CRISPR

BIOLOGY AND APPLICATIONS

CRISPR

BIOLOGY AND APPLICATIONS

EDITED BY

Rodolphe Barrangou
Department of Food,
Bioprocessing, and
Nutrition Sciences
North Carolina State University
Raleigh, North Carolina

Erik J. Sontheimer
RNA Therapeutics Institute
University of Massachusetts
Chan Medical School
Worcester, Massachusetts

Luciano A. Marraffini
Howard Hughes Medical Institute
Chevy Chase, Maryland
Laboratory of Bacteriology
The Rockefeller University New York,
New York

ASM
PRESS
Washington, DC

WILEY

Limit of Liability/Disclaimer of Warranty
While the publisher and author have used their best efforts in preparing this book, they make no representations or warranties with respect to the accuracy of completeness of the contents of this book and specifically disclaim any implied warranties or merchantability of fitness for a particular purpose. No warranty may be created or extended by sales representatives or written sales materials. The publisher is not providing legal, medical, or other professional services. Any reference herein to any specific commercial products, procedures, or services by trade name, trademark, manufacturer, or otherwise does not constitute or imply endorsement, recommendation, or favored status by the American Society for Microbiology (ASM). The views and opinions of the author(s) expressed in this publication do not necessarily state or reflect those of ASM, and they shall not be used to advertise or endorse any product.

Editorial Correspondence:
ASM Press, 1752 N Street, NW, Washington, DC 20036-2904, USA

Registered Offices:
John Wiley & Sons, Inc., 111 River Street, Hoboken, NJ 07030, USA

For details of our global editorial offices, customer services, and more information about Wiley products, visit us at www.wiley.com.

Wiley also publishes its books in a variety of electronic formats and by print-on-demand. Some content that appears in standard print versions of this book may not be available in other formats.

Library of Congress Cataloging-in-Publication Data has been applied for
ISBN 9781683670377 (Hardback); ISBN 9781683670384 (Adobe PDF);
ISBN 9781683673613 (e-Pub)

Cover design and illustration: Owen Design Co.
Interior design by: Susan Brown Schmidler
Illustrations: Patrick Lane, ScEYEnce Studios

Set in 11/13.5pt Sabon LT Std by Straive, Chennai, India

SKY10036075_091622

Contents

Contributors

Omar O. Abudayyeh
McGovern Institute for Brain
    Research at MIT
Massachusetts Institute of Technology
Cambridge, MA 02139

Scott Bailey
Department of Biochemistry and
    Molecular Biology
Department of Biophysics and
    Biophysical Chemistry
Bloomberg School of Public Health
Johns Hopkins University
Baltimore, MD 21205

Rodolphe Barrangou
CRISPRlab
Department of Food, Bioprocessing and
    Nutrition Sciences
North Carolina State University
Raleigh, NC 27695

Morgan Quinn Beckett
Department of Biochemistry and
    Molecular Biology
Department of Biophysics and
    Biophysical Chemistry
Bloomberg School of Public Health
Johns Hopkins University
Baltimore, MD 21205

David Bikard
Synthetic Biology Group
Department of Microbiology
Institut Pasteur
Paris 75015, France

Joseph Bondy-Denomy
Department of Microbiology & Immunology
University of California, San Francisco
San Francisco, CA 94143

Aroa Rey Campa
Department of Microbiology and
    Immunology
Bio-Protection Research Centre
University of Otago
Dunedin 9054, New Zealand

Crystal Chen
Department of Chemical Engineering
Stanford University
Stanford, CA 94305

Mariia Y. Cherepkova
Department of Biosystems Science and
    Engineering
ETH Zurich
4058 Basel, Switzerland

Jonathan D. D'Gama
Department of Microbiology
Harvard Medical School
Division of Infectious Diseases
Brigham & Women's Hospital
Boston, MA 02115

Peter C. Fineran
Department of Microbiology and Immunology
Bio-Protection Research Centre
University of Otago
Dunedin 9054
New Zealand

Jonathan S. Gootenberg
McGovern Institute for Brain
  Research at MIT
Massachusetts Institute of Technology
Cambridge, MA 02139

Uri Gophna
The Shmunis School of Biomedicine and
  Cancer Research
Tel Aviv University
Tel Aviv, Israel, 69978

Sutharsan Govindarajan
Department of Microbiology & Immunology
University of California, San Francisco
San Francisco, CA 94143

Patrick D. Hsu
Department of Bioengineering
Innovative Genomics Institute
Center for Computational Biology
University of California, Berkeley
Berkeley, CA 94720

Tautvydas Karvelis
Institute of Biotechnology
Life Sciences Center
Vilnius University
Vilnius, Lithuania LT-10257

Eugene V. Koonin
National Center for Biotechnology
  Information
National Library of Medicine
Bethesda, MD 20894

Peter Lotfy
Biological and Biomedical Sciences
  PhD Program
Harvard Medical School
Division of Gastroenterology, Hepatology,
  and Nutrition
Boston Children's Hospital
Boston, MA 02115
Broad Institute of MIT and Harvard
Cambridge, MA 02142

Kira S. Makarova
National Center for Biotechnology
  Information
National Library of Medicine
Bethesda, MD 20894

Luciano A. Marraffini
Laboratory of Bacteriology
Howard Hughes Medical Institute
The Rockefeller University
New York, NY 10065

Jasprina N. Noordermeer
Department of Bioengineering
Stanford University
Stanford, CA 94305

Joseph S. Park
Department of Microbiology
Harvard Medical School
Division of Infectious Diseases
Brigham & Women's Hospital
Boston, MA 02115

Randall J. Platt
Department of Biosystems Science and
  Engineering
ETH Zurich
4058 Basel
Department of Chemistry
University of Basel
4003, Basel
Switzerland

Lei S. Qi
Department of Chemical and Systems Biology
ChEM-H Institute
Stanford University
Stanford, CA 94305

Anita Ramachandran
Department of Biochemistry and
  Molecular Biology
Bloomberg School of Public Health
Johns Hopkins University
Baltimore, MD 21205

Avery Roberts
CRISPRlab
Department of Food, Bioprocessing and
  Nutrition Sciences
North Carolina State University
Raleigh, NC 27695

Justen Russell
Synthetic Biology Group
Department of Microbiology
Institut Pasteur
Paris 75015, France

Virginijus Siksnys
Institute of Biotechnology
Life Sciences Center
Vilnius University
Vilnius, Lithuania LT-10257

Leah M. Smith
Department of Microbiology and
    Immunology
University of Otago
Dunedin 9054, New Zealand

Erik J. Sontheimer
RNA Therapeutics Institute
University of Massachusetts Chan
    Medical School
Worcester, MA 01605

Tanmay Tanna
Department of Biosystems Science and
    Engineering
ETH Zurich
4058 Basel, Switzerland

John van der Oost
Wageningen University & Research
Laboratory of Microbiology
6708WE Wageningen
The Netherlands

Matthew K. Waldor
Department of Microbiology
Harvard Medical School
Division of Infectious Diseases
Brigham & Women's Hospital
Howard Hughes Medical Institute
Boston, MA 02115

Yuri I. Wolf
National Center for Biotechnology Information
National Library of Medicine
Bethesda, MD 20894

Jenny Y. Zhang
Department of Microbiology & Immunology
University of California, San Francisco
San Francisco, CA 94143

Preface

The rise and popularity of CRISPR have been associated with technology underpinning genome manipulation. Gene editing represents, by many measures, one of the most important achievements of modern science. The alteration the genetic makeup of an organism, so often at the center of science fiction classics (Ridley Scott's *Blade Runner* and Steven Spielberg's *Jurassic Park*, for example), is now a reality. Little did the scientific community imagine that the best tool to do this, Cas9, would originate from the investigation of how bacteria and archaea fend off phages and plasmids that invade them. Microbiologists were probably the least surprised about this development, as bacteria have repeatedly provided a plethora of molecular gadgets with highly useful biotechnological applications. Most notable among these are restriction enzymes. About 40 years before the Cas9 revolution in gene editing, the harnessing of the specific DNA cleavage properties of another bacterial defense system, restriction-modification, was the central catalyst for another revolution in biomedical sciences: recombinant DNA. Both CRISPR and restriction-modification are testaments of the importance of basic science, microbiology, and bacterial defenses for the advancement of biotechnology.

While ahead of Cas9 there is an exciting future of genetic engineering and gene therapy, behind Cas9 there is the complex and fascinating world of CRISPR immunity. The first part of this book provides extensive coverage of this unique world. It opens with a brief recollection of the historical events and seminal work that led to the discovery of CRISPR-Cas systems and their function in adaptive immunity, as well as the features that they have in common (chapter 1) and a description of their tremendous diversity (chapter 2). This is followed by in-depth explanations of the different molecular mechanisms of immunity behind this diversity (chapters 3 to 7), their impact on the ecology and evolution of prokaryotic populations (chapter 8), the countermeasures developed by phages to evade CRISPR defenses (chapter 9), and the regulation of expression of CRISPR loci (chapter 10). In the second part, the book focuses on the biotechnological applications of CRISPR-Cas systems and their derived effectors: the widespread gene editing (chapter 11) and gene regulation (chapter 12) techniques, the use of CRISPR screens in microbiology research (chapter 13), the repurposing of CRISPR-Cas systems to attack the bacterial chromosome as programmable antimicrobials (chapter 14), the exploitation of CRISPR immunity to protect bacterial cultures of industrial value from phage predation (chapter 15), and, last but not least, the development of molecular recording circuits for synthetic biology applications (chapter 16).

Such a comprehensive collection of CRISPR chapters would have been impossible without the generous commitment and invaluable contributions of all the authors. We are grateful for the privileged participation of founders of the CRISPR field (Koonin, Makarova, Siksnys, and van der Oost), technical innovators and high-profile contributors (Bailey, Gophna, Fineran, Waldor, and Bikard) and rising stars that have advanced and will continue to advance the field and bring it to new heights (Abudayyeh, Bondy-Denomy, Gootenberg, Hsu, Platt, and Qi). We are personally grateful for their contribution to this book, their collaborative efforts and spirit over the years, and their commitment to the readership.

We also thank the American Society for Microbiology (ASM), not only for publishing this book but also for their constant support of the CRISPR field, especially for featuring numerous microbiological studies of CRISPR-Cas systems for over 15 years and recognizing early on the potential interest in and impact of the CRISPR literature. Indeed, the ASM has also provided financial support for many of the annual CRISPR meetings organized by researchers of the field. In addition, the Society has organized innumerable symposia and sessions covering the different aspects of CRISPR immunity at its annual general meeting and also at local meetings.

Finally, we thank our families. Every endeavor we take on as scientists, such as attending week-long meetings afar, writing grants 3 days before the deadline, and embarking in the publication of a book with more than a dozen different authors, sends us away from home and keeps our minds preoccupied with our work. We are indebted to them for their constant understanding, support, and love.

RODOLPHE BARRANGOU
ERIK J. SONTHEIMER
LUCIANO A. MARRAFFINI

Diversity of CRISPR-Cas Systems

CRISPR-Cas Systems: Core Features and Common Mechanisms

CHAPTER

Rodolphe Barrangou[1], Luciano A. Marraffini[2,3], and Erik J. Sontheimer[4]

[1]CRISPRlab, Department of Food, Bioprocessing and Nutrition Sciences, North Carolina State University, Raleigh, NC, 27695
[2]Laboratory of Bacteriology, The Rockefeller University, New York, NY, 10065
[3]Howard Hughes Medical Institute, The Rockefeller University, New York, NY, 10065
[4]RNA Therapeutics Institute, University of Massachusetts Chan Medical School, Worcester, MA, 01605

Introduction

CRISPR loci were first described as a group of short, repetitive sequences downstream of the *iap* gene in *Escherichia coli* K-12 (1) and were later also identified in many additional bacterial and archaeal genomes (2). As opposed to other DNA repeats that are adjacent to one another, CRISPR repeats were found to be distinctively separated by similarly short sequences of unknown origin. These sequences, known as "spacers," were shown to match regions of the genomes of bacteriophages and plasmids that invade prokaryotes (3, 4, 5) (known as the target or protospacer), leading to the hypothesis that CRISPR loci could be linked to infection by these elements. In addition, a conserved set of CRISPR-associated (*cas*) genes coding for domains harbored by proteins that participate in transactions among nucleic acids (e.g., nucleases, helicases, and integrases) were often found to flank the CRISPR repeat-spacer arrays (6). Finally, the isolation and sequencing of apparent non-protein-coding RNAs from the archaeon *Archaeoglobus fulgidus* revealed that spacer sequences were transcribed and processed into small RNAs (7). All of these observations, along with in-depth bioinformatic analyses of prokaryotic genomes, allowed Koonin and Makarova to synthesize and articulate the first model for CRISPR loci as a prokaryotic defense system (8). Specifically, they proposed an RNA interference (RNAi)-based immunity, in which the small RNAs derived from the spacers would be used by Cas nucleases to find and destroy complementary transcripts, and suggested a mechanism with functional analogies to eukaryotic RNAi.

This initial phase of CRISPR studies, carried out largely *in silico*, generated intriguing and testable predictions that spawned a period of experimentation to test the bioinformatic hypotheses. In the first such study, Barrangou and colleagues showed

CRISPR: Biology and Applications, First Edition. Edited by Rodolphe Barrangou, Erik J. Sontheimer, and Luciano A. Marraffini.
© 2022 American Society for Microbiology. DOI: 10.1128/9781683673798.ch01

that CRISPR-Cas systems provide spacer-specific immunity against bacteriophages in *Streptococcus thermophilus* (9). More importantly, the work also produced the astonishing finding that immunity is acquired: upon infection, new spacers that match a region of the genome of the invading phage are inserted into the CRISPR array. This process, the hallmark of CRISPR immunity, is known in the field as "adaptation," as it enables the bacterial host to adapt to the environmental stress imposed by the virus. The authors also showed by gene knockout that *cas9* (originally known as *cas5* or *csn1*) is necessary for immunity and that *cas4* is required for immunization, implicating *cas* genes in CRISPR-encoded resistance. In a second study, Andersson and Banfield used metagenomic analysis of CRISPR spacer sequences to assemble viral genomes present in natural acidophilic biofilms (10). The reconstituted genomes revealed extensive recombination events that disrupted phage genome matches with the CRISPR spacers, a result showing the presence of an active arms race between CRISPR systems and phages, further supporting the role of these loci in prokaryotic defense. Later in the same year, the van der Oost group provided experimental evidence that immunity is guided by the small RNAs derived from spacers, known as CRISPR RNAs (crRNAs) (11). These researchers constructed *Escherichia coli* strains with individual deletions of the *cas* genes to identify the endoribonuclease responsible for the generation of crRNAs from a large CRISPR precursor transcript (the pre-crRNA), Cas6 (originally known as CseE). The mutant strain lacking *cas6* was unable to produce crRNAs and also failed at providing CRISPR immunity against lambda phage, demonstrating the essentiality of protospacer-complementary crRNAs for defense. The reliance of CRISPR-Cas system on this processed RNA guide also established "crRNA biogenesis" as a critical mechanistic step for immunity. Finally, Marraffini and Sontheimer showed that CRISPR loci prevent the transfer of conjugative plasmids in *Staphylococcus epidermidis* in a spacer-specific manner (12), demonstrating that CRISPR-Cas immunity extends beyond phage defense. More importantly, in part through the insertion of a self-splicing intron within the target sequence, which disrupts the DNA target but not its transcript (splicing reconstitutes the original sequence), they demonstrated that immunity can operate by targeting the DNA, as distinct from the RNA, of the invader. This finding not only contrasted with the original prediction that CRISPR-Cas systems have an RNAi-like mechanism but also prompted the proposal that the CRISPR machinery could be repurposed as an RNA-programmable, sequence-specific DNA cleavage system, perhaps even beyond prokaryotes (12). Altogether, these early mechanistic studies collectively established CRISPR-Cas as adaptive, DNA-encoded (heritable), RNA-mediated, nucleic acid-targeting immune systems (Fig. 1.1).

After these seminal studies, CRISPR research quickly accelerated. Explorations of the molecular mechanisms of immunity, as well as deep and comprehensive identification and bioinformatic analyses of CRISPR loci deposited in GenBank, led to the recognition of different CRISPR-Cas types and subtypes (13, 14, 15). Interestingly, each of the experimental studies described above investigated a different CRISPR-Cas system: the *S. thermophilus*, *E. coli*, and *S. epidermidis* experiments focused on type II-A, I-E, and III-A systems, respectively. However, the findings of these foundational studies are still relevant today because they uncovered fundamental aspects of the CRISPR-Cas immune response that are common to all types. While the chapters that follow present the details and unique features of each different CRISPR-Cas type (with the exception of type IV, about which relatively little is known), below we describe some of the unifying features of the CRISPR mechanism of action.

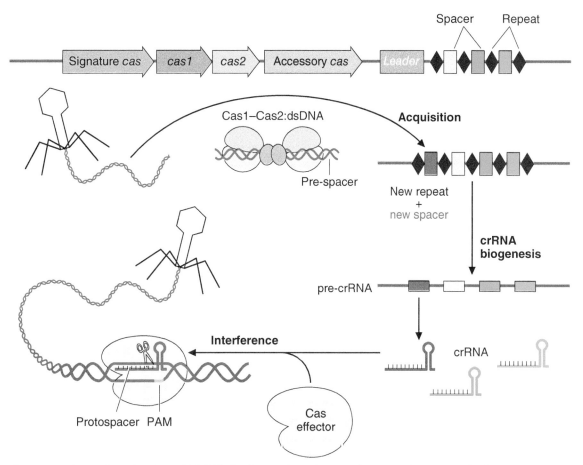

Figure 1.1 General mechanism of CRISPR-Cas immunity. The typical CRISPR locus (top) harbors the CRISPR array of repeats and spacers flanked by a set of *cas* genes. This set contains *cas1* and *cas2*, which are almost universally conserved and participate in the spacer acquisition process, as well as signature and accessory *cas* genes that determine the CRISPR type and are usually involved in the later stages of the CRISPR-Cas immune response, crRNA biogenesis and interference. During spacer acquisition, short pieces of the invader's DNA are integrated into the CRISPR array, a process that is accompanied by the duplication of a repeat sequence. During crRNA biogenesis, the CRISPR array is transcribed and processed into small RNA guides known as CRISPR RNAs (crRNAs). Finally, crRNAs guide the Cas effector complex to its complementary target nucleic acid. Here, we show the mechanism for types I, II and V CRISPR-Cas systems, where the crRNA recognizes complementary DNA target sequences that are flanked by a protospacer adjacent motif (PAM). The effector complexes of types III and VI systems use the crRNA to find complementary transcripts of the invader.

The CRISPR Array

Across all CRISPR-Cas systems, CRISPR loci are defined by their arrays of short (~40-bp) DNA repeats separated by similarly short spacer sequences. Common to the great majority of CRISPR loci is the presence of a degenerate repeat at the end of the array, with a handful of nucleotides that differ from the rest of the repeats, typically at the 3′ edge of the very last repeat (6). The repeat sequences, however, vary greatly from one locus to another, without any apparent pattern that correlates with the different CRISPR-Cas types (though with some variation that correlates with the coevolutionary dynamics between the CRISPR repeat sequences and the Cas proteins that interact with them). Indeed, it is possible to classify CRISPR loci based on their repeat sequences (16), and this can be useful for the study of "orphan" arrays (uncommon loci that lack associated *cas* genes) or arrays found in unassembled, fragmented genomes (missing their associated *cas* genes). The number of repeats also varies substantially, with many arrays containing just a few and others a few hundred. To date, the record holder is the bacterium *Haliangium ochraceum* strain DSM 14365, which harbors 588 repeats (and 587 spacers) (17).

Spacer Acquisition

As mentioned above, one of the hallmarks of CRISPR immunity is that new spacers can be acquired upon infection from the genome of the invader (9). This creates a molecular memory of the pathogen that is used to recognize and neutralize it during subsequent attacks. New spacers are added in a polarized manner adjacent to the first repeat of the locus, also known as the "leader" end due to the presence of an A/T-rich sequence that immediately precedes this repeat. The insertion of a new spacer results in the duplication of the first repeat. This polarized form of integration creates a temporal, iterative record of infections suffered by the host, and this "recording" has been used to trace the evolutionary history of bacteria and archaea (5). Spacer acquisition thus enables an adaptive, genetically encoded, and heritable immune response that is shared across all CRISPR-Cas types.

Most CRISPR-Cas types acquire spacer sequences from DNA molecules. The exceptions are type III systems, which can convert RNA into cDNA before its integration into the CRISPR array (18), effectively capturing RNA molecules as DNA spacers (see chapter 5). In both type I and II systems, free double-stranded DNA (dsDNA) ends are the preferred substrates for the CRISPR adaptation machinery (19, 20). For both systems, the host's recombination nuclease RecBCD increases the efficiency of spacer acquisition at the free dsDNA ends. As a consequence, hot spots of acquisition are generated between these ends and the first *chi* site that stops RecBCD degradation (19, 20). Although not yet proven experimentally, it is believed that along its path of degradation from the free dsDNA end to the *chi* site, random and spontaneous disengagement of RecBCD from its substrate will presumably expose a new free dsDNA end from which spacers can be generated and acquired (21). This mode of action has important consequences for the CRISPR pathway. First, since prokaryotic chromosomes are circular but many of their invaders inject a linear DNA genome with a free dsDNA end, it provides a means to distinguish "self" from "foreign" DNA and avoid autoimmunity. Second, it allows the amplification of the CRISPR-Cas immune response through a process known as "priming" (22, 23, 24). At least in type I and II systems, the dsDNA break introduced by the Cas nucleases can lead to spacer acquisition from the free dsDNA ends that are generated (24, 25). This is particularly important to counteract the rise of escaper phages containing target mutations that prevent efficient targeting and cleavage by Cas effector nucleases. While these mutations lead to the overall failure of CRISPR immunity, in many cases this still allows a relatively low level of target cleavage, sufficient to create new free dsDNA ends and thus trigger the acquisition of new spacers to neutralize phage escapers.

At the molecular level, spacer acquisition is akin to transposon integration, although the inserted DNA is much smaller than typical transposons. The integration reaction is mediated by *cas1* and *cas2*, the only *cas* genes universally conserved across all CRISPR-Cas types (15). Cas1 and Cas2 form an integrase complex consisting of two distal Cas1 dimers bridged by a Cas2 dimer (26, 27). This complex is loaded with a prespacer DNA (the sequence that will become a spacer after integration into the CRISPR array), usually harboring 3′ overhangs, presumably originating from dsDNA ends. The mechanisms of prespacer loading are not completely elucidated and seem to be different for diverse types (27, 28). In contrast, at least in all studies to date, the integration mechanism is universal, involving the nucleophilic attack of the 3′-OH group of the prespacer overhang on the minus (bottom) strand of the first repeat, distal to the leader (Fig. 1.1). This leads to the formation of a half-site intermediate and it is followed by a second nucleophilic attack of the other 3′-OH overhang on the plus (top) strand at the leader-repeat border. The fully integrated spacer is flanked by single-stranded repeat sequences, which are converted into double-stranded sequences, probably by a DNA polymerase gap-filling activity, to finalize the complete duplication of the repeat. Finally, there is the mechanism of selection of the first repeat for integration varies across types. In type II systems a leader-anchoring sequence promotes polarized spacer acquisition (29, 30). In contrast, type I spacer integration typically requires the bacterial integration host factor (31), which binds to the leader sequence and induces a sharp DNA bend, required for the type I Cas1-Cas2 integrase to catalyze the first integration reaction at the leader-repeat border.

CRISPR Targets

Spacer sequences can be used to find matching sequences in the large databases of DNA sequences and thus infer the targets of the CRISPR-Cas immune response. It has been estimated that only ~7% of the known spacers have homologous sequences in the databases (26,364 out of 363,460 unique spacers in reference 32), an observation that probably reflects the lack of sequence information from a majority of the prokaryotic genetic elements in general and from viruses in particular. Most of this small fraction of spacers (~96%) matched the genomes of either bacteriophages or prophages. The rest matched to plasmids (~3%), CRISPR-*cas* loci (>1%, especially *cas3*), and prokaryotic DNA that could not be identified as part of mobile genetic elements (>1%) (32). This distribution highlights the fundamental function of CRISPR-Cas systems in the defense against prokaryotic viruses (9). Similarly to phages, plasmids are a major category of prokaryotic mobile genetic elements (33), but it is not completely clear why they represent the second most abundant CRISPR target. On one hand, plasmids can carry addiction modules, such as toxin/antitoxin systems, that are costly to the host (34), which would benefit from their elimination via CRISPR immunity. On the other hand, the transfer of plasmids can be also beneficial for the recipient organism, for example, antibiotic-resistant plasmids in bacterial pathogens (35). At the population level, the limits imposed by CRISPR immunity to both phage infection and plasmid conjugation, two of the major routes of horizontal gene transfer (36), have often led to the hypothesis that CRISPR-Cas systems represent a barrier to the evolution of bacteria and archaea (12, 37). However, theoretical modeling (38) and computational estimation of the rates of horizontal gene transfer (39) have challenged this idea.

Although matched by a minority of spacers, *cas* genes represent one of the most interesting targets of CRISPR immunity. The simplest explanation for this observation is that CRISPR-Cas loci are occasionally encoded by both phage (40, 41) and plasmid (42, 43) genomes, and therefore, *cas* spacers may appear as part of the general CRISPR response against these elements. But more complex scenarios are also conceivable. For instance, some *Escherichia coli* strains carry an orphan CRISPR locus consisting of a single spacer targeting *cas* genes present in the type I-F systems of other strains (44). Importantly, the sequences of the repeats flanking the spacer are recognized and processed into a crRNA by the type I-F Cascade complex. Therefore, this orphan CRISPR can repel invasion of mobile elements that carry the type I-F *cas* locus (44), which upon entry into the host will produce a targeting complex that will turn against their own genome. This example reveals the presence of evolutionary forces that use CRISPR as a barrier against the horizontal transfer of CRISPR-*cas* loci themselves. Finally, the presence of spacers against genes that belong to core (i.e. nonmobile) prokaryotic genomic regions suggests a direct role for CRISPR against the horizontal transfer of such genes, as opposed to preventing their spread indirectly, via the attack of phages and plasmids, the main vehicles for genetic exchange (36). Supporting this role, it has been reported that such spacers can prevent the natural transformation of both Gram-positive (45) and Gram-negative (46) bacteria. In addition, new spacers matching archaeal housekeeping genes can be acquired during interspecies mating (47), a process in which cells fuse to form a diploid state containing the full genetic repertoire of both parental cells to facilitate genetic exchange and recombination (48). Although not yet completely clear, these spacers could limit subsequent gene transfer, constraining evolution and thus facilitating archaeal speciation.

RNA-Guided Nucleases

A second hallmark of the CRISPR-Cas immune response is the employment of programmable RNA-guided nucleases. The crRNA generated from the transcription and processing of the CRISPR array is loaded into Cas proteins to form a ribonucleoprotein complex that is able to scan nucleic acids for a sequence complementary to the crRNA guide. After successful base-pairing, the target sequence is cut, with different mechanisms of cleavage for different types. For example, the crRNA is loaded into the Cas nuclease itself in types II, III, V, and VI, resulting in cleavage near or within the target sequence itself (49, 50, 51, 52, 53, 54). However, in type I systems, the crRNA is part of the Cascade complex, which lacks nuclease activity (11). This complex uses the crRNA to find the target and then recruits the helicase/nuclease Cas3, which goes on to degrade the DNA in the vicinity of the target sequence (55).

A major distinction between crRNA-guided Cas nucleases is their substrates. Types I, II, and V recognize dsDNA, a process that requires the formation of an R loop, a three-stranded nucleic acid structure composed of the target DNA:crRNA hybrid and the displaced noncomplementary DNA strand (56, 57, 58, 59, 60, 61). Recently, this form of DNA targeting has been shown (in some type I and type V instances) to specify sites of transposon insertion rather than DNA destruction (62, 63). For either of these functional outcomes, stable R-loop formation is not possible without the presence of a flanking sequence known as the protospacer adjacent motif (PAM) (53, 64, 65, 66). The PAM requirement is thought to fulfill two roles during DNA targeting. First, it facilitates the recognition of the target sequence within the invader genome. The crRNA-guided nucleases use a PAM recognition domain (60, 67, 68, 69) to scan DNA for the presence of the motif, and R-loop formation begins only when the protein-PAM interaction occurs (70, 71). Annealing of the spacer-target nucleotides immediately flanking the PAM is critical for the success of R-loop initial formation, and the presence of mismatches in this region (known as the "seed sequence") prevents targeting (64, 66). This mechanism limits the number of nonproductive attempts at R-loop formation, thus enabling rapid interrogation and identification of potential target sites within large DNA molecules. It is also suspected of providing binding energy that can be used to drive initial DNA duplex unwinding. A second function for the PAM is to avoid self-targeting of the CRISPR locus, which contains the target sequence but is flanked by a CRISPR repeat rather than a PAM. Indeed, given that the crRNA guides are fully complementary to the spacer DNA, in the absence of a PAM requirement, the spacer would be recognized as a target. However, since the spacer sequence is flanked by a repeat in the CRISPR array, which does not contain a PAM, self-targeting is not possible.

Type III and VI CRISPR-Cas systems, in contrast, use the crRNA guide to find complementary RNA transcripts produced by the invader (52, 54, 72). In these systems, target recognition triggers diverse nonspecific nuclease activities. During type III immunity, base-pairing of the crRNA with a target transcript activates the DNase activity of the Cas10 subunit (73) to destroy the template DNA. In addition, the Palm domain of Cas10 is activated and converts ATP into cyclic oligoadenylates that act as second messengers to bind and activate Csm6/Csx1 (74, 75), nonspecific RNases that induce a growth arrest of the host to assist the CRISPR defense (76, 77). Interestingly, this self-targeting is somewhat reminiscent of abortive infection, preventing infected cells from acting as viral factories. In type VI systems, target recognition by the crRNA guide within Cas13 triggers a nonspecific RNase activity of this nuclease (54, 72), also inducing a dormant state in the host that prevents the propagation of the infecting phage (78). In both types, the target RNA is cleaved shortly after its recognition, resulting in the inactivation of the nonspecific nuclease activities (54, 72, 73), presumably to prevent the irreversible destruction of the host cell. As opposed to DNA targeting systems, the crRNA-guided endoribonucleases do not appear to require a specific motif, i.e., PAM, to cleave the target RNA (79). It is believed that this may be due to the relatively easier mode of target finding (single-stranded target, no need for unwinding and R-loop formation, and high copy number of a target transcript compared to its DNA template) and also to the presence of an intrinsic limitation on self-targeting. This limitation arises because the CRISPR locus generates crRNAs but not complementary transcripts that would trigger self-immunity: "self-target" RNAs would be produced only through transcription of the CRISPR array in the opposite direction to the transcription of the crRNAs. Some level of this transcription, however, could be possible even if accidental, and both type III and VI crRNA-guided nucleases are inhibited upon recognition of self-targets. Inhibition relies on the annealing of the repeat sequence downstream of the target (the "antitag") that is complementary to the portion of the repeat sequence harbored upstream of the spacer sequence in the crRNA (the "tag") (80, 81). As a consequence of this mechanism, bona fide type III and VI targets are never flanked by antitag sequences.

Evasion of CRISPR Immunity

All bacterial and archaeal immune systems are engaged in an arms race with phages and plasmids and are therefore subject to subversion by these invaders. A great proportion of phages carry anti-CRISPR inhibitors (Acrs) (82), most of which specifically bind to and prevent the function of the crRNA-guided nucleases of each different type (covered in chapter 9). In addition to Acrs, mutations present in the phage or plasmid population that affect the recognition of the target by the Cas nucleases allow evasion of the immunity imparted by DNA-targeting CRISPR-Cas systems. As explained above, these mutations are present in either the PAM or seed sequences (64, 66). However, depending on the genetic function of the target sequence, some escape mutations are not viable, for example, if they change the structure or activity of an essential phage or plasmid gene. Therefore, some spacers mediate less "escapable," and thus more robust, immunity than others (83, 84). Importantly, spacer acquisition generates a population of bacteria that harbor numerous different spacer sequences (10, 85, 86, 87, 88). This diversity prevents the rise of escapers (89), as most phages in the population will harbor mutations that only enable escape from the immunity provided by one or two spacer sequences but eventually will infect a host cell harboring a spacer they cannot evade. Interestingly, it has been shown that phage and plasmid recombination systems can induce the accumulation of escape mutations within the targets of type I and II Cas nucleases (90, 91). Presumably, these systems can facilitate the introduction of target mutations through their involvement in the repair of the cleaved DNA. Finally, in contrast to DNA targeting CRISPR-Cas systems, the growth arrest generated during both type III and VI CRISPR immunity limits the possibilities of genetic escape (78, 79). Although there is a low number of phages within the viral population carrying target mutations that could enable escape from recognition and targeting by Cas nucleases, these mutant phages eventually end up infecting a host in which a previous wild-type phage triggered a growth arrest. These cells are inhospitable for the propagation of any phage, including those with mutant targets. Accordingly, the diversity of the spacer repertoire seems to be less important to counteract the rise of escaper phages during type III and VI immunity (79), an observation that could have an evolutionary correlation with the very low frequencies of spacer acquisition in these systems.

Conclusions

In summary, all CRISPR-Cas systems share a common general mechanism centered on the CRISPR-*cas* locus. The CRISPR array is used to store genetic information from prokaryotic invaders, mostly phages and plasmids, through the extraction and integration of a short DNA spacer sequence from their genome. This information is passed to the Cas nucleases in the form of a crRNA guide, which is used to find and destroy specific, invasive nucleic acids. CRISPR-Cas systems are locked in a coevolutionary arms race with their targets, which have evolved a series of countermeasures to evade immunity. There are, however, many different CRISPR-Cas systems, with different features and mechanisms. The chapters that follow describe in detail the wonderful CRISPR diversity that exists in nature.

References

1. Ishino Y, Shinagawa H, Makino K, Amemura M, Nakata A. 1987. Nucleotide sequence of the *iap* gene, responsible for alkaline phosphatase isozyme conversion in *Escherichia coli*, and identification of the gene product. *J Bacteriol* **169**:5429–5433.

2. Mojica FJ, Díez-Villaseñor C, Soria E, Juez G. 2000. Biological significance of a family of regularly spaced repeats in the genomes of Archaea, Bacteria and mitochondria. *Mol Microbiol* **36**:244–246.

3. Bolotin A, Quinquis B, Sorokin A, Ehrlich SD. 2005. Clustered regularly interspaced short palindrome repeats (CRISPRs) have spacers of extrachromosomal origin. *Microbiology (Reading)* **151**:2551–2561.

4. Mojica FJM, Díez-Villaseñor C, García-Martínez J, Soria E. 2005. Intervening sequences of regularly spaced prokaryotic repeats derive from foreign genetic elements. *J Mol Evol* **60**:174–182.

5. Pourcel C, Salvignol G, Vergnaud G. 2005. CRISPR elements in *Yersinia pestis* acquire new repeats by preferential uptake of bacteriophage DNA, and provide additional tools for evolutionary studies. *Microbiology (Reading)* **151**:653–663.

6. Jansen R, Embden JD, Gaastra W, Schouls LM. 2002. Identification of genes that are associated with DNA repeats in prokaryotes. *Mol Microbiol* **43**:1565–1575.

7. Tang T-H, Bachellerie J-P, Rozhdestvensky T, Bortolin M-L, Huber H, Drungowski M, Elge T, Brosius J, Hüttenhofer A. 2002. Identification of 86 candidates for small non-messenger RNAs from the archaeon *Archaeoglobus fulgidus*. *Proc Natl Acad Sci U S A* **99**:7536–7541.

8. Makarova KS, Grishin NV, Shabalina SA, Wolf YI, Koonin EV. 2006. A putative RNA-interference-based immune system in prokaryotes: computational analysis of the predicted enzymatic machinery, functional analogies with eukaryotic RNAi, and hypothetical mechanisms of action. *Biol Direct* **1**:7 http://dx.doi.org/10.1186/1745-6150-1-7.

9. Barrangou R, Fremaux C, Deveau H, Richards M, Boyaval P, Moineau S, Romero DA, Horvath P. 2007. CRISPR provides acquired resistance against viruses in prokaryotes. *Science* **315**:1709–1712.

10. Andersson AF, Banfield JF. 2008. Virus population dynamics and acquired virus resistance in natural microbial communities. *Science* **320**:1047–1050.

11. Brouns SJJ, Jore MM, Lundgren M, Westra ER, Slijkhuis RJH, Snijders APL, Dickman MJ, Makarova KS, Koonin EV, van der Oost J. 2008. Small CRISPR RNAs guide antiviral defense in prokaryotes. *Science* **321**:960–964.

12. Marraffini LA, Sontheimer EJ. 2008. CRISPR interference limits horizontal gene transfer in staphylococci by targeting DNA. *Science* **322**:1843–1845.

13. Makarova KS, Haft DH, Barrangou R, Brouns SJJ, Charpentier E, Horvath P, Moineau S, Mojica FJM, Wolf YI, Yakunin AF, van der Oost J, Koonin EV. 2011. Evolution and classification of the CRISPR-Cas systems. *Nat Rev Microbiol* **9**:467–477.

14. Makarova KS, Wolf YI, Alkhnbashi OS, Costa F, Shah SA, Saunders SJ, Barrangou R, Brouns SJJ, Charpentier E, Haft DH, Horvath P, Moineau S, Mojica FJM, Terns RM, Terns MP, White MF, Yakunin AF, Garrett RA, van der Oost J, Backofen R, Koonin EV. 2015. An updated evolutionary classification of CRISPR-Cas systems. *Nat Rev Microbiol* **13**:722–736.

15. Makarova KS, Wolf YI, Iranzo J, Shmakov SA, Alkhnbashi OS, Brouns SJJ, Charpentier E, Cheng D, Haft DH, Horvath P, Moineau S, Mojica FJM, Scott D, Shah SA, Siksnys V, Terns MP, Venclovas Č, White MF, Yakunin AF, Yan W, Zhang F, Garrett RA, Backofen R, van der Oost J, Barrangou R, Koonin EV. 2020. Evolutionary classification of CRISPR-Cas systems: a burst of class 2 and derived variants. *Nat Rev Microbiol* **18**:67–83.

16. Nethery MA, Korvink M, Makarova KS, Wolf YI, Koonin EV, Barrangou R. 2021. CRISPRclassify: repeat-based classification of CRISPR loci. *CRISPR J* **4**:558–574.

17. Pourcel C, Touchon M, Villeriot N, Vernadet J-P, Couvin D, Toffano-Nioche C, Vergnaud G. 2020. CRISPRCasdb a successor of CRISPRdb containing CRISPR arrays and cas genes from complete genome sequences, and tools to download and query lists of repeats and spacers. *Nucleic Acids Res* **48**:D535–D544.

18. Silas S, Mohr G, Sidote DJ, Markham LM, Sanchez-Amat A, Bhaya D, Lambowitz AM, Fire AZ. 2016. Direct CRISPR spacer acquisition from RNA by a natural reverse transcriptase-Cas1 fusion protein. *Science* **351**:aad4234.

19. Levy A, Goren MG, Yosef I, Auster O, Manor M, Amitai G, Edgar R, Qimron U, Sorek R. 2015. CRISPR adaptation biases explain preference for acquisition of foreign DNA. *Nature* **520**:505–510.

20. Modell JW, Jiang W, Marraffini LA. 2017. CRISPR-Cas systems exploit viral DNA injection to establish and maintain adaptive immunity. *Nature* **544**:101–104.

21. Jakhanwal S, Cress BF, Maguin P, Lobba MJ, Marraffini LA, Doudna JA. 2021. A CRISPR-Cas9-integrase complex generates precise DNA fragments for genome integration. *Nucleic Acids Res* **49**:3546–3556.

22. Datsenko KA, Pougach K, Tikhonov A, Wanner BL, Severinov K, Semenova E. 2012. Molecular memory of prior infections activates the CRISPR/Cas adaptive bacterial immunity system. *Nat Commun* **3**:945.

23. Swarts DC, Mosterd C, van Passel MWJ, Brouns SJJ. 2012. CRISPR interference directs strand specific spacer acquisition. *PLoS One* **7**:e35888 http://dx.doi.org/10.1371/journal.pone.0035888.

24. Nussenzweig PM, McGinn J, Marraffini LA. 2019. Cas9 cleavage of viral genomes primes the acquisition of new immunological memories. *Cell Host Microbe* **26**:515–526.e6.

25. Semenova E, Savitskaya E, Musharova O, Strotskaya A, Vorontsova D, Datsenko KA, Logacheva MD, Severinov K. 2016. Highly efficient primed spacer acquisition from targets destroyed by the *Escherichia coli* type I-E CRISPR-Cas interfering complex. *Proc Natl Acad Sci U S A* **113**:7626–7631.

26. Nuñez JK, Kranzusch PJ, Noeske J, Wright AV, Davies CW, Doudna JA. 2014. Cas1-Cas2 complex formation mediates spacer acquisition during CRISPR-Cas adaptive immunity. *Nat Struct Mol Biol* **21**:528–534.

27. Xiao Y, Ng S, Nam KH, Ke A. 2017. How type II CRISPR-Cas establish immunity through Cas1-Cas2-mediated spacer integration. *Nature* **550**:137–141.

28. Rollie C, Graham S, Rouillon C, White MF. 2018. Prespacer processing and specific integration in a type I-A CRISPR system. *Nucleic Acids Res* **46**:1007–1020.

29. McGinn J, Marraffini LA. 2016. CRISPR-Cas systems optimize their immune response by specifying the site of spacer integration. *Mol Cell* **64**:616–623.

30. Wright AV, Doudna JA. 2016. Protecting genome integrity during CRISPR immune adaptation. *Nat Struct Mol Biol* **23**:876–883.

31. Nuñez JK, Bai L, Harrington LB, Hinder TL, Doudna JA. 2016. CRISPR immunological memory requires a host factor for specificity. *Mol Cell* **62**:824–833.

32. Shmakov SA, Sitnik V, Makarova KS, Wolf YI, Severinov KV, Koonin EV. 2017. The CRISPR spacer space is dominated by sequences from species-specific mobilomes. *mBio* **8**:e01397-17 http://dx.doi.org/10.1128/mBio.01397-17.

33. Smillie C, Garcillán-Barcia MP, Francia MV, Rocha EPC, de la Cruz F. 2010. Mobility of plasmids. *Microbiol Mol Biol Rev* **74**:434–452.

34. Yang QE, Walsh TR. 2017. Toxin-antitoxin systems and their role in disseminating and maintaining antimicrobial resistance. *FEMS Microbiol Rev* **41**:343–353.

35. Weigel LM, Clewell DB, Gill SR, Clark NC, McDougal LK, Flannagan SE, Kolonay JF, Shetty J, Killgore GE, Tenover FC. 2003. Genetic analysis of a high-level vancomycin-resistant isolate of *Staphylococcus aureus*. *Science* **302**:1569–1571.

36. Thomas CM, Nielsen KM. 2005. Mechanisms of, and barriers to, horizontal gene transfer between bacteria. *Nat Rev Microbiol* **3**:711–721.

37. Wheatley RM, MacLean RC. 2021. CRISPR-Cas systems restrict horizontal gene transfer in *Pseudomonas aeruginosa*. *ISME J* **15**:1420–1433.

38. Westra ER, Levin BR. 2020. It is unclear how important CRISPR-Cas systems are for protecting natural populations of bacteria against infections by mobile genetic elements. *Proc Natl Acad Sci U S A* **117**:27777–27785.

39. Gophna U, Kristensen DM, Wolf YI, Popa O, Drevet C, Koonin EV. 2015. No evidence of inhibition of horizontal gene transfer by CRISPR-Cas on evolutionary timescales. *ISME J* **9**:2021–2027.

40. Minot S, Sinha R, Chen J, Li H, Keilbaugh SA, Wu GD, Lewis JD, Bushman FD. 2011. The human gut virome: inter-individual variation and dynamic response to diet. *Genome Res* **21**:1616–1625.

41. Seed KD, Lazinski DW, Calderwood SB, Camilli A. 2013. A bacteriophage encodes its own CRISPR/Cas adaptive response to evade host innate immunity. *Nature* **494**:489–491.

42. Godde JS, Bickerton A. 2006. The repetitive DNA elements called CRISPRs and their associated genes: evidence of horizontal transfer among prokaryotes. *J Mol Evol* **62**:718–729.

43. Millen AM, Horvath P, Boyaval P, Romero DA. 2012. Mobile CRISPR/Cas-mediated bacteriophage resistance in *Lactococcus lactis*. *PLoS One* **7**:e51663 http://dx.doi.org/10.1371/journal.pone.0051663.

44. Almendros C, Guzmán NM, García-Martínez J, Mojica FJM. 2016. Anti-*cas* spacers in orphan CRISPR4 arrays prevent uptake of active CRISPR-Cas I-F systems. *Nat Microbiol* **1**:16081 http://dx.doi.org/10.1038/nmicrobiol.2016.81.

45. Bikard D, Hatoum-Aslan A, Mucida D, Marraffini LA. 2012. CRISPR interference can prevent natural transformation and virulence acquisition during in vivo bacterial infection. *Cell Host Microbe* **12**:177–186.

46. Zhang Y, Heidrich N, Ampattu BJ, Gunderson CW, Seifert HS, Schoen C, Vogel J, Sontheimer EJ. 2013. Processing-independent CRISPR RNAs limit natural transformation in *Neisseria meningitidis*. *Mol Cell* **50**:488–503.

47. Turgeman-Grott I, Joseph S, Marton S, Eizenshtein K, Naor A, Soucy SM, Stachler A-E, Shalev Y, Zarkor M, Reshef L, Altman-Price N, Marchfelder A, Gophna U. 2019. Pervasive acquisition of CRISPR memory driven by inter-species mating of archaea can limit gene transfer and influence speciation. *Nat Microbiol* **4**:177–186.

48. Naor A, Lapierre P, Mevarech M, Papke RT, Gophna U. 2012. Low species barriers in halophilic archaea and the formation of recombinant hybrids. *Curr Biol* **22**:1444–1448.

49. Garneau JE, Dupuis M-È, Villion M, Romero DA, Barrangou R, Boyaval P, Fremaux C, Horvath P, Magadán AH, Moineau S. 2010. The CRISPR/Cas bacterial immune system cleaves bacteriophage and plasmid DNA. *Nature* **468**:67–71.

50. Gasiunas G, Barrangou R, Horvath P, Siksnys V. 2012. Cas9-crRNA ribonucleoprotein complex mediates specific DNA cleavage for adaptive immunity in bacteria. *Proc Natl Acad Sci U S A* **109**:E2579–E2586.

51. Jinek M, Chylinski K, Fonfara I, Hauer M, Doudna JA, Charpentier E. 2012. A programmable dual-RNA-guided DNA endonuclease in adaptive bacterial immunity. *Science* **337**:816–821.

52. Hale CR, Zhao P, Olson S, Duff MO, Graveley BR, Wells L, Terns RM, Terns MP. 2009. RNA-guided RNA cleavage by a CRISPR RNA-Cas protein complex. *Cell* **139**:945–956.

53. Zetsche B, Gootenberg JS, Abudayyeh OO, Slaymaker IM, Makarova KS, Essletzbichler P, Volz SE, Joung J, van der Oost J, Regev A, Koonin EV, Zhang F. 2015. Cpf1 is a single RNA-guided endonuclease of a class 2 CRISPR-Cas system. *Cell* **163**:759–771.

54. Abudayyeh OO, Gootenberg JS, Konermann S, Joung J, Slaymaker IM, Cox DBT, Shmakov S, Makarova KS, Semenova E, Minakhin L, Severinov K, Regev A, Lander ES, Koonin EV, Zhang F. 2016. C2c2 is a single-component programmable RNA-guided RNA-targeting CRISPR effector. *Science* **353**:aaf5573 http://dx.doi.org/10.1126/science.aaf5573.

55. Westra ER, van Erp PBG, Künne T, Wong SP, Staals RHJ, Seegers CLC, Bollen S, Jore MM, Semenova E, Severinov K, de Vos WM, Dame RT, de Vries R, Brouns SJJ, van der Oost J. 2012. CRISPR immunity relies on the consecutive binding and degradation of negatively supercoiled invader DNA by Cascade and Cas3. *Mol Cell* **46**:595–605.

56. Nishimasu H, Ran FA, Hsu PD, Konermann S, Shehata SI, Dohmae N, Ishitani R, Zhang F, Nureki O. 2014. Crystal structure of Cas9 in complex with guide RNA and target DNA. *Cell* **156**:935–949.

57. Zhao H, Sheng G, Wang J, Wang M, Bunkoczi G, Gong W, Wei Z, Wang Y. 2014. Crystal structure of the RNA-guided immune surveillance Cascade complex in *Escherichia coli*. *Nature* **515**:147–150.

58. Jackson RN, Golden SM, van Erp PBG, Carter J, Westra ER, Brouns SJJ, van der Oost J, Terwilliger TC, Read RJ, Wiedenheft B. 2014. Structural biology. Crystal structure of the CRISPR RNA-guided surveillance complex from *Escherichia coli*. *Science* **345**:1473–1479.

59. Mulepati S, Héroux A, Bailey S. 2014. Structural biology. Crystal structure of a CRISPR RNA-guided surveillance complex bound to a ssDNA target. *Science* **345**:1479–1484.

60. Yamano T, Nishimasu H, Zetsche B, Hirano H, Slaymaker IM, Li Y, Fedorova I, Nakane T, Makarova KS, Koonin EV, Ishitani R, Zhang F, Nureki O. 2016. Crystal structure of Cpf1 in complex with guide RNA and target DNA. *Cell* **165**:949–962.

61. Jiang F, Taylor DW, Chen JS, Kornfeld JE, Zhou K, Thompson AJ, Nogales E, Doudna JA. 2016. Structures of a CRISPR-Cas9 R-loop complex primed for DNA cleavage. *Science* **351**:867–871.

62. Klompe SE, Vo PLH, Halpin-Healy TS, Sternberg SH. 2019. Transposon-encoded CRISPR-Cas systems direct RNA-guided DNA integration. *Nature* **571**:219–225.

63. Strecker J, Ladha A, Gardner Z, Schmid-Burgk JL, Makarova KS, Koonin EV, Zhang F. 2019. RNA-guided DNA insertion with CRISPR-associated transposases. *Science* **365**:48–53.

64. Deveau H, Barrangou R, Garneau JE, Labonté J, Fremaux C, Boyaval P, Romero DA, Horvath P, Moineau S. 2008. Phage response to CRISPR-encoded resistance in *Streptococcus thermophilus*. *J Bacteriol* **190**:1390–1400.

65. Mojica FJM, Díez-Villaseñor C, García-Martínez J, Almendros C. 2009. Short motif sequences determine the targets of the prokaryotic CRISPR defence system. *Microbiology (Reading)* **155**:733–740.

66. Semenova E, Jore MM, Datsenko KA, Semenova A, Westra ER, Wanner B, van der Oost J, Brouns SJJ, Severinov K. 2011. Interference by clustered regularly interspaced short palindromic repeat (CRISPR) RNA is governed by a seed sequence. *Proc Natl Acad Sci U S A* **108**:10098–10103.

67. Sashital DG, Wiedenheft B, Doudna JA. 2012. Mechanism of foreign DNA selection in a bacterial adaptive immune system. *Mol Cell* **46**:606–615.

68. Anders C, Niewoehner O, Duerst A, Jinek M. 2014. Structural basis of PAM-dependent target DNA recognition by the Cas9 endonuclease. *Nature* **513**:569–573.

69. Hochstrasser ML, Taylor DW, Bhat P, Guegler CK, Sternberg SH, Nogales E, Doudna JA. 2014. CasA mediates Cas3-catalyzed target degradation during CRISPR RNA-guided interference. *Proc Natl Acad Sci U S A* **111**:6618–6623.

70. Sternberg SH, Redding S, Jinek M, Greene EC, Doudna JA. 2014. DNA interrogation by the CRISPR RNA-guided endonuclease Cas9. *Nature* **507**:62–67.

71. Redding S, Sternberg SH, Marshall M, Gibb B, Bhat P, Guegler CK, Wiedenheft B, Doudna JA, Greene EC. 2015. Surveillance and processing of foreign DNA by the *Escherichia coli* CRISPR-Cas system. *Cell* **163**:854–865.

72. East-Seletsky A, O'Connell MR, Knight SC, Burstein D, Cate JHD, Tjian R, Doudna JA. 2016. Two distinct RNase activities of CRISPR-C2c2 enable guide-RNA processing and RNA detection. *Nature* **538**:270–273.

73. Kazlauskiene M, Tamulaitis G, Kostiuk G, Venclovas Č, Siksnys V. 2016. Spatiotemporal control of type III-A CRISPR-Cas immunity: coupling DNA degradation with the target RNA recognition. *Mol Cell* **62**:295–306.

74. Kazlauskiene M, Kostiuk G, Venclovas Č, Tamulaitis G, Siksnys V. 2017. A cyclic oligonucleotide signaling pathway in type III CRISPR-Cas systems. *Science* **357**:605–609.

75. Niewoehner O, Garcia-Doval C, Rostøl JT, Berk C, Schwede F, Bigler L, Hall J, Marraffini LA, Jinek M. 2017. Type III CRISPR-Cas systems produce cyclic oligoadenylate second messengers. *Nature* **548**:543–548.

76. Jiang W, Samai P, Marraffini LA. 2016. Degradation of phage transcripts by CRISPR-associated RNases enables type III CRISPR-Cas immunity. *Cell* **164**:710–721.

77. Rostøl JT, Marraffini LA. 2019. Non-specific degradation of transcripts promotes plasmid clearance during type III-A CRISPR-Cas immunity. *Nat Microbiol* **4**:656–662.

78. Meeske AJ, Nakandakari-Higa S, Marraffini LA. 2019. Cas13-induced cellular dormancy prevents the rise of CRISPR-resistant bacteriophage. *Nature* **570**:241–245.

79. Pyenson NC, Gayvert K, Varble A, Elemento O, Marraffini LA. 2017. Broad targeting specificity during bacterial type III CRISPR-Cas immunity constrains viral escape. *Cell Host Microbe* **22**:343–353.e3.

80. Marraffini LA, Sontheimer EJ. 2010. Self versus non-self discrimination during CRISPR RNA-directed immunity. *Nature* **463**:568–571.

81. Meeske AJ, Marraffini LA. 2018. RNA guide complementarity prevents self-targeting in type VI CRISPR systems. *Mol Cell* **71**:791–801.e3.

82. Bondy-Denomy J, Pawluk A, Maxwell KL, Davidson AR. 2013. Bacteriophage genes that inactivate the CRISPR/Cas bacterial immune system. *Nature* **493**:429–432.

83. Paez-Espino D, Morovic W, Sun CL, Thomas BC, Ueda K, Stahl B, Barrangou R, Banfield JF. 2013. Strong bias in the bacterial CRISPR elements that confer immunity to phage. *Nat Commun* **4**:1430 http://dx.doi.org/10.1038/ncomms2440.

84. Pyenson NC, Marraffini LA. 2020. Co-evolution within structured bacterial communities results in multiple expansion of CRISPR loci and enhanced immunity. *eLife* **9**:e53078 http://dx.doi.org/10.7554/eLife.53078.

85. Sun CL, Barrangou R, Thomas BC, Horvath P, Fremaux C, Banfield JF. 2013. Phage mutations in response to CRISPR diversification in a bacterial population. *Environ Microbiol* **15**:463–470.

86. Savitskaya E, Semenova E, Dedkov V, Metlitskaya A, Severinov K. 2013. High-throughput analysis of type I-E CRISPR/Cas spacer acquisition in *E. coli*. *RNA Biol* **10**:716–725.

87. Kieper SN, Almendros C, Behler J, McKenzie RE, Nobrega FL, Haagsma AC, Vink JNA, Hess WR, Brouns SJJ. 2018. Cas4 facilitates PAM-compatible spacer selection during CRISPR adaptation. *Cell Rep* **22**:3377–3384.

88. Shiimori M, Garrett SC, Graveley BR, Terns MP. 2018. Cas4 nucleases define the PAM, length, and orientation of DNA fragments integrated at CRISPR loci. *Mol Cell* **70**:814–824.e6.

89. van Houte S, Ekroth AKE, Broniewski JM, Chabas H, Ashby B, Bondy-Denomy J, Gandon S, Boots M, Paterson S, Buckling A, Westra ER. 2016. The diversity-generating benefits of a prokaryotic adaptive immune system. *Nature* **532**:385–388.

90. Roy D, Huguet KT, Grenier F, Burrus V. 2020. IncC conjugative plasmids and SXT/R391 elements repair double-strand breaks caused by CRISPR-Cas during conjugation. *Nucleic Acids Res* **48**:8815–8827.

91. Hossain AA, McGinn J, Meeske AJ, Modell JW, Marraffini LA. 2021. Viral recombination systems limit CRISPR-Cas targeting through the generation of escape mutations. *Cell Host Microbe* **29**:1482–1495.e12.

CHAPTER

Evolutionary Classification of CRISPR-Cas Systems

Kira S. Makarova, Yuri I. Wolf, and Eugene V. Koonin

National Center for Biotechnology Information, National Library of Medicine, Bethesda, MD 20894

Introduction

CRISPR-Cas systems are bacterial and archaeal adaptive immune systems. The mechanisms and biology of CRISPR are discussed in many recent reviews (1–7) and other chapters of this book, so here we provide only bare essentials as the context for the discussion of CRISPR-Cas diversity, classification, and evolution.

The CRISPR immune response includes three principal phases: adaptation, expression, and interference. In the adaptation phase, the CRISPR effector, a distinct complex of Cas proteins, binds to a target DNA, often after recognizing a short sequence motif known as the protospacer-adjacent motif (PAM), and excises a fragment of the target DNA, the protospacer. The adaptation complex inserts the protospacer DNA into the 5′ end of a CRISPR array, where it becomes a spacer, which is accompanied by duplication of the 5′-terminal repeat. Some CRISPR systems use an alternative mechanism of adaptation, spacer acquisition from RNA via reverse transcription, which is catalyzed by a reverse transcriptase (RT) encoded in the CRISPR-*cas* locus.

In the expression phase, the CRISPR array is typically transcribed as a single transcript, known as the pre-CRISPR RNA (pre-crRNA), that is processed into mature crRNAs. Each crRNA consists of a spacer sequence and portions of the flanking repeats. In different CRISPR variants, the pre-crRNA is processed by a dedicated multiprotein Cas complex, by a single, multidomain effector Cas protein, or by RNases encoded by genes other than *cas*.

In the interference phase, the crRNA typically remains bound to the effector Cas protein complex (or single effector protein) and functions as the guide to recognize the protospacer (or a closely similar sequence) in an invading genome of a virus or plasmid. The nucleic acid bound to the crRNA is then cleaved and inactivated by a Cas nuclease (or nucleases) that can be either a subunit (domain) of the effector or a different protein recruited at the interference stage. We only present here the briefest, oversimplified overview of CRISPR-Cas functionality that is necessary to discuss the evolutionary classification of CRISPR, omitting all the (fascinating) details.

Similar to other defense systems, archaeal and bacterial CRISPR systems show remarkable diversity of Cas protein sequences, gene composition, and architecture of

CRISPR: Biology and Applications, First Edition. Edited by Rodolphe Barrangou, Erik J. Sontheimer, and Luciano A. Marraffini.
© 2022 American Society for Microbiology. DOI: 10.1128/9781683673798.ch02

the genomic loci (1, 3, 8–13). Our knowledge of this diversity is continuously expanding through screening of the ever-growing genomic and metagenomic databases. To keep pace with this expansion, a robust classification of CRISPR systems based on evolutionary relationships is essential for the progress of CRISPR research but presents formidable challenges due to the lack of universal markers and the fast evolution of the CRISPR-*cas* loci (14). Therefore, the three consecutive versions of CRISPR-Cas classification employed a multipronged approach that combined comparison of the gene compositions of CRISPR-Cas systems and locus architectures clustering by sequence similarity and phylogenetic analysis of the slowly evolving Cas proteins, such as Cas1 (15–17).

The latest classification, published in 2020, includes 2 classes, 6 types, and 33 subtypes; two or more variants are distinguished within some of the subtypes (17). The two CRISPR-Cas classes are distinguished by the fundamentally different architectures of the effector modules involved in crRNA processing and interference (Fig. 2.1). The classification of CRISPR-Cas systems is linked to the classification of *cas* genes that currently consists of 13 core families (Fig. 2.1A and B). Class 1 systems have effector modules composed of multiple Cas proteins, some of which form crRNA-binding complexes (such as the Cascade complex in type I systems) that, with contributions from additional Cas proteins, mediate pre-crRNA processing and interference. In contrast, the effector moieties of class 2 systems are single, multidomain crRNA-binding proteins (such as Cas9 in type II) that combine all activities required for interference and, in many variants, those involved in pre-crRNA processing as well (Fig. 2.1B).

In recent years, the known diversity of CRISPR-Cas systems has dramatically expanded, thanks to the advances of genomics and metagenomics complemented by dedicated efforts to identify new variants of CRISPR with an extended range of activities that potentially could be harnessed for new applications (18–27). Given the greater

Figure 2.1 The two classes of CRISPR-Cas systems, their modular organization, and Cas protein nomenclature (**A**) Generic organization of class 1 and class 2 CRISPR-Cas loci. Class 1 CRISPR-Cas systems have effector modules composed of multiple Cas proteins that form a crRNA-binding complex and function together in target binding and processing. Class 2 systems have a single, multidomain crRNA-binding protein that is functionally analogous to the entire effector complex of class 1. All core Cas genes are indicated. The most common ancillary components that are, however, missing in some subtypes and variants are indicated by dashed outlines. (**B**) Functional modules of CRISPR-Cas systems. The scheme shows the typical relationships between genetic, structural, and functional organizations of the 6 types of CRISPR-Cas systems. Protein names follow the current nomenclature (17). An asterisk indicates the putative small subunit that is often fused to the large subunit in several type I subtypes. The most common ancillary components that are, however, missing in some subtypes and variants are indicated by dashed outlines. Cas6 is shown with a thin solid outline for type I because it is dispensable in some, but not most, systems and with a dashed line for type III because most of these apparently use the Cas6 protein provided in *trans* by other CRISPR-Cas loci. The three colors for Cas9, Cas10, Cas12, and Cas13 indicate that these proteins contribute to different stages of the CRISPR-Cas response. The CARF and HEPN domain proteins are the most common sensors and effectors, respectively, in the type III ancillary modules, but several alternative sensors and effectors have been identified as well (47, 94). Most of the ring nucleases are a distinct variety of CARF domain proteins that cleave cyclic oligo(**A**) produced by Cas10 and thus control the indiscriminate RNase activity of the HEPN domain of Csx6 (54). (**C**) The hierarchical *cas* gene nomenclature scheme and the evidence used for CRISPR-Cas classification exemplified by subtype VI-B systems. Gene neighborhood analysis allows unambiguous classification of this system as class 2. Motif search and profile-profile comparison of HEPN domains results in its classification as type VI. However, PSI-BLAST searches do not detect sequence similarity to any of the previously identified type VI effector proteins. Moreover, these loci encompass distinct ancillary genes, supporting their classification as a separate subtype (VI-B). The phylogenetic tree of Cas13b contains two strongly supported branches that are associated with distinct ancillary genes. Accordingly, subtype VI-B is subdivided into two variants (19). The figure panels are modified from reference 17.

utility of class 2 CRISPR for applications thanks to the compactness of the respective effectors (above all, Cas9), such efforts have been primarily directed at the discovery of new class 2 subtypes and variants, resulting in the identification of the RNA-targeting type VI (encompassing Cas13) and multiple previously unknown subtypes of type V systems (encompassing Cas12). Furthermore, unexpectedly, it has been recently shown that

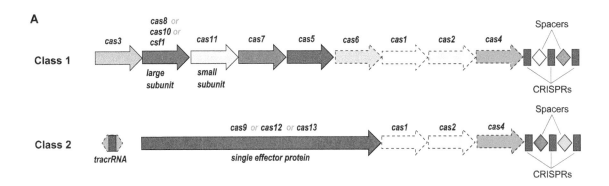

type V CRISPR effectors evolved from transposon-encoded TnpB nucleases on multiple, independent occasions, resulting in a large pool of type V variants that have been gradually elevated to the subtype status as they were studied experimentally (9, 17, 19). Another remarkable development was the discovery of several class 1 and class 2 CRISPR-Cas variants that appear to lack the RNA-guided targeted cleavage activity and thus either mediate adaptive immunity via mechanisms that are substantially different from those of typical CRISPR systems or perform completely different functions (28–30). Such derived CRISPR-Cas systems include type IV, several variants of type I, and at least one type V variant, and they are often encoded by mobile genetic elements (MGEs) (30–32). Following predictions made on the basis of comparative genome analysis, it has been demonstrated experimentally that type I and type V CRISPR-Cas variants encoded by Tn7-like transposons mediate crRNA-dependent DNA transposition (33, 34). The origin of some of these derived forms from particular class 1 and class 2 subtypes is readily identifiable, but nevertheless, they present a challenge from the point of view of CRISPR-Cas classification. The third group of important recent findings expanding CRISPR diversity is the identification of numerous gene families associated with specific variants of CRISPR-Cas systems, particularly, those of type III systems, and implicated in signal transduction and regulatory roles (28–30, 35, 36).

In this chapter, we review the current state and future challenges of CRISPR-Cas classification systems, the emerging picture of the diverse routes of the origin and evolution of the different CRISPR variants, and the evolutionary connections between CRISPR and other defense and signal transduction systems of prokaryotes.

Principles of CRISPR-Cas System Classification

There are no universal genes in CRISPR-Cas systems, which rules out a straightforward, comprehensive phylogenetic approach to classification modeled on the global taxonomy of cellular life forms that is based on phylogenetic analysis of universal genes, such as those encoding rRNAs or ribosomal proteins. Moreover, even in the case of widely conserved *cas* genes, such as *cas1*, phylogenetic analysis is of limited use for classification because of frequent module shuffling among the CRISPR loci (14, 37, 38). Therefore, the computational strategy for CRISPR classification relies on multiple criteria, including identification of signature genes that can be used to define CRISPR-Cas types, subtypes, and variants and comparison of gene compositions and organizations of genes in CRISPR-*cas* loci, as well as sequence similarity-based clustering and phylogenetic analysis of *cas* genes that are sufficiently highly conserved in subsets of CRISPR-Cas systems to be phylogenetically informative. The available experimental data were taken into consideration as well (15–17, 39).

For the latest (2020) version of the CRISPR classification (17), 566 amino acid sequence profiles representing all identified variants of the 13 core *cas* genes (Table 2.1), some components of effector complexes that remain uncharacterized, and reliably identified ancillary CRISPR-linked genes were compared to the protein sequences from 13,116 complete archaeal and bacterial genomes that were available at the NCBI as of 1 March 2019, using PSI-BLAST (40). After extensive manual curation of the results of this search, a collection of 7,915 reliably identified CRISPR-*cas* loci was obtained. Based on the presence of signature *cas* genes, sequence similarity between Cas proteins, phylogenies of the conserved Cas proteins (including Cas1 and effector proteins for individual types and subtypes), and comparison of the loci organization, these loci were included into previously identified subtypes or else assigned to new subtypes.

Table 2.1 The core proteins of CRISPR-Cas systems

Family	Function and structure	References for further reading
Cas1	Metal-dependent deoxyribonuclease that functions as the integrase during adaptation. Typically forms complex with Cas2, but Cas1 from type V-C does not require Cas2 for spacer integration. Unique fold with two domains: N-terminal β-stranded domain and catalytic C-terminal α-helical domain.	66, 97–99
Cas2	Some are RNases specific to U-rich regions, others are double-stranded DNases, but some are apparently inactivated; form a tight complex with Cas1 and appear to perform a structural role during adaptation. Belong to RRM (ferredoxin) fold.	98, 100–102
Cas3	ssDNA metal-dependent deoxyribonuclease (HD domain) and helicase; required for interference in type I systems. In some systems, the two domains encoded separately and designated Cas3′ (helicase) and Cas″ (HD domain).	103–106
Cas4	PD-(DE)xK superfamily nuclease with four conserved cysteines coordinating one [4Fe-4S] or [2Fe-2S] cluster; cleaves ssDNA in the 5′–3′ direction or both directions. A component of the adaptation complexes in many subtypes, assisting in precise protospacer processing and PAM selection.	37, 107–110
Cas5	Subunit of Cascade complex interacting with large subunit and Cas7 subunit and binding the 5′-handle of crRNA. In the subtype I-C system, Cas5 is the ribonuclease that replaces the Cas6 function. Two domains of RRM (ferredoxin) fold; the C-terminal domain is deteriorated in many Cas5 proteins of type I systems.	111–115
Cas6	Metal-independent endoribonuclease that generates crRNAs. Two domains of RRM (ferredoxin) fold; RAMP superfamily.	38, 116–119
Cas7	Subunit of Cascade complexes binding crRNA; often present in Cascade complexes in several copies. In type III systems, cleaves target RNA. Single RRM (ferredoxin) fold with subdomains; RAMP superfamily.	55, 62, 114, 115, 120
Cas8abcefg, (large subunit)	Subunit of class 1 Cascade complex, involved in PAM recognition. Typically, multidomain proteins; most of the distinct families have no sequence and structure similarity but perform the same function in the respective effector complexes.	114, 121–124
Cas9	Type II effector protein, programmable DNA nuclease containing RuvC and HNH nuclease domains, which are involved in the cleavage of the target DNA. Cas9 is required to generate crRNA and to cleave the target DNA. Function in complex tracrRNA encoded in the same locus; additionally, Cas9 contributes to adaptation, in particular by recognizing the PAM motif. Cas9 has several subdomains, including RuvC and HNH nuclease domains, and adopts a bilobed general structure.	4, 67, 68, 125–129
Cas10 large subunit	Large subunit of most type III effector complexes. Active Cas10 synthetizes cycle oligoadenylates that are signaling molecule-activating multiple ancillary proteins. Two domains homologous to Palm domain polymerases and cyclases; both belong to RRM (ferredoxin) fold; Zn finger-containing domain and C-terminal α-helical domain. Many are fused to HD nuclease domain.	35, 36, 65, 114, 115, 130–135
Cas11 (Cse2, Csa5, Csm2, Cmr5) small subunit	Small, mostly α-helical protein, subunit of class 1 Cascade complexes; interacts with crRNA. In many type I systems is fused to large subunit Cas8. Cse2 has two α-helical bundle-like domains; Cmr5 has a domain matching the N-terminal domain of Cse1 and Csa5 has a domain matching the C-terminal domain of Cse2. Distinct family have low or no sequence similarity but perform the same function in class 1 effector complexes.	5, 114, 136–138
Cas12	Type V effector proteins, programmable DNA nuclease. Cas12 proteins contain an active RuvC-like domain responsible for the cleavage of both strands of the target DNA. Several Cas12 also contain a subdomain involved in processing of pre-crRNA, but other type V loci, similarly to type II, encode a tracrRNA that is involved in the processing of pre-crRNA along with the housekeeping bacterial RNase III. Cas12 adopts a bilobed shape and contains has several subdomains, including an RuvC nuclease domain and often an OB-fold domain involved in pre-RNA cleavage.	21, 68, 69, 72, 139
Cas13	Type VI effectors, programmable RNA nuclease. All Cas13 proteins contains two HEPN superfamily RNase domains. Cas13a and Cas13d are active in processing of pre-crRNA. Once activated by the RNA target recognition, Cas13 becomes a nonspecific RNase that appears to be toxic for the cell, inducing dormancy. Cas13 are multidomain proteins adopting a bilobed shape.	18, 23–25, 73

Classification of CRISPR-Cas Systems

Class 1 and Its Derivatives

Class 1 is defined by effector modules consisting of several *cas* genes encoding Cas proteins that function as subunits of effector complexes or as standalone enzymes. The 2020 version of class 1 includes CRISPR-Cas systems of types I, III, and IV, with 16 subtypes, 4 of which—subtypes III-E, III-F, IV-B, and IV-C—were new compared to the previous (2015) version of the classification (16) (Fig. 2.2). The class 1 systems share a common overall organization of the effector complexes, which implies common ancestry. The scaffold of these complexes is formed by multiple, paralogous proteins, Cas5 and Cas7, that contain highly divergent versions of the RNA recognition motif (RRM) domain (Fig. 2.1B). Members of the third family of RRM domain proteins, Cas6, are standalone ribonucleases that in most class 1 systems are responsible for processing of pre-crRNA. In addition to the RRM proteins (historically known as repeat-associated mysterious proteins [RAMPs]), class 1 effector complexes typically contain the so-called large and small subunits, which provide for a uniform gross architecture but differ substantially as discussed below (Fig. 2.2).

Type I

Type I CRISPR-Cas systems are the most diverse and most abundant ones in the microbial world. The signature gene for type I is *cas3*, which encodes a single-stranded DNA (ssDNA)-stimulated helicase that in most type I systems is fused to an N-terminal HD nuclease domain, but in some subtypes and variants, the helicase and the nuclease are encoded by separate genes. Type I consists of 7 subtypes (I-A to I-G), each with distinct features, gene compositions, and operonic organizations (Fig. 2.2). The prototype, probably, ancestral organization of type I systems seems to comprise 8 genes, *cas1* to *cas8*. This prototypical gene arrangement persists in the expansive subtype I-B, whereas in other subtypes, various genes were lost and rearranged. In subtypes I-C, I-D, I-E, and I-F, all *cas* genes typically comprise a single operon that combines the adaptation and effector modules. In the other type I subtypes, the *cas* genes are distributed between two or more operons. The subtypes within type I lack unique gene signatures, but most possess distinct, defining features that often represent partial disruption or modification of the ancestral gene complement. Thus, subtype I-A appears to be a derivative of I-B, in which Cas3 is split into distinct helicase and nuclease moieties, and Cas8 is also split into two distinct proteins, which comprise derived large and small subunits of the effector complex (Fig. 2.2). Subtype I-C also seems to be a partially degraded derivative of I-B, in which Cas6 was lost, apparently, being functionally replaced by Cas5 (Fig. 2.2). Subtype I-D stands apart from the other type I systems in that the HD nuclease domain involved in the target DNA cleavage is relocated from Cas3 to the large effector complex subunit, designated Cas10d (Fig. 2.2). Thus, the large subunit architecture in subtype I-D resembles that in type III systems, although the rest of the Cas10d sequence shows no similarity to the polymerase-cyclase domains of Cas10 (see below). In subtype I-E, similarly to I-A, there are two distinct genes encoding the large (Cas8e) and the small (Cse2) subunits of the effector complex. Subtype I-F is characterized by the fusion of the *cas2* and *cas3* genes, which reflects involvement of Cas3 in adaptation, although neither the helicase nor the nuclease activity is required (41). Finally, subtype I-G encompasses two unique gene fusions. The first is *cas1-cas4*, which similarly reflects an accessory role of Cas4 in adaptation, and the second one is *cas5-cas6*, which most likely doubles as a ribonuclease cleaving the pre-crRNA and a component of the effector complex. A notable feature of type I CRISPR-Cas is the extreme variability of the large subunit

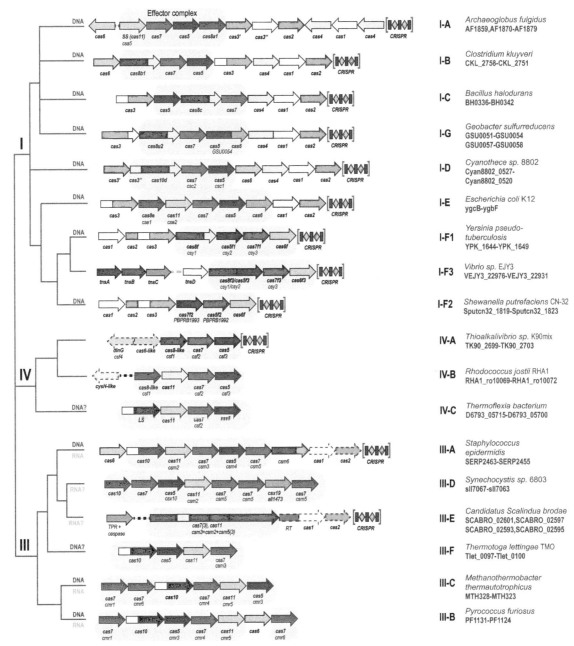

Figure 2.2 Updated classification of class 1 CRISPR-Cas systems The diagram schematically shows representative (typical) CRISPR-cas loci of each class 1 subtype and selected distinct variants, with the dendrogram on the left showing the likely evolutionary relationships between the types and subtypes. The column on the right indicates the organism and the corresponding gene range. Homologous genes are color-coded and identified by family name. The gene names follow the previous classification (17). Where both a systematic name and a legacy name are commonly used, the legacy name is given under the systematic name. The small subunit is encoded by *csm2*, *cmr5*, *cse2*, *csa5*, and several additional families of homologous genes that are collectively designated *cas11*. The adaptation module genes *cas1* and *cas2* are dispensable in subtype III-A and subtype III-B (dashed lines). Gene regions colored cream represent the HD nuclease domain; the HD domain in Cas10 is distinct from the HD domain of Cas3. Functionally uncharacterized genes are shown in gray. The pink shading shows the effector module. Most of the subtype III-B, III-C, III-E, and III-F loci as well as IV-B and IV-C loci lack CRISPR arrays and are shown accordingly, although for each of the type III subtypes, exceptions have been detected. CHAT, protease domain of the caspase family; RT, reverse transcriptase; TPR, tetratricopeptide repeats. The figure is modified from reference 17.

of the effector complex, Cas8. In most cases, there is no detectable sequence similarity between Cas8 proteins of different subtypes and even variants within the same subtype, such as I-B. Due to this extreme variability, Cas8 proteins are not suitable signatures for subtype differentiation within type I. Phylogenetic trees of two conserved genes of type I systems, *cas1* and *cas3*, are only partially compatible with the subtype classification, suggestive of multiple gene exchanges between different type I variants, in accord with the paradigm of modular evolution of CRISPR-Cas systems (14, 16, 42).

Derived Variants of Type I and Exaptation of Type I Systems for Nondefense Functions

Apart from the fully functional adaptive immune systems, several derived variants of type I have been discovered that have been shown or predicted to be recruited for functions distinct from defense (these are typical cases of exaptation, that is, recruitment of functional systems for new functions that are biologically unrelated but mechanistically similar to the original ones, a general evolutionary phenomenon defined by Gould and Vrba [43]). Numerous Tn7-like transposons encompass partially degraded subtype I-B, I-F, and I-E CRISPR loci that lack Cas3 and thus are incapable of interference (30–32). Phylogenetic analysis of conserved transposase subunits indicates that these CRISPR systems were captured by transposons on multiple, independent occasions, suggesting that they perform a function that is beneficial or even essential for the transposons (31). It has been first predicted and then shown experimentally that the I-B and I-F transposon-encoded CRISPR systems mediate RNA-guided transposition, a phenomenon that has not been previously described (31, 34, 44, 45). The effector complex of the transposon-associated CRISPR systems (named CAST [CRISPR-associated transposase]) (33) employs the guide crRNA to bind to the unique protospacer sequence in the target DNA and to deliver the transposase, which inserts the transposon in the vicinity of the recognized protospacer.

The CASTs are the best characterized but not the only derived variants of type I CRISPR-Cas systems. Many species of *Streptomyces* encode a reduced subtype I-E system that consists of *cas5*, *cas6*, *cas7*, and *cas8* genes which colocalize with a gene coding for a STAND NTPase, a putative programmed cell death (PCD) effector (29). Given the absence of Cas3, an adaptation module or a CRISPR array, this module is not a typical CRISPR system but, rather, can be predicted to function as a distinct defense mechanism, in conjunction with the STAND NTPase. Although the CASTs and other derived forms of type I systems substantially depart from the established subtypes both structurally and functionally, their evolutionary origin from these subtypes is unequivocal. Therefore, at least, the current solution is to classify these derived forms not as new subtypes but rather as variants with the established ones (Fig. 2.2). In particular, the CASTs are regarded as variants within subtypes I-B, I-E, and I-F.

Type III

Type III CRISPR-Cas systems are similar to type I in many features of the effector modules, suggesting a common origin (5, 9). The unique signature of type III is Cas10, the large subunit of the effector complex that contains two polymerase-cyclase Palm domains, one of which contains the typical catalytic sites and is active, whereas the other one is inactivated. All type III systems also encompass the small effector complex subunit Cas11, one Cas5 protein, and typically, several paralogous Cas7 proteins (Fig. 2.2). Cas6 is present only in some of these systems, whereas others can probably utilize crRNAs processed by Cas6 associated with co-occurring type I systems (46).

Type III loci are overall less abundant among prokaryotes, especially bacteria, than type I systems (see below), but they show a surprising diversity of gene composition and locus organization. So far, 5 subtypes have been established, but even within some of these, there is remarkable variability (Fig. 2.2). In particular, and in a sharp contrast with type I systems, many type III loci include diverse ancillary genes, implying a far greater functional versatility (29, 47). Unlike type I systems, which typically encompass both the adaptation and the effector modules, only subtype III-A (with some exceptions) and a minority of subtype III-B systems possess both modules. The rest of the type III systems lack the adaptation module and rely on other systems, typically of type I, for spacer acquisition. In many of those type III systems that include the adaptation module, it contains an extra component, an RT domain that is typically fused to Cas1 and, in some variants, to Cas6 (38). The RT provides for spacer acquisition by reverse transcription of virus mRNAs (48).

The functionality of type III CRISPR-Cas systems is more complex than that of type I systems. Unlike type I systems, which only shred the target DNA, type III systems cleave both DNA, via the HD domain that is typically fused to Cas10, and RNA transcripts of the target genome, via the RNase activity of one of the Cas7 subunits. Furthermore, most of the type III systems encompass a built-in signaling pathway that triggers indiscriminate RNA cleavage upon target recognition (35, 36, 47). This pathway depends on the polymerase activity of Cas10, which doubles as the structural component (large subunit) of the effector complex and is induced to catalyze the synthesis of cyclic oligo(A) (cOA) when this complex binds the target. The produced cOA binds to the sensor CARF (CRISPR-associated Rossmann fold) domain of the Csm6 protein or other CARF domain-containing proteins (these proteins are classified as accessory and have no Cas numbers but are present in the great majority of the type III systems) and activates nonspecific RNA cleavage by the HEPN nuclease domain of the same protein, which results in cell dormancy or death. The CARF domains present in different variants of type III systems show considerable sequence and structural diversity, implying variations in the signaling pathways that might use different signal oligonucleotides (47). The effector domains of these cOA-activated CRISPR-associated proteins also vary: although HEPN is most prevalent, other nucleases, such as those of the restriction endonuclease fold, and in some cases other enzymes, for example, proteases, are also induced by cOA and apparently cause cell dormancy (49). Thus, the cOA-activated signaling pathway displays substantial combinatorial complexity, the functional implications of which remain to be explored (47). The promiscuous RNase activity of HEPN nucleases is costly for the cell, so type III systems possess a dedicated mechanism to control the damage, whereby cOA is cleaved by a RING nuclease that is encoded either within the CRISPR locus itself or elsewhere in the genome (50–54). Some of the RING nucleases are distinct variants of the CARF domain that not only bind but cleave cOA, whereas others are enzymes of different families (50–54). The specifics of the coupling between immunity and dormancy/PCD in the type III-A CRISPR response are likely to differ depending on the lifestyles of the respective bacteria and archaea and remain to be experimentally characterized.

The components of the cOA pathway are detectable in virtually all type III CRISPR-Cas systems that include a (predicted) enzymatically active Cas10 protein, primarily subtypes III-A, III-B, and III-D. In contrast, in subtypes III-C and III-F and some variants of subtype III-D, the polymerase active site of Cas10 is disrupted by substitution of the catalytic amino acid residues, so that Cas10 only functions in its structural capacity, and the CARF-HEPN protein is missing. Subtype III-E systems lack both Cas10 and the CARF-HEPN protein. Thus, these subtypes of type III function in a simpler manner than the complete type III systems of subtypes III-A and III-B, without the coupling between immunity and dormancy/ PCD via the cOA pathway.

Derived Variants of Type III

As discussed in detail below, the ancestral form of type III, in all likelihood, encompassed the complete cOA pathway, whereas all variants lacking this functionality are partially degraded, derived forms. Evolution of derived type III systems variants is particularly apparent in subtypes III-D and III-E, through the comparison of the locus organization and phylogenetic analysis of Cas7 and Cas5. The subtype-defining III-D1 variant shows a typical type III organization of the effector module, with the intact, presumably, catalytically active Cas10 and the uncharacterized protein Csx19, a putative component of the effector complex (Fig. 2.3). Thus, the III-D1 systems can be confidently predicted to encompass the

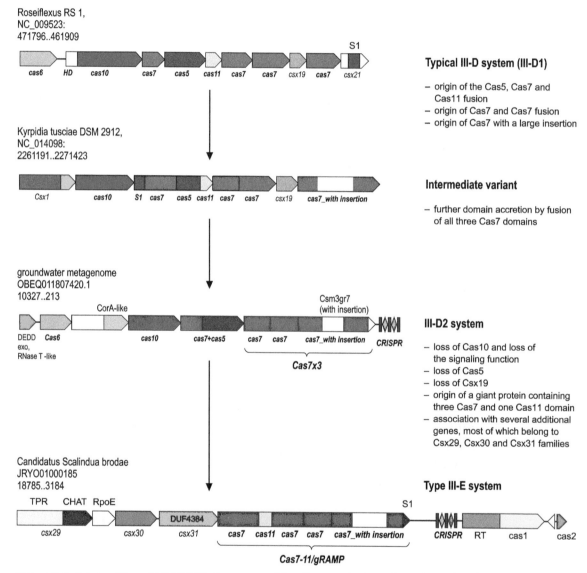

Figure 2.3 Origin of type III-E CRISPR-Cas systems The diagram depicts a hypothetical evolutionary path from a typical, fully functional subtype III-D systems to the highly derived subtype III-E. The proposed evolutionary stages are illustrated using particular loci from selected bacterial genomes. For each locus, the species name, GenBank accession number, and genomic coordinates are indicated above the scheme of the operonic organization. Homologous genes are color-coded and identified by the corresponding family name following the previous classification (18). For additional genes in each locus, abbreviated domain names are indicated. Key evolutionary events are described to the right of the schematics. DEDD exo, exonuclease of the DEDD family. The figure is modified from reference 55.

cOA pathway. Like many other type III systems, those of the III-D1 variant often lack the adaptation module. The III-D2 variant clearly is a derivative of III-D1, as indicated by the fusion of 3 Cas7 proteins and an additional fusion of Cas7 with Cas5 (Fig. 2.3). Also, the III-D2 systems occasionally associate with *cas1*, *cas6*, and/or the membrane transporter CorA (see below) and, in some cases, also proteins containing CARF or WYL domains and HEPN or other effector domains. Subtype III-E is a derivative of III-D2, in which 4 Cas7 subunits are fused with Cas11 into a single, giant effector protein (dubbed Cas7-11, for the component domains). Apart from this gene fusion, the III-E systems have lost Cas5, Csx19, Cas10, and, accordingly, the cOA pathway (Fig. 2.3). Instead, the III-E systems have gained 4 ancillary genes, including the gene for caspase-like protease Csx29 (17). In accordance with the loss of Cas10, the III-E systems have been shown to lack any DNA cleavage capacity, and given the loss of the cOA pathway, they show no indiscriminate RNA cleavage. The only activities the III-E systems retain are pre-crRNA processing and crRNA-guided cleavage of target RNA, which are both catalyzed by distinct active sites of the Cas7 domains of the effector protein (55, 56).

With regard to the classification of CRISPR-Cas systems, subtype III-E could appear problematic because the organization of its effector module, which comprises a single, multidomain protein, blurs the formal distinction between class 1 and class 2 (Fig. 2.1B). Indeed, as pointed out above and discussed in detail below, such an organization of the effector complex is the signature of class 2 systems. However, subtype III-E clearly evolved from typical type III systems (Fig. 2.3), which justifies its classification within class 1, even if a footnote now has to be added to the definitions of the two classes of CRISPR-Cas.

Some distinct variants of type III could become new subtypes when more diversity and/or more structural and functional information becomes available. In particular, a distinct type III variant was identified in archaea of the order *Sulfolobales* (represented by the locus YN1551_RS11700-YN1551_RS11720 from *Sulfolobus islandicus*) that is distinguished by extremely diverged versions of Cas10 and Cas5 and a unique, uncharacterized predicted component of the effector module, Csx26. Another distinct type III system, so far detected only in the archaeon *Ignisphaera aggregans* (locus Igag_0607..Igag_0623), includes several proteins with no detectable similarity to any known Cas or ancillary CRISPR-linked proteins.

Additional Genes Linked to Type III CRISPR-Cas Systems

As mentioned above, systematic analysis of the CRISPR-*cas* neighborhoods revealed the association of over 100 additional genes with different type III subtypes and variants, far more such CRISPR-linked genes than could be identified for any other type (17). The involvement of these ancillary genes in the activity of the respective CRISPR-Cas systems remains to be experimentally characterized, but some themes can be inferred from the predictions from sequence and structure analyses. The predominant among such functional themes are the apparent membrane association of many type III systems and connections to various forms of signal transduction (17, 28, 29). In particular, the most common among these additional CRISPR-linked genes is *corA*, which encodes a divalent cation channel and is present in many subtype III-B and III-D loci (6, 29). Many variants of subtypes III-A, III-B, and III-D contain a gene encoding a membrane protein containing a SAVED domain (a distinct variety of CARF domains) and often also a Lon family protease. Other CRISPR-linked genes encode uncharacterized membrane proteins (29, 47, 57, 58). In terms of CRISPR-Cas classification, the ancillary genes are too narrowly spread to serve as signatures of subtypes, although some can help differentiate variants. Furthermore, there is no established, coherent classification of the ancillary genes that are often shared with other defense systems.

Type IV

Sequence and structure comparisons suggest that type IV systems were derived from type III by reductive evolution (Fig. 2.2). Type IV systems include Cas7 and Cas5 but lack Cas10 (or Cas8) and instead encompass a much smaller protein, Csf1, which does not contain any recognizable enzymatic domains but has been predicted and later experimentally demonstrated to perform the role of the large subunit of the effector complex (39, 59). The type IV systems are classified into three subtypes, which differ in *cas* operon organization (17). Two additional subtypes have been proposed, but their status remains uncertain (60, 61). The most common subtypes, IV-A and IV-B, share the typical Cas5-Cas7-Csf1 organization, where in subtype IV-C, the predicted large subunit does not share any sequence similarity with Csf1 but contains an N-terminal HD nuclease domain related to the HD domain of Cas10 (17). This feature, along with the topology of the phylogenetic tree of Cas7, in which subtype IV-C comprises the basal branch to the rest of type IV, suggests that IV-C is the intermediate stage of evolution of type IV from type III (60). Comparisons of the solved structures of the effector complexes support the affinity of types IV and III (55, 62), although the similarity is too low to trace type IV to any specific type III subtype.

The subtypes within type IV display considerable variability of gene composition (60, 61). Remarkably, type IV core genes display different phylogenetic tree topologies, suggesting that shuffling of the components of the effector complexes is common among these systems (60). This makes classification of type IV systems especially challenging. Typically, however, subtypes IV-B and IV-C encompass predicted Cas11, the small subunit of the effector complex, whereas subtype IV-A systems lack this protein. Furthermore, and again in parallel with type III, different type IV subtypes include distinct accessory proteins, such as DinG helicase in subtype IV-A and an inactivated homolog of APS/PAPS reductase CysH in subtype IV-B and some subtype IV-A systems (60). These genes are strongly associated with type IV loci, suggesting distinct functionalities that remain uncharacterized, although it has been shown that DinG is required for subtype IV-A system interference activity against plasmids (63).

All type IV systems identified so far are located on plasmids, integrative conjugative elements, prophages, and some free phages (60, 61). Moreover, (nearly) all protospacer matches for the spacers in CRISPR arrays of subtypes IV-A and IV-C are in genes encoding proteins of the plasmid conjugative machinery (60, 61). Thus, the primary role of type IV systems, most likely, is in competition among MGEs, in particular, plasmid exclusion. The molecular mechanisms of type IV, however, probably differ from the typical CRISPR mode of action, given that except for the rare subtype IV-C, they all lack recognizable nuclease domains. It appears likely that subtype IV-A and IV-B systems inhibit plasmid replication without cleaving the target DNA but rather by blocking the movement of the replication fork or transcription. There seem to be major, functionally relevant differences between subtypes IV-A and IV-C, on the one hand, and subtype IV-B, on the other hand. The IV-A effector complexes bind crRNA similarly to other CRISPR systems (59), whereas subtype IV-B effectors bind heterogeneous small RNAs via a filament formed by Cas7 and Cas11 subunits (62), suggestive of a distinct yet unknown mechanism. Irrespective of the molecular details, type IV systems are another case of exaptation of partially degraded derivatives of CRISPR-Cas systems for function distinct from adaptive immunity. The functional and evolutionary parallels between type IV systems derived from type III and the CASTs derived from type I are clear and prominent.

Extremely Derived Class 1 Variants

Some derived CRISPR-Cas variants are so distant from the organization of any of the established types and subtypes that their very status as *bona fide* CRISPR-Cas systems becomes questionable. A case in point is a recently described variant found in many halo-archaea that encompasses only highly divergent forms of Cas5 and Cas7 (haloarchaeal RAMPs [HRAMPs]) along with an uncharacterized conserved protein and various nucleases (64). Recently, the structure of the uncharacterized protein of these systems has been predicted and identified as an extremely derived Cas10-like protein (65). Along similar lines, Asgard archaea encode several highly derived CRISPR-Cas variants that resemble HRAMPs in the Cas protein composition and include a unique large protein containing a diverged Cas1 domain, along with distinct variants of Cas5 (a fusion with an HD nuclease) and Cas7 as well as additional nucleases (66). The functions of these extremely derived systems are unknown, and given the lack of adjacent CRISPR arrays, it is unclear whether their activity is guide RNA dependent. If these derived systems are shown to function via CRISPR-Cas-like mechanisms, they might qualify as distinct types, given the drastic change of the Cas protein repertoire and recruitment of additional genes not found in the established types of CRISPR-Cas.

On the whole, evolution of derived variants that lack the interference capacity and have been shown or predicted to perform functions distinct from adaptive immunity is a pervasive trend in the evolution of class 1 CRISPR-Cas systems. More highly divergent CRISPR-Cas derivatives are likely to be discovered as the genomic databases rapidly grow, especially through the advances of metagenomics and single-cell genomics. Experimental study of these unusual systems is likely to become an important research direction.

Class 2

Class 2 consists of CRISPR-Cas types II, V, and VI (Fig. 2.4). The defining feature of class 2 systems is that their effector modules comprise a single, large, multidomain protein, such as Cas9 in type II, which has become the principal tool of the new generation of genome editing methods. Class 2 systems are far less common among bacteria than those of class 1 and are nearly absent in archaea. However, since 2015, concerted efforts have been made to discover new class 2 variants in genomics and metagenomic data, in part considering the potential that these systems might have as genome editing tools with unique features. As a result, two new CRISPR types (V and VI) were identified, and the number of subtypes is steadily expanding (reference 17 and references therein). Assignment of subtypes for class 2 systems presents a challenge because of the uniform domain architecture of the respective effector proteins and limited variation of locus architecture. The current general practice is to assign a new subtype for variants in which the effector proteins do not show significant sequence similarity to those of any of the already-established subtypes in BLAST searches; the presence of additional accessory genes and organization of the adaptation module are also taken into account.

Type II

Among the class 2 systems, type II is by far the most abundant among bacteria. The effector protein Cas9 has a unique domain architecture, which is the signature of type II, with an HNH nuclease inserted into the RuvC-like nuclease domain. Each of the two nuclease moieties of Cas9 is responsible for the cleavage of one strand of the target DNA, with the strand complementary to the guide RNA cleaved by HNH and the second strand by the RuvC-like nuclease. Another distinctive feature of type II (shared also with some of the

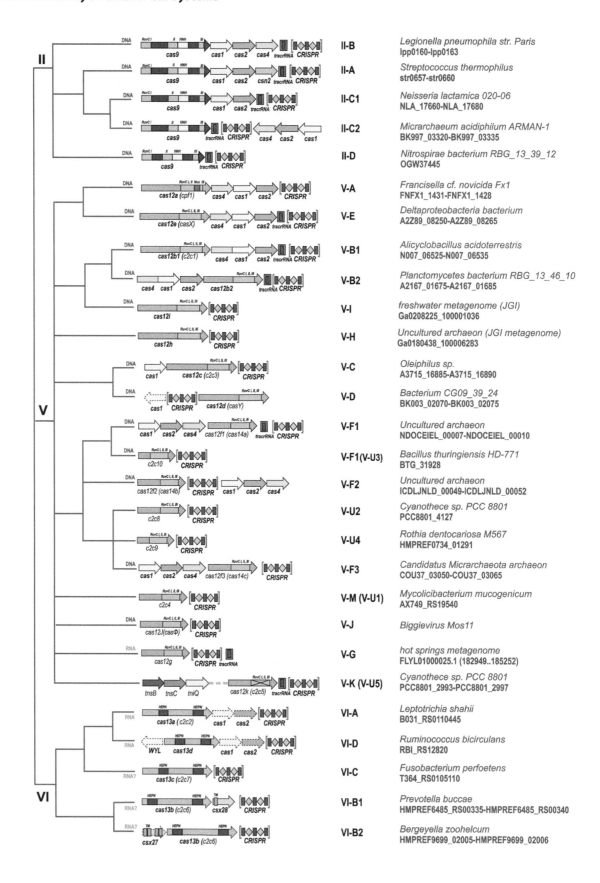

Figure 2.4 Updated classification of class 2 CRISPR-Cas systems The schematic shows representative (typical) CRISPR-cas loci of each class 2 subtype and selected distinct variants, with the dendrogram on the left showing the likely evolutionary relationships between the types and subtypes. The column on the right indicates the organism and the corresponding gene range. Homologous genes are color-coded and identified by a family name following the previous classification (18). Where both a systematic name and a legacy name are commonly used, the legacy name is given under the systematic name. The gray shading of different hues shows the two levels of classification, subtypes and variants. The adaptation module genes *cas1* and *cas2* are present in only a subset of the subtype V-D, VI-A, and VI-D loci and are accordingly shown by dashed lines. The WYL domain-encoding genes and *csx27* genes are also dispensable and shown by dashed lines. Additional genes encoding components of the interference module, such as tracrRNA, are shown. The domains of the effector proteins are color-coded: RuvC-like nuclease, yellow; HNH nuclease, green; HEPN RNase, purple; and transmembrane domains, blue. The figure is modified from reference 17.

type V subtypes as specified below) is the involvement of a distinct species of RNA, the *trans*-acting crRNA (tracrRNA), in the processing of the pre-crRNA as well as in interference (Fig. 2.1B).

Type II systems show far less variability in gene composition and locus organization than those of class 1. Nevertheless, there are three type II subtypes with their distinct signature genes. Subtype II-A is distinguished by an additional gene, *csn2*, which encodes an inactivated P-loop ATPase involved in spacer acquisition (67). Two variants of subtype II-A are differentiated by the presence of either a long or a short form of Csn2. Subtype II-B loci lack *csn2* but encode the Cas4 nuclease, which is also involved in adaptation (67). Subtype II-C has a minimal gene complement, consisting of the *cas1, cas2,* and *cas9* genes only (67). A recent comprehensive analysis of Cas9 homologs revealed a candidate for a new type II subtype, II-D (68). This system is characterized by a group of small Cas9 proteins (~700 amino acids) and a CRISPR array but is not associated with any other known *cas* genes (Fig. 2.4).

Type V

Type V systems resemble type II in the overall simple organization of the CRISPR-*cas* loci, which encode the large effector protein, in many cases also the adaptation module, and sometimes additional, ancillary proteins. Moreover, the effectors of types II and V share the RuvC-like nuclease domain. Otherwise, however, the domain architectures of the effector proteins differ between the two types. Unlike Cas9, with its two nuclease domains, each responsible for the cleavage of one strand of the target DNA, the type V effectors, Cas12 proteins, contain only the RuvC-like nuclease domain that consecutively cleaves the two strands of the target (69). Also in contrast to the relatively uniform type II, type V systems, although comparatively rare in bacteria, are highly diverse with respect to the size and architecture of the effector proteins as well as molecular mechanisms, including target specificity. Due to this diversity, 12 subtypes of type V have already been established, and many more are expected to be recognized (17). These new type V variants originate from a heterogeneous assemblage of loci that encompass RuvC-like domain containing proteins that are smaller than typical Cas12 and show high sequence similarity to TnpB proteins encoded by IS605-like transposons (68). The TnpB proteins have been recently shown to function as RNA-guided nucleases (68, 70), and their association with CRISPR apparently evolved on multiple, independent occasions, as suggested by phylogenetic analysis of TnpB and Cas1 proteins (19, 68). Different variants of type V seem to correspond to different stages of CRISPR effector evolution from TnpB (see discussion in the next section). Thus, there are likely numerous such variants that potentially would merit the subtype status, which is a challenge for the practical task of CRISPR classification. Originally, the type V variants that seem to have relatively recently evolved from TnpB were collectively designated subtype V-U, but once functionally characterized, some have been reclassified as new subtypes. These variants show notable differences in interference specificity. Thus, Cas12f (originally designated Cas14) has been shown to cleave ssDNA (21), although subsequently, a double-stranded DNA cleavage activity has been reported as well, whereas Cas12g is an RNA-guided RNase with additional, collateral RNase and ssDNase activities (22). Notably, the subtypes of type V differ with respect to the mechanisms of pre-crRNA processing. Some, like type II systems, employ tracrRNA and RNase III as the processing enzyme, for example, subtype V-B (18, 71). In others, for example, subtype V-A, pre-crRNA is processed by the effector protein itself (72).

One of the former V-U variants, V-U5, subsequently renamed subtype V-K, contains an apparently inactivated RuvC-like nuclease domain, as indicated by the replacement of essential catalytic residues, and is encoded by cyanobacterial Tn7-like transposons (19, 30). The role of the inactivated nuclease Cas12k in RNA-guided transposition has been experimentally demonstrated, analogously to the interference-defective variants of subtypes I-B, I-E, and I-F described above (33). Thus, subtype V-K is another case of independent exaptation of a CRISPR-Cas systems for the same nondefense role. It appears likely that more such inactivated and repurposed variants of class 2 systems will be discovered in further searches and metagenomic sequences.

Type VI

Type VI is the only known variety of CRISPR-Cas systems that target exclusively RNA. Type VI effectors (Cas13) are unrelated to the type II and type V effectors, contain two HEPN RNase domains, and apparently target transcripts of invading DNA genomes, although they protect bacteria against RNA phages under laboratory conditions and might function in this capacity in some bacteria (18, 24). Cas13 proteins additionally show collateral, nonspecific RNase activity that is activated by target recognition and induces dormancy in virus-infected bacteria (73). Furthermore, similarly to subtype V-A and some other type V subtypes, Cas13 itself is responsible for the processing of the pre-crRNA.

Four subtypes of type VI systems have been recognized (Fig. 2.4). Two variants of subtype VI-B encompass distinct membrane proteins, in addition to Cas13 (19). Most of the subtype VI-D systems include accessory proteins containing a WYL domain, a regulator of the Cas13 activity (23). Subtypes VI-A and VI-D sometimes carry an adaptation module, whereas the rest of type VI systems lack one and thus apparently depend on other types of CRISPR-Cas systems for spacer acquisition, as demonstrated for one of the VI-B systems that employs a subtype II-C adaptation machinery (74). Overall, type VI shows much less diversity than type V, but nevertheless, discovery of additional subtypes appears likely.

Distribution of Different Variants of CRISPR-Cas Systems Among Bacteria and Archaea

The different types and subtypes of CRISPR-Cas systems are nonuniformly distributed among bacterial and archaeal phyla (17). In the course of the latest iteration of CRISPR-Cas classification, a census of 13,116 complete prokaryote genomes identified CRISPR-*cas* loci in a substantial majority of archaea (276 of 324 genomes [85.2%]), including nearly all hyperthermophiles (89 of 92 genomes [96.7%]), but only in about 40% of bacteria (5,412 of 12,792 genomes [42.3%]). Individual CRISPR-Cas classes, types, and subtypes show clear occurrence trends. Class 2 is virtually exclusive to bacteria, with only a few instances in mesophilic archaea. In part, the absence of class 2 in archaea can be explained by the absence of RNase III, the pan-bacterial enzyme that is responsible for pre-crRNA processing in type II and some subtypes of type V, which together comprise a substantial majority of the class 2 systems (75, 76). In contrast, the genomes of hyperthermophilic Crenarchaeota are notably enriched for type III systems of class 1, and more generally, type III is highly prevalent in thermophiles. Overall, and in most groups of bacteria and archaea, class 1 is much more abundant than class 2, but there are exceptions, for example, the bacterial phylum *Tenericutes*, in which only class 2 systems have been identified so far (Fig. 2.5). In some bacterial phyla that consist mostly of symbiotic and parasitic organisms,

	Type I	Type III	Type IV	Type II	Type V	Type VI	Partial	None
Euryarchaeota	0.454246707	0.181921762	0.003226258	0	0.012051092	0	0.249561237	0.098992944
Crenarchaeota	0.352721593	0.39647819	0	0	0	0	0.225457867	0.025342351
Thaumarchaeota	0.266548731	0.162495042	0	0	0	0	0.417566102	0.153390125
other Archaea	0.27746252	0.09688994	0	0	0	0	0.34333808	0.28230946
Acidobacteria	0.361167822	0.238111255	0	0	0	0	0.077306986	0.323413937
Aquificae	0.423928482	0.210749524	0	0	0.026165873	0	0.314426127	0.024729995
Bacteroidetes	0.142911157	0.086353031	0	0.203671333	0.006872379	0.040908599	0.117455163	0.401828338
Chlorobi	0.468334091	0.340518225	0	0	0	0	0.191147684	0
Fusobacteria	0.260195859	0.186475317	0	0.198371834	0.121738652	0.070422786	0.127822834	0.034972718
Chlamydiae	0.110463157	0	0	0	0	0	0	0.889536843
Planctomycetes	0.437459577	0.164168461	0.040225727	0.125230098	0.061195016	0	0.11408116	0.05763996
Verrucomicrobia	0.175003642	0.122737394	0	0.169690411	0.047779057	0	0.095558113	0.389231382
Alphaproteobacteria	0.181144332	0.02263982	0.001755339	0.062072393	0.001573258	0.003153582	0.047703857	0.67995742
Betaproteobacteria	0.235784806	0.061926929	0.005951858	0.082192412	0.001096798	0	0.101088285	0.511958912
Gammaproteobacteria	0.319464834	0.061375438	0.003377891	0.017294053	0.006416276	0	0.102465481	0.489606027
Oligoflexia	0	0	0	0.13082689	0	0	0	0.86947311
Deltaproteobacteria	0.430474569	0.165555374	0.007550137	0	0.007503251	0	0.252992443	0.135924226
Epsilonproteobacteria	0.164616898	0.090443393	0	0.26967838	0	0.004284385	0.07307512	0.397901825
other Proteobacteria	0	0	0.038969911	0	0.106657558	0	0.287484511	0.56668802
Spirochaetes	0.334233151	0.071412516	0.038256806	0.08816374	0	0	0.233619446	0.234314341
Actinobacteria	0.355842877	0.061963762	0.016161607	0.03432967	0.020885186	0	0.082134763	0.428682135
Chloroflexi	0.36012324	0.421120579	0	0	0.009743717	0	0.147133916	0.061878548
Cyanobacteria	0.281747888	0.294593165	0	0	0.133164428	0	0.197418331	0.093076188
Deinococcus-Thermus	0.437499448	0.254045222	0	0	0.018899517	0	0.143866058	0.145689754
Bacilli	0.228554143	0.068248076	0.004132532	0.164160084	0.003331204	0.003731718	0.087964563	0.439877681
Clostridia	0.495445454	0.225096401	0.007213958	0.023629766	0.035475777	0.003716288	0.092726133	0.116696224
Erysipelotrichia	0.253212746	0.126606373	0	0.360017871	0	0	0	0.26016301
Negativicutes	0.326549757	0.290631325	0	0.191638706	0	0	0.125774338	0.065405873
Tissierellia	0.528539553	0.230128461	0	0.063440344	0	0	0.046729353	0.131162289
Tenericutes	0	0	0	0.213577252	0	0	0.022145832	0.764276916
Thermotogae	0.375688828	0.475454063	0	0	0	0	0.128228206	0.020628904
other Bacteria	0.285308321	0.196953668	0	0.09832999	0.012581807	0	0.07450966	0.332316554

Figure 2.5 Distribution of the 6 types of CRISPR-Cas systems in the major archaeal and bacterial phyla The heat map shows the weighted fraction (between 0 and 1.0) of the genomes in each of the major archaeal and bacterial phyla, in which CRISPR-Cas systems of the respective type were detected. "Partial" or "None" indicates CRISPR-*cas* loci that could not be assigned to any of the known types. The figure is modified from reference 17.

such as *Chlamydia* (Fig. 2.4) or the Candidate Phyla Radiation, CRISPR-Cas systems are rare (20, 77, 78). This, however, is not a general rule: almost all archaea in the DPANN superphylum that also appear to be mostly, if not exclusively, symbionts or parasites carry CRISPR-Cas systems. The majority of type VI systems, in particular, all those of the most abundant subtype (VI-B), so far have been identified in bacterial genomes of the phyla *Bacteroidetes* and *Fusobacteria* (Fig. 2.4).

The biological causes of the nonuniform spread of CRISPR-Cas systems remain to be elucidated. Considering the propensity of the CRISPR-*cas* loci for horizontal gene transfer, their loss or retention in prokaryotic genomes most likely is determined by the trade-off between the fitness cost that is due, mostly, to autoimmunity and curtailment of horizontal gene transfer and by the advantages of defense provided by adaptive immunity (79–84). The benefits of adaptive immunity most likely depend on the abundance and diversity of viruses in the habitats of different groups of bacteria and archaea as well as the biology of host-parasite interactions (85, 86). The evolutionary dynamics of CRISPR-Cas likely depend, to a large extent, on the interactions between CRISPR-Cas and DNA repair systems, such as that of double-strand break repair (87). Characterization of the factors that determine CRISPR persistence and loss in bacteria and archaea can be expected to become a major direction of CRISPR-Cas research.

A Brief Evolutionary History of CRISPR

The principal theme of this chapter is the diversity of CRISPR-Cas systems and its classification. Evidently, however, this diversity is shaped by CRISPR-Cas origins and evolution, and comparative analysis of CRISPR-Cas systems has led to multiple insights into their evolution, so much so that a fairly complete picture has emerged, albeit with different degrees of confidence in its different elements and with details missing. So we present a brief overview of our current understanding of the evolutionary history of CRISPR-Cas.

Three main trends seem to have shaped the evolution of CRISPR-Cas: (i) exaptation of components of various MGEs for the key functionalities of the CRISPR-Cas systems, (ii) accretion of protein domains, yielding versatile, elaborate CRISPR effectors, and (iii) extensive module recombination and domain rearrangements in Cas proteins. The first observation revealing the MGE connection was the discovery of a distinct variety of self-synthesizing transposons, in which the transposase is a homolog of Cas1 (these transposons were accordingly dubbed casposons, and the transposase was named casposase) (88). In retrospect, recruitment of a transposase for CRISPR adaptation does not appear surprising because the reactions involved in spacer insertion into a CRISPR array and in transposon insertion into its homing site are mechanistically and chemically similar. Exaptation of transposases seems to be a general route of evolution of adaptive immune systems and other mechanisms of "natural genome engineering," as demonstrated by the recruitment of the RAG1-RAG2 transposase for the role of VD(J) recombinase in the vertebrate adaptive immune system and the Piggy-BAC transposase in ciliate macronuclear genome maturation (89). Some of the casposons, in addition to the casposase, encode an endonuclease homologous to Cas4, which is a component of the adaptation module in several CRISPR-Cas subtypes (see above). Homologs of Cas2 so far have not been identified in casposons. Cas2 is a homolog of the VapD nuclease found in numerous TA systems and probably was acquired by the emerging CRISPR systems from a TA module, directly or via a casposon.

As already mentioned above, the origins of the effector proteins of types II and V can be readily traced to nucleases encoded by IS*605*-like transposons (17, 68). The most common and evolutionarily successful type II systems are monophyletic, having evolved from transposon-encoded, RNA-guided nucleases known as IscB (68, 90). The IscB proteins possess the distinctive domain architecture shared with Cas9, whereby the HNH nuclease domain is inserted within the RuvC-like domain, and phylogenetic analysis of the RuvC-like domain supports the ancestral relationship. In contrast, Cas12 proteins are clearly polyphyletic, having evolved from different branches of the TnpB tree on multiple, independent occasions (19). Apparently, given the broad spread of TnpB across bacterial and archaeal genomes, and their pronounced mobility, insertion of a TnpB-encoding transposon near a CRISPR array produces a functional type V system with relative ease. The recruitment of TnpB and IscB for the effector function in class 2 CRISPR-Cas systems seems to occur so readily because both are RNA-guided nucleases, with the guide RNA (dubbed OMEGA) encoded in the same loci (68). The OMEGA RNA might have been the ancestor of the tracrRNA associated with type II and many of the type V systems (68, 70).

The type VI effector, Cas13, seems to originate from a toxin-antitoxin (TA) or abortive infection (Abi) modules that includes a HEPN RNase as the toxin moiety. The low sequence conservation between the two HEPN domains of Cas13 suggests likely origin by fusion of two distinct toxins, rather than by duplication (17). The TA modules differ from transposons in that they lack transposases or other proteins mediating active mobility. Nevertheless, they exhibit high mobility and can be considered a distinct type of MGE that typically piggyback on plasmids and persist by rendering the host bacteria or archaea addictive to both the TA itself and the carrier plasmid (91).

The evolution of class 2 CRISPR effectors from components of different MGEs followed the same path of accretion of structural elements yielding the large effector proteins (typically, about 1,000 amino acids) from the much smaller MGE proteins (17). During such evolution, the effector proteins become specifically adopted for crRNA, PAM recognition, and target accommodation, whereas the sequence similarity to the ancestral proteins is quickly eroded.

The origin of the class 1 effector modules is not as immediately apparent as that of class 2 effectors, but tracing evolutionary connections of the signal transduction machinery of type III yields credible clues (Fig. 2.6). Search of prokaryote genomes for homologs of Cas10 led to the identification of a putative operon that consists of a gene encoding a small Cas10 homolog consisting of the polymerase domain alone and a protein that consists of fused CARF and HEPN domains (58). So far, this module has not been characterized experimentally, but the domain composition leads to the prediction that it is an Abi (abortive infection) system that, after cOA synthesis by the Cas10 homolog is triggered by virus infection, induces dormancy/PCD through the nonspecific RNase activity of HEPN. The simplicity and compactness of this putative ABI module imply that it is the ancestor of the type III effectors, as opposed to the reverse direction of evolution. Subsequent evolution involved major complexification of type III effector modules through the accretion of additional proteins and domains, such as the HD nuclease domain of Cas10 (5). The second key evolutionary process seems to have involved serial duplication of the RRM domain of Cas10 that gave rise to the entire superfamily of RRM-containing Cas proteins, namely, Cas5, Cas6, and Cas7 (collectively referred to as RAMPs) (5). Thus, type III CRISPR systems appear to have been the first to evolve in class 1, and type I systems are a subsequent derivation, in which the cOA signaling circuit and hence the coupling of target recognition and cleavage with dormancy/PCD were lost (17). Thus, a major trend in the evolution of class 1 CRISPR systems, after the initial phase of gene and domain, apparently, was functional reduction and simplification, which yielded type I and the even further degraded type IV and other derived forms discussed above (Fig. 2.5).

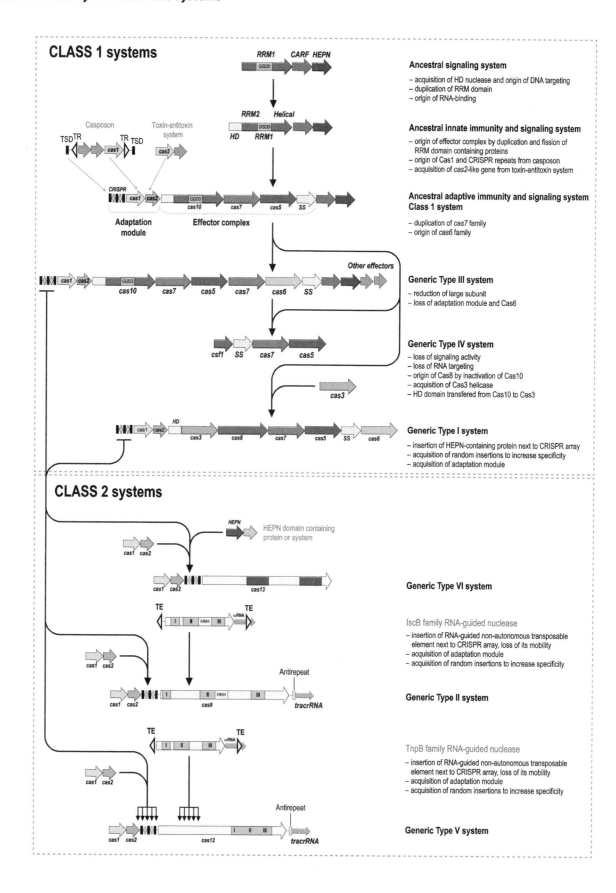

Figure 2.6 Origins and evolution of CRISPR-Cas systems Shown is a hypothetical scenario of the origin of CRISPR-Cas systems from an ancestral signaling system (possibly an abortive infection defense system [Abi]). This putative ancestral Abi module shares a cyclic oligo(A) polymerase Palm domain (RNA recognition motif [RRM] fold) with Cas10 and is proposed to function analogously to type III CRISPR-Cas systems. Specifically, cyclic oligo(A) molecules synthesized in response to virus infection bind to the CARF domain of the second protein in this system, resulting in activation of the RNase activity of the HEPN domain, which induces dormancy through indiscriminate RNA cleavage. This putative ancestral Abi module would give rise to the type III-like CRISPR-Cas effector module via the duplication of the RRM domain, with subsequent inactivation of one of the copies (the two RRM domains are designated RRM1 and RRM2). The ancestral class 1 CRISPR-Cas system is inferred to have evolved through the merger of two modules, the adaptation module, including the CRISPR repeats, derived from a casposon, and the type III-like effector module likely derived from the ancestral Abi system. The subsequent acquisition of the HD nuclease domain by the effector module provided for RNA-guided DNA cleavage. Inactivation of the oligo(A) polymerase domain in the effector complex or, possibly, replacement of Cas10 by an unrelated protein and acquisition of the Cas3 helicase led to the emergence of type I systems, which lack the cyclic oligo(A)-dependent signaling pathway and exclusively cleave double-stranded DNA (dsDNA). Class 2 systems of type II and different subtypes of type V appear to have evolved independently by recruitment of distinct TnpB nucleases that are encoded by IS*605*-like transposable elements. Type VI likely originated from an RNA-cleaving, HEPN domain-containing abortive infection or TA system. Some CRISPR-Cas systems, such as type IV and Tn7-linked systems I-F3 and V-K, were subsequently recruited by mobile genetic elements and lost their interference capacity along with the original defense function. The key evolutionary events are described to the right of the images. The typical CRISPR-*cas* operon organization is shown for each CRISPR-Cas subtype and selected, distinct variants. Homologous genes are color-coded and identified by a family name following the previous classification (18). The multiforking arrows indicate events that have been inferred to have occurred on multiple, independent occasions during the evolution of CRISPR-Cas systems. GGDD, key catalytic motif of the cyclase/polymerase domain of Cas10 that is involved in the synthesis of cyclic oligo(A) signaling molecules; TR, terminal repeats; TSD, target site duplication, the likely source of the ancestral repeats (95). The figure is modified from reference 96.

To complete the theme of MGE contributions to CRISPR evolution, it is worth pointing out that the RT associated with the adaptation modules of many type III systems derives from yet another class of MGEs, group II introns, which are bacterial retrotransposons (92).

To summarize, the CRISPR-Cas systems of adaptive immunity appear to have evolved via convergence of several distinct routes of evolution, all of which involved recruitment of components of MGEs. Specifically, at least 4 distinct varieties of MGEs were involved: (i) casposons that gave rise to the CRISPR adaptation module; (ii) IS*605*-like transposons, which became the ancestors of type II and V effector modules; (iii) TA/Abi modules, from which both class 1 and type VI effectors apparently evolved; and (iv) group II introns, from which the RT was recruited to become part of the type III adaptation modules. The previous scenarios for CRISPR evolution assumed that class 1 evolved first and class 2 came about later, via substitution of TnpB/IscB derivatives for the class 2 effector modules. However, the latest observation on the modular evolution of CRISPR-Cas systems and the multiple, independent occasions of MGE recruitment suggest independent evolution of different types of CRISPR-Cas by combination of adaptation and effector modules as the more parsimonious interpretation. It is important to emphasize that the evolutionary connection between CRISPR-Cas and MGEs is a two-way street as captured in the "guns for hire" concept (93). Complementary to the recruitment of MGE components for multiple functions in CRISPR immunity, CRISPR-Cas systems or their parts were recruited to function in MGE reproduction on multiple occasions. The CAST complexes of Tn7-like transposons discussed above are a typical example of this route of evolution.

Conclusions and Outlook

The most abundant types and subtypes of CRISPR-Cas systems are already known; thus, the general framework of the current classification can be expected to stand the test of time (17). However, the discovery of relatively rare but functionally diverse and evolutionarily informative variants, such as the multiple type V subtypes, continues, and many more probably remain to be identified, especially as diverse environments are probed by metagenomics and single-cell genomics. The more distinct of these variants will undoubtedly become new subtypes, but so far, no new types have been discovered since type VI. Under the current criteria, to be classified as a distinct type, a CRISPR-Cas variant has to possess an effector module that is unrelated (or extremely distantly related) to those of the known types. Some new types might remain to be discovered, but it appears obvious that these would be relatively rare among prokaryotes and/or highly specialized.

The consecutive versions of CRISPR-Cas classification, although based on explicitly defined criteria, were constructed in a more or less *ad hoc* mode that involved intensive expert curation. This approach has considerable advantages but at this stage appears to be unsustainable in the face of the accumulation of enormous amounts of genomic and especially metagenomics sequences containing diverse CRISPR-*cas* loci. It appears that developing automated classification methods based on machine learning approaches is indispensable for the future of CRISPR-Cas study.

The main trends in the evolution of CRISPR-Cas systems, in particular, their multifaceted evolutionary links to MGE, seem to have been established, although the search for specific ancestors of individual types, subtypes, and *cas* genes will undoubtedly continue. However, the grand challenge for the future is to identify the biological factors underlying the evolutionary dynamics of CRISPR-Cas and in particular leading to the distinctly nonuniform distribution among bacteria and archaea.

Acknowledgments

K.S.M., Y.I.W., and E.V.K. are supported by the Intramural Research Program of the National Institutes of Health of the United States (National Library of Medicine).

References

1. Mohanraju P, Makarova KS, Zetsche B, Zhang F, Koonin EV, van der Oost J. 2016. Diverse evolutionary roots and mechanistic variations of the CRISPR-Cas systems. *Science* 353:aad5147.

2. Jackson SA, McKenzie RE, Fagerlund RD, Kieper SN, Fineran PC, Brouns SJ. 2017. CRISPR-Cas: adapting to change. *Science* 356:eaal5056.

3. Barrangou R, Horvath P. 2017. A decade of discovery: CRISPR functions and applications. *Nat Microbiol* 2:17092.

4. Jiang F, Doudna JA. 2017. CRISPR-Cas9 structures and mechanisms. *Annu Rev Biophys* 46:505–529.

5. Koonin EV, Makarova KS. 2019. Origins and evolution of CRISPR-Cas systems. *Philos Trans R Soc Lond B Biol Sci* 374:20180087.

6. Faure G, Makarova KS, Koonin EV. 2019. CRISPR-Cas: complex functional networks and multiple roles beyond adaptive immunity. *J Mol Biol* 431:3–20.

7. McGinn J, Marraffini LA. 2019. Molecular mechanisms of CRISPR-Cas spacer acquisition. *Nat Rev Microbiol* 17:7–12.

8. Koonin EV, Makarova KS, Wolf YI. 2017. Evolutionary genomics of defense systems in archaea and bacteria. *Annu Rev Microbiol* 71:233–261.

9. Koonin EV, Makarova KS, Zhang F. 2017. Diversity, classification and evolution of CRISPR-Cas systems. *Curr Opin Microbiol* 37:67–78.

10. Ishino Y, Krupovic M, Forterre P. 2018. History of CRISPR-Cas from encounter with a mysterious repeated sequence to genome editing technology. *J Bacteriol* 200:e00580-17.

11. Hille F, Charpentier E. 2016. CRISPR-Cas: biology, mechanisms and relevance. *Philos Trans R Soc Lond B Biol Sci* 371:20150496.

12. Wright AV, Nuñez JK, Doudna JA. 2016. Biology and applications of CRISPR systems: harnessing nature's toolbox for genome engineering. *Cell* 164:29–44.

13. Klompe SE, Sternberg SH. 2018. Harnessing "a billion years of experimentation": the ongoing exploration and exploitation of CRISPR-Cas immune systems. *CRISPR J* 1:141–158.

14. Makarova KS, Wolf YI, Koonin EV. 2018. Classification and nomenclature of CRISPR-Cas systems: where from here? *CRISPR J* 1:325–336.

15. Makarova KS, Haft DH, Barrangou R, Brouns SJ, Charpentier E, Horvath P, Moineau S, Mojica FJ, Wolf YI, Yakunin AF, van der Oost J, Koonin EV. 2011. Evolution and classification of the CRISPR-Cas systems. *Nat Rev Microbiol* 9:467–477.

16. Makarova KS, Wolf YI, Alkhnbashi OS, Costa F, Shah SA, Saunders SJ, Barrangou R, Brouns SJ, Charpentier E, Haft DH, Horvath P, Moineau S, Mojica FJ, Terns RM, Terns MP, White MF, Yakunin AF, Garrett RA, van der Oost J, Backofen R, Koonin EV. 2015. An updated evolutionary classification of CRISPR-Cas systems. *Nat Rev Microbiol* 13:722–736.

17. Makarova KS, Wolf YI, Iranzo J, Shmakov SA, Alkhnbashi OS, Brouns SJJ, Charpentier E, Cheng D, Haft DH, Horvath P, Moineau S, Mojica FJM, Scott D, Shah SA, Siksnys V, Terns MP, Venclovas Č, White MF, Yakunin AF, Yan W, Zhang F, Garrett RA, Backofen R, van der Oost J, Barrangou R, Koonin EV. 2020. Evolutionary classification of CRISPR-Cas systems: a burst of class 2 and derived variants. *Nat Rev Microbiol* 18:67–83.

18. Shmakov S, Abudayyeh OO, Makarova KS, Wolf YI, Gootenberg JS, Semenova E, Minakhin L, Joung J, Konermann S, Severinov K, Zhang F, Koonin EV. 2015. Discovery and functional characterization of diverse class 2 CRISPR-Cas systems. *Mol Cell* 60:385–397.

19. Shmakov S, Smargon A, Scott D, Cox D, Pyzocha N, Yan W, Abudayyeh OO, Gootenberg JS, Makarova KS, Wolf YI, Severinov K, Zhang F, Koonin EV. 2017. Diversity and evolution of class 2 CRISPR-Cas systems. *Nat Rev Microbiol* 15:169–182.

20. Burstein D, Harrington LB, Strutt SC, Probst AJ, Anantharaman K, Thomas BC, Doudna JA, Banfield JF. 2017. New CRISPR-Cas systems from uncultivated microbes. *Nature* 542:237–241.

21. Harrington LB, Burstein D, Chen JS, Paez-Espino D, Ma E, Witte IP, Cofsky JC, Kyrpides NC, Banfield JF, Doudna JA. 2018. Programmed DNA destruction by miniature CRISPR-Cas14 enzymes. *Science* 362:839–842.

22. Yan WX, Hunnewell P, Alfonse LE, Carte JM, Keston-Smith E, Sothiselvam S, Garrity AJ, Chong S, Makarova KS, Koonin EV, Cheng DR, Scott DA. 2019. Functionally diverse type V CRISPR-Cas systems. *Science* 363:88–91.

23. Yan WX, Chong S, Zhang H, Makarova KS, Koonin EV, Cheng DR, Scott DA. 2018. Cas13d is a compact RNA-targeting type VI CRISPR effector positively modulated by a WYL-domain-containing accessory protein. *Mol Cell* 70:327–339.e5.

24. Abudayyeh OO, Gootenberg JS, Konermann S, Joung J, Slaymaker IM, Cox DB, Shmakov S, Makarova KS, Semenova E, Minakhin L, Severinov K, Regev A, Lander ES, Koonin EV, Zhang F. 2016. C2c2 is a single-component programmable RNA-guided RNA-targeting CRISPR effector. *Science* 353:aaf5573.

25. Smargon AA, Cox DBT, Pyzocha NK, Zheng K, Slaymaker IM, Gootenberg JS, Abudayyeh OA, Essletzbichler P, Shmakov S, Makarova KS, Koonin EV, Zhang F. 2017. Cas13b is a type VI-B CRISPR-associated RNA-guided RNase differentially regulated by accessory proteins Csx27 and Csx28. *Mol Cell* 65:618–630.e7.

26. Murugan K, Babu K, Sundaresan R, Rajan R, Sashital DG. 2017. The revolution continues: newly discovered systems expand the CRISPR-Cas toolkit. *Mol Cell* 68:15–25.

27. Stella S, Alcón P, Montoya G. 2017. Class 2 CRISPR-Cas RNA-guided endonucleases: Swiss army knives of genome editing. *Nat Struct Mol Biol* 24:882–892.

28. Shah SA, Alkhnbashi OS, Behler J, Han W, She Q, Hess WR, Garrett RA, Backofen R. 2019. Comprehensive search for accessory proteins encoded with archaeal and bacterial type III CRISPR-cas gene cassettes reveals 39 new cas gene families. *RNA Biol* 16:530–542.

29. Shmakov SA, Makarova KS, Wolf YI, Severinov KV, Koonin EV. 2018. Systematic prediction of genes functionally linked to CRISPR-Cas systems by gene neighborhood analysis. *Proc Natl Acad Sci U S A* 115:E5307–E5316.

30. Faure G, Shmakov SA, Yan WX, Cheng DR, Scott DA, Peters JE, Makarova KS, Koonin EV. 2019. CRISPR-Cas in mobile genetic elements: counter-defence and beyond. *Nat Rev Microbiol* 17:513–525.

31. Peters JE, Makarova KS, Shmakov S, Koonin EV. 2017. Recruitment of CRISPR-Cas systems by Tn7-like transposons. *Proc Natl Acad Sci U S A* 114:E7358–E7366.

32. Koonin EV, Makarova KS. 2017. Mobile genetic elements and evolution of CRISPR-Cas systems: all the way there and back. *Genome Biol Evol* 9:2812–2825.

33. Strecker J, Ladha A, Gardner Z, Schmid-Burgk JL, Makarova KS, Koonin EV, Zhang F. 2019. RNA-guided DNA insertion with CRISPR-associated transposases. *Science* 365:48–53.

34. Klompe SE, Vo PLH, Halpin-Healy TS, Sternberg SH. 2019. Transposon-encoded CRISPR-Cas systems direct RNA-guided DNA integration. *Nature* 571:219–225.

35. Kazlauskiene M, Kostiuk G, Venclovas Č, Tamulaitis G, Siksnys V. 2017. A cyclic oligonucleotide signaling pathway in type III CRISPR-Cas systems. *Science* 357:605–609.

36. Niewoehner O, Garcia-Doval C, Rostøl JT, Berk C, Schwede F, Bigler L, Hall J, Marraffini LA, Jinek M. 2017. Type III CRISPR-Cas systems produce cyclic oligoadenylate second messengers. *Nature* 548:543–548.

37. Hudaiberdiev S, Shmakov S, Wolf YI, Terns MP, Makarova KS, Koonin EV. 2017. Phylogenomics of Cas4 family nucleases. *BMC Evol Biol* 17:232.

38. Mohr G, Silas S, Stamos JL, Makarova KS, Markham LM, Yao J, Lucas-Elío P, Sanchez-Amat A, Fire AZ, Koonin EV, Lambowitz AM. 2018. A reverse transcriptase-Cas1 fusion protein contains a Cas6 domain required for both CRISPR RNA biogenesis and RNA spacer acquisition. *Mol Cell* 72:700–714.e8.

39. Makarova KS, Aravind L, Wolf YI, Koonin EV. 2011. Unification of Cas protein families and a simple scenario for the origin and evolution of CRISPR-Cas systems. *Biol Direct* 6:38 http://dx.doi.org/10.1186/1745-6150-6-38.

40. Altschul SF, Madden TL, Schäffer AA, Zhang J, Zhang Z, Miller W, Lipman DJ. 1997. Gapped BLAST and PSI-BLAST: a new generation of protein database search programs. *Nucleic Acids Res* 25:3389–3402.

41. Fagerlund RD, Wilkinson ME, Klykov O, Barendregt A, Pearce FG, Kieper SN, Maxwell HWR, Capolupo A, Heck AJR, Krause KL, Bostina M, Scheltema RA, Staals RHJ, Fineran PC. 2017. Spacer capture and integration by a type I-F Cas1-Cas2-3 CRISPR adaptation complex. *Proc Natl Acad Sci U S A* 114:E5122–E5128.

42. Makarova KS, Wolf YI, Koonin EV. 2013. The basic building blocks and evolution of CRISPR-CAS systems. *Biochem Soc Trans* 41:1392–1400.

43. Gould SJ, Vrba ES. 1982. Exaptation—a missing term in science of form. *Paleobiology* 8:4–15.

44. Petassi MT, Hsieh SC, Peters JE. 2020. Guide RNA categorization enables target site choice in Tn7-CRISPR-Cas transposons. *Cell* 183:1757–1771.e18.

45. Saito M, Ladha A, Strecker J, Faure G, Neumann E, Altae-Tran H, Macrae RK, Zhang F. 2021. Dual modes of CRISPR-associated transposon homing. *Cell* 184:2441–2453.e18.

46. Majumdar S, Zhao P, Pfister NT, Compton M, Olson S, Glover CV, III, Wells L, Graveley BR, Terns RM, Terns MP. 2015. Three CRISPR-Cas immune effector complexes coexist in *Pyrococcus furiosus*. *RNA* 21:1147–1158.

47. Makarova KS, Timinskas A, Wolf YI, Gussow AB, Siksnys V, Venclovas Č, Koonin EV. 2020. Evolutionary and functional classification of the CARF domain superfamily, key sensors in prokaryotic antivirus defense. *Nucleic Acids Res* 48:8828–8847.

48. Silas S, Mohr G, Sidote DJ, Markham LM, Sanchez-Amat A, Bhaya D, Lambowitz AM, Fire AZ. 2016. Direct CRISPR spacer acquisition from RNA by a natural reverse transcriptase-Cas1 fusion protein. *Science* 351:aad4234.

49. McMahon SA, Zhu W, Graham S, Rambo R, White MF, Gloster TM. 2020. Structure and mechanism of a type III CRISPR defence DNA nuclease activated by cyclic oligoadenylate. *Nat Commun* 11:500.

50. Athukoralage JS, Graham S, Grüschow S, Rouillon C, White MF. 2019. A type III CRISPR ancillary ribonuclease degrades its cyclic oligoadenylate activator. *J Mol Biol* 431:2894–2899.

51. Athukoralage JS, Graham S, Rouillon C, Grüschow S, Czekster CM, White MF. 2020. The dynamic interplay of host and viral enzymes in type III CRISPR-mediated cyclic nucleotide signalling. *eLife* 9:e55852.

52. Athukoralage JS, McMahon SA, Zhang C, Grüschow S, Graham S, Krupovic M, Whitaker RJ, Gloster TM, White MF. 2020. An anti-CRISPR viral ring nuclease subverts type III CRISPR immunity. *Nature* 577:572–575.

53. Athukoralage JS, McQuarrie S, Grüschow S, Graham S, Gloster TM, White MF. 2020. Tetramerisation of the CRISPR ring nuclease Csx3 facilitates cyclic oligoadenylate cleavage. bioRxiv.

54. Athukoralage JS, Rouillon C, Graham S, Grüschow S, White MF. 2018. Ring nucleases deactivate type III CRISPR ribonucleases by degrading cyclic oligoadenylate. *Nature* 562:277–280.

55. Özcan A, Krajeski R, Ioannidi E, Lee B, Gardner A, Makarova KS, Koonin EV, Abudayyeh OO, Gootenberg JS. 2021. Programmable RNA targeting with the single-protein CRISPR effector Cas7-11. *Nature* 597:720–725.

56. van Beljouw SPB, Haagsma AC, Rodríguez-Molina A, van den Berg DF, Vink JNA, Brouns SJJ. 2021. The gRAMP CRISPR-Cas effector is an RNA endonuclease complexed with a caspase-like peptidase. *Science* 373:1349–1353.

57. Lowey B, Whiteley AT, Keszei AFA, Morehouse BR, Mathews IT, Antine SP, Cabrera VJ, Kashin D, Niemann P, Jain M, Schwede F, Mekalanos JJ, Shao S, Lee ASY, Kranzusch PJ. 2020. CBASS immunity uses CARF-related effectors to sense 3′-5′- and 2′-5′-linked cyclic oligonucleotide signals and protect bacteria from phage infection. *Cell* 182:38–49.e17.

58. Burroughs AM, Zhang D, Schäffer DE, Iyer LM, Aravind L. 2015. Comparative genomic analyses reveal a vast, novel network of nucleotide-centric systems in biological conflicts, immunity and signaling. *Nucleic Acids Res* 43:10633–10654.

59. Özcan A, Pausch P, Linden A, Wulf A, Schühle K, Heider J, Urlaub H, Heimerl T, Bange G, Randau L. 2019. Type IV CRISPR RNA processing and effector complex formation in *Aromatoleum aromaticum*. *Nat Microbiol* 4:89–96.

60. Moya-Beltrán A, Makarova KS, Acuña LG, Wolf YI, Covarrubias PC, Shmakov SA, Silva C, Tolstoy I, Johnson DB, Koonin EV, Quatrini R. 2021. Evolution of type IV CRISPR-Cas systems: insights from CRISPR loci in integrative conjugative elements of *Acidithiobacillia*. *CRISPR J* 4:656–672.

61. Pinilla-Redondo R, Mayo-Muñoz D, Russel J, Garrett RA, Randau L, Sørensen SJ, Shah SA. 2020. Type IV CRISPR-Cas systems are highly diverse and involved in competition between plasmids. *Nucleic Acids Res* 48:2000–2012.

62. Zhou Y, Bravo JPK, Taylor HN, Steens JA, Jackson RN, Staals RHJ, Taylor DW. 2021. Structure of a type IV CRISPR-Cas ribonucleoprotein complex. iScience 24:102201.

63. Crowley VM, Catching A, Taylor HN, Borges AL, Metcalf J, Bondy-Denomy J, Jackson RN. 2019. A type IV-A CRISPR-Cas system in *Pseudomonas aeruginosa* mediates RNA-guided plasmid interference *in vivo*. *CRISPR J* 2:434–440.

64. Makarova KS, Karamycheva S, Shah SA, Vestergaard G, Garrett RA, Koonin EV. 2019. Predicted highly derived class 1 CRISPR-Cas system in Haloarchaea containing diverged Cas5 and Cas7 homologs but no CRISPR array. *FEMS Microbiol Lett* 366:fnz079.

65. Kolesnik MV, Fedorova I, Karneyeva KA, Artamonova DN, Severinov KV. 2021. Type III CRISPR-Cas systems: deciphering the most complex prokaryotic immune system. *Biochemistry (Mosc)* 86:1301–1314.

66. Makarova KS, Wolf YI, Shmakov SA, Liu Y, Li M, Koonin EV. 2020. Unprecedented diversity of unique CRISPR-Cas-related systems and Cas1 homologs in Asgard archaea. *CRISPR J* 3:156–163.

67. Chylinski K, Makarova KS, Charpentier E, Koonin EV. 2014. Classification and evolution of type II CRISPR-Cas systems. *Nucleic Acids Res* 42:6091–6105.

68. Altae-Tran H, Kannan S, Demircioglu FE, Oshiro R, Nety SP, McKay LJ, Dlakić M, Inskeep WP, Makarova KS, Macrae RK, Koonin EV, Zhang F. 2021. The widespread IS200/IS605 transposon family encodes diverse programmable RNA-guided endonucleases. *Science* 374:57–65.

69. Swarts DC, van der Oost J, Jinek M. 2017. Structural basis for guide RNA processing and seed-dependent DNA targeting by CRISPR-Cas12a. *Mol Cell* 66:221–233.e4.

70. Karvelis T, Druteika G, Bigelyte G, Budre K, Zedaveinyte R, Silanskas A, Kazlauskas D, Venclovas Č, Siksnys V. 2021. Transposon-associated TnpB is a programmable RNA-guided DNA endonuclease. *Nature* 599:692–696.

71. Strecker J, Jones S, Koopal B, Schmid-Burgk J, Zetsche B, Gao L, Makarova KS, Koonin EV, Zhang F. 2019. Engineering of CRISPR-Cas12b for human genome editing. *Nat Commun* 10:212.

72. Fonfara I, Richter H, Bratovič M, Le Rhun A, Charpentier E. 2016. The CRISPR-associated DNA-cleaving enzyme Cpf1 also processes precursor CRISPR RNA. *Nature* 532:517–521.

73. Meeske AJ, Nakandakari-Higa S, Marraffini LA. 2019. Cas13-induced cellular dormancy prevents the rise of CRISPR-resistant bacteriophage. *Nature* 570:241–245.

74. Hoikkala V, Ravantti J, Díez-Villaseñor C, Tiirola M, Conrad RA, McBride MJ, Moineau S, Sundberg LR. 2021. Cooperation between different CRISPR-Cas types enables adaptation in an RNA-targeting system. *mBio* 12:e03338-20.

75. Garrett RA, Vestergaard G, Shah SA. 2011. Archaeal CRISPR-based immune systems: exchangeable functional modules. *Trends Microbiol* 19:549–556.

76. Charpentier E, Richter H, van der Oost J, White MF. 2015. Biogenesis pathways of RNA guides in archaeal and bacterial CRISPR-Cas adaptive immunity. *FEMS Microbiol Rev* 39:428–441.

77. Dudek NK, Sun CL, Burstein D, Kantor RS, Aliaga Goltsman DS, Bik EM, Thomas BC, Banfield JF, Relman DA. 2017. Novel microbial diversity and functional potential in the marine mammal oral microbiome. *Curr Biol* 27:3752–3762.e6.

78. Castelle CJ, Brown CT, Anantharaman K, Probst AJ, Huang RH, Banfield JF. 2018. Biosynthetic capacity, metabolic variety and unusual biology in the CPR and DPANN radiations. *Nat Rev Microbiol* 16:629–645.

79. Levin BR. 2010. Nasty viruses, costly plasmids, population dynamics, and the conditions for establishing and maintaining CRISPR-mediated adaptive immunity in bacteria. *PLoS Genet* 6:e1001171.

80. Iranzo J, Lobkovsky AE, Wolf YI, Koonin EV. 2013. Evolutionary dynamics of the prokaryotic adaptive immunity system CRISPR-Cas in an explicit ecological context. *J Bacteriol* 195:3834–3844.

81. Iranzo J, Lobkovsky AE, Wolf YI, Koonin EV. 2015. Immunity, suicide or both? Ecological determinants for the combined evolution of anti-pathogen defense systems. *BMC Evol Biol* 15:43.

82. García-Martínez J, Maldonado RD, Guzmán NM, Mojica FJM. 2018. The CRISPR conundrum: evolve and maybe die, or survive and risk stagnation. *Microb Cell* 5:262–268.

83. Gurney J, Pleška M, Levin BR. 2019. Why put up with immunity when there is resistance: an excursion into the population and evolutionary dynamics of restriction-modification and CRISPR-Cas. *Philos Trans R Soc Lond B Biol Sci* 374:20180096.

84. van Houte S, Ekroth AK, Broniewski JM, Chabas H, Ashby B, Bondy-Denomy J, Gandon S, Boots M, Paterson S, Buckling A, Westra ER. 2016. The diversity-generating benefits of a prokaryotic adaptive immune system. *Nature* 532:385–388.

85. Weinberger AD, Wolf YI, Lobkovsky AE, Gilmore MS, Koonin EV. 2012. Viral diversity threshold for adaptive immunity in prokaryotes. *mBio* 3:e00456-12.

86. Westra ER, van Houte S, Oyesiku-Blakemore S, Makin B, Broniewski JM, Best A, Bondy-Denomy J, Davidson A, Boots M, Buckling A. 2015. Parasite exposure drives selective evolution of constitutive versus inducible defense. *Curr Biol* 25:1043–1049.

87. Bernheim A, Bikard D, Touchon M, Rocha EPC. 2019. A matter of background: DNA repair pathways as a possible cause for the sparse distribution of CRISPR-Cas systems in bacteria. *Philos Trans R Soc Lond B Biol Sci* 374:20180088.

88. Krupovic M, Makarova KS, Forterre P, Prangishvili D, Koonin EV. 2014. Casposons: a new superfamily of self-synthesizing DNA transposons at the origin of prokaryotic CRISPR-Cas immunity. *BMC Biol* 12:36.

89. Kapitonov VV, Koonin EV. 2015. Evolution of the RAG1-RAG2 locus: both proteins came from the same transposon. *Biol Direct* 10:20.

90. Kapitonov VV, Makarova KS, Koonin EV. 2015. ISC, a novel group of bacterial and archaeal DNA transposons that encode Cas9 homologs. *J Bacteriol* 198:797–807.

91. Harms A, Brodersen DE, Mitarai N, Gerdes K. 2018. Toxins, targets, and triggers: an overview of toxin-antitoxin biology. *Mol Cell* 70:768–784.

92. Toro N, Martínez-Abarca F, Mestre MR, González-Delgado A. 2019. Multiple origins of reverse transcriptases linked to CRISPR-Cas systems. *RNA Biol* 16:1486–1493.

93. Koonin EV, Makarova KS, Wolf YI, Krupovic M. 2020. Evolutionary entanglement of mobile genetic elements and host defence systems: guns for hire. *Nat Rev Genet* 21:119–131.

94. Koonin EV, Makarova KS. 2018. Discovery of oligonucleotide signaling mediated by CRISPR-associated polymerases solves two puzzles but leaves an enigma. *ACS Chem Biol* 13:309–312.

95. Koonin EV, Krupovic M. 2015. Evolution of adaptive immunity from transposable elements combined with innate immune systems. *Nat Rev Genet* 16:184–192.

96. Koonin EV, Makarova KS. 2022. Evolutionary plasticity and functional versatility of CRISPR systems. *PLoS Biol* 20:e3001481.

97. Hickman AB, Dyda F. 2015. The casposon-encoded Cas1 protein from *Aciduliprofundum boonei* is a DNA integrase that generates target site duplications. *Nucleic Acids Res* 43:10576–10587.

98. Nuñez JK, Kranzusch PJ, Noeske J, Wright AV, Davies CW, Doudna JA. 2014. Cas1-Cas2 complex formation mediates spacer acquisition during CRISPR-Cas adaptive immunity. *Nat Struct Mol Biol* 21:528–534.

99. Rollie C, Schneider S, Brinkmann AS, Bolt EL, White MF. 2015. Intrinsic sequence specificity of the Cas1 integrase directs new spacer acquisition. *eLife* 4:e08716.

100. Beloglazova N, Brown G, Zimmerman MD, Proudfoot M, Makarova KS, Kudritska M, Kochinyan S, Wang S, Chruszcz M, Minor W, Koonin EV, Edwards AM, Savchenko A, Yakunin AF. 2008. A novel family of sequence-specific endoribonucleases associated with the clustered regularly interspaced short palindromic repeats. *J Biol Chem* 283:20361–20371.

101. Samai P, Smith P, Shuman S. 2010. Structure of a CRISPR-associated protein Cas2 from *Desulfovibrio vulgaris*. *Acta Crystallogr Sect F Struct Biol Cryst Commun* 66:1552–1556.

102. Dixit B, Ghosh KK, Fernandes G, Kumar P, Gogoi P, Kumar M. 2016. Dual nuclease activity of a Cas2 protein in CRISPR-Cas subtype I-B of *Leptospira interrogans*. *FEBS Lett* 590:1002–1016.

103. Beloglazova N, Petit P, Flick R, Brown G, Savchenko A, Yakunin AF. 2011. Structure and activity of the Cas3 HD nuclease MJ0384, an effector enzyme of the CRISPR interference. *EMBO J* 30:4616–4627.

104. Huo Y, Nam KH, Ding F, Lee H, Wu L, Xiao Y, Farchione MD, Jr, Zhou S, Rajashankar K, Kurinov I, Zhang R, Ke A. 2014. Structures of CRISPR Cas3 offer mechanistic insights into Cascade-activated DNA unwinding and degradation. *Nat Struct Mol Biol* 21:771–777.

105. Sinkunas T, Gasiunas G, Fremaux C, Barrangou R, Horvath P, Siksnys V. 2011. Cas3 is a single-stranded DNA nuclease and ATP-dependent helicase in the CRISPR/Cas immune system. *EMBO J* 30:1335–1342.

106. Jackson RN, Lavin M, Carter J, Wiedenheft B. 2014. Fitting CRISPR-associated Cas3 into the helicase family tree. *Curr Opin Struct Biol* 24:106–114.

107. Lemak S, Nocek B, Beloglazova N, Skarina T, Flick R, Brown G, Joachimiak A, Savchenko A, Yakunin AF. 2014. The CRISPR-associated Cas4 protein Pcal_0546 from *Pyrobaculum calidifontis* contains a [2Fe-2S] cluster: crystal structure and nuclease activity. *Nucleic Acids Res* 42:11144–11155.

108. Zhang J, Kasciukovic T, White MF. 2012. The CRISPR associated protein Cas4 is a 5′ to 3′ DNA exonuclease with an iron-sulfur cluster. *PLoS One* 7:e47232.

109. Lee H, Dhingra Y, Sashital DG. 2019. The Cas4-Cas1-Cas2 complex mediates precise prespacer processing during CRISPR adaptation. *eLife* 8:e44248.

110. Shiimori M, Garrett SC, Graveley BR, Terns MP. 2018. Cas4 nucleases define the PAM, length, and orientation of DNA fragments integrated at CRISPR loci. *Mol Cell* 70:814–824.e6.

111. Nam KH, Haitjema C, Liu X, Ding F, Wang H, DeLisa MP, Ke A. 2012. Cas5d protein processes pre-crRNA and assembles into a cascade-like interference complex in subtype I-C/Dvulg CRISPR-Cas system. *Structure* 20:1574–1584.

112. Reeks J, Naismith JH, White MF. 2013. CRISPR interference: a structural perspective. *Biochem J* 453:155–166.

113. Shao Y, Cocozaki AI, Ramia NF, Terns RM, Terns MP, Li H. 2013. Structure of the Cmr2-Cmr3 subcomplex of the Cmr RNA silencing complex. *Structure* 21:376–384.

114. Liu TY, Doudna JA. 2020. Chemistry of class 1 CRISPR-Cas effectors: binding, editing, and regulation. *J Biol Chem* 295:14473–14487.

115. Mogila I, Kazlauskiene M, Valinskyte S, Tamulaitiene G, Tamulaitis G, Siksnys V. 2019. Genetic dissection of the type III-A CRISPR-Cas system Csm complex reveals roles of individual subunits. *Cell Rep* 26:2753–2765.e4.

116. Carte J, Pfister NT, Compton MM, Terns RM, Terns MP. 2010. Binding and cleavage of CRISPR RNA by Cas6. *RNA* 16:2181–2188.

117. Carte J, Wang R, Li H, Terns RM, Terns MP. 2008. Cas6 is an endoribonuclease that generates guide RNAs for invader defense in prokaryotes. *Genes Dev* 22:3489–3496.

118. Niewoehner O, Jinek M, Doudna JA. 2014. Evolution of CRISPR RNA recognition and processing by Cas6 endonucleases. *Nucleic Acids Res* 42:1341–1353.

119. Wang R, Preamplume G, Terns MP, Terns RM, Li H. 2011. Interaction of the Cas6 riboendonuclease with CRISPR RNAs: recognition and cleavage. *Structure* 19:257–264.

120. Zhu X, Ye K. 2015. Cmr4 is the slicer in the RNA-targeting Cmr CRISPR complex. *Nucleic Acids Res* 43:1257–1267.

121. Brouns SJ, Jore MM, Lundgren M, Westra ER, Slijkhuis RJ, Snijders AP, Dickman MJ, Makarova KS, Koonin EV, van der Oost J. 2008. Small CRISPR RNAs guide antiviral defense in prokaryotes. *Science* 321:960–964.

122. Jore MM, Lundgren M, van Duijn E, Bultema JB, Westra ER, Waghmare SP, Wiedenheft B, Pul U, Wurm R, Wagner R, Beijer MR, Barendregt A, Zhou K, Snijders AP, Dickman MJ, Doudna JA, Boekema EJ, Heck AJ, van der Oost J, Brouns SJ. 2011. Structural basis for CRISPR RNA-guided DNA recognition by Cascade. *Nat Struct Mol Biol* 18:529–536.

123. van der Oost J, Westra ER, Jackson RN, Wiedenheft B. 2014. Unravelling the structural and mechanistic basis of CRISPR-Cas systems. *Nat Rev Microbiol* 12:479–492.

124. Wang B, Zhang T, Yin J, Yu Y, Xu W, Ding J, Patel DJ, Yang H. 2021. Structural basis for self-cleavage prevention by tag:anti-tag pairing complementarity in type VI Cas13 CRISPR systems. *Mol Cell* 81:1100–1115.e5.

125. Heler R, Samai P, Modell JW, Weiner C, Goldberg GW, Bikard D, Marraffini LA. 2015. Cas9 specifies functional viral targets during CRISPR-Cas adaptation. *Nature* 519:199–202.

126. Wei Y, Terns RM, Terns MP. 2015. Cas9 function and host genome sampling in type II-A CRISPR-Cas adaptation. *Genes Dev* 29:356–361.

127. Cong L, Zhang F. 2015. Genome engineering using CRISPR-Cas9 system. *Methods Mol Biol* 1239:197–217.

128. Deltcheva E, Chylinski K, Sharma CM, Gonzales K, Chao Y, Pirzada ZA, Eckert MR, Vogel J, Charpentier E. 2011. CRISPR RNA maturation by trans-encoded small RNA and host factor RNase III. *Nature* 471:602–607.

129. Jinek M, Chylinski K, Fonfara I, Hauer M, Doudna JA, Charpentier E. 2012. A programmable dual-RNA-guided DNA endonuclease in adaptive bacterial immunity. *Science* 337:816–821.

130. Elmore JR, Sheppard NF, Ramia N, Deighan T, Li H, Terns RM, Terns MP. 2016. Bipartite recognition of target RNAs activates DNA cleavage by the type III-B CRISPR-Cas system. *Genes Dev* 30:447–459.

131. Jung TY, An Y, Park KH, Lee MH, Oh BH, Woo E. 2015. Crystal structure of the Csm1 subunit of the Csm complex and its single-stranded DNA-specific nuclease activity. *Structure* 23:782–790.

132. Rouillon C, Zhou M, Zhang J, Politis A, Beilsten-Edmands V, Cannone G, Graham S, Robinson CV, Spagnolo L, White MF. 2013. Structure of the CRISPR interference complex CSM reveals key similarities with cascade. *Mol Cell* 52:124–134.

133. Samai P, Pyenson N, Jiang W, Goldberg GW, Hatoum-Aslan A, Marraffini LA. 2015. Co-transcriptional DNA and RNA cleavage during type III CRISPR-Cas immunity. *Cell* 161:1164–1174.

134. Zhu X, Ye K. 2012. Crystal structure of Cmr2 suggests a nucleotide cyclase-related enzyme in type III CRISPR-Cas systems. *FEBS Lett* 586:939–945.

135. Sofos N, Feng M, Stella S, Pape T, Fuglsang A, Lin J, Huang Q, Li Y, She Q, Montoya G. 2020. Structures of the Cmr-β complex reveal the regulation of the immunity mechanism of type III-B CRISPR-Cas. *Mol Cell* 79:741–757.e7.

136. Daume M, Plagens A, Randau L. 2014. DNA binding properties of the small cascade subunit Csa5. *PLoS One* 9:e105716.

137. Reeks J, Graham S, Anderson L, Liu H, White MF, Naismith JH. 2013. Structure of the archaeal Cascade subunit Csa5: relating the small subunits of CRISPR effector complexes. *RNA Biol* 10:762–769.

138. Sakamoto K, Agari Y, Agari K, Yokoyama S, Kuramitsu S, Shinkai A. 2009. X-ray crystal structure of a CRISPR-associated RAMP superfamily protein, Cmr5, from *Thermus thermophilus* HB8. *Proteins* 75:528–532

139. Garcia-Doval C, Jinek M. 2017. Molecular architectures and mechanisms of class 2 CRISPR-associated nucleases. *Curr Opin Struct Biol* 47:157–166.

Molecular Mechanisms of Type I CRISPR-Cas Systems

John van der Oost

Wageningen University & Research, Laboratory of Microbiology, 6708WE, Wageningen, The Netherlands

Introduction

The diversity of bacterial and archaeal CRISPR-Cas systems is overwhelming. Based on the most recent classification (1), CRISPR-Cas class 1 includes types I, III, and IV. All class 1 systems have typical multisubunit complexes, often with a Cas nuclease, either stand-alone (Cas3 in type I) or tightly associated with the Cascade-like complex (Cas10 in type III). In contrast, class 2 (type II [Cas9], type V [Cas12], and type VI [Cas13]) systems all consist of a single, large, multidomain nuclease that binds a CRISPR RNA (crRNA) guide.

Many fundamental discoveries of relevant details of the CRISPR-Cas mechanism have been described in a series of seminal studies of type I CRISPR-Cas systems. The very first encounter of the CRISPR-Cas system was the unexpected finding of an unusual repetitive sequence in an intergenic region of the genome of *Escherichia coli*. Subsequently, many molecular details of general features were first described for the *E. coli* system (now called subtype I-E), including all three mechanistic stages: adaptation (selection and processing of protospacers as well as their integration as new spacers in the CRISPR array), expression (transcription and maturation of the crRNA guides), and interference (RNA-guided recognition and cleavage of target DNA). In addition, examples of anti-CRISPR systems were originally discovered in viruses counteracting type I-F systems. Moreover, structural details of the complete set of type I-E CRISPR-associated Cas proteins were initially revealed: the spacer acquisition nuclease-integrase complex (Cas1-Cas2), the guide-binding and target search complex (Cascade: Cas5-Cas6-Cas7-Cas8-Cas11), and the target cleaving helicase-nuclease (Cas3). In this chapter, these key discoveries and relevant mechanistic features are described, and recently established applications of type I complexes are discussed. For more details, excellent reviews are available that provide in-depth descriptions of CRISPR-Cas mechanisms and applications (2–5).

CRISPR: Biology and Applications, First Edition. Edited by Rodolphe Barrangou, Erik J. Sontheimer, and Luciano A. Marraffini.
© 2022 American Society for Microbiology. DOI: 10.1128/9781683673798.ch03

CRISPR-Cas Mechanism

Cascade-crRNA Complex

The first biochemical study of CRISPR-Cas was executed in my Bacterial Genetics group in Wageningen, The Netherlands (6). As a model, we selected the CRISPR-Cas system of *E. coli* (class 1/type I-E): a CRISPR array that consists of alternating repeats (29 bp) and spacers (32 bp), linked to a gene cluster (*cas3-cas8-cas11-cas7-cas5-cas6-cas1-cas2* [Fig. 3.1A]) that encodes eight Cas proteins. The CRISPR array together with different combinations of *cas* genes were transplanted from one *E. coli* strain (K-12) to a CRISPR-deficient *E. coli* strain [BL21(DE3)]. Small crRNAs appeared to be bound by a C̲RISPR-a̲ssociated c̲omplex for a̲nti-viral d̲efense (Cascade, consisting of 5 different subunits, Cas5-Cas6-Cas7-Cas8-Cas11) (Fig. 3.1B). It was demonstrated that the CRISPR array is transcribed as a long precursor CRISPR RNA (pre-crRNA) that in the presence of the Cascade complex is processed through cleavage at a specific site in the repeat sequence by a Cascade-associated ribonuclease (Cas6). The resulting mature crRNA guides consist of a 5′ handle (8 repeat-derived nucleotides), the central guide [derived from the variable spacer; 32 nucleotides (nt)], and a 3′ handle (21 repeat-derived nucleotides, including a hairpin structure) (Fig. 3.1B).

Next, we analyzed the composition and architecture of Cascade. A low-resolution "seahorse-like" shape of the complex was revealed by negative-stain electron microscopy (EM), whereas native mass spectrometry analysis revealed a 405-kDa complex with an uneven stoichiometry (Cas5-Cas6-Cas7-Cas8-Cas11 = 1-2-6-1-1) (7–9) (Fig. 3.1Di). This unique composition of the 11-subunit *E. coli* Cascade complex was confirmed by a higher-resolution cryo-EM structure (10) (Fig. 3.1Dii). Moreover, the latter structure demonstrated that the crRNA guide was positioned in a groove along on the surface the 6-subunit Cas7 backbone, with the crRNA ends at opposite sides of the Cascade structure: the hook-like 5′ handle is bound to the Cas5 subunit on one end (seahorse tail), and the 3′ handle with its hairpin is anchored by the Cas6 subunit (seahorse head [Fig. 3.1Dii]). Also, structures have been obtained of the Cascade-crRNA complex with single-stranded targets. A cryo-EM structure with an RNA target (complementary to the spacer of the crRNA guide) revealed a major conformational change of the Cascade backbone (10). This structural rearrangement upon target binding was confirmed in crystal structures of the *E. coli* Cascade-crRNA complex, with and without a complementary single-stranded DNA target (11–13) (Fig. 3.1D). Moreover, these structures revealed that the guide sequence base-pairs with the target DNA strand in a pseudo-A-form configuration, with five segments of 5 bp, each time interrupted by a flipped-out base of the guide (10, 12) (Fig. 3.1B and D). In contrast, structures of class 2 nucleases, Cas9 and Cas12, revealed uninterrupted base-pairing between guide RNA and target DNA, resulting in a helical conformation. The unusual guide-target base-pairing by Cascade relates to the firm binding of both guide ends to Cascade subunits, as well as to the binding of the guide and the Cas7 subunits along the Cascade backbone (4).

To test the *in vivo* role of the Cascade complex in antiviral defense, the complex was coexpressed with a designed CRISPR, resulting in mature crRNA guides with spacer sequences complementary to gene fragments of phage lambda (double-stranded DNA [dsDNA] virus that specifically uses *E. coli* as a host). However, the Cascade-crRNA system appeared insufficient to protect *E. coli* from infection by this virus (6).

RNA-Guided DNA Interference

As the Cascade-crRNA complex on its own appeared unable to provide immunity to *E. coli* BL21(DE3), a next attempt was to transplant the Cascade complex together with Cas3, the

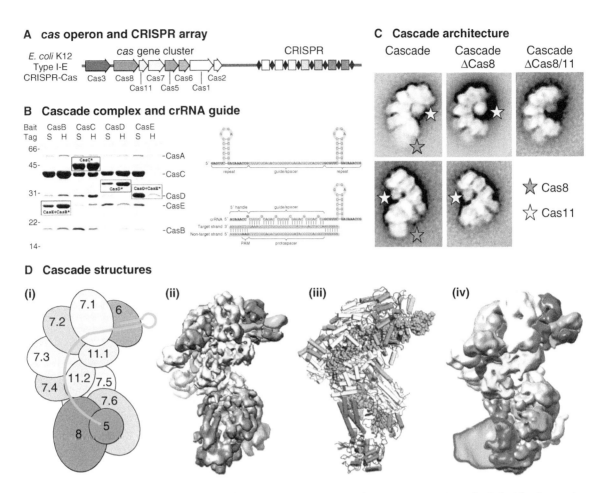

Figure 3.1 Cascade structure. (**A**) *E. coli* type I-E *cas* operon and CRISPR array. (**B**) The left side shows that affinity chromatography of Strep (**S**) and His (**H**) tag variants of five Cas proteins results in robust copurification of Cas8-Cas11-Cas7-Cas5-Cas6 (CasABCDE) and hence that they are subunits of the Cascade complex. The right side shows a repeat-spacer-repeat fragment of the precursor crRNA (pre-crRNA) (top). The mature crRNA with a central spacer-derived guide and repeat-derived flanks (5' handle and 3' hairpin) (bottom) allows for base-pairing with the target strand of a dsDNA fragment that contains an appropriate PAM sequence, resulting in displacement of the nontarget strand and forming an R-loop structure. (**C**) First insights in Cascade architecture by negative-stain EM of Cascade, Cascade without Cas8, and Cascade without Cas8/Cas11. (**D**) Cascade structures (similar color schemes): (i) model of Cascade and crRNA (green), (ii) cryo-EM structure of Cascade with crRNA (green), (iii) crystal structure of Cascade with crRNA (green) and DNA target (orange), and (iv) cryo-EM structure of Cascade with crRNA, dsDNA, and Cas3 (blue).

predicted helicase/nuclease. This time we were successful: coexpression of the 6 *cas* genes encoding Cascade and Cas3, together with the aforementioned synthetic CRISPR array (with 4 anti-lambda spacers) in the *E. coli* host, appeared to be the right combination to neutralize an attack by phage lambda. Apparently, the nucleic acid sequences of the infecting virus were targeted specifically by the 4 complementary crRNA guides, resulting in robust immunity (6) (Fig. 3.2A).

The next important question to be answered was whether the virus-associated target was a cRNA transcript (analogous to eukaryotic RNA interference) or rather a cDNA strand. As mentioned above, structures of the Cascade-crRNA complex that firmly binds cRNA or cDNA target strands after *in vitro* assembly were later obtained. However, to reveal whether the Cascade-crRNA/Cas3 system targets RNA or DNA

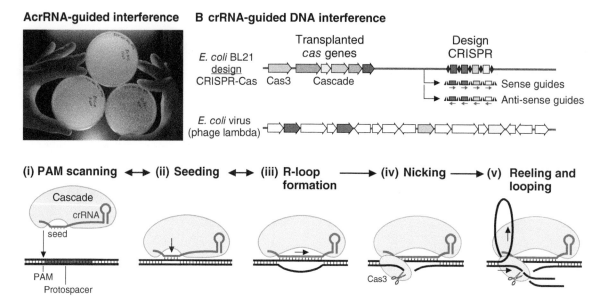

Figure 3.2 RNA-guided DNA interference by Cascade and Cas3. (A) The Eureka moment, when we demonstrated that the combination of Cas3 and Cascade loaded a designed crRNA and resulted in immunization against phage lambda (6), hence allowing for specific and programmable DNA interference (photo by Stan J. J. Brouns). **(B)** Experimental design to demonstrate crRNA-guided DNA interference. In a CRISPR-deficient *E. coli* host strain [BL21(DE3)] the depicted components were transplanted on different plasmids carrying genes encoding Cascade and Cas3, as well as designed CRISPRs with 4 spacers (either in the sense or in the antisense orientation) corresponding to 4 genes of phage lambda. The engineered *E. coli* strains were exposed to phage lambda, after which the mixture was spread on agar plates. After overnight incubation, plaques indicated attack by the phage, and the absence of plaques indicated successful protection of the host by the right combination of CRISPR-Cas components. We found that both sense and antisense guides could provide protection (as depicted in panel A) and concluded that the target was the double-stranded DNA of phage lambda. **(C)** The stepwise mechanism of Cascade/Cas3-based DNA interference: (i) PAM scanning, (ii) seed attack, (iii) R-loop formation, (iv) nicking of the displaced nontarget strand, and (v) reeling of the DNA resulting in occasional nicking of the nontarget strand and looping of the target strand.

in vivo, two parallel experiments were performed in which the functionality of RNA guides was tested in the two possible orientations, sense and antisense (Fig. 3.2B). In the case of an RNA target, only the guides with a complementary antisense orientation would be functional, as in eukaryotic RNA interference. On the other hand, in the case of a DNA target, both orientations would work, as each of them can base-pair with a different strand. The outcome of the experiment, yet another milestone in CRISPR research, clearly demonstrated that both guide orientations were functional in neutralizing the viral attack. Hence, it was concluded that DNA is the target of the CRISPR-Cas system (6) (Fig. 3.2B). Interestingly, previous analysis of the virus-derived sequences that were integrated into CRISPR of the *Streptococcus* Cas9 system revealed that irrespective of their orientation in the virus genes (sense/antisense), they were functional in supporting specific immunity) (14). This observation was in perfect agreement with the proposed DNA targeting.

The aforementioned claim that DNA was the specific target of the type I CRISPR-Cas system (6) was indeed confirmed by experiments in which the DNA cleavage was demonstrated directly. In the presence of the Cascade complex, severe DNA damage of the displaced nontarget DNA strand is caused by the helicase/nuclease activity of Cas3 (7, 15–18) (Fig. 3.2Ci). In the next paragraph, the stepwise interference process is described in more detail, highlighting the different checkpoints between DNA surveillance and DNA destruction, which make sure that only the viral DNA will be targeted and not a similar sequence on the host genome.

Self/Non-Self-Target Selection

After the discovery of adaptive immunity by a type II system (14), it was observed that despite the presence of the antivirus spacers, a viral infection occasionally resulted in bacterial infections, as monitored by plaques in bacterial lawns after growth on agar plates (19). It turned out that the escape was a consequence of mutations in the viral genomes, either within the targeted sequence (protospacer) or just adjacent to the targeted sequence (19). Using a bioinformatics approach, these motifs were detected also in other CRISPR-Cas systems, and they were called protospacer-adjacent motifs (PAM) (20). Indeed, in all well-studied DNA-targeting CRISPR-Cas systems (types I, II, and V), PAMs have since been defined (20–22). The PAM is crucial for CRISPR-Cas targeting, both as a first checkpoint for self/non-self-discrimination in biological defense and as a requirement for target selection in biotechnological applications (see below). The molecular basis of self/non-self-discrimination by Cascade has been revealed by a structural analysis, initially of the isolated Cas8 subunit (23) and later by the complete Cascade complex (24). It was demonstrated that the PAM (e.g., 5′-TTG on the nontarget strand [Fig. 3.1B and 3.2Ci]) is recognized in the minor groove of the double helix by the Cas8 subunit. The minor groove DNA recognition explains the observation that a single Cascade complex can recognize several distinct PAM sequences (25). PAM recognition causes severe DNA bending around Cas8 (26), leading to local DNA unwinding (27). This partially melted state of the double helix is stabilized by Cas8, which upon PAM binding moves an amino acid side chain as a wedge between the two DNA strands, right next to the PAM (24) (Fig. 3.2Ci).

This partially melted structure of the double helix downstream of the PAM is referred to as the "seed bubble." The seed sequence has been identified as part of the guide (6 to 8 nt at the 5′ end of the spacer sequence) where no mismatches are tolerated in base-pairing with the protospacer target, and as such, it is a second checkpoint for accurate target selection (21, 28) (Fig. 3.2Cii). Apart from its role in type I systems, this seed region in the crRNA guide appears to be a general feature in all DNA-targeting CRISPR-Cas systems, including type II (29, 30) and type V (22). The seed sequence is generally preordered with its bases exposed (31), to allow for efficient base-pairing with the complementary target strand (27). In the case of perfect base-pairing, the unzipping of the dsDNA is extended from the seed bubble to the end of the 32-bp protospacer, resulting in an R-loop structure in which the target strand base-pairs with the crRNA guide in the typical segmented manner (Fig. 3.2Ciii). Downstream of the seed, the requirement for perfect base-pairing is slightly more relaxed (32), but too many mismatches will cause reversal of the strand displacement process and abortion of the DNA interference (33). The displaced nontarget strand in its turn is guided along a parallel path on the Cascade surface, and the R-loop structure is stabilized by locking this strand behind the Cse2 dimer (27).

After complete R-loop formation, when all quality control selection criteria have been met, the stage is set for the final step of the interference process: recruitment and activation of Cas3. Initial *in vitro* analysis has revealed that Cas3 has 3′-to-5′ nuclease activity (17, 18). In addition, structural analyses confirmed earlier predictions (34) that Cas3 is composed of an ATP-dependent SF2-like helicase domain and an HD-like endonuclease domain (15, 16). Cas3 is specifically recruited by the Cascade/crRNA complex due to the aforementioned conformational change, which occurs only in the case of maximal guide-target base-pairing (the R-loop state), to avoid cleaving partially complementary targets. Using cryo-EM, snapshots of the entire interference process have been obtained. Binding of Cas3 to the complex relies on direct physical interactions with both the nontarget strand and the Cas8 subunit, and the Cas3 nuclease domain attacks the nontarget strand at an exposed region (35). Using single-molecule imaging, Cas3 has been demonstrated to remain tightly associated with a Cascade/R-loop complex, after which it nicks the nontarget strand (Fig. 3.2Civ). While holding on to the Cascade complex, the Cas3-helicase activity results in local unwinding of the double helix target DNA (36, 37). After melting, reeling of the nontarget strand occurs, directing the single-stranded DNA (ssDNA) to the Cas3 nuclease domain and resulting in occasional nicking and releasing of ssDNA fragments. Indeed, *in vitro* nicking by Cascade-associated Cas3 has been demonstrated to result in 30- to 100-nt ssDNA fragments (38). Concomitantly, as a result of the reeling mechanism, the noncleaved target strand is put aside as a growing loop (37) (Fig. 2Cv). In a prokaryotic cell, this ssDNA loop is a potential target for other endonucleases, resulting in even more damage of the targeted DNA, neutralizing an invasion by a virus or plasmid. Additional support for the reeling mechanism has been provided by a cryo-EM structure of the Cascade-crRNA/dsDNA/Cas3 complex after nicking of the nontarget strand, revealing retraction of the nontarget strand in the helicase channel (35). In terms of safety, the catalytic features of Cas3 (discontinuous helicase activity and relatively inefficient nuclease activity) combined with its tight interaction with Cascade have been proposed to further ensure controlled degradation of target DNA only (36).

Altogether, the multistep process of recognition events ensures well-controlled DNA targeting and degradation of foreign DNA by Cascade and Cas3 (Fig. 3.2C). Given the strict requirement for a PAM to allow targeting, it did not come as a surprise that this feature is also taken into account in the protospacer selection process, as described in the next section.

Spacer Acquisition

The initial identification of Cascade and Cas3 as the key players in the expression and the interference mechanism of the type I system (6) led to the conclusion that Cas1 and Cas2 are not involved in these processes and to the proposal that they rather may play a role in CRISPR adaptation (39). Indeed, this has been demonstrated by functional transplantation of the *cas1* and *cas2* genes from *E. coli* K-12 to the aforementioned CRISPR-deficient *E. coli* BL21(DE3) strain and revealing acquisition of new spacers at the leader end of the CRISPR array (40, 41). When a phage or plasmid has not been encountered before, this process is called "naive acquisition." In addition, spacer acquisition in type I systems may also proceed via "primed acquisition" when a previously obtained spacer has lost its functionality due to an escape phage with a mutated PAM or target sequence (42). Despite the escape, however, priming of new spacer acquisition against that phage occurs because Cascade can still detect the target. Probably because of incomplete R-loop formation, Cas3 does not nick the DNA. Instead, Cas3 may be released from Cascade, migrating along the invader DNA where, in collaboration with Cas1 and Cas2, prespacer fragments are generated from adjacent DNA sequences (37).

A series of complementary studies have revealed that CRISPR adaptation proceeds in a stepwise manner (43). Initially, generation of ssDNA and dsDNA fragments of invading nucleic acids occurs. During naive adaptation, this first step is catalyzed by non-Cas systems, either by endonucleases of a restriction-modification defense system (44) or by RecBCD during repair of DNA damage caused by a stalled replication fork (45). During the more efficient primed adaptation, the aforementioned Cas3 nuclease activity is responsible for generation of ssDNA fragments for spacer acquisition (37, 38). Although these predigestion steps are not essential, they may substantially enhance the efficiency of the overall acquisition process.

The proposed role of Cas1 and Cas2 in spacer acquisition has been confirmed *in vitro*, both by biochemical analyses (46, 47) and by structural studies (48, 49). The two proteins are very well conserved in all CRISPR-Cas types. In the *E. coli* type I system, they form a complex with an uneven stoichiometry: Cas1-Cas2 = 4-2 (Fig. 3.3A and B) (46, 47). This complex has the potential to capture the DNA fragments that are generated during the previous step of the acquisition process. Recent single-molecule analyses have revealed that the Cas1-Cas2 complex selects precursors of prespacers from DNA in various forms, including ssDNA and partial duplexes. The outcome of the eventual selection depends both on the size of the DNA strand(s) and on the presence of a PAM sequence (5′-CTT) (Fig. 3.3Ci). Cytoplasmic exonucleases (e.g., DNA polymerase III and ExoT [50]) are responsible for uneven trimming of the Cas1-Cas2-loaded DNA, generating mature prespacers of a suitable size for integration, typically with a central double-stranded part (23 bp) and 3′ extensions at both sides (Fig. 3.3B, Ci, and Cii). Cas1-Cas2 protects the PAM-containing sequence (CTTN) from being degraded, which results in the production of asymmetrically trimmed prespacers. Typically, the sizes of the extensions are 8 nt at the PAM end (NNNNCTTN-3′) and 5 nt at the other end, allowing for integration in a specific orientation. For inserting the prespacer fragment, the CRISPR array is attacked at the leader end of the CRISPR array. In the type I-E system of *E. coli*, the integration host factor (IHF), well known for its role in integration of plasmids in the host chromosome, has been demonstrated to guide the protospacer-loaded Cas1-Cas2 complex to the leader end of a CRISPR array (51, 52). Most likely, binding of IHF to a recognition site in the leader sequence and the subsequent sharp bending of the CRISPR DNA lead to destabilization of the double helix of the first repeat, allowing for attack by the Cas1-Cas2-loaded prespacer Fig. 3.3B and Cii). The hydroxyl groups at the 3′ overhangs of the prespacers are responsible for a site-specific nucleophilic attack of both strands of the leader-proximal repeat of the CRISPR array, initially at the leader site (Fig. 3.3Cii), and then, after a final processing step of the PAM extension (leaving a C at the 3′ end), an attack occurs at the spacer site (Fig. 3.3Ciii) (46, 47, 50). In some CRISPR-Cas systems, Cas4 has been demonstrated to associate with Cas1 and Cas2, playing an important role in PAM detection and maturation of the prespacer DNA (53, 54). After melting of the targeted repeat (Fig. 3.3Civ to Cv), the single-stranded repeats flank the new spacer (Fig. 3.3Cvi). The acquisition process is completed by second-strand synthesis and ligation of the repeat flanks by a non-Cas polymerase and a ligase (47) (Fig. 3Cvi to Cvii). Note that in the case of type I-E CRISPR adaptation, the leader-proximal end of the spacers always have a GC base pair (of which the C is derived from the integrated prespacer and the G ends up in the transcribed crRNA). For that reason, the G is generally considered to be part of the repeat (5′ handle: AUAAACCC*G*), consistent with its lack of involvement in base-pairing with the target DNA (Fig. 1B).

Applications

Reflecting its role as a defense system, the type I-E Cascade/Cas3 system has initially been used to inhibit phage infection (6), to target plasmids (55), and to avoid conjugation of plasmids (25). In addition, a wide spectrum of DNA editing applications has been developed for different subtypes of type I systems in recent years. As described below, these approaches are based on Cascade, either alone or in combination with different DNA-modifying enzymes.

A Cas1-Cas2 complex

B IHF-assisted spacer integration

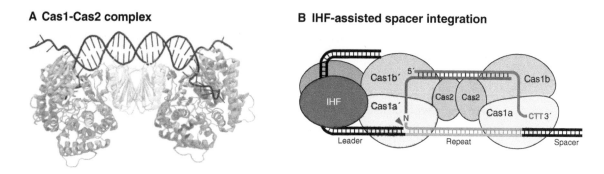

C Integration of spacer precursor in CRISPR

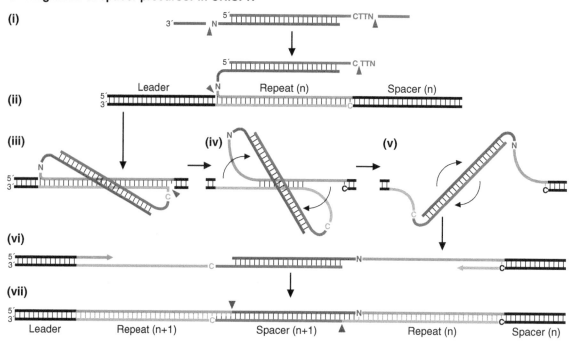

Figure 3.3 CRISPR adaptation. (**A**) Crystal structure of the prespacer-loaded Cas1-Cas2 integration complex. (**B**) Cartoon of the first nucleophilic attack ("half-site integration intermediate") at the leader site (L-site) by the 3′ non-PAM extension of the prespacer loaded on the Cas1-Cas2 complex. In *E. coli*, destabilization of the leader-proximal repeat through binding of the integration host factor (IHF) to the leader contributes to integration of new spacers at the leader-proximal end of the CRISPR. (**C**) Stepwise mechanism of Cas1-Cas2-based spacer integration: (i) double-stranded, PAM-containing DNA fragments are bound by the Cas1-Cas2 complex, after which trimming of the 3′ extensions (by non-Cas exonucleases, red arrowheads) results in an asymmetrical prespacer; (ii) attack of the L-site by the 3′ non-PAM extension and trimming of the 3′ PAM containing extension (red arrowheads); (iii) attack of the S-site by the 3′ PAM-derived extension (red arrowhead); (iv to v) melting of the repeat; (vi) second-strand synthesis by non-Cas DNA polymerase (orange and blue arrows); (vii) covalent linking of gaps by non-Cas ligase (blue arrowhead).

Genome Deletions

The initial insights of the RNA-guided DNA interference mechanism of the type-I CRISPR-Cas system have provided the basis for a range of genome applications. The first applications used Cascade/Cas3 for the immunization of bacteria, either naturally or synthetically. Natural immunization has been established through the exposure of bacterial hosts to phages in a subset of the infected cells, resulting in acquisition of phage-derived CRISPR spacers and hence specific immunity in these adapted bacteria. As described above, synthetic immunization has been achieved by supplementing a host bacterium with the genes encoding Cascade and Cas3 as well as with a designed CRISPR array with one or more spacers complementary to a certain phage, resulting in a recombinant host with robust resistance to this phage (6).

Using Cascade/Cas3 for targeting of intracellular plasmids may result in curing of these plasmids (55), whereas targeting of bacterial chromosomes often results in cell death due to the absence of efficient repair systems (56) (Fig. 3.4A). Escaping the toxicity of chromosome targeting has been demonstrated to occur through spontaneous mutations (deletions or substitutions in *cas* genes, CRISPR spacers, and/or PAM) that result in disrupted CRISPR functionality. In addition, chromosomal targeting by subtype I-E (*E. coli* [55]) and I-F systems (*Pectobacterium atrosepticum* [57]) has been demonstrated to result in escape by large deletions (>50 kb) of target sequences; in the absence of nonhomologous end joining (NHEJ), it has been proposed that an alternative, micro-homology (1- to 9-bp)-mediated end-joining (A-EJ/MMEJ) system is involved (58) (Fig. 3.4A). The type I-E Cascade/Cas3 system has also been used in human cells to generate major (unidirectional) deletions up to 200 kb in length (59, 60). Recently, also, deletions in genomes of *Pseudomonas aeruginosa* and *E. coli* have been generated with a heterologous type I-C system (61).

Genome editing by homologous recombination generally relies on repair templates and the cellular homology-directed repair (HDR) system. A repair template is a DNA fragment that consists of left and right flanks that are identical to the flanks of the chromosomal target site, surrounding the editing sequence. As the efficiency of the HDR system of most organisms is relatively low, selection of the desired recombinants increases the chance of success. A very efficient HDR-coupled selection strategy involves a CRISPR-based counterselection approach that specifically targets wild-type sequences, in order to enrich for recombinant clones (29). The counterselection approach works well because repair by end-joining systems (NHEJ and MMEJ) in many prokaryotes is rather inefficient, and hence, dsDNA breaks in bacterial and archaeal chromosomes frequently result in cell death. After the successful genome editing by Cas9 and Cas12a, several type I systems have been repurposed similarly. Cascade/Cas3 complexes with designed crRNA guides have been used in combination with appropriate repair templates (Fig. 3.4A). Again, the repair fragments allow for homologous recombination at the target locus, whereas the Cascade/Cas3 system is used for generating toxic double-strand breaks in case no recombination occurred. Using this approach, the native type I-A system of *Sulfolobus* has been used for generating specific deletions and substitutions (62). Likewise, specific modifications have been introduced in *Clostridium* with native type I-B Cascade/Cas3 (63).

Transcriptional Control

Compared to the "scalpel precision" of class 2 systems (blunt-end or sticky-end double-strand breaks by Cas9 or Cas12, respectively), DNA targeting by natural type I systems is more like "shredder destruction." For that reason, several applications have been developed based on only the Cascade system, in the absence of the Cas3 shredder. The targeting specificity of the type IE Cascade-crRNA complex was used to develop a CRISPR inhibition tool, i.e., to block the transcription of target genes through blocking of the RNA polymerase (Fig. 3.4B). Targeting of Cascade to a promoter sequence leads to strong inhibition of the transcription of the downstream gene, whereas targeting of the coding sequence results in substantially lower silencing efficiency. The system has been used to successfully knock down several genes simultaneously (64, 65). In addition, type I-E CRISPR-Cas systems have recently been functionally expressed in nuclei of human and plant cells. Through

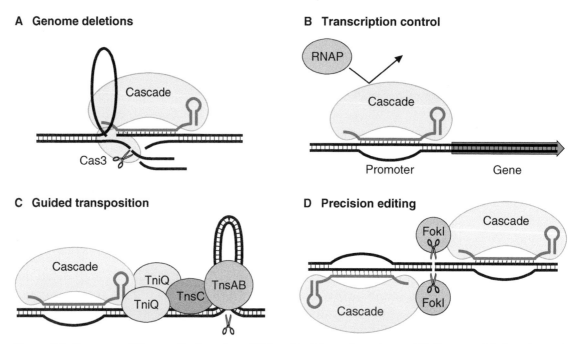

Figure 3.4 Genome editing applications by crRNA-guided Cascade variants. (A) Genome deletions by native or heterologous Cascade and Cas3. (B) Transcriptional control either by CRISPR inhibition, i.e., establishing a guide-specific roadblock of Cascade to inhibit initiation or elongation of the RNA polymerase (RNAP), or by CRISPR activation (not shown), i.e., enhancing the recruitment of RNAP through fusing of RNAP-binding domains to Cascade. (C) Guided transposition using natural transposons. (D) Precision editing by generating double-strand breaks through FokI-Cascade dimers, followed by repair-associated edits.

fusion of activation and repression domains to Cascade, modulated expression of target genes has been established (66, 67).

Guided Transposition

Recently, a unique type I-F Cascade complex has been characterized (68). Whereas CRISPR-Cas systems generally protect archaeal and bacterial hosts from invasions by mobile genetic elements, examples have been described in which CRISPR systems have been captured by plasmids and viruses (69–71). In addition, type I-F3 and type V-K CRISPR-Cas systems have been discovered to be associated with Tn7-like transposons (1, 72). The crRNA-guided transposition of both variants has recently been characterized (type I-F [68] and type V-K [73]). Programmable transposition of the IF type mini-Cascade/crRNA complex requires four transposition proteins: TniQ and TrsABC (Fig. 3.4C). Integration of up to 10 kb of donor/cargo DNA occurs 50 to 60 bp downstream of the targeted DNA sequence. Interestingly, this RNA-guided integrase is unique in that it allows for site-specific integration and that it not only operates independently of a distinct nuclease, but also it requires neither supply of repair templates nor involvement of a homology-directed repair system. Recent structural studies have revealed a Cascade-like complex associated with a dimer of TniQ (reviewed in reference 74).

Precision Editing

When the destructive Cas3 nuclease activity was revealed (17, 18), an alternative strategy was proposed to allow for more precise Cascade-based genome editing (75). Inspired by synthetic approaches developed for the zinc finger nucleases and transcription activator-like effector nucleases (reviewed in reference 76), a design was made of dimeric Cascade/crRNA-complex in which each complex was fused to the nuclease domain of the FokI restriction enzyme, resulting in a central DNA cleaving FokI dimer flanked with a DNA recognition Cascade/crRNA complex at both sides (75) (Fig. 4D).

Initially, a fusion of FokI was generated at the N terminus of Cas8 of the *E. coli* Cascade complex. After successful generation of double-strand breaks in a target DNA fragment, the design of the system was optimized by testing different linker sequences between FokI and Cas8, as well as different spacing between both protospacers. Subsequently, the best-performing design was transplanted into mammalian cells. First, FokI-Cascade complexes with nuclear localization signals were obtained from *E. coli* and transfected as ribonucleo-protein (RNP) complexes into multiple human cell lines. Next, plasmid-based systems have been established for expression in human cells, initially with six plasmids, five of which contain a gene encoding one of the Cascade subunits and one which has the gene of the crRNA guide. When this DNA delivery method resulted in editing efficiencies comparable with that of RNP delivery, a 2-plasmid system was developed by fusing the 5 FokI-Cascade encoding genes with intermediate 2A linkers. During translation, these linkers allow for ribosome skipping that results in the production of separate Cascade subunits. Surprisingly, the editing efficiencies of the 2-plasmid system were in the same range as those of the aforementioned delivery methods. Last but not least, the functionality of FokI-Cascade complexes from dozens of bacteria has been tested in human cells using the 2-plasmid expression system. Eventually, the best RNA-guided FokI-Cascade performance was obtained when using a type I-E Cascade complex from *Pseudomonas aeruginosa*, in multiple human cell lines with high specificity and knockout (insertion/deletion) efficiencies of up to 50%.

Conclusion and Outlook

Of the six different types of CRISPR-Cas systems, type I systems are the most abundant in nature. A series of studies on type I Cascade-like systems has resulted in groundbreaking discoveries on general mechanistic features of CRISPR-Cas and in relevant insights that have contributed substantially to the development of CRISPR-Cas systems as unprecedented genome editing tools. The highlights include the discovery of (i) CRISPR arrays (type I-E; [77]) and *cas* genes (types I and III [78]), (ii) maturation of CRISPR-derived RNA precursors yielding functional crRNA guides (type I-E [6]), (iii) DNA interference by crRNA-guided Cascade/Cas3 (type I-E [6, 17]), (iv) spacer acquisition by Cas1 and Cas2 (type I-E [40, 41]), (v) functional transplantation of Cas proteins (type I-E [6, 41]), and, last but not least, (vi) the fact that designing CRISPR with dedicated spacers allows for (multiplex) targeting of any DNA (type I-E [6]). Moreover, crystal and cryo-EM structures of all functional modules (CRISPR adaptation, CRISPR expression, and CRISPR interference) have been revealed first for type I.

A wide range of applications has been developed for type I Cascade, initially mainly for immunizing bacteria and archaea. However, because of the destructive mode of action of the Cas3 nuclease, as well as complications related to the heterologous expression of the multisubunit Cascade complex, it took many years to transplant Cascade into mammalian cells. In the meantime, successful genome editing has been accomplished with an astonishing pace with the DNA-targeting class 2 systems, initially for type II and more recently for type V. Recently, type I systems have successfully been repurposed for dedicated genome editing applications in eukaryotic cells, including cells from yeasts, plants, and humans. Examples of Cascade-based gene disruption, modulation of gene expression, and guided transposition have been reported. It is anticipated that the spectacular expansion of the

CRISPR-Cas toolbox will continue and that also natural and synthetic variants of type I systems will contribute to the development of functionalities for applications in genome engineering and diagnostics.

Looking back, I can only conclude that it has been an immense pleasure and a great privilege to have been active in the CRISPR field from the very beginning. This is a beautiful example of how fundamental research on a bacterial antivirus system has been repurposed as a generic genetic tool with unprecedented impact in basic biology as well as in applied areas ranging from microbial and plant biotechnology to human gene therapy.

References

1. Makarova KS, Wolf YI, Iranzo J, Shmakov SA, Alkhnbashi OS, Brouns SJJ, Charpentier E, Cheng D, Haft DH, Horvath P, Moineau S, Mojica FJM, Scott D, Shah SA, Siksnys V, Terns MP, Venclovas Č, White MF, Yakunin AF, Yan W, Zhang F, Garrett RA, Backofen R, van der Oost J, Barrangou R, Koonin EV. 2020. Evolutionary classification of CRISPR-Cas systems: a burst of class 2 and derived variants. *Nat Rev Microbiol* **18**:67–83.

2. Hidalgo-Cantabrana C, Barrangou R. 2020. Characterization and applications of type I CRISPR-Cas systems. *Biochem Soc Trans* **48**:15–23.

3. McGinn J, Marraffini LA. 2019. Molecular mechanisms of CRISPR-Cas spacer acquisition. *Nat Rev Microbiol* **17**:7–12.

4. Mohanraju P, Makarova KS, Zetsche B, Zhang F, Koonin EV, van der Oost J. 2016. Diverse evolutionary roots and mechanistic variations of the CRISPR-Cas systems. *Science* **353**:aad5147.

5. Sontheimer EJ, Barrangou R. 2015. The bacterial origins of the CRISPR genome-editing revolution. *Hum Gene Ther* **26**:413–424.

6. Brouns SJ, Jore MM, Lundgren M, Westra ER, Slijkhuis RJ, Snijders AP, Dickman MJ, Makarova KS, Koonin EV, van der Oost J. 2008. Small CRISPR RNAs guide antiviral defense in prokaryotes. *Science* **321**:960–964.

7. Jore MM, Lundgren M, van Duijn E, Bultema JB, Westra ER, Waghmare SP, Wiedenheft B, Pul U, Wurm R, Wagner R, Beijer MR, Barendregt A, Zhou K, Snijders AP, Dickman MJ, Doudna JA, Boekema EJ, Heck AJ, van der Oost J, Brouns SJ. 2011. Structural basis for CRISPR RNA-guided DNA recognition by Cascade. *Nat Struct Mol Biol* **18**:529–536.

8. Quax TE, Wolf YI, Koehorst JJ, Wurtzel O, van der Oost R, Ran W, Blombach F, Makarova KS, Brouns SJ, Forster AC, Wagner EG, Sorek R, Koonin EV, van der Oost J. 2013. Differential translation tunes uneven production of operon-encoded proteins. *Cell Rep* **4**:938–944.

9. van Duijn E, Barbu IM, Barendregt A, Jore MM, Wiedenheft B, Lundgren M, Westra ER, Brouns SJ, Doudna JA, van der Oost J, Heck AJ. 2012. Native tandem and ion mobility mass spectrometry highlight structural and modular similarities in clustered-regularly-interspaced short-palindromic-repeats (CRISPR)-associated protein complexes from *Escherichia coli* and *Pseudomonas aeruginosa*. *Mol Cell Proteomics* **11**:1430–1441.

10. Wiedenheft B, Lander GC, Zhou K, Jore MM, Brouns SJJ, van der Oost J, Doudna JA, Nogales E. 2011b. Structures of the RNA-guided surveillance complex from a bacterial immune system. *Nature* **477**:486–489.

11. Jackson RN, Golden SM, van Erp PB, Carter J, Westra ER, Brouns SJ, van der Oost J, Terwilliger TC, Read RJ, Wiedenheft B. 2014. Structural biology. Crystal structure of the CRISPR RNA-guided surveillance complex from Escherichia coli. *Science* **345**:1473–1479.

12. Mulepati S, Héroux A, Bailey S. 2014. Structural biology. Crystal structure of a CRISPR RNA-guided surveillance complex bound to a ssDNA target. *Science* **345**:1479–1484.

13. Zhao H, Sheng G, Wang J, Wang M, Bunkoczi G, Gong W, Wei Z, Wang Y. 2014. Crystal structure of the RNA-guided immune surveillance Cascade complex in *Escherichia coli*. *Nature* **515**:147–150.

14. Barrangou R, Fremaux C, Deveau H, Richards M, Boyaval P, Moineau S, Romero DA, Horvath P. 2007. CRISPR provides acquired resistance against viruses in prokaryotes. *Science* **315**:1709–1712.

15. Gong B, Shin M, Sun J, Jung CH, Bolt EL, van der Oost J, Kim JS. 2014. Molecular insights into DNA interference by CRISPR-associated nuclease-helicase Cas3. *Proc Natl Acad Sci U S A* **111**:16359–16364.

16. Huo Y, Nam KH, Ding F, Lee H, Wu L, Xiao Y, Farchione MD, Jr, Zhou S, Rajashankar K, Kurinov I, Zhang R, Ke A. 2014. Structures of CRISPR Cas3 offer mechanistic insights into Cascade-activated DNA unwinding and degradation. *Nat Struct Mol Biol* **21**:771–777.

17. Sinkunas T, Gasiunas G, Fremaux C, Barrangou R, Horvath P, Siksnys V. 2011. Cas3 is a single-stranded DNA nuclease and ATP-dependent helicase in the CRISPR/Cas immune system. *EMBO J* **30**:1335–1342.

18. Westra ER, van Erp PB, Künne T, Wong SP, Staals RH, Seegers CL, Bollen S, Jore MM, Semenova E, Severinov K, de Vos WM, Dame RT, de Vries R, Brouns SJ, van der Oost J. 2012. CRISPR immunity relies on the consecutive binding and degradation of negatively supercoiled invader DNA by Cascade and Cas3. *Mol Cell* **46**:595–605.

19. Deveau H, Barrangou R, Garneau JE, Labonté J, Fremaux C, Boyaval P, Romero DA, Horvath P, Moineau S. 2008. Phage response to CRISPR-encoded resistance in *Streptococcus thermophilus*. *J Bacteriol* **190**:1390–1400.

20. Mojica FJM, Díez-Villaseñor C, García-Martínez J, Almendros C. 2009. Short motif sequences determine the targets of the prokaryotic CRISPR defence system. *Microbiology (Reading)* **155**:733–740.

21. Semenova E, Jore MM, Datsenko KA, Semenova A, Westra ER, Wanner B, van der Oost J, Brouns SJ, Severinov K. 2011. Interference by clustered regularly interspaced short palindromic repeat (CRISPR) RNA is governed by a seed sequence. *Proc Natl Acad Sci U S A* **108**:10098–10103.

22. Zetsche B, Gootenberg JS, Abudayyeh OO, Slaymaker IM, Makarova KS, Essletzbichler P, Volz SE, Joung J,

van der Oost J, Regev A, Koonin EV, Zhang F. 2015. Cpf1 is a single RNA-guided endonuclease of a class 2 CRISPR-Cas system. *Cell* **163**:759–771.

23. Sashital DG, Wiedenheft B, Doudna JA. 2012. Mechanism of foreign DNA selection in a bacterial adaptive immune system. *Mol Cell* **46**:606–615.

24. Hayes RP, Xiao Y, Ding F, van Erp PB, Rajashankar K, Bailey S, Wiedenheft B, Ke A. 2016. Structural basis for promiscuous PAM recognition in type I-E Cascade from *E. coli. Nature* **530**:499–503.

25. Westra ER, Semenova E, Datsenko KA, Jackson RN, Wiedenheft B, Severinov K, Brouns SJ. 2013. Type I-E CRISPR-cas systems discriminate target from non-target DNA through base pairing-independent PAM recognition. *PLoS Genet* **9**:e1003742.

26. Hochstrasser ML, Taylor DW, Bhat P, Guegler CK, Sternberg SH, Nogales E, Doudna JA. 2014. CasA mediates Cas3-catalyzed target degradation during CRISPR RNA-guided interference. *Proc Natl Acad Sci U S A* **111**:6618–6623.

27. Xiao Y, Luo M, Hayes RP, Kim J, Ng S, Ding F, Liao M, Ke A. 2017. Structure basis for directional R-loop formation and substrate handover mechanisms in type I CRISPR-Cas system. *Cell* **170**:48–60.e11.

28. Wiedenheft B, van Duijn E, Bultema JB, Waghmare SP, Zhou K, Barendregt A, Westphal W, Heck AJ, Boekema EJ, Dickman MJ, Doudna JA. 2011. RNA-guided complex from a bacterial immune system enhances target recognition through seed sequence interactions. *Proc Natl Acad Sci U S A* **108**:10092–10097.

29. Jiang W, Bikard D, Cox D, Zhang F, Marraffini LA. 2013. RNA-guided editing of bacterial genomes using CRISPR-Cas systems. *Nat Biotechnol* **31**:233–239.

30. Jinek M, Chylinski K, Fonfara I, Hauer M, Doudna JA, Charpentier E. 2012. A programmable dual-RNA-guided DNA endonuclease in adaptive bacterial immunity. *Science* **337**:816–821.

31. Künne T, Swarts DC, Brouns SJ. 2014. Planting the seed: target recognition of short guide RNAs. *Trends Microbiol* **22**:74–83.

32. Fineran PC, Gerritzen MJ, Suárez-Diez M, Künne T, Boekhorst J, van Hijum SA, Staals RH, Brouns SJ. 2014. Degenerate target sites mediate rapid primed CRISPR adaptation. *Proc Natl Acad Sci U S A* **111**:E1629–E1638.

33. Jung C, Hawkins JA, Jones SK, Jr, Xiao Y, Rybarski JR, Dillard KE, Hussmann J, Saifuddin FA, Savran CA, Ellington AD, Ke A, Press WH, Finkelstein IJ. 2017. Massively parallel biophysical analysis of CRISPR-Cas complexes on next generation sequencing chips. *Cell* **170**:35–47.e13.

34. Makarova KS, Grishin NV, Shabalina SA, Wolf YI, Koonin EV. 2006. A putative RNA-interference-based immune system in prokaryotes: computational analysis of the predicted enzymatic machinery, functional analogies with eukaryotic RNAi, and hypothetical mechanisms of action. *Biol Direct* **1**:7.

35. Xiao Y, Luo M, Dolan AE, Liao M, Ke A. 2018. Structure basis for RNA-guided DNA degradation by Cascade and Cas3. *Science* **361**:6397 http://dx.doi.org/10.1126/science.aat0839.

36. Loeff L, Brouns SJJ, Joo C. 2018. Repetitive DNA Reeling by the Cascade-Cas3 complex in nucleotide unwinding steps. *Mol Cell* **70**:385–394.e3.

37. Redding S, Sternberg SH, Marshall M, Gibb B, Bhat P, Guegler CK, Wiedenheft B, Doudna JA, Greene EC. 2015. Surveillance and processing of foreign DNA by the *Escherichia coli* CRISPR-Cas System. *Cell* **163**:854–865.

38. Künne T, Kieper SN, Bannenberg JW, Vogel AI, Miellet WR, Klein M, Depken M, Suarez-Diez M, Brouns SJ. 2016. Cas3-derived target DNA degradation fragments fuel primed CRISPR adaptation. *Mol Cell* **63**:852–864.

39. van der Oost J, Jore MM, Westra ER, Lundgren M, Brouns SJ. 2009. CRISPR-based adaptive and heritable immunity in prokaryotes. *Trends Biochem Sci* **34**:401–407.

40. Swarts DC, Mosterd C, van Passel MW, Brouns SJ. 2012. CRISPR interference directs strand specific spacer acquisition. *PLoS One* **7**:e35888 http://dx.doi.org/10.1371/journal.pone.0035888.

41. Yosef I, Goren MG, Qimron U. 2012. Proteins and DNA elements essential for the CRISPR adaptation process in *Escherichia coli. Nucleic Acids Res* **40**:5569–5576.

42. Datsenko KA, Pougach K, Tikhonov A, Wanner BL, Severinov K, Semenova E. 2012. Molecular memory of prior infections activates the CRISPR/Cas adaptive bacterial immunity system. *Nat Commun* **3**:945 http://dx.doi.org/10.1038/ncomms1937.

43. Jackson SA, McKenzie RE, Fagerlund RD, Kieper SN, Fineran PC, Brouns SJ. 2017. CRISPR-Cas: adapting to change. *Science* **356**:6333 http://dx.doi.org/10.1126/science.aal5056.

44. Dupuis ME, Villion M, Magadán AH, Moineau S. 2013. CRISPR-Cas and restriction-modification systems are compatible and increase phage resistance. *Nat Commun* **4**:2087 http://dx.doi.org/10.1038/ncomms3087.

45. Levy A, Goren MG, Yosef I, Auster O, Manor M, Amitai G, Edgar R, Qimron U, Sorek R. 2015. CRISPR adaptation biases explain preference for acquisition of foreign DNA. Nature. **520**:505–510.

46. Arslan Z, Hermanns V, Wurm R, Wagner R, Pul Ü. 2014. Detection and characterization of spacer integration intermediates in type I-E CRISPR-Cas system. *Nucleic Acids Res* **42**:7884–7893.

47. Nuñez JK, Lee AS, Engelman A, Doudna JA. 2015. Integrase-mediated spacer acquisition during CRISPR-Cas adaptive immunity. *Nature* **519**:193–198.

48. Nuñez JK, Harrington LB, Kranzusch PJ, Engelman AN, Doudna JA. 2015. Foreign DNA capture during CRISPR-Cas adaptive immunity. *Nature* **527**:535–538.

49. Wang J, Li J, Zhao H, Sheng G, Wang M, Yin M, Wang Y. 2015. Structural and mechanistic basis of PAM-dependent spacer acquisition in CRISPR-Cas systems. *Cell* **163**:840–853.

50. Kim S, Loeff L, Colombo S, Jergic S, Brouns SJJ, Joo C. 2020. Selective loading and processing of prespacers for precise CRISPR adaptation. *Nature* **579**:141–145.

51. Nuñez JK, Bai L, Harrington LB, Hinder TL, Doudna JA. 2016. CRISPR immunological memory requires a host factor for specificity. *Mol Cell* **62**:824–833.

52. Wright AV, Liu JJ, Knott GJ, Doxzen KW, Nogales E, Doudna JA. 2017. Structures of the CRISPR genome integration complex. *Science* **357**:1113–1118.

53. Kieper SN, Almendros C, Behler J, McKenzie RE, Nobrega FL, Haagsma AC, Vink JNA, Hess WR, Brouns SJJ. 2018. Cas4 facilitates PAM-compatible spacer selection during CRISPR adaptation. *Cell Rep* **22**:3377–3384.

54. Shiimori M, Garrett SC, Graveley BR, Terns MP. 2018. Cas4 nucleases define the PAM, length, and orientation of DNA fragments integrated at CRISPR loci. *Mol Cell* **70**:814–824.e6.

55. Caliando BJ, Voigt CA. 2015. Targeted DNA degradation using a CRISPR device stably carried in the host genome. *Nat Commun* **6**:6989.

56. Gomaa AA, Klumpe HE, Luo ML, Selle K, Barrangou R, Beisel CL. 2014. Programmable removal of bacterial strains by use of genome-targeting CRISPR-Cas systems. *mBio* **5**:e00928-13.

57. Vercoe RB, Chang JT, Dy RL, Taylor C, Gristwood T, Clulow JS, Richter C, Przybilski R, Pitman AR, Fineran PC. 2013. Cytotoxic chromosomal targeting by CRISPR/Cas systems can reshape bacterial genomes and expel or remodel pathogenicity islands. *PLoS Genet* **9**:e1003454 http://dx.doi.org/10.1371/journal.pgen.1003454.

58. Chayot R, Montagne B, Mazel D, Ricchetti M. 2010. An end-joining repair mechanism in *Escherichia coli. Proc Natl Acad Sci U S A* **107**:2141–2146.

59. Cameron P, Coons MM, Klompe SE, Lied AM, Smith SC, Vidal B, Donohoue PD, Rotstein T, Kohrs BW, Nyer DB, Kennedy R, Banh LM, Williams C, Toh MS, Irby MJ, Edwards LS, Lin CH, Owen ALG, Künne T, van der Oost J, Brouns SJJ, Slorach EM, Fuller CK, Gradia S, Kanner SB, May AP, Sternberg SH. 2019. Harnessing type I CRISPR-Cas systems for genome engineering in human cells. *Nat Biotechnol* **37**:1471–1477.

60. Dolan AE, Hou Z, Xiao Y, Gramelspacher MJ, Heo J, Howden SE, Freddolino PL, Ke A, Zhang Y. 2019. Introducing a spectrum of long-range genomic deletions in human embryonic stem cells using type I CRISPR-Cas. *Mol Cell* **74**: 936–950.

61. Csörgő B, León LM, Chau-Ly IJ, Vasquez-Rifo A, Berry JD, Mahendra C, Crawford ED, Lewis JD, Bondy-Denomy J. 2020. A compact Cascade-Cas3 system for targeted genome engineering. *Nat Methods* **17**:1183–1190.

62. Liu G, She Q, Garrett RA. 2016. Diverse CRISPR-Cas responses and dramatic cellular DNA changes and cell death in pKEF9-conjugated *Sulfolobus* species. *Nucleic Acids Res* **44**:4233–4242.

63. Pyne ME, Bruder MR, Moo-Young M, Chung DA, Chou CP. 2016. Harnessing heterologous and endogenous CRISPR-Cas machineries for efficient markerless genome editing in *Clostridium. Sci Rep* **6**:25666.

64. Luo ML, Mullis AS, Leenay RT, Beisel CL. 2015. Repurposing endogenous type I CRISPR-Cas systems for programmable gene repression. *Nucleic Acids Res* **43**:674–681.

65. Rath D, Amlinger L, Hoekzema M, Devulapally PR, Lundgren M. 2015. Efficient programmable gene silencing by Cascade. *Nucleic Acids Res* **43**:237–246.

66. Pickar-Oliver A, Black JB, Lewis MM, Mutchnick KJ, Klann TS, Gilcrest KA, Sitton MJ, Nelson CE, Barrera A, Bartelt LC, Reddy TE, Beisel CL, Barrangou R, Gersbach CA. 2019. Targeted transcriptional modulation with type I CRISPR-Cas systems in human cells. *Nat Biotechnol* **37**:1493–1501.

67. Young JK, Gasior SL, Jones S, Wang L, Navarro P, Vickroy B, Barrangou R. 2019. The repurposing of type I-E CRISPR-Cascade for gene activation in plants. *Commun Biol* **2**:383.

68. Klompe SE, Vo PLH, Halpin-Healy TS, Sternberg SH. 2019. Transposon-encoded CRISPR-Cas systems direct RNA-guided DNA integration. *Nature* **571**:219–225.

69. Makarova KS, Wolf YI, Alkhnbashi OS, Costa F, Shah SA, Saunders SJ, Barrangou R, Brouns SJ, Charpentier E, Haft DH, Horvath P, Moineau S, Mojica FJ, Terns RM, Terns MP, White MF, Yakunin AF, Garrett RA, van der Oost J, Backofen R, Koonin EV. 2015. An updated evolutionary classification of CRISPR-Cas systems. *Nat Rev Microbiol* **13**:722–736.

70. McDonald ND, Regmi A, Morreale DP, Borowski JD, Boyd EF. 2019. CRISPR-Cas systems are present predominantly on mobile genetic elements in *Vibrio* species. *BMC Genomics* **20**:105.

71. Seed KD, Lazinski DW, Calderwood SB, Camilli A. 2013. A bacteriophage encodes its own CRISPR/Cas adaptive response to evade host innate immunity. *Nature* **494**:489–491.

72. Peters JE, Makarova KS, Shmakov S, Koonin EV. 2017. Recruitment of CRISPR-Cas systems by Tn7-like transposons. *Proc Natl Acad Sci U S A* **114**:E7358–E7366.

73. Strecker J, Ladha A, Gardner Z, Schmid-Burgk JL, Makarova KS, Koonin EV, Zhang F. 2019. RNA-guided DNA insertion with CRISPR-associated transposases. *Science* **365**:48–53.

74. van der Oost J, Mougiakos I. 2020. First structural insights into CRISPR-Cas-guided DNA transposition. *Cell Res* **30**:193–194.

75. Brouns SJ, van der Oost J. 4 July 2013. Modified Cascade ribonucleoproteins and uses thereof. Patent WO2013098244 (original filing, December 2011).

76. van der Oost J. 2013. Molecular biology. New tool for genome surgery. *Science* **339**:768–770.

77. Ishino Y, Shinagawa H, Makino K, Amemura M, Nakata A. 1987. Nucleotide sequence of the *iap* gene, responsible for alkaline phosphatase isozyme conversion in *Escherichia coli*, and identification of the gene product. *J Bacteriol* **169**:5429–5433.

78. Makarova KS, Aravind L, Grishin NV, Rogozin IB, Koonin EV. 2002. A DNA repair system specific for thermophilic Archaea and bacteria predicted by genomic context analysis. *Nucleic Acids Res* **30**:482–496.

Molecular Mechanisms of Type II CRISPR-Cas Systems

Tautvydas Karvelis and Virginijus Siksnys

Institute of Biotechnology, Life Sciences Center, Vilnius University, Vilnius, Lithuania LT-10257

A Brief Overview of Type II CRISPR-Cas Systems

CRISPR-Cas systems emerged as prokaryotic defense systems against foreign nucleic acids (1). They are very diverse and vary by gene arrangement, molecular mechanisms, and the type of targeted nucleic acids (2). Despite the remarkable diversity, the CRISPR-Cas "firewall" buildup can be broadly divided into three stages. At the adaptation stage, the fragments of the foreign nucleic acid are captured and inserted into the CRISPR array as new spacers. The spacer integration process is mediated mainly by Cas1 and Cas2 proteins and is conserved across different CRISPR Cas systems. Next, the CRISPR region is transcribed into pre-*crispr* RNA (pre-crRNA) and further processed to produce mature crRNA. Finally, crRNA and Cas proteins combine into an effector complex that provides interference against invading nucleic acids. The interference mechanism differs significantly among the CRISPR-Cas systems. Based on the composition of the effector complex, CRISPR-Cas systems are grouped into two classes and six types (2). Class 1 systems use multiprotein complexes for interference and are further subdivided into types I, III, and IV. Class 2 effectors utilize single nuclease protein (as exemplified for type II by Cas9, type V by Cas12, and type VI by Cas13) for nucleic acid destruction.

Among the CRISPR-Cas systems identified in bacteria and archaea, ~10% belong to the class 2 type II systems, which are almost exclusively found in bacteria (3, 4). Type II systems are further classified into three subtypes based on Cas9 protein sequence similarities and *cas* gene context (Fig. 4.1A). Type II-A systems contain the *cas9*, *cas1*, *cas2*, and *csn2* genes in the CRISPR locus and are most abundant (~55% of type II systems identified to date). In the type II-B systems, the *csn2* gene is replaced by *cas4*, while most of the type II-C systems contain only *cas9*, *cas1*, and *cas2* (2–4).

Discovery of Type II CRISPR-Cas Systems

The paradigm that CRISPR-Cas systems function as an antiviral defense system in bacteria first emerged by studying phage-host interactions in the *Streptococcus thermophilus* DGCC7710 strain, which possesses four different CRISPR-Cas (CRISPR1 to -4) systems representing three major CRISPR-Cas types (5). Phage challenge experiments revealed that in response to lytic phage infection, most of the cells die; however, a

CRISPR: Biology and Applications, First Edition. Edited by Rodolphe Barrangou, Erik J. Sontheimer, and Luciano A. Marraffini.
© 2022 American Society for Microbiology. DOI: 10.1128/9781683673798.ch04

few bacteriophage-insensitive mutants (BIMs) were recovered. Sequencing of the survivors revealed that these clones were able to incorporate phage-derived spacers into the CRISPR1 locus of the type II system. The 100% identity of the spacer to the phage genome ensured resistance to the cognate phage, while single nucleotide mutations in the spacer region abolished protection against the invading phage. To provide proof of the concept that spacer sequences inserted in the CRISPR array play a key role in phage resistance, Barrangou et al. designed a set of elegant experiments. They first demonstrated that spacers from phage-resistant bacteria inserted into the CRISPR array of the phage-sensitive *S. thermophilus* strain immunized bacteria against phage. On the other hand, they showed that the deletion of the spacers from the CRISPR region in the BIM strain made it susceptible to phage infection. Additionally, genetic dissection of two of the four CRISPR-associated genes revealed that the gene encoding Cas9 protein (formerly termed Cas5 or Csn1) containing putative RuvC and HNH nuclease domains (6, 7) was required for phage interference, while Csn2 protein (named Cas7 by Barrangou et al.) was mandatory for the spacer acquisition. Taken together, these experiments suggested for the first time that Cas9 is an important player in CRISPR-Cas immunity and set the stage for the mechanistic characterization of type II systems.

The Biology of Type II CRISPR-Cas Systems

Adaptation

Adaptation is the first step in CRISPR-Cas immunity, during which new spacers are acquired from the invading foreign nucleic acid and inserted into the CRISPR array (Fig. 4.1B). Cas1 and Cas2 proteins, which are conserved across all CRISPR-Cas types, play a key role in the spacer integration (8, 9). In the prototypical type I-E CRISPR-Cas system from *Escherichia coli*, these proteins form a heterocomplex with $Cas1_4$-$Cas2_2$ stoichiometry (10–14). This complex binds and positions a double-stranded DNA (dsDNA) prespacer to enable the nucleophilic attack of the 3′-hydroxyl end at the leader-repeat junction. The integration reaction is carried out by the Cas1 protein; however, the integration host factor contributes to the target specificity directing integration to the leader-repeat junction. Finally, the integration intermediate is repaired by the host factors, resulting in a new spacer insertion and repeat duplication. The universal conservation of the Cas1 and Cas2 proteins implies that the spacer integration mechanism is shared across different CRISPR-Cas systems. Indeed, the Cas1-Cas2 complex from the type II-A CRISPR-Cas system can integrate short DNA duplexes into the CRISPR region *in vitro* (14), similar to the *E. coli* Cas1-Cas2 complex. However, all four Cas proteins—Cas9, Cas1, Cas2, and Csn2—encoded in the type II-A CRISPR-Cas locus are required for adaptation *in vivo*, suggesting a possible role for Cas9 and Csn2 proteins in the spacer capture step (15–18). Although Cas9 nucleolytic activity is not required for adaptation *in vivo*, mutations in the protospacer adjacent motif (PAM) recognition domain result in PAM-independent spacer acquisition, confirming that Cas9 is required for insertion of spacers with a correct PAM sequence (15, 17). These findings collectively suggest that Cas9 protein plays an additional role beyond interference and contributes to the prespacer selection during adaptation, at least in type II-A systems. The role of the Csn2 protein that is required for adaptation *in vivo* is obscure; however, the protein forms a complex with Cas1-Cas2, suggesting a possible function in the spacer capture step (16, 18).

In type II-B systems, Csn2 is replaced by Cas4, which in the type I systems forms a complex with Cas1-Cas2 and contributes to spacer processing and integration (19, 20).

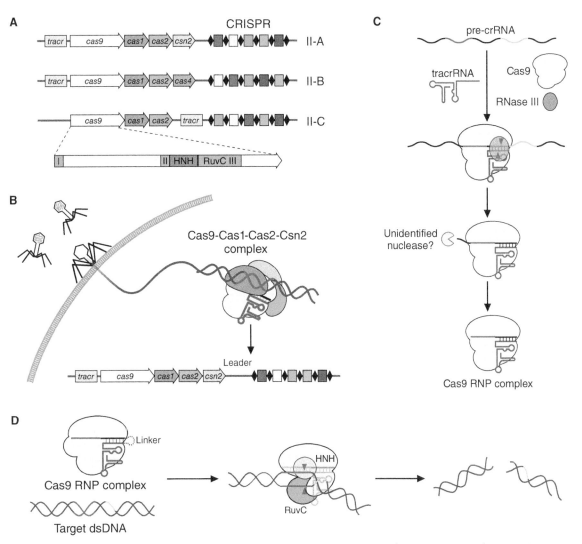

Figure 4.1 Biology of type II CRISPR-Cas systems. (A) Type II CRISPR-Cas locus structure. The *cas9* signature gene, encoding Cas9 nuclease, is conserved throughout all subtypes. **(B)** Adaptation. In type II-A systems, fragments of the invading DNA are captured by the Cas9-Cas1-Cas2-Csn2 adaptation complex, followed by spacer processing and integration into the CRISPR locus. **(C)** crRNA biogenesis. In type II systems, *trans*-encoded tracrRNA pairs with the repeat region of pre-crRNA in the presence of Cas9. The RNA duplex stretch is recognized and cleaved by endogenous RNase III. Further processing by unidentified nuclease of the crRNA 5′ end results in the ~42-nt mature form of crRNA. **(D)** DNA interference. The Cas9 RNP complex, guided by RNA (crRNA:tracrRNA duplex or engineered sgRNA), binds to the matching PAM-flanked DNA target, forming an R-loop. After R-loop formation, Cas9 nuclease uses HNH and RuvC active sites to cleave the target (complementary to gRNA) and nontarget (displaced) strands, respectively, to introduce DSBs.

It remains to be established whether Cas4 plays a similar role during adaptation in type II-B systems. On the other hand, in most type II-C systems, both Csn2 and Cas4 proteins are absent, suggesting that Cas1 and Cas2 might be sufficient for spacer capture and integration (21); however, it remains to be established whether host-specific factors contribute to this process.

crRNA Biogenesis

In the crRNA maturation stage, the CRISPR region is transcribed into the long pre-crRNA, which is further processed into shorter mature crRNAs. In most of the CRISPR-Cas systems, this process is carried out by a dedicated CRISPR endoribonuclease, Cas6 (or Cas5), or the effector protein (Cas13 and some Cas12 variants) (22–26). However, in the type II systems, crRNA maturation proceeds through a unique pathway. Profiling of *Streptococcus pyogenes* CRISPR-Cas locus expression using differential RNA sequencing (27) revealed two RNA species: ~42-nucleotide (nt) crRNAs, consisting of a 20-nt spacer-derived 5′ guide sequence and 22-nt repeat-derived 3′ sequence, and an additional RNA molecule that showed partial complementarity to the crRNA.

tracrRNA

Transactivating crRNA (tracrRNA) contained a sequence stretch, termed anti-repeat, with almost perfect complementarity to CRISPR repeats, suggesting base-pairing between the two RNA species. The experiments confirmed the formation of a tracrRNA and pre-crRNA partial duplex that is further cleaved by endogenous RNase III (Fig. 4.1C). Subsequent processing of the 5′ end of crRNA by unknown nuclease generates the mature ~42-nt crRNA and ~75-nt tracrRNA. Further genetic deletion experiments revealed that alongside to Cas9 and crRNA, both tracrRNA and RNase III are required for *in vivo* DNA interference by the type II-A CRISPR-Cas system (27). It was shown later that tracrRNA is an indispensable element for the catalytic activity of type II-A systems (28, 29). Interestingly, the crRNA maturation pathway established for the type II-C system from *Neisseria meningitidis* shows a variation on type II crRNA biogenesis. In contrast to the "canonical" type II pre-crRNA processing pathway discussed above, in *N. meningitidis,* crRNA transcription initiates in the spacer region from internal promoters within the repeats, resulting in 5′ crRNA ends that are not further processed (30). While repeat and anti-repeat sequences still form a duplex that is cleaved by RNase III, inactivation of the *rnc* gene, encoding RNase III, did not compromise the interference. Similarly, the heterologous *S. thermophilus* CRISPR-Cas system containing a minimal repeat-spacer-repeat unit in the CRISPR array provided plasmid interference in *E. coli* in the absence of RNase III (28), implying that RNase III is primarily required for dicing of long pre-crRNA into a set of crRNAs containing a single spacer. The tracrRNA requirement has been demonstrated for some Cas12 effectors (31–36); however, it still remains to be established if tracrRNA in these systems is involved in crRNA maturation.

Interference

PAM Dependence

Early experiments revealed that a short sequence motif (PAM) adjacent to the 3′ end of targeted sequences (6) is required for the immunity provided by the type II CRISPR-Cas systems (37). Specifically, in the CRISPR1 and CRISPR3 systems of *S. thermophilus*, phage interference was dependent on the presence of the 5′-NNAGAAW-3′ (where W represents A or T) and 5′-NGGNG-3′ PAM sequences, respectively (37, 38). Phages that escaped the CRISPR1-Cas immunity contained the mutations either in sequences complementary to the spacers (protospacers) or in PAM. By studying plasmid interference by the type II CRISPR1 system in the *S. thermophilus* DGCC7710 strain, Garneau et al. were able to capture a linear DNA fragment resulting from blunt-end cleavage by Cas9 nuclease 3 nt upstream of the PAM, providing direct evidence that the type II CRISPR-Cas system targets DNA (39).

Cas9: A Sole Protein Required for Interference

After the initial experiments, demonstrating antiviral defense of the type II CRISPR-Cas system (5) and identification of DNA cleavage requirements *in vivo*, the research focus shifted to the molecular mechanism of interference provided by these systems. Sapranaus-kas et al. were the first to demonstrate that the entire *S. thermophilus* CRISPR3-Cas locus is an independent module that can be transferred to distant *E. coli* bacteria and provide interference against invading DNA in the heterologous host (40). Most importantly, they showed that Cas9 is the only protein in the CRISPR-Cas locus required for interference and that mutations in Cas9 HNH or RuvC active sites abolished CRISPR-Cas immunity.

Cas9: Programmable RNA-Guided DNA Endonuclease

In vivo studies of type II CRISPR-Cas systems (see above) set the stage for *in vitro* reconstitution of Cas9-mediated immunity. In 2012, two independent studies demonstrated DNA cleavage by Cas9 in biochemical experiments (41, 42). *S. thermophilus* CRISPR3 (St3) Cas9 ribonucleoprotein (RNP) complex purified from *E. coli* cells (41) and *S. pyogenes* Cas9 (SpCas9) complex assembled from purified components *in vitro* (42) efficiently generated double-strand breaks (DSBs) 3 nt upstream of the PAM sequence, consistent with *in vivo* experiments (39). The cleavage reaction required both crRNA and tracrRNA that could be fused into a single guide RNA (sgRNA) to streamline the system (Fig. 4.1D) (42). Cas9 RNP complex recognizes the DNA target through Watson-Crick base-pairing between the ~20 nt of crRNA and complementary DNA strand forming an R-loop. Both studies showed that by simply changing the crRNA sequence, Cas9 protein can be reprogrammed to cleave a desired DNA target (41, 42). Mutational analysis confirmed the importance of the HNH and RuvC active sites of Cas9 predicted bioinformatically. Further studies revealed that HNH cleaves DNA target strand complementary to crRNA, while RuvC cuts the displaced nontarget DNA strand (Fig. 4.1D). Consequently, mutations in either HNH or RuvC active sites converted Cas9 endonuclease into a nickase (Cas9n) or catalytically dead Cas9 variant (dCas9) (41, 42).

Taken together, biochemical experiments revealed that in type II CRISPR-Cas systems, Cas9 functions as an RNA-guided DNA nuclease that cuts DNA in a PAM-dependent manner. A simple modular organization of the Cas9 RNP complex, where specificity for DNA targets is encoded by guide RNA (gRNA) consisting of crRNA:tracrRNA duplex or sgRNA and the cleavage is performed by the Cas9 protein, paved the way for targeted genome modification. Indeed, by altering the gRNA sequence within the Cas9 RNP complex, Cas9 can be directed to precisely cut DNA nearly at any site in the genome.

Adaptation of Cas9 Nuclease for *In Vivo* Genome Modification

The principles of the current genome editing technology were established almost three decades ago, when it was shown that DSBs greatly facilitated DNA repair in yeast (43) and mammalian (44) cells that can occur by nonhomologous end joining (NHEJ) or homology-directed repair (HDR) (45). During NHEJ, broken DNA strands are joined; however, small insertions and deletions (indels) are generated at the cleavage site. HDR, on the other hand, precisely repairs DSBs by replacing the broken region using a homologous template. Therefore, repair by NHEJ often leads to frameshifts, resulting in a premature stop codon and gene knockout. Alternatively, by utilizing the HDR pathway, precise mutations, deletions, or insertions from a single-stranded DNA (ssDNA) or dsDNA donor template can be introduced.

Three major nuclease scaffolds were adapted for the generation of DSBs at desired sites: homing endonucleases from bacterial mobile genetic elements (46), zinc finger nucleases derived from eukaryotic transcription factors (47–50), and transcription activator-like effector nucleases from *Xanthomonas* bacteria (51–54). Although they enabled precise genome modification in cells, reprogramming of their specificity for different targets remained a major challenge.

In 2012, after successful reconstitution of Cas9 endonuclease activity *in vitro*, both Gasiunas et al. and Jinek et al. realized the transformative potential of Cas9 in genome editing applications. Jinek et al. wrote, "We propose an alternative methodology based on RNA-programmed Cas9 that could offer considerable potential for gene-targeting and genome-editing applications" (42), while Gasiunas et al. came to the conclusion that "taken together, these findings pave the way for the development of unique molecular tools for RNA-directed DNA surgery" (41). Although some reports raised concerns about whether bacterial Cas9 protein can introduce DSBs in large and complex eukaryotic genomes (55, 56), the skepticism was soon dispersed when the first studies demonstrated that SpCas9 complex could be effectively used as a genome editing tool in human (57–59), mouse (59, 60), zebrafish (61, 62), yeast (63), bacterial (64), plant (65), and other cells. Soon these studies were expanded to germline cells to produce modified organisms ranging from mice (66) to primates (67).

Repurposing Cas9 nuclease activity to induce targeted DSBs in genomic DNA is just one way to use this technology. dCas9 complex can be utilized as a programmable DNA binding scaffold (Fig. 4.2) to sterically hinder RNA polymerase binding sites and down-regulate gene expression in bacterial and mammalian cells (68, 69). dCas9 fusion with effector domains enabled targeting of these proteins to specific DNA loci to induce or repress gene expression (70–72), modify histones or DNA bases (73–76), and track DNA

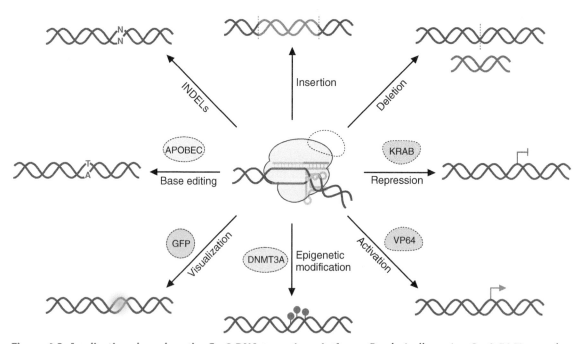

Figure 4.2 Applications based on the Cas9 DNA targeting platform. Catalytically active Cas9 RNP complex can be used for targeted DSB generation in chromosomal DNA. DSB repair by NHEJ results in insertions/deletions (indels) and gene knockouts. DSB repair by the homology-directed repair (HDR) pathway, which requires a homologous DNA template, can be used to accurately edit genomes in various applications (precise gene replacement or insertion, precise mutations, etc.). Additionally, catalytically inactive Cas9 (dCas9) can be fused with effector domains for targeted gene activation/repression, epigenetic modification, DNA visualization *in vivo*, and other applications. The use of dCas9 or Cas9n (nickase) variants fused with AID or APOBEC cytosine deaminases allows targeted C conversion to T, while fusions with engineered TadA allows A-to-G conversion in precise base editing applications.

location *in vivo* (77). Furthermore, the PAM providing DNA oligonucleotides enabled Cas9 targeting and degradation of RNA both *in vitro* and *in vivo* (78, 79) and RNA tracking in cells (80).

Cas9 was also adapted in genetic screens for systematic analysis of gene function. Cas9- or dCas9-based loss-of-function gene screens were developed to identify genes essential for the cell survival phenotype (81–85). Gain-of-function screens were designed using dCas9 tethered to transcriptional activation factors (85).

dCas9 or Cas9n variants were instrumental in the development of base editing technology that does not require DSB formation for targeted genome modification. In this case, dCas9 or Cas9n is tethered to ssDNA deaminases that are able to generate transition mutations by changing C to T or A to G in the targeted sequence (86). DNA target sequence recognition by dCas9 or Cas9n creates an R-loop (87) displacing the nontarget strand as the ssDNA that becomes a target for deaminases. Cas9 fusions with AID or APOBEC cytosine deaminases (76, 88) or engineered TadA adenine deaminase (89) were used for the C-to-T and A-to-G conversions, respectively. Coupling of both deaminases with a single Cas9 protein resulted in dual-function base editors that were able to edit cytosine and adenine residues simultaneously (90, 91).

To expand the scope of base editing to other base conversions (e.g., transversion mutations) or precise insertions and deletions, an alternative DNA prime editing technology has been developed in the lab of David Liu. Tethering of nCas9 protein to reverse transcriptase, which is programmed by a template encoded at the 3′ end of prime editing gRNA (pegRNA), allowed introduction of precise insertions, deletions, and all 12 possible base-to-base edits at the targeted site (92). Overall, the base and prime editing technologies look very promising for clinical applications since they do not rely on the generation of DSBs, which trigger HDR or NHEJ repair.

Mechanism of DNA Cleavage by Cas9 Nuclease

Cas9 structures solved in different states significantly improved our understanding of structural and molecular mechanisms of RNA-guided DNA targeting and cleavage (93–105). The structure of the apo-SpCas9 protein (1,368 amino acids [aa]) revealed a bilobed architecture comprised of nuclease (NUC) and recognition (REC) lobes with a positively charged channel at the interface (Fig. 4.3, 1st step) (100). The NUC lobe contains the HNH and split RuvC domains involved in DNA cleavage and the PAM-interacting (PI) domain. In the apo-Cas9, the PI domain is disordered, assuming an inactive state. In the binary Cas9-gRNA complex, the gRNA is accommodated in the positively charged groove at the interface between the REC and NUC lobes (93, 95, 100–102). The repeat:anti-repeat duplex of the gRNA makes sequence-dependent interactions with Cas9, providing a structural explanation for the observed coevolution between tracrRNA and Cas9 sequences (106). Cas9-gRNA interactions also preorder the 20-nt spacer sequence into an A-form conformation that facilitates the Watson-Crick base-pairing between the RNA guide and target DNA strand. In the SpCas9-gRNA complex, the PI domain undergoes structural rearrangement and R1333 and R1335 residues become prepositioned for 5′-NGG-3′ PAM recognition. All these structural changes occurring upon gRNA binding enable formation of a functional Cas9-gRNA complex (Fig. 4.3, 2nd step) that searches for PAM sequences in DNA by three-dimensional (3D) collisions and 2D diffusion (107–109) (Fig. 4.3, 3rd step). PAM sequence is recognized through sequence-specific protein-DNA interactions (Fig. 4.3, 4th step) that trigger local DNA distortion to facilitate initiation of R-loop formation (99). R-loop formation (Fig. 4.3, 5th step) is directional starting at the seed sequence proximal

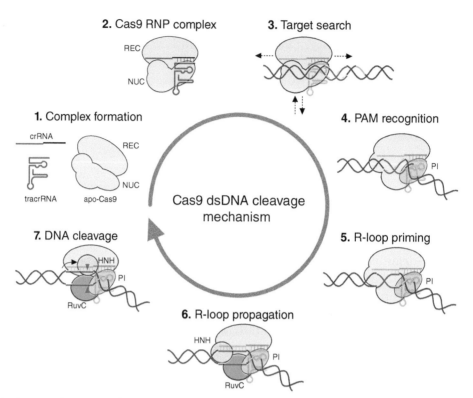

Figure 4.3 Cas9 nuclease DNA cleavage mechanism. Apo-Cas9 protein consists of nuclease (NUC) and recognition (REC) lobes with a positively charged interface between them (1st step). The gRNA binding drives the apo-Cas9 nuclease from the inactive state to the DNA target recognition-competent complex (2nd step). During the DNA target search, Cas9 complex interrogates double-stranded DNA (dsDNA) for PAM sequences through 3D collisions/2D diffusion (3rd step). Upon collision with the PAM, the Cas9 complex makes sequence-specific protein-DNA contacts (4th step) and bends the DNA helix by ~30°, providing the structural distortion required for R-loop formation initiation. Successful R-loop initiation (priming) at the PAM-proximal seed sequence (5th step) and subsequent propagation (6th step) to the PAM distal end result in HNH domain movement towards scissile phosphate and RuvC domain activation to enable concerted DNA cleavage (7th step).

to PAM and propagating to the PAM distal end (87). A perfect or near-perfect complementarity between the spacer sequence and DNA is required for DNA cleavage, with some mismatches tolerated at the PAM distal end but inhibitory in the 10- to 12-nt seed sequence proximal to PAM (42, 59, 64, 87, 109). Structural analyses of Cas9 in different states imply conformational flexibility of the HNH domain. R-loop formation drives the repositioning and reorientation of the HNH domain into a catalytically competent conformation prepositioned for complementary strand cleavage in the guide:target heteroduplex (93, 95, 99, 101, 102). Förster resonance energy transfer (FRET) based experiments confirmed HNH domain movement that is coordinated with RuvC domain activation for the noncomplementary strand cleavage (Fig. 4.3, 7th step) (105, 110). Interestingly, in the postcleavage state, due to the structural rearrangements of the REC domain, Cas9 makes multiple DNA contacts with the cleaved DNA, providing a structural explanation for Cas9 inhibition by reaction products (109).

Structures of Cas9 orthologs from *Staphylococcus aureus* (Sa) (1,053 aa), *Francisella novicida* (Fn) (1,629 aa), *Campylobacter jejuni* (Cj) (984 aa), *Corynebacterium diphtheriae* (Cd) (1,084 aa) and *Streptococcus thermophilus* CRISPR1 (St1) (1,121 aa) (95, 97, 102–104) revealed that in spite of the size differences, Cas9 proteins retain bilobed (REC and NUC) structure. The major differences were observed in the REC lobe, responsible for

RNA repeat:anti-repeat duplex binding, and the C-terminal domain, involved in PAM recognition. These differences explain the distinct PAM specificities (SaCas9, 5'-NNGRRT-3'; FnCas9, 5'-NGG-3'; CjCas9, 5'-NNNVRYM-3'; CdCas9, 5'-NNRHHHY-3'; and St1, 5'-NNRGAA-3'; R represents A or G, V represents a base not T, Y represents C or T, M represents A or C, and H represents a base not G). On the other hand, the structural similarity of the NUC lobes of different orthologs suggests conserved DNA recognition and cleavage mechanisms across the Cas9 protein family.

Cas9 Challenges and Limitations

Since the first demonstration that SpCas9 can be used for genome editing, it still remains the most widely used CRISPR nuclease (111). High SpCas9 activity across different cell types and the requirement for relatively short PAM sequence (5'-NGG-3') enabled myriad genome editing applications in different cells and organisms. The CRISPR-Cas9 system is now rapidly advancing into clinics for the treatment of human diseases. It is a great tool but it still has some limitations that need to be overcome. The first challenge is off-target activity that leads to the unintended cleavage at nontargeted sites and undesired mutations in the chromosomal DNA. Off-target sites containing a few nucleotide mismatches between the gRNA and target site sequences or substitutions in the PAM sequence are prone to cleavage by SpCas9 (112–115). Next, the SpCas9 requirement for the 5'-NGG-3' PAM sequence may limit the targeting space in allele-specific modifications (116–118) or base editing applications (86). Next, there are difficulties with the delivery of Cas9 into tissues or cells to achieve therapeutic effects. Adeno-associated virus (AAV) often is a preferable delivery vector but has a limited DNA cargo capacity. Therefore, simultaneous packaging of SpCas9 (4,107 bp) and a gRNA into AAV becomes challenging (119–121). Furthermore, the preexisting immunity against SpCas9 and SaCas9 proteins in humans may add an extra layer of complexity for *in vivo* genome editing applications (122). Therefore, CRISPR nucleases that are more specific, show diverse PAM requirements, and are compact are highly desirable. To address this issue, several approaches have been developed, including structure-guided rational design, directed evolution, and exploration of Cas9 ortholog space (Fig. 4.4).

Cas9 Proteins with Increased Specificity

Early attempts to identify possible off-targets were limited to the cleavage analysis of nontarget sites that show similarities to the target sequence (112–115). These experiments revealed that SpCas9 can tolerate up to 5 individual mismatches between gRNA and DNA. Mismatches in the PAM sequence were also tolerated, resulting in the cleavage of DNA targets flanked by noncanonical PAM sequences.

To get a comprehensive genome-wide profile of possible Cas9 off-targets, a number of unbiased methods were developed, including whole-genome sequencing (123, 124), high-throughput genome-wide translocation sequencing (125), Digenome-seq (126, 127), integrase-defective lentiviral vector capture (128–130), GUIDE-seq (genome-wide unbiased identification of DSBs enabled by sequencing) (131), BLESS (break labeling, enrichment on streptavidin, and next-generation sequencing) (132–134), BLISS (break labeling *in situ* and sequencing) (135), CIRCLE-seq (circularization for *in vitro* reporting of cleavage effects by sequencing) (136), SITE-seq (selective enrichment and identification of tagged genomic DNA ends by sequencing) (137), CLEAVE-seq (138), and CHANGE-seq (139). These studies revealed a relatively comprehensive picture of Cas9 editing specificity that is mostly dependent on the gRNA sequence.

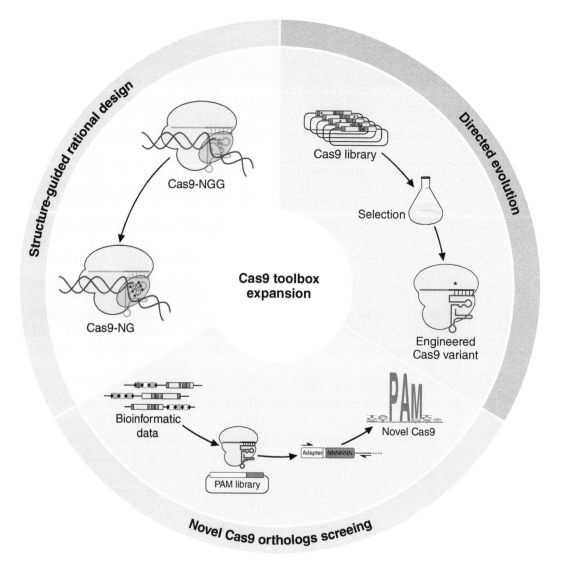

Figure 4.4 Strategies for expanding the Cas9 toolbox. The major experimental approaches used for the development of the expanded Cas9 toolkit can be broadly categorized into structure-guided rational design, directed evolution, and novel Cas9 ortholog screening. Both structure-guided mutagenesis and directed evolution coupled with different screening strategies resulted in engineered Cas9 protein variants with increased specificity and altered PAM recognition. Bioinformatic database mining for novel Cas9 proteins followed by screening resulted in Cas9 proteins that differ in size, specificity, and PAM sequence requirements.

Multiple attempts were made to improve Cas9 fidelity through protein engineering. Structure-based rational engineering approaches generated SpCas9-HF1 (high-fidelity variant 1) (140) and eSpCas9 1.1 (enhanced SpCas9 version 1.1) (134) variants retaining efficient on-target cleavage but reduced off-target activity. In these variants, residues responsible for nonspecific interactions with complementary (target) and noncomplementary DNA strands, respectively, were replaced with alanine. A different rationale was applied to engineer HypaCas9 (hyperaccurate Cas9) (141). Mutations of amino acid residues involved in DNA:RNA heteroduplex binding in the REC domain constrained the HNH domain in the inactive state on the mismatched targets, increasing the DNA cleavage specificity.

Directed-evolution approaches were also employed for generation of SpCas9 variants with increased specificity. Combination of positive- and negative-selection strategies in

E. coli or yeast cells for SpCas9 protein library screening enabled selection of high-fidelity Cas9 variants (EvoCas9, SniperCas9, and HiFiCas9) that retained high on-target activity and showed decreased off-target cleavage (142–144).

Alternatively, mining of the natural diversity of Cas9 protein family could identify Cas9 orthologs with increased specificity. DSB capture after DNA cleavage by SpCas9 and SaCas9 nucleases revealed that SaCas9 introduced fewer DSBs at off-target sites (133). Furthermore, St3 and St1Cas9 nucleases targeting the same sequence were less prone than SpCas9 to off-target cleavage (145). *N. meningitidis* Cas9 orthologs (Nm1 and Nm2) showed lower off-target cleavage activity than SpCas9 (146, 147). These studies suggest that more Cas9 orthologs with improved specificity profiles can be discovered.

Cas9 Variants with Expanded Targeting Range

Although Cas9 proteins can be reprogrammed by gRNA to target virtually any DNA sequence, PAM sequence requirement limits the targeting space. The 5′-NGG-3′ PAM sequence recognized by SpCas9 is theoretically found every 8 bp in the human genome; however, PAM density and distribution in the genome are uneven, creating problems when targeting at specific sites is required, e.g., to increase efficiency of HDR (148), for correction of mutations by base editing (86), or for single allele modification (116–118). To overcome these limitations, multiple efforts were directed to engineer Cas9 nucleases with altered PAM specificity or mining the natural diversity of Cas9 orthologs.

The first SpCas9 variants with altered PAM recognition were developed by directed evolution generating 5′-NGA-3′ (VQR variant), 5′-NGAG-3′ (EQR), and 5′-NGCG-3′ (VRER) PAM specificities. Using the same approach, the relaxed 5′-NNNRRT-3′ PAM variant was obtained for SaCas9, which recognizes 5′-NNGRRT-3′ PAM (149). Later, the phage-assisted continuous evolution approach yielded an xCas9 variant that recognizes a range of 5′-NG-3′, 5′-GAA-3′, and 5′-GAT-3′ PAM sequences (150) and SpCas9 variants specific for non-G 5′-NRRH-3′, 5′-NRCH-3′, and 5′-NRTH-3′ PAMs (151).

Using a rational structure-guided approach, FnCas9 variants with relaxed 5′-YG-3′ (RHA variant) specificity were generated (95). Subsequent attempts to relax PAM recognition for SpCas9 generated SpCas9 variants recognizing 5′-NG-3′, 5′-NR-3′, and 5′-NY-3′ PAMs (152, 153) that show very simple PAM requirement.

Exploration of Cas9 diversity provides an alternative approach to expand the DNA targeting space of Cas9 nucleases. The PAM specificity of the first Cas9 orthologs was predicted bioinformatically (38, 154, 155); however, this approach is limited due to the relatively low number of available phage genomes. To overcome this problem, high-throughput PAM screens using PAM libraries were developed to interrogate the PAM specificity of novel Cas9 orthologs (156). This strategy allowed establishment of PAM specificity for a number of Cas9 orthologs, including SaCas9 (5′-NNGRRT-3′), Nm1Cas9 (5′-NNNNGATT-3′), Nm2Cas9 (5′-NNNNCC-3′), *Brevibacillus laterosporus* Cas9 (5′-NNNNCNDD-3′, where D is A, T, or G), FnCas9 (5′-NGG-3′), CjCas9 (5′-NNNNRYAC-3′), *Streptococcus canis* Cas9 (5′-NNG-3′), and others (95, 133, 146, 147, 157–160). Recently, systematic exploration of the Cas9 diversity using high-throughput PAM screens produced a catalog of Cas9 orthologs (161) that includes 79 Cas9 variants recognizing the entire spectrum of T-, A-, C-, and G-rich PAM sequences, ranging from single nucleotides to 5-nt sequence strings. Additional characterization of a subset of Cas9 orthologs also revealed biochemical diversity, including different ranges of temperature dependence, cleavage pattern, and optimal spacer length. Therefore, the natural diversity of Cas9 proteins offers a rich source of novel tools for the Cas9-based genome editing toolbox.

Small Cas9 Orthologs for Viral Delivery

AAVs show great promise for Cas9 *in vivo* delivery for human therapeutic applications (120). However, due to the limited capacity of AAV DNA cargo, the simultaneous packaging of the SpCas9 and gRNA encoding sequences (~4.2 kb) into AAV is challenging. Therefore, smaller Cas9 orthologs with different PAM sequences are highly desired to expand *in vivo* applications. Search efforts for more compact Cas9 nucleases resulted in the discovery of SaCas9 (1,053 aa), Nm1Cas9 (1,082 aa), Nm2Cas9 (1,082 aa), and CjCas9 (984 aa) variants that have been efficiently delivered into cells using AAV vectors (133, 147, 160, 162). Moreover, the SaCas9 nuclease delivered by AAV entered phase 1/2 clinical trials for treatment of Leber congenital amaurosis 10, which causes blindness (https://clinicaltrials.gov/ct2/show/NCT03872479), while SpCas9 is used in phase 1/2 trials for *ex vivo* treatment of β-thalassemia and sickle cell disease (https://clinicaltrials.gov/ct2/show/NCT03655678 and https://clinicaltrials.gov/ct2/show/NCT03745287). Nevertheless, more compact CRISPR nucleases with an expanded targeting range and increased specificity are still highly desirable to facilitate advancement of *in vivo* genome editing applications into the clinic.

Concluding Remarks

Almost 15 years ago, basic research on molecular mechanisms of type II CRISPR-Cas systems opened a new era in molecular biology by repurposing the antiviral defense system of bacteria into a versatile tool for genome modification. This period is marked by unprecedented scientific progress ranging from the fundamental discoveries resulting in deciphering of the molecular mechanism of type II CRISPR-Cas systems to the first clinical trials. Novel CRISPR-Cas9 nucleases with increased fidelity, diverse PAM requirements, and smaller size are expanding the CRISPR-Cas toolbox, paving the way to more fascinating discoveries and applications across different fields of biology and medicine.

References

1. Koonin EV, Makarova KS, Wolf YI. 2017. Evolutionary genomics of defense systems in archaea and bacteria. *Annu Rev Microbiol* **71**:233–261.

2. Makarova KS, Wolf YI, Iranzo J, Shmakov SA, Alkhnbashi OS, Brouns SJJ, Charpentier E, Cheng D, Haft DH, Horvath P, Moineau S, Mojica FJM, Scott D, Shah SA, Siksnys V, Terns MP, Venclovas Č, White MF, Yakunin AF, Yan W, Zhang F, Garrett RA, Backofen R, van der Oost J, Barrangou R, Koonin EV. 2020. Evolutionary classification of CRISPR-Cas systems: a burst of class 2 and derived variants. *Nat Rev Microbiol* **18**:67–83.

3. Makarova KS, Wolf YI, Alkhnbashi OS, Costa F, Shah SA, Saunders SJ, Barrangou R, Brouns SJJ, Charpentier E, Haft DH, Horvath P, Moineau S, Mojica FJM, Terns RM, Terns MP, White MF, Yakunin AF, Garrett RA, van der Oost J, Backofen R, Koonin EV. 2015. An updated evolutionary classification of CRISPR-Cas systems. *Nat Rev Microbiol* **13**:722–736.

4. Shmakov S, Smargon A, Scott D, Cox D, Pyzocha N, Yan W, Abudayyeh OO, Gootenberg JS, Makarova KS, Wolf YI, Severinov K, Zhang F, Koonin EV. 2017. Diversity and evolution of class 2 CRISPR-Cas systems. *Nat Rev Microbiol* **15**:169–182.

5. Barrangou R, Fremaux C, Deveau H, Richards M, Boyaval P, Moineau S, Romero DA, Horvath P. 2007. CRISPR provides acquired resistance against viruses in prokaryotes. *Science* **315**:1709–1712.

6. Bolotin A, Quinquis B, Sorokin A, Ehrlich SD. 2005. Clustered regularly interspaced short palindrome repeats (CRISPRs) have spacers of extrachromosomal origin. *Microbiology (Reading)* **151**:2551–2561.

7. Makarova KS, Grishin NV, Shabalina SA, Wolf YI, Koonin EV. 2006. A putative RNA-interference-based immune system in prokaryotes: computational analysis of the predicted enzymatic machinery, functional analogies with eukaryotic RNAi, and hypothetical mechanisms of action. *Biol Direct* **1**:7.

8. Jackson SA, McKenzie RE, Fagerlund RD, Kieper SN, Fineran PC, Brouns SJJ. 2017. CRISPR-Cas: adapting to change. *Science* **356**: eaal5056.

9. Sasnauskas G, Siksnys V. 2020. CRISPR adaptation from a structural perspective. *Curr Opin Struct Biol* **65**:17–25.

10. Nuñez JK, Kranzusch PJ, Noeske J, Wright AV, Davies CW, Doudna JA. 2014. Cas1-Cas2 complex formation mediates spacer acquisition during CRISPR-Cas adaptive immunity. *Nat Struct Mol Biol* **21**:528–534.

11. Nuñez JK, Harrington LB, Kranzusch PJ, Engelman AN, Doudna JA. 2015. Foreign DNA capture during CRISPR-Cas adaptive immunity. *Nature* **527**:535–538.

12. Nuñez JK, Lee ASY, Engelman A, Doudna JA. 2015. Integrase-mediated spacer acquisition during CRISPR-Cas adaptive immunity. *Nature* **519**:193–198.

13. Nuñez JK, Bai L, Harrington LB, Hinder TL, Doudna JA. 2016. CRISPR immunological memory requires a host factor for specificity. *Mol Cell* **62**:824–833.

14. Wright AV, Doudna JA. 2016. Protecting genome integrity during CRISPR immune adaptation. *Nat Struct Mol Biol* **23**:876–883.

15. Heler R, Samai P, Modell JW, Weiner C, Goldberg GW, Bikard D, Marraffini LA. 2015. Cas9 specifies functional viral targets during CRISPR-Cas adaptation. *Nature* 519:199–202.

16. Ka D, Jang DM, Han BW, Bae E. 2018. Molecular organization of the type II-A CRISPR adaptation module and its interaction with Cas9 via Csn2. *Nucleic Acids Res* 46:9805–9815.

17. Wei Y, Terns RM, Terns MP. 2015. Cas9 function and host genome sampling in type II-A CRISPR-Cas adaptation. *Genes Dev* 29:356–361.

18. Wilkinson M, Drabavicius G, Silanskas A, Gasiunas G, Siksnys V, Wigley DB. 2019. Structure of the DNA-bound spacer capture complex of a type II CRISPR-Cas system. *Mol Cell* 75:90–101.e5.

19. Lee H, Zhou Y, Taylor DW, Sashital DG. 2018. Cas4-dependent prespacer processing ensures high-fidelity programming of CRISPR arrays. *Mol Cell* 70:48–59.e5.

20. Shiimori M, Garrett SC, Graveley BR, Terns MP. 2018. Cas4 nucleases define the PAM, length, and orientation of DNA fragments integrated at CRISPR loci. *Mol Cell* 70:814–824.e6.

21. He Y, Wang M, Liu M, Huang L, Liu C, Zhang X, Yi H, Cheng A, Zhu D, Yang Q, Wu Y, Zhao X, Chen S, Jia R, Zhang S, Liu Y, Yu Y, Zhang L. 2018. Cas1 and Cas2 from the type II-C CRISPR-Cas system of Riemerella anatipestifer are required for spacer acquisition. *Front Cell Infect Microbiol* 8:195.

22. Charpentier E, Richter H, van der Oost J, White MF. 2015. Biogenesis pathways of RNA guides in archaeal and bacterial CRISPR-Cas adaptive immunity. *FEMS Microbiol Rev* 39:428–441.

23. East-Seletsky A, O'Connell MR, Knight SC, Burstein D, Cate JHD, Tjian R, Doudna JA. 2016. Two distinct RNase activities of CRISPR-C2c2 enable guide-RNA processing and RNA detection. *Nature* 538:270–273.

24. Fonfara I, Richter H, Bratovič M, Le Rhun A, Charpentier E. 2016. The CRISPR-associated DNA-cleaving enzyme Cpf1 also processes precursor CRISPR RNA. *Nature* 532:517–521.

25. Hochstrasser ML, Doudna JA. 2015. Cutting it close: CRISPR-associated endoribonuclease structure and function. *Trends Biochem Sci* 40:58–66.

26. Pausch P, Al-Shayeb B, Bisom-Rapp E, Tsuchida CA, Li Z, Cress BF, Knott GJ, Jacobsen SE, Banfield JF, Doudna JA. 2020. CRISPR-CasΦ from huge phages is a hypercompact genome editor. *Science* 369:333–337.

27. Deltcheva E, Chylinski K, Sharma CM, Gonzales K, Chao Y, Pirzada ZA, Eckert MR, Vogel J, Charpentier E. 2011. CRISPR RNA maturation by trans-encoded small RNA and host factor RNase III. *Nature* 471:602–607.

28. Karvelis T, Gasiunas G, Miksys A, Barrangou R, Horvath P, Siksnys V. 2013. crRNA and tracrRNA guide Cas9-mediated DNA interference in Streptococcus thermophilus. *RNA Biol* 10:841–851.

29. Chylinski K, Le Rhun A, Charpentier E. 2013. The tracrRNA and Cas9 families of type II CRISPR-Cas immunity systems. *RNA Biol* 10:726–737.

30. Zhang Y, Heidrich N, Ampattu BJJ, Gunderson CWW, Seifert HSS, Schoen C, Vogel J, Sontheimer EJJ. 2013. Processing-independent CRISPR RNAs limit natural transformation in Neisseria meningitidis. *Mol Cell* 50:488–503.

31. Harrington LB, Burstein D, Chen JS, Paez-Espino D, Ma E, Witte IP, Cofsky JC, Kyrpides NC, Banfield JF, Doudna JA. 2018. Programmed DNA destruction by miniature CRISPR-Cas14 enzymes. *Science* 362:839–842.

32. Karvelis T, Bigelyte G, Young JK, Hou Z, Zedaveinyte R, Budre K, Paulraj S, Djukanovic V, Gasior S, Silanskas A, Venclovas Č, Siksnys V. 2020. PAM recognition by miniature CRISPR-Cas12f nucleases triggers programmable double-stranded DNA target cleavage. *Nucleic Acids Res* 48:5016–5023.

33. Liu J-J, Orlova N, Oakes BL, Ma E, Spinner HB, Baney KLM, Chuck J, Tan D, Knott GJ, Harrington LB, Al-Shayeb B, Wagner A, Brötzmann J, Staahl BT, Taylor KL, Desmarais J, Nogales E, Doudna JA. 2019. CasX enzymes comprise a distinct family of RNA-guided genome editors. *Nature* 566:218–223.

34. Strecker J, Jones S, Koopal B, Schmid-Burgk J, Zetsche B, Gao L, Makarova KS, Koonin EV, Zhang F. 2019. Engineering of CRISPR-Cas12b for human genome editing. *Nat Commun* 10:212.

35. Teng F, Cui T, Feng G, Guo L, Xu K, Gao Q, Li T, Li J, Zhou Q, Li W. 2018. Repurposing CRISPR-Cas12b for mammalian genome engineering. *Cell Discov* 4:63.

36. Yan WX, Hunnewell P, Alfonse LE, Carte JM, Keston-Smith E, Sothiselvam S, Garrity AJ, Chong S, Makarova KS, Koonin EV, Cheng DR, Scott DA. 2019. Functionally diverse type V CRISPR-Cas systems. *Science* 363:88–91.

37. Deveau H, Barrangou R, Garneau JE, Labonté J, Fremaux C, Boyaval P, Romero DA, Horvath P, Moineau S. 2008. Phage response to CRISPR-encoded resistance in Streptococcus thermophilus. *J Bacteriol* 190:1390–1400.

38. Horvath P, Romero DA, Coûté-Monvoisin A-C, Richards M, Deveau H, Moineau S, Boyaval P, Fremaux C, Barrangou R. 2008. Diversity, activity, and evolution of CRISPR loci in Streptococcus thermophilus. *J Bacteriol* 190:1401–1412.

39. Garneau JE, Dupuis M-È, Villion M, Romero DA, Barrangou R, Boyaval P, Fremaux C, Horvath P, Magadán AH, Moineau S. 2010. The CRISPR/Cas bacterial immune system cleaves bacteriophage and plasmid DNA. *Nature* 468:67–71.

40. Sapranauskas R, Gasiunas G, Fremaux C, Barrangou R, Horvath P, Siksnys V. 2011. The Streptococcus thermophilus CRISPR/Cas system provides immunity in Escherichia coli. *Nucleic Acids Res* 39:9275–9282.

41. Gasiunas G, Barrangou R, Horvath P, Siksnys V. 2012. Cas9-crRNA ribonucleoprotein complex mediates specific DNA cleavage for adaptive immunity in bacteria. *Proc Natl Acad Sci USA* 109:E2579–E2586.

42. Jinek M, Chylinski K, Fonfara I, Hauer M, Doudna JA, Charpentier E. 2012. A programmable dual-RNA-guided DNA endonuclease in adaptive bacterial immunity. *Science* 337:816–821.

43. Rudin N, Sugarman E, Haber JE. 1989. Genetic and physical analysis of double-strand break repair and recombination in Saccharomyces cerevisiae. *Genetics* 122:519–534.

44. Rouet P, Smih F, Jasin M. 1994. Introduction of double-strand breaks into the genome of mouse cells by expression of a rare-cutting endonuclease. *Mol Cell Biol* 14:8096–8106.

45. Wyman C, Kanaar R. 2006. DNA double-strand break repair: all's well that ends well. *Annu Rev Genet* 40:363–383.

46. Smith J, Grizot S, Arnould S, Duclert A, Epinat J-C, Chames P, Prieto J, Redondo P, Blanco FJ, Bravo J, Montoya G, Pâques F, Duchateau P. 2006. A combinatorial approach to create artificial homing endonucleases cleaving chosen sequences. *Nucleic Acids Res* 34:e149.

47. Bibikova M, Carroll D, Segal DJ, Trautman JK, Smith J, Kim Y-G, Chandrasegaran S. 2001. Stimulation of homologous recombination through targeted cleavage by chimeric nucleases. *Mol Cell Biol* 21:289–297.

48. Bibikova M, Golic M, Golic KG, Carroll D. 2002. Targeted chromosomal cleavage and mutagenesis in Drosophila using zinc-finger nucleases. *Genetics* 161:1169–1175.

49. Bibikova M, Beumer K, Trautman JK, Carroll D. 2003. Enhancing gene targeting with designed zinc finger nucleases. *Science* 300:764.

50. Kim YG, Cha J, Chandrasegaran S. 1996. Hybrid restriction enzymes: zinc finger fusions to Fok I cleavage domain. *Proc Natl Acad Sci USA* 93:1156–1160.

51. Boch J, Scholze H, Schornack S, Landgraf A, Hahn S, Kay S, Lahaye T, Nickstadt A, Bonas U. 2009. Breaking the code of DNA binding specificity of TAL-type III effectors. *Science* 326:1509–1512.

52. Christian M, Cermak T, Doyle EL, Schmidt C, Zhang F, Hummel A, Bogdanove AJ, Voytas DF. 2010. Targeting DNA double-strand breaks with TAL effector nucleases. *Genetics* 186:757–761.

53. Miller JC, Tan S, Qiao G, Barlow KA, Wang J, Xia DF, Meng X, Paschon DE, Leung E, Hinkley SJ, Dulay GP, Hua KL, Ankoudinova I, Cost GJ, Urnov FD, Zhang HS, Holmes MC, Zhang L, Gregory PD,

Rebar EJ. 2011. A TALE nuclease architecture for efficient genome editing. *Nat Biotechnol* **29:**143–148.

54. Moscou MJ, Bogdanove AJ. 2009. A simple cipher governs DNA recognition by TAL effectors. *Science* **326:**1501.

55. Barrangou R. 2012. RNA-mediated programmable DNA cleavage. *Nat Biotechnol* **30:**836–838.

56. Carroll D. 2012. A CRISPR approach to gene targeting. *Mol Ther* **20:**1658–1660.

57. Jinek M, East A, Cheng A, Lin S, Ma E, Doudna J. 2013. RNA-programmed genome editing in human cells. *eLife* **2:**e00471.

58. Mali P, Yang L, Esvelt KM, Aach J, Guell M, DiCarlo JE, Norville JE, Church GM. 2013. RNA-guided human genome engineering via Cas9. *Science* **339:**823–826.

59. Cong L, Ran FA, Cox D, Lin S, Barretto R, Habib N, Hsu PD, Wu X, Jiang W, Marraffini LA, Zhang F. 2013. Multiplex genome engineering using CRISPR/Cas systems. *Science* **339:**819–823.

60. Wang H, Yang H, Shivalila CS, Dawlaty MM, Cheng AW, Zhang F, Jaenisch R. 2013. One-step generation of mice carrying mutations in multiple genes by CRISPR/Cas-mediated genome engineering. *Cell* **153:**910–918.

61. Hwang WY, Fu Y, Reyon D, Maeder ML, Tsai SQ, Sander JD, Peterson RT, Yeh J-RJ, Joung JK. 2013. Efficient genome editing in zebrafish using a CRISPR-Cas system. *Nat Biotechnol* **31:**227–229.

62. Chang N, Sun C, Gao L, Zhu D, Xu X, Zhu X, Xiong J-W, Xi JJ. 2013. Genome editing with RNA-guided Cas9 nuclease in zebrafish embryos. *Cell Res* **23:**465–472.

63. DiCarlo JE, Norville JE, Mali P, Rios X, Aach J, Church GM. 2013. Genome engineering in Saccharomyces cerevisiae using CRISPR-Cas systems. *Nucleic Acids Res* **41:**4336–4343.

64. Jiang W, Bikard D, Cox D, Zhang F, Marraffini LA. 2013. RNA-guided editing of bacterial genomes using CRISPR-Cas systems. *Nat Biotechnol* **31:**233–239.

65. Nekrasov V, Staskawicz B, Weigel D, Jones JDG, Kamoun S. 2013. Targeted mutagenesis in the model plant Nicotiana benthamiana using Cas9 RNA-guided endonuclease. *Nat Biotechnol* **31:**691–693.

66. Shen B, Zhang J, Wu H, Wang J, Ma K, Li Z, Zhang X, Zhang P, Huang X. 2013. Generation of gene-modified mice via Cas9/RNA-mediated gene targeting. *Cell Res* **23:**720–723.

67. Niu Y, Shen B, Cui Y, Chen Y, Wang J, Wang L, Kang Y, Zhao X, Si W, Li W, Xiang APP, Zhou J, Guo X, Bi Y, Si C, Hu B, Dong G, Wang H, Zhou Z, Li T, Tan T, Pu X, Wang F, Ji S, Zhou Q, Huang X, Ji W, Sha J. 2014. Generation of gene-modified cynomolgus monkey via Cas9/RNA-mediated gene targeting in one-cell embryos. *Cell* **156:**836–843.

68. Qi LS, Larson MH, Gilbert LA, Doudna JA, Weissman JS, Arkin AP, Lim WA. 2013. Repurposing CRISPR as an RNA-guided platform for sequence-specific control of gene expression. *Cell* **152:**1173–1183.

69. Bikard D, Marraffini LA. 2013. Control of gene expression by CRISPR-Cas systems. *F1000Prime Rep* **5:**47.

70. Perez-Pinera P, Kocak DD, Vockley CM, Adler AF, Kabadi AM, Polstein LR, Thakore PI, Glass KA, Ousterout DG, Leong KW, Guilak F, Crawford GE, Reddy TE, Gersbach CA. 2013. RNA-guided gene activation by CRISPR-Cas9-based transcription factors. *Nat Methods* **10:**973–976.

71. Maeder ML, Linder SJ, Cascio VM, Fu Y, Ho QH, Joung JK. 2013. CRISPR RNA-guided activation of endogenous human genes. *Nat Methods* **10:**977–979.

72. Gilbert LA, Larson MH, Morsut L, Liu Z, Brar GA, Torres SE, Stern-Ginossar N, Brandman O, Whitehead EH, Doudna JA, Lim WA, Weissman JS, Qi LS. 2013. CRISPR-mediated modular RNA-guided regulation of transcription in eukaryotes. *Cell* **154:**442–451.

73. Hilton IB, D'Ippolito AM, Vockley CM, Thakore PI, Crawford GE, Reddy TE, Gersbach CA. 2015. Epigenome editing by a CRISPR-Cas9-based acetyltransferase activates genes from promoters and enhancers. *Nat Biotechnol* **33:**510–517.

74. Vojta A, Dobrinić P, Tadić V, Bočkor L, Korać P, Julg B, Klasić M, Zoldoš V. 2016. Repurposing the CRISPR-Cas9 system for targeted DNA methylation. *Nucleic Acids Res* **44:**5615–5628.

75. Kearns NA, Pham H, Tabak B, Genga RM, Silverstein NJ, Garber M, Maehr R. 2015. Functional annotation of native enhancers with a Cas9-histone demethylase fusion. *Nat Methods* **12:**401–403.

76. Komor AC, Kim YB, Packer MS, Zuris JA, Liu DR. 2016. Programmable editing of a target base in genomic DNA without double-stranded DNA cleavage. *Nature* **533:**420–424.

77. Chen B, Gilbert LA, Cimini BA, Schnitzbauer J, Zhang W, Li G-W, Park J, Blackburn EH, Weissman JS, Qi LS, Huang B. 2013. Dynamic imaging of genomic loci in living human cells by an optimized CRISPR/Cas system. *Cell* **155:**1479–1491.

78. Batra R, Nelles DA, Pirie E, Blue SM, Marina RJ, Wang H, Chaim IA, Thomas JD, Zhang N, Nguyen V, Aigner S, Markmiller S, Xia G, Corbett KD, Swanson MS, Yeo GW. 2017. Elimination of toxic microsatellite repeat expansion RNA by RNA-targeting Cas9. *Cell* **170:**899–912.e10.

79. O'Connell MRL, Oakes BL, Sternberg SH, East-Seletsky A, Kaplan M, Doudna JA. 2014. Programmable RNA recognition and cleavage by CRISPR/Cas9. *Nature* **516:**263–266.

80. Nelles DA, Fang MY, O'Connell MR, Xu JL, Markmiller SJ, Doudna JA, Yeo GW. 2016. Programmable RNA tracking in live cells with CRISPR/Cas9. *Cell* **165:**488–496.

81. Shalem O, Sanjana NE, Hartenian E, Shi X, Scott DA, Mikkelson T, Heckl D, Ebert BL, Root DE, Doench JG, Zhang F. 2014. Genome-scale CRISPR-Cas9 knockout screening in human cells. *Science* **343:**84–87.

82. Wang T, Wei JJ, Sabatini DM, Lander ES. 2014. Genetic screens in human cells using the CRISPR-Cas9 system. *Science* **343:**80–84.

83. Koike-Yusa H, Li Y, Tan E-P, Velasco-Herrera MC, Yusa K. 2014. Genome-wide recessive genetic screening in mammalian cells with a lentiviral CRISPR-guide RNA library. *Nat Biotechnol* **32:**267–273.

84. Zhou Y, Zhu S, Cai C, Yuan P, Li C, Huang Y, Wei W. 2014. High-throughput screening of a CRISPR/Cas9 library for functional genomics in human cells. *Nature* **509:**487–491.

85. Gilbert LA, Horlbeck MA, Adamson B, Villalta JE, Chen Y, Whitehead EH, Guimaraes C, Panning B, Ploegh HL, Bassik MC, Qi LS, Kampmann M, Weissman JS. 2014. Genome-scale CRISPR-mediated control of gene repression and activation. *Cell* **159:**647–661.

86. Rees HA, Liu DR. 2018. Base editing: precision chemistry on the genome and transcriptome of living cells. *Nat Rev Genet* **19:**770–788.

87. Szczelkun MD, Tikhomirova MS, Sinkunas T, Gasiunas G, Karvelis T, Pschera P, Siksnys V, Seidel R. 2014. Direct observation of R-loop formation by single RNA-guided Cas9 and Cascade effector complexes. *Proc Natl Acad Sci USA* **111:**9798–9803.

88. Nishida K, Arazoe T, Yachie N, Banno S, Kakimoto M, Tabata M, Mochizuki M, Miyabe A, Araki M, Hara KY, Shimatani Z, Kondo A. 2016. Targeted nucleotide editing using hybrid prokaryotic and vertebrate adaptive immune systems. *Science* **353:**aaf8729.

89. Gaudelli NM, Komor AC, Rees HA, Packer MS, Badran AH, Bryson DI, Liu DR. 2017. Programmable base editing of A•T to G•C in genomic DNA without DNA cleavage. *Nature* **551:**464–471.

90. Grünewald J, Zhou R, Lareau CA, Garcia SP, Iyer S, Miller BR, Langner LM, Hsu JY, Aryee MJ, Joung JK. 2020. A dual-deaminase CRISPR base editor enables concurrent adenine and cytosine editing. *Nat Biotechnol* **38:**861–864.

91. Zhang X, Zhu B, Chen L, Xie L, Yu W, Wang Y, Li L, Yin S, Yang L, Hu H, Han H, Li Y, Wang L, Chen G, Ma X, Geng H, Huang W, Pang X, Yang Z, Wu Y, Siwko S, Kurita R, Nakamura Y, Yang L, Liu M, Li D. 2020. Dual base editor catalyzes both cytosine and adenine base conversions in human cells. *Nat Biotechnol* **38:**856–860.

92. Anzalone AV, Randolph PB, Davis JR, Sousa AA, Koblan LW, Levy JM, Chen PJ, Wilson C, Newby GA, Raguram A, Liu DR. 2019. Search-and-replace genome editing without double-strand breaks or donor DNA. *Nature* **576:**149–157.

93. Anders C, Niewoehner O, Duerst A, Jinek M. 2014. Structural basis of PAM-dependent target DNA recognition by the Cas9 endonuclease. *Nature* 513:569–573.

94. Anders C, Bargsten K, Jinek M. 2016. Structural plasticity of PAM recognition by engineered variants of the RNA-guided endonuclease Cas9. *Mol Cell* 61:895–902.

95. Hirano H, Gootenberg JS, Horii T, Abudayyeh OO, Kimura M, Hsu PD, Nakane T, Ishitani R, Hatada I, Zhang F, Nishimasu H, Nureki O. 2016. Structure and engineering of Francisella novicida Cas9. *Cell* 164:950–961.

96. Hirano S, Nishimasu H, Ishitani R, Nureki O. 2016. Structural basis for the altered PAM specificities of engineered CRISPR-Cas9. *Mol Cell* 61:886–894.

97. Hirano S, Abudayyeh OO, Gootenberg JS, Horii T, Ishitani R, Hatada I, Zhang F, Nishimasu H, Nureki O. 2019. Structural basis for the promiscuous PAM recognition by Corynebacterium diphtheriae Cas9. *Nat Commun* 10:1968.

98. Jiang F, Zhou K, Ma L, Gressel S, Doudna JA. 2015. A Cas9-guide RNA complex preorganized for target DNA recognition. *Science* 348:1477–1481.

99. Jiang F, Taylor DW, Chen JS, Kornfeld JE, Zhou K, Thompson AJ, Nogales E, Doudna JA. 2016. Structures of a CRISPR-Cas9 R-loop complex primed for DNA cleavage. *Science* 351:867–871.

100. Jinek M, Jiang F, Taylor DW, Sternberg SH, Kaya E, Ma E, Anders C, Hauer M, Zhou K, Lin S, Kaplan M, Iavarone AT, Charpentier E, Nogales E, Doudna JA. 2014. Structures of Cas9 endonucleases reveal RNA-mediated conformational activation. *Science* 343:1247997.

101. Nishimasu H, Ran FA, Hsu PD, Konermann S, Shehata SI, Dohmae N, Ishitani R, Zhang F, Nureki O. 2014. Crystal structure of Cas9 in complex with guide RNA and target DNA. *Cell* 156:935–949.

102. Nishimasu H, Cong L, Yan WX, Ran FA, Zetsche B, Li Y, Kurabayashi A, Ishitani R, Zhang F, Nureki O. 2015. Crystal structure of Staphylococcus aureus Cas9. *Cell* 162:1113–1126.

103. Yamada M, Watanabe Y, Gootenberg JS, Hirano H, Ran FA, Nakane T, Ishitani R, Zhang F, Nishimasu H, Nureki O. 2017. Crystal structure of the minimal Cas9 from Campylobacter jejuni reveals the molecular diversity in the CRISPR-Cas9 systems. *Mol Cell* 65:1109–1121.e3.

104. Zhang Y, Zhang H, Xu X, Wang Y, Chen W, Wang Y, Wu Z, Tang N, Wang Y, Zhao S, Gan J, Ji Q. 2020. Catalytic-state structure and engineering of Streptococcus thermophilus Cas9. *Nat Catal* 3:813–823.

105. Zhu X, Clarke R, Puppala AK, Chittori S, Merk A, Merrill BJ, Simonović M, Subramaniam S. 2019. Cryo-EM structures reveal coordinated domain motions that govern DNA cleavage by Cas9. *Nat Struct Mol Biol* 26:679–685.

106. Faure G, Shmakov SA, Makarova KS, Wolf YI, Crawley AB, Barrangou R, Koonin EV. 2019. Comparative genomics and evolution of trans-activating RNAs in class 2 CRISPR-Cas systems. *RNA Biol* 16:435–448.

107. Globyte V, Lee SH, Bae T, Kim JS, Joo C. 2019. CRISPR/Cas9 searches for a protospacer adjacent motif by lateral diffusion. *EMBO J* 38:e99466.

108. Knight SC, Xie L, Deng W, Guglielmi B, Witkowsky LB, Bosanac L, Zhang ET, El Beheiry M, Masson J-B, Dahan M, Liu Z, Doudna JA, Tjian R. 2015. Dynamics of CRISPR-Cas9 genome interrogation in living cells. *Science* 350:823–826.

109. Sternberg SH, Redding S, Jinek M, Greene EC, Doudna JA. 2014. DNA interrogation by the CRISPR RNA-guided endonuclease Cas9. *Nature* 507:62–67.

110. Sternberg SH, LaFrance B, Kaplan M, Doudna JA. 2015. Conformational control of DNA target cleavage by CRISPR-Cas9. *Nature* 527:110–113.

111. Doudna JA. 2020. The promise and challenge of therapeutic genome editing. *Nature* 578:229–236.

112. Fu Y, Foden JA, Khayter C, Maeder ML, Reyon D, Joung JK, Sander JD. 2013. High-frequency off-target mutagenesis induced by CRISPR-Cas nucleases in human cells. *Nat Biotechnol* 31:822–826.

113. Hsu PD, Scott DA, Weinstein JA, Ran FA, Konermann S, Agarwala V, Li Y, Fine EJ, Wu X, Shalem O, Cradick TJ, Marraffini LA, Bao G, Zhang F. 2013. DNA targeting specificity of RNA-guided Cas9 nucleases. *Nat Biotechnol* 31:827–832.

114. Mali P, Aach J, Stranges PB, Esvelt KM, Moosburner M, Kosuri S, Yang L, Church GM. 2013. CAS9 transcriptional activators for target specificity screening and paired nickases for cooperative genome engineering. *Nat Biotechnol* 31:833–838.

115. Pattanayak V, Lin S, Guilinger JP, Ma E, Doudna JA, Liu DR. 2013. High-throughput profiling of off-target DNA cleavage reveals RNA-programmed Cas9 nuclease specificity. *Nat Biotechnol* 31:839–843.

116. Courtney DG, Moore JE, Atkinson SD, Maurizi E, Allen EHA, Pedrioli DML, McLean WHI, Nesbit MA, Moore CBT. 2016. CRISPR/Cas9 DNA cleavage at SNP-derived PAM enables both in vitro and in vivo KRT12 mutation-specific targeting. *Gene Ther* 23:108–112.

117. György B, Nist-Lund C, Pan B, Asai Y, Karavitaki KD, Kleinstiver BP, Garcia SP, Zaborowski MP, Solanes P, Spataro S, Schneider BL, Joung JK, Géléoc GSG, Holt JR, Corey DP. 2019. Allele-specific gene editing prevents deafness in a model of dominant progressive hearing loss. *Nat Med* 25:1123–1130.

118. Li Y, Mendiratta S, Ehrhardt K, Kashyap N, White MA, Bleris L. 2016. Exploiting the CRISPR/Cas9 PAM constraint for single-nucleotide resolution interventions. *PLoS One* 11:e0144970.

119. Lino CA, Harper JC, Carney JP, Timlin JA. 2018. Delivering CRISPR: a review of the challenges and approaches. *Drug Deliv* 25:1234–1257.

120. Wang D, Zhang F, Gao G. 2020. CRISPR-based therapeutic genome editing: strategies and in vivo delivery by AAV vectors. *Cell* 181:136–150.

121. Wu Z, Yang H, Colosi P. 2010. Effect of genome size on AAV vector packaging. *Mol Ther* 18:80–86.

122. Charlesworth CT, Deshpande PS, Dever DP, Camarena J, Lemgart VT, Cromer MK, Vakulskas CA, Collingwood MA, Zhang L, Bode NM, Behlke MA, Dejene B, Cieniewicz B, Romano R, Lesch BJ, Gomez-Ospina N, Mantri S, Pavel-Dinu M, Weinberg KI, Porteus MH. 2019. Identification of preexisting adaptive immunity to Cas9 proteins in humans. *Nat Med* 25:249–254.

123. Smith C, Gore A, Yan W, Abalde-Atristain L, Li Z, He C, Wang Y, Brodsky RA, Zhang K, Cheng L, Ye Z. 2014. Whole-genome sequencing analysis reveals high specificity of CRISPR/Cas9 and TALEN-based genome editing in human iPSCs. *Cell Stem Cell* 15:12–13.

124. Veres A, Gosis BS, Ding Q, Collins R, Ragavendran A, Brand H, Erdin S, Cowan CA, Talkowski ME, Musunuru K. 2014. Low incidence of off-target mutations in individual CRISPR-Cas9 and TALEN targeted human stem cell clones detected by whole-genome sequencing. *Cell Stem Cell* 15:27–30.

125. Frock RL, Hu J, Meyers RM, Ho Y-J, Kii E, Alt FW. 2015. Genome-wide detection of DNA double-stranded breaks induced by engineered nucleases. *Nat Biotechnol* 33:179–186.

126. Kim D, Bae S, Park J, Kim E, Kim S, Yu HR, Hwang J, Kim J-I, Kim J-S. 2015. Digenome-seq: genome-wide profiling of CRISPR-Cas9 off-target effects in human cells. *Nat Methods* 12:237–243.

127. Kim D, Kim S, Kim S, Park J, Kim J-S. 2016. Genome-wide target specificities of CRISPR-Cas9 nucleases revealed by multiplex Digenome-seq. *Genome Res* 26:406–415.

128. Osborn MJ, Webber BR, Knipping F, Lonetree C-L, Tennis N, DeFeo AP, McElroy AN, Starker CG, Lee C, Merkel S, Lund TC, Kelly-Spratt KS, Jensen MC, Voytas DF, von Kalle C, Schmidt M, Gabriel R, Hippen KL, Miller JS, Scharenberg AM, Tolar J, Blazar BR. 2016. Evaluation of TCR gene editing achieved by TALENs, CRISPR/Cas9, and megaTAL nucleases. *Mol Ther* 24:570–581.

129. Gabriel R, Lombardo A, Arens A, Miller JC, Genovese P, Kaeppel C, Nowrouzi A, Bartholomae CC, Wang J, Friedman G, Holmes MC, Gregory PD, Glimm H, Schmidt M, Naldini L, von Kalle C. 2011. An unbiased genome-wide analysis of zinc-finger nuclease specificity. *Nat Biotechnol* 29:816–823.

130. Wang X, Wang Y, Wu X, Wang J, Wang Y, Qiu Z, Chang T, Huang H, Lin R-J, Yee J-K. 2015. Unbiased detection of off-target cleavage by CRISPR-Cas9 and TALENs using integrase-defective lentiviral vectors. *Nat Biotechnol* 33:175–178.

131. Tsai SQ, Zheng Z, Nguyen NT, Liebers M, Topkar VV, Thapar V, Wyvekens N, Khayter C, Iafrate AJ, Le LP, Aryee MJ, Joung JK. 2015. GUIDE-seq enables genome-wide profiling of off-target cleavage by CRISPR-Cas nucleases. *Nat Biotechnol* 33:187–197.

132. Crosetto N, Mitra A, Silva MJ, Bienko M, Dojer N, Wang Q, Karaca E, Chiarle R, Skrzypczak M, Ginalski K, Pasero P, Rowicka M, Dikic I. 2013. Nucleotide-resolution DNA double-strand break mapping by next-generation sequencing. *Nat Methods* 10:361–365.

133. Ran FA, Cong L, Yan WX, Scott DA, Gootenberg JS, Kriz AJ, Zetsche B, Shalem O, Wu X, Makarova KS, Koonin EV, Sharp PA, Zhang F. 2015. In vivo genome editing using Staphylococcus aureus Cas9. *Nature* 520:186–191.

134. Slaymaker IM, Gao L, Zetsche B, Scott DA, Yan WX, Zhang F. 2016. Rationally engineered Cas9 nucleases with improved specificity. *Science* 351:84–88.

135. Yan WX, Mirzazadeh R, Garnerone S, Scott D, Schneider MW, Kallas T, Custodio J, Wernersson E, Li Y, Gao L, Federova Y, Zetsche B, Zhang F, Bienko M, Crosetto N. 2017. BLISS is a versatile and quantitative method for genome-wide profiling of DNA double-strand breaks. *Nat Commun* 8:15058.

136. Tsai SQ, Nguyen NT, Malagon-Lopez J, Topkar VV, Aryee MJ, Joung JK. 2017. CIRCLE-seq: a highly sensitive in vitro screen for genome-wide CRISPR-Cas9 nuclease off-targets. *Nat Methods* 14:607–614.

137. Cameron P, Fuller CK, Donohoue PD, Jones BN, Thompson MS, Carter MM, Gradia S, Vidal B, Garner E, Slorach EM, Lau E, Banh LM, Lied AM, Edwards LS, Settle AH, Capurso D, Llaca V, Deschamps S, Cigan M, Young JK, May AP. 2017. Mapping the genomic landscape of CRISPR-Cas9 cleavage. *Nat Methods* 14:600–606.

138. Young J, Zastrow-Hayes G, Deschamps S, Svitashev S, Zaremba M, Acharya A, Paulraj S, Peterson-Burch B, Schwartz C, Djukanovic V, Lenderts B, Feigenbutz L, Wang L, Alarcon C, Siksnys V, May G, Chilcoat ND, Kumar S. 2019. CRISPR-Cas9 editing in maize: systematic evaluation of off-target activity and its relevance in crop improvement. *Sci Rep* 9:6729.

139. Lazzarotto CR, Malinin NL, Li Y, Zhang R, Yang Y, Lee G, Cowley E, He Y, Lan X, Jividen K, Katta V, Kolmakova NG, Petersen CT, Qi Q, Strelcov E, Maragh S, Krenciute G, Ma J, Cheng Y, Tsai SQ. 2020. CHANGE-seq reveals genetic and epigenetic effects on CRISPR-Cas9 genome-wide activity. *Nat Biotechnol* 38:1317–1327.

140. Kleinstiver BP, Pattanayak V, Prew MS, Tsai SQ, Nguyen NT, Zheng Z, Joung JK. 2016. High-fidelity CRISPR-Cas9 nucleases with no detectable genome-wide off-target effects. *Nature* 529:490–495.

141. Chen JS, Dagdas YS, Kleinstiver BP, Welch MM, Sousa AA, Harrington LB, Sternberg SH, Joung JK, Yildiz A, Doudna JA. 2017. Enhanced proofreading governs CRISPR-Cas9 targeting accuracy. *Nature* 550:407–410.

142. Casini A, Olivieri M, Petris G, Montagna C, Reginato G, Maule G, Lorenzin F, Prandi D, Romanel A, Demichelis F, Inga A, Cereseto A. 2018. A highly specific SpCas9 variant is identified by in vivo screening in yeast. *Nat Biotechnol* 36:265–271.

143. Lee JK, Jeong E, Lee J, Jung M, Shin E, Kim YH, Lee K, Jung I, Kim D, Kim S, Kim JS. 2018. Directed evolution of CRISPR-Cas9 to increase its specificity. *Nat Commun* 9:3048.

144. Vakulskas CA, Dever DP, Rettig GR, Turk R, Jacobi AM, Collingwood MA, Bode NM, McNeill MS, Yan S, Camarena J, Lee CM, Park SH, Wiebking V, Bak RO, Gomez-Ospina N, Pavel-Dinu M, Sun W, Bao G, Porteus MH, Behlke MA. 2018. A high-fidelity Cas9 mutant delivered as a ribonucleoprotein complex enables efficient gene editing in human hematopoietic stem and progenitor cells. *Nat Med* 24:1216–1224.

145. Müller M, Lee CM, Gasiunas G, Davis TH, Cradick TJ, Siksnys V, Bao G, Cathomen T, Mussolino C. 2016. Streptococcus thermophilus CRISPR-Cas9 systems enable specific editing of the human genome. *Mol Ther* 24:636–644.

146. Lee CM, Cradick TJ, Bao G. 2016. The Neisseria meningitidis CRISPR-Cas9 system enables specific genome editing in mammalian cells. *Mol Ther* 24:645–654.

147. Edraki A, Mir A, Ibraheim R, Gainetdinov I, Yoon Y, Song CQ, Cao Y, Gallant J, Xue W, Rivera-Pérez JA, Sontheimer EJ. 2019. A compact, high-accuracy Cas9 with a dinucleotide PAM for in vivo genome editing. *Mol Cell* 73:714–726.e4.

148. Paquet D, Kwart D, Chen A, Sproul A, Jacob S, Teo S, Olsen KM, Gregg A, Noggle S, Tessier-Lavigne M. 2016. Efficient introduction of specific homozygous and heterozygous mutations using CRISPR/Cas9. *Nature* 533:125–129.

149. Kleinstiver BP, Prew MS, Tsai SQ, Nguyen NT, Topkar VV, Zheng Z, Joung JK. 2015. Broadening the targeting range of Staphylococcus aureus CRISPR-Cas9 by modifying PAM recognition. *Nat Biotechnol* 33:1293–1298.

150. Hu JH, Miller SM, Geurts MH, Tang W, Chen L, Sun N, Zeina CM, Gao X, Rees HA, Lin Z, Liu DR. 2018. Evolved Cas9 variants with broad PAM compatibility and high DNA specificity. *Nature* 556:57–63.

151. Miller SM, Wang T, Randolph PB, Arbab M, Shen MW, Huang TP, Matuszek Z, Newby GA, Rees HA, Liu DR. 2020. Continuous evolution of SpCas9 variants compatible with non-G PAMs. *Nat Biotechnol* 38:471–481.

152. Nishimasu H, Shi X, Ishiguro S, Gao L, Hirano S, Okazaki S, Noda T, Abudayyeh OO, Gootenberg JS, Mori H, Oura S, Holmes B, Tanaka M, Seki M, Hirano H, Aburatani H, Ishitani R, Ikawa M, Yachie N, Zhang F, Nureki O. 2018. Engineered CRISPR-Cas9 nuclease with expanded targeting space. *Science* 361:1259–1262.

153. Walton RT, Christie KA, Whittaker MN, Kleinstiver BP. 2020. Unconstrained genome targeting with near-PAMless engineered CRISPR-Cas9 variants. *Science* 368:290–296.

154. Fonfara I, Le Rhun A, Chylinski K, Makarova KS, Lécrivain AL, Bzdrenga J, Koonin EV, Charpentier E. 2014. Phylogeny of Cas9 determines functional exchangeability of dual-RNA and Cas9 among orthologous type II CRISPR-Cas systems. *Nucleic Acids Res* 42:2577–2590.

155. Mojica FJM, Díez-Villaseñor C, García-Martínez J, Almendros C. 2009. Short motif sequences determine the targets of the prokaryotic CRISPR defence system. *Microbiology (Reading)* 155:733–740.

156. Karvelis T, Gasiunas G, Siksnys V. 2017. Methods for decoding Cas9 protospacer adjacent motif (PAM) sequences: a brief overview. *Methods* 121-122:3–8.

157. Chatterjee P, Jakimo N, Jacobson JM. 2018. Minimal PAM specificity of a highly similar SpCas9 ortholog. *Sci Adv* 4:eaau0766.

158. Chatterjee P, Jakimo N, Lee J, Amrani N, Rodríguez T, Koseki SRT, Tysinger E, Qing R, Hao S, Sontheimer EJ, Jacobson J. 2020. An engineered ScCas9 with broad PAM range and high specificity and activity. *Nat Biotechnol* 38:1154–1158.

159. Karvelis T, Gasiunas G, Young J, Bigelyte G, Silanskas A, Cigan M, Siksnys V. 2015. Rapid characterization of CRISPR-Cas9 protospacer adjacent motif sequence elements. *Genome Biol* 16:253.

160. Kim E, Koo T, Park SW, Kim D, Kim K, Cho HY, Song DW, Lee KJ, Jung MH, Kim S, Kim JH, Kim JH, Kim JS. 2017. In vivo genome editing with a small Cas9 orthologue derived from Campylobacter jejuni. *Nat Commun* 8:14500.

161. Gasiunas G, Young JK, Karvelis T, Kazlauskas D, Urbaitis T, Jasnauskaite M, Grusyte MM, Paulraj S, Wang P-H, Hou Z, Dooley SK, Cigan M, Alarcon C, Chilcoat ND, Bigelyte G, Curcuru JL, Mabuchi M, Sun Z, Fuchs RT, Schildkraut E, Weigele PR, Jack WE, Robb GB, Venclovas Č, Siksnys V. 2020. A catalogue of biochemically diverse CRISPR-Cas9 orthologs. *Nat Commun* **11:**5512.

162. Ibraheim R, Song CQ, Mir A, Amrani N, Xue W, Sontheimer EJ. 2018. All-in-one adeno-associated virus delivery and genome editing by Neisseria meningitidis Cas9 in vivo. *Genome Biol* **19:**137.

Mechanism of Type III CRISPR-Cas Immunity

Luciano A. Marraffini[1,2]

[1]Laboratory of Bacteriology, The Rockefeller University, New York, NY 10065
[2]Howard Hughes Medical Institute, The Rockefeller University, New York, NY 10065

Introduction

One of the first things to notice when looking at the table of contents of this book is the many different types of CRISPR-Cas systems it covers. Type III systems are second in abundance across bacteria and archaea (1) but first in terms of complexity. In this chapter, I review the many different strategies and mechanisms of nucleic acid recognition and degradation that make the molecular biology of these systems so intricate. I also cover how new spacers are incorporated into type III CRISPR arrays. This is another unique aspect of these systems, which can integrate RNA spacer sequences in between repeats and then convert them to DNA using a reverse transcriptase (RT) associated with the Cas1 integrase. The complexity of targeting limits the development of type III systems to achieve programmable destruction of nucleic acids. Biotechnological applications are very scarce compared to those of type I, II, V, and VI CRISPR-Cas systems and are not described in this chapter.

Type III CRISPR-Cas Immunity Requires Target Transcription

The study of type III-B immunity against plasmids in *Sulfolobus islandicus* provided the first hint that targeting by this system requires transcription (2). After transformation of different plasmids that contained targets downstream of the *pyrF* promoter, plasmid destruction was achieved only when the target sequence (matching the spacer) was located in the bottom strand (template strand for transcription from the *pyrF* promoter), not when it was located in the top strand (nontemplate strand). To demonstrate that transcription was responsible for this asymmetrical targeting, an arabinose-inducible promoter was added downstream of the targets to drive transcription in the opposite direction to the *pyrF* promoter. This resulted in efficient immunity against targets located in the top strand, which was converted into the template strand for transcription originating from the arabinose-inducible promoter.

Because spacer arrays in *S. islandicus* are shared by three different CRISPR-Cas systems (one type I and two type III), the above-described results could not unequivocally be attributed to type III immunity. Decisive demonstration of transcription-dependent targeting, as well as its role in immunological tolerance of mobile genetic elements, was

CRISPR: Biology and Applications, First Edition. Edited by Rodolphe Barrangou, Erik J. Sontheimer, and Luciano A. Marraffini.
© 2022 American Society for Microbiology. DOI: 10.1128/9781683673798.ch05

obtained after investigation of the type III immune response against temperate phages in staphylococci (3). When the *Staphylococcus epidermidis* type III-A CRISPR-Cas system was programmed to target the lytic region of a lambda-like staphylococcal phage, immunity prevented lytic propagation of the phage but not its lysogenization. This result suggested that the presence of a target sequence, the same in both the replicating phage and the integrated prophage, is not sufficient to license type III-A CRISPR-Cas immunity. A major difference between these two phage life-forms is their transcription: phages actively transcribe their genome during the lytic cycle but repress most of it in the prophage state. Indeed, the targeting of the region responsible for maintaining the lysogenic state that is constantly transcribed from the prophage did prevent the formation of lysogens. But the key evidence came from the analysis of phages that escape type III-A targeting: instead of containing target sequence mutations (which are the primary mode of escape in other CRISPR-Cas types), they harbored promoter mutations that abrogated target transcription. This study also found the strand asymmetry of targeting; i.e., efficient immunity was achieved only when the sequence matching the spacer was located in the bottom strand. More importantly, it showed a biological role for the transcription-dependent type III targeting: immunological tolerance. As is the case for the mammalian immune system, it is fundamental to distinguish between harmful pathogens and beneficial commensals. Since they commonly carry accessory genes that increase the fitness of the host (4), prophages can be considered a commensal of bacteria. Therefore, the targeting mechanism of type III-A systems that attacks the lytic (harmful) phage but not the (beneficial) prophage equips the host with an immunological tolerance that is not possible with the other CRISPR types (I, II, and V).

crRNA Biogenesis

The two studies described above provided the foundational evidence for the requirement of active target transcription during type III CRISPR-Cas immunity. Subsequent research elucidated the molecular details of its mechanism. Type III systems encode an RNA-guided complex formed by the signature protein Cas10 and either Csm2-Csm5 or Cmr2-Cmr5 in type III-A or III-B, respectively (5–7) (Fig. 5.1A). This complex is loaded with crispr RNAs (crRNAs) derived from the CRISPR array (Fig. 5.1B). The array is first transcribed as a long precursor containing alternate repeats and spacers. Type III repeats form stem-loop structures that are recognized by the endoribonuclease Cas6, also encoded in the type III locus and cleaved downstream of this secondary structure (8–10). Cleavage occurs at every repeat and liberates an intermediate crRNA containing ~8 nucleotides (nt) of the 3′ end of the repeat sequence (known as the crRNA "tag"), followed by the full spacer and by the 5′ end of the repeat, including the stem-loop. The intermediate crRNA is transferred from Cas6 to the Cas10 complex (11) by a mechanism that most likely involves a direct but transient physical interaction, evidenced by the copurification of trace amounts of Cas6 with the complex (6).

Once in the Cas10 complex, the intermediate crRNA undergoes maturation at the 3′ end. This process eliminates the repeat stem-loop structure and parts of the spacer at 6-nt intervals (10). It is believed that Csm3/Cmr4, which form the backbone of the complex and directly interact with the crRNA (7, 12–14), define the length of the intervals of maturation (15), possibly by protecting 5-nt stretches and leaving the 6 nt exposed for the attack of polynucleotide phosphorylase (PNPase), a host nuclease recruited by Csm5 for crRNA processing (16). The crRNA tag is locked into the Csm4/Cmr3-Cas10 subunits (17, 18), and the engineering of longer crRNA tags results in the shortening of the 3′ end of the mature crRNA by the same extension of the tag addition (10). Therefore, it is believed that

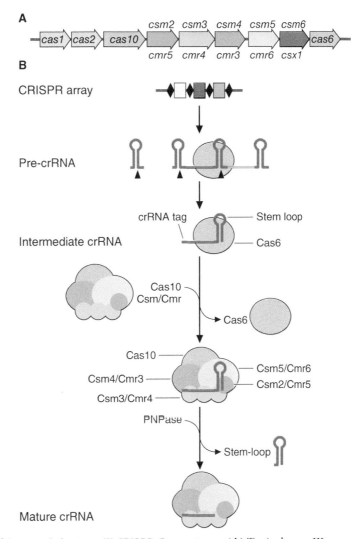

Figure 5.1 CrRNA biogenesis by type III CRISPR-Cas systems. (**A**) Typical type III *cas* operon; III-A and III-B homologs are indicated at the top and bottom, respectively, of each gene. (**B**) Different steps of crRNA maturation: transcription of the CRISPR array (black diamonds, repeats; colored boxes, spacers) to generate the crRNA precursor (pre-crRNA), cleavage of the precursor by Cas6 at the base of the stem-loop structure (black triangle) to generate the intermediate crRNA, transfer of the intermediate from Cas6 to the Cas10-Csm/Cmr complex, and degradation of the stem-loop structure of the intermediate (presumably by PNPase) to generate the mature crRNA within the complex.

Csm3/Cmr4 and PNPase establish a ruler mechanism for crRNA maturation that establishes the length of the mature crRNA from the 5′ end of the intermediate crRNA.

RNA Recognition

A breakthrough in the understanding of the molecular mechanism of type III targeting was the determination that the Cas10-Csm/Cmr complexes use their crRNA guide to recognize complementary RNA molecules (5) (Fig. 5.2A). In the context of infection, these molecules are the invader's transcripts, which explains the observations that target transcription is required for immunity and that only spacers with homology to the nontemplate strand provide efficient targeting (2, 3). Only these spacers generate crRNA guides that are complementary to the transcript.

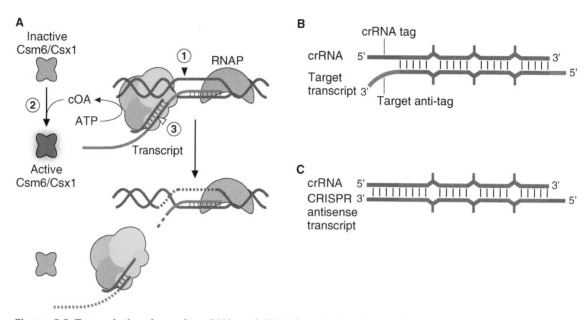

Figure 5.2 Transcription-dependent DNA and RNA degradation during the type III CRISPR-Cas immune response. (A) Upon target transcription by RNA polymerase (RNAP), the crRNA guide within the Cas10-Csm/Cmr complex is able to find a complementary invader's transcript. Target recognition triggers three events: the HD domain of Cas10 is activated and degrades ssDNA (black triangle), presumably generated by the transcription bubble (1); the palm domain of Cas10 catalyzes the oligomerization of ATP into a cyclic oligoadenylate (cOA), which activates the Csm6/Csx1 nonspecific RNase (2); and the target RNA is cleaved by the Csm3/Cmr4 subunit of the complex, which disrupts the base-pairing between crRNA and transcript and deactivates both domains of Cas10 (3). **(B)** Within the Cas10 complex, the crRNA-target RNA duplex adopts a discontinuous helical structure with every 6th base being flipped in the spacer region. Critical for the activation of type III immunity is the lack of base-pairing between the sequences upstream of the spacer (crRNA tag) and downstream of the protospacer (target anti-tag). **(C)** Antisense transcription of the CRISPR array generates a protospacer complementary to the crRNA spacer. This target does not activate Cas10 due to the additional complementarity between the crRNA tag and the CRISPR repeat sequences.

Presumably, the type III complex is able to bind the target transcript as soon as it comes out of the RNA polymerase (19) (Fig. 5.2A). Interactions between the 3' end of the mature crRNA and the target are important for the initial binding (20–22). Structural analysis revealed an overall architecture similar to the Cascade complex of type I CRISPR-Cas systems (13, 23), with a helical protein arrangement that positions the crRNA for RNA substrate binding (14, 17, 18, 23). Cas10 complexes display a central double-helical core of three Csm3/Cmr4 and two Csm2/Cmr5 subunits bound to the Cas10-Csm4/Cmr3 subcomplex at one end and by a Csm5/Cmr6 subunit at the opposite end (Fig. 5.1B). The crRNA passes through the entire complex, with the 5' end tag being capped by the Cas10-Csm4/Cmr3 subunits, the spacer sequence bound by the Csm3/Cmr4 and Csm2/Cmr5 subunits, and the 3' end interacting with the Csm5/Cmr6 subunit. This structure suggests that the crRNA plays an essential role in its assembly (6). Upon target binding, the Cas10 complex undergoes a conformational change that generates a wider binding channel that is large enough to harbor the crRNA-target RNA duplex (17, 18, 24). The guide-target RNA duplex adopts a discontinuous helical structure with every 6th base being flipped in the spacer region (Fig. 5.2B). The duplex is stabilized by the stacking interactions and hydrogen bonds formed between phosphate groups and the different subunits (14, 17, 18, 23).

DNA Degradation

The structural change produced after target recognition by the crRNA impacts most dramatically the Cas10 subunit, which undergoes rapid conformational fluctuations characteristic of the activation of its two domains (24). The signature protein of type III systems carries out two activities that are fundamental to confer immunity to the host: DNA degradation and adenine oligomerization (1). The HD domain of Cas10 performs nonspecific single-stranded DNA (ssDNA) degradation upon target binding (25–27). In the context of transcription, it was determined that upon generation of the target RNA by RNA polymerase, the HD domain can cleave the nontemplate, but not the template, strand (19). This observation has led to the hypothesis that Cas10 is able to degrade the ssDNA of the transcription bubble (Fig. 5.2A).

As is the case for other CRISPR-Cas systems, DNA degradation is fundamental for immunity (28). However, efficient DNA destruction is achieved only when the target is strongly transcribed (29). Presumably as a consequence of this, type III-A immunity against temperate staphylococcal phages relies on Cas10 degradation of phage DNA when the target is located in early-expressed, but not late-expressed, genes (30). This is because high phage transcription is achieved early in the lytic cycle, when the phage genome undergoes bidirectional replication. During the late stages of infection, however, the genome experiences rolling-circle replication, producing long concatemers that are transcribed inefficiently.

Activation of CARF Accessory Proteins by cOA Second Messengers

In addition to the HD domain, Cas10 possesses a polymerase/cyclase palm domain whose presence was recognized early on, first through bioinformatic analysis (31) and then through structural homology (32). Recently, this domain was found to be responsible for the conversion of ATP into cyclic oligoadenylates (cOA) (33, 34). The palm domain relies on its GGDD motif to directly catalyze 5'–3' phosphodiester formation between ATP molecules and generate a cyclic molecule containing 2 to 6 adenosine units, depending on the species (Fig. 5.2A). Similarly to the ssDNase activity of Cas10, the cyclase activity is triggered by target RNA recognition.

cOA act as second messengers to activate type III CRISPR-Cas accessory proteins and enhance the immune response provided by these systems. These accessory proteins contain a CRISPR-associated Rossman fold (CARF) domain, predicted to bind nucleotides for regulatory purposes (35, 36). Indeed, recent experimental evidence showed that the most common CARF-containing proteins, the type III-associated Csm6 and Csx1, are able to bind the cOA molecule and activate a second, RNase catalytic domain (33, 34, 37–39) (Fig. 5.2A). Csm6 is a nonspecific RNase (30, 40) that uses an HEPN (higher eukaryotes and prokaryotes nuclease) domain to degrade host and invader transcripts indiscriminately (29). Csm6/Csx1 activity is fundamental for type III immunity (6, 41) when the DNA degradation mediated by the Cas10 HD domain is inefficient, such as is the case of poorly transcribed targets within plasmids (29) or during the targeting of late-expressed phage genes (30). In these situations, activation of Csm6/Csx1 can lead to a temporary growth arrest (29) that is relieved when the invader's DNA is cleared off the host cell through Cas10 ssDNase activity; i.e., the target transcript and therefore the cOA second messenger are not produced anymore. It remains to be elucidated how the massive degradation of transcripts provides immunity. In studies with target plasmids, it was shown that the presence of Csm6 is required for the rapid disappearance of the plasmid DNA upon target transcription (29). This has led to the hypothesis that the degradation of host and plasmid transcripts that are required for plasmid replication prevents further accumulation of target genomes, which can be then cleared off by the inefficient Cas10 ssDNase activity (29). In support of this hypothesis, absence of Csm6 could be chemically complemented with erythromycin, a global translation inhibitor that would also lead to the depletion of important plasmid replication factors. Finally, the activation of Csm6 has been shown to prevent the rise of escaper phage containing target mutations (30), and therefore, it is an important contributor to the robustness of the type III-A CRISPR-Cas immune response.

Inactivation of the Type III CRISPR-Cas Immune Response

Different from most other CRISPR types that specifically target DNA (type I, II, and V), the type III CRISPR-Cas immune response is built upon nonspecific degradation of nucleic acids: ssDNA by the HD domain of Cas10 and ssRNA by the HEPN domain of Csm6/Csx1. Both of these activities are triggered by the binding of the crRNA guide to the target transcript, and if continued for long periods of time, they could jeopardize the survival of the host. To avoid the potential toxicity of indiscriminate ssDNA and ssRNA degradation, type III systems have mechanisms to neutralize both of these activities. First, the activating transcript is cleaved by the Csm3/Cmr4 subunits of the Cas10 complex (5, 42, 43) (Fig. 5.2A). Cleavage occurs within the target sequence at 6-nt intervals, a spacing that is determined by the number of Csm3/Cmr4 molecules in the Cas10 complex. Structural analysis revealed that thumb-like β-hairpin domains of each Csm3/Cmr4 subunit intercalate between segments of duplexed crRNA:target RNA, flipping every 6th nucleotide of the spacer region and bending the duplex to form an extended A-form double helix (Fig. 5.2B). This distortion places ssRNA target in proximity to the catalytic residues from the Csm3/Cmr4 subunit, positioning it for productive cleavage (14, 17, 18). This rearrangement of the target nucleic acid is very similar to that occurring in the type I Cascade complex (44).

Cleavage of the target transcript instantly inactivates both domains of Cas10 (Fig. 5.2A). This stops DNA degradation and cOA production immediately (26, 27); however, Csm6/Csx1 nucleases already bound to the second messenger cannot be neutralized by this route. Instead, rapid inactivation of the nuclease is achieved through the cleavage of the cOA by the CARF domain itself of *Thermococcus onnurineus* Csm6 (45). Therefore, it seems that the RNase activity of the type III accessory proteins is short lived and needs a constant supply of inducer to maintain it. Interestingly, in the archaeon *Sulfolobus solfataricus*, two ring nucleases encoded outside of the type III locus were identified as the external inactivators of Csx1 (46). Both of these nucleases contain a CARF domain with catalytic residues that are absent from the CARF domain of Csx1 that are required to cut the cOA ring. It remains to be determined which mechanism, *cis* or *trans* cOA cleavage, is more common to inactivate Csm6/Csx1.

Finally, a potential "autoimmune" reaction of type III systems also needs to be neutralized. While most of the transcription of the CRISPR array is carried out in one direction, spurious transcription from the opposite direction would generate a transcript that is complementary to the crRNA guide and therefore would activate the type III CRISPR-Cas immune response in the absence of a target invader, against the host. Indeed, some type III loci are transcribed bidirectionally (47). However, while most invader targets share homology only with the spacer region of the crRNA, the transcript generated by antisense transcription of the CRISPR array is also complementary to the tag region (Fig. 5.2C). This extended complementarity prevents a type III immune response (48) by blocking the activation of both catalytic domains of Cas10 (26, 27, 49). Single-molecule experiments showed that base-pairing between the 8 nucleotides of the crRNA tag and the 8 corresponding nucleotides at the 3′ flanking sequence of the target (the "anti-tag") locks Cas10 into a static structure that is incapable of catalytic activity (24). In line with this result, structural analysis of a Cas10 complex bound to RNA substrates with and without tag:anti-tag complementarity showed that specific residues of Cas10 interact with the anti-tag to induce a conformational change that allosterically activates DNA cleavage and cOA generation (17, 18). In the presence of complementarity, the crRNA tag sequesters away the target anti-tag, preventing these interactions and thus Cas10 activation. Only Cas10 is inhibited; Csm3 is able to cleave the target RNA at the proto-spacer region (27, 42).

Target Sequence Requirements: The PAM

CRISPR types that recognize DNA targets (I, II, and V) need to avoid the cleavage of the spacer sequences in the CRISPR array, which could be targeted by the crRNA that they encode. This challenge has been solved by adding a second requirement for DNA cleavage besides complementarity with the crRNA: the presence of a protospacer adjacent motif (PAM) flanking the region of complementarity (50, 51). This motif is not present in the repeat sequence that flanks the spacers in the CRISPR array, and self-targeting is prevented. Although useful to avoid self-targeting, the PAM requirement also represents an Achilles heel of CRISPR-Cas systems, as invader populations frequently have individuals carrying mutations in the PAM that allow them to escape immunity (52). Because they recognize transcripts, type III systems do not face this self-targeting problem (targeting of the CRISPR array can happen only if it is transcribed bidirectionally; see above). Therefore, these systems are not under pressure to evolve PAM-dependent targeting. If this is the case, then mutations in the flanking sequences of the invader's target should not enable escape from type III immunity.

This hypothesis has been tested in different subtypes *in vivo* and *in vitro*, with different results. In *Staphylococcus epidermidis*, which harbors a single type III-A locus, early studies showed that individual mutations in the flanking regions of the *nes* target in the staphylococcal conjugative plasmid pG0400 do not alter anticonjugation immunity (48). A more comprehensive search for a PAM was performed by transforming a library of target plasmids harboring all possible sequence combinations (1,024 different sequences) on the five base pairs in the anti-tag, i.e., immediately upstream of the protospacer (53). With the exception of a few sequences that created complementarity with the crRNA tag, most of the sequences licensed type III-A immunity, demonstrating the absence of a motif necessary for targeting *in vivo*.

A similar experiment was performed to evaluate the requirement of a PAM for the targeting of the type III-B system of *Pyrococcus furiosus* (25). In addition to this locus, this archaeon harbors two other *cas* operons that encode two type I systems, I-A and I-G, all sharing and using ~200 different crRNAs encoded by 7 CRISPR arrays (54). Target plasmid libraries containing all the possible combinations of the three base pairs immediately upstream of the protospacer were transformed in a strain with complete deletions of the type I *cas* loci. In addition to combinations that create tag:anti-tag complementarity, several combinations (NGN, NNG, and NAA) abrogated immunity (25). This result was further corroborated *in vitro* by testing the ability of RNA targets with different 5′ end tri-nucleotide flanking sequences to activate DNA degradation. It was found that only targets with the motifs found to license type III-B immunity *in vivo* were able to trigger Cas10's HD domain. In contrast, cleavage of the target RNA was independent of the 5′ end flanking sequence. These results established the existence of an "RNA PAM" (rPAM) which is required in *P. furiosus* for productive target recognition.

How is it possible for these two type III CRISPR-Cas systems to have different targeting rules? One explanation for this discrepancy could be related to the co-occurrence of the type III-B with the type I-A and I-G loci in *P. furiosus*, whose PAM sequence space is markedly similar to that of the rPAM (25). It is conceivable that all three systems present in this organism have coevolved targeting mechanisms that resulted in (i) the necessity of type I systems for PAM recognition to avoid self-immunity, (ii) the acquisition of spacers matching protospacers flanked by such PAMs, (iii) the co-option of these spacers by the type III-B system, and (iv), finally, the adoption of a similar requirement in the form of the rPAM by the type III-B system. Indeed, the coresidence of type I and III systems, with type III systems using spacers acquired by the type I acquisition machinery to neutralize escapers (see below), has been shown to occur in *Marinomonas mediterranea* (55) and *Sulfolobus solfataricus* (56). Whether the rPAM requirement is limited to *P. furiosus* remains an open question. Although not studied as comprehensively as the *P. furiosus* flanking sequences, *S. solfataricus*, another archaeon that harbors both a type I-A and a type III-B system, failed in experiments to show any specific targeting requirements at the 5′ flank of the protospacer (57).

Target Sequence Requirements: Protospacer Mutations

Mismatches between the crRNA and the protospacer sequence in the target transcript can interfere with four steps of the type III immunity mechanism: the binding of the Cas10 complex to its target, the activation of Cas10's HD domain, the synthesis of cOA and Csm6 activation, and the cleavage of the target transcript. Although the exact effect of target mutations on each of these events has not been determined yet, different studies that explored the consequences of mismatches between the crRNA and the protospacer RNA have all come to the same general conclusion: type III CRISPR-Cas systems are much more tolerant of target mutations than other CRISPR types.

For example, high tolerance to mismatches was demonstrated for the *S. epidermidis* type III-A whole CRISPR-Cas immune response against plasmids *in vivo* (53). In this study, three target plasmid libraries were transformed into staphylococci, containing either the base complementary to the crRNA or its reverse complement mutation at each position of the first or second 10-nt region of the target that follows the anti-tag sequence. It was found that no single mutation in any of these regions could prevent immunity. Then it was investigated how many mismatches, regardless of their position in the 10-nt region tested, were necessary to abrogate plasmid transformation. It was shown that between 4 and 6 mismatches with the crRNA were required. These results, however, could not determine which individual step of the type III-A targeting mechanism (transcript recognition, Cas10 HD, and/or palm domain activation) is affected by the accumulation of mutations. Consistent with the tolerance to plasmid target mutations, the anti-phage immunity of the *S. epidermidis* type III-A system is also able to endure several transcript:crRNA mismatches (3, 53).

Several studies have investigated the impact of mismatches on the cleavage of the protospacer RNA *in vitro*. One of these evaluated the binding and cleavage of truncated protospacers (43). It was found that the presence of deletions of the 5′ or 3′ end of the protospacer did not affect target cleavage by the Csm complex of *Streptococcus thermophilus*; however, deletion of the 5′ end significantly abrogated target binding. Another study evaluated the importance of point mutations in the target (42), revealing that neither single-nucleotide nor 5-nt (consecutive) mismatches with the crRNA affected cleavage by the Csm complex of *Thermus thermophilus*.

The effect of protospacer mutations on Cas10 ssDNase activity was also investigated (27). The presence of dinucleotide mismatches in the 5′ end or middle region of the crRNA, but not in the 3′ end, prevented ssDNA degradation by the Csm complex of *Streptococcus thermophilus*. Similar results were obtained for the Cmr complex of *Thermotoga maritima* (49), where mismatches in the first 5 nt of the 5′ end of the crRNA prevented ssDNA cleavage but not binding. When different combinations of mismatches were evaluated, it was found that up to three mismatches could be introduced without significant impact on ssDNA degradation. As reported for the Csm complex of *Streptococcus thermophilus*, truncations of the 5′ end of the protospacer RNA drastically affected its binding to the crRNA within the complex and therefore prevented DNA destruction. To date, very little is known about the effect of mismatches between the crRNA guide and the protospacer RNA on the adenylate cyclase of Cas10; future studies will certainly address this knowledge gap.

Robustness of Type III CRISPR-Cas Immune Response

The success of any immune response is measured by the ability of the pathogen to escape it. Phages and plasmids can certainly escape all types of CRISPR immunity, but they do so much less often when the host carries a type III system. One study (53) compared the immunity provided by the CRISPR types III-A (*S. epidermidis*) and II-A (*Streptococcus pyogenes*), both expressed in the same host (*Staphylococcus aureus*) and programmed with spacers that recognize the same targets in the same phage. When the target was located in an intergenic region dispensable for the lytic propagation of the virus, escapers of both immune systems were found, but the frequency of escape was 2 orders of magnitude higher for type II-A than for type III-A. Mutant viruses were also qualitatively different: whereas phages evaded type II-A targeting via PAM single-nucleotide mutations, evasion of type III-A immunity required a deletion of the intergenic region targeted. When the target was in an essential gene, however, phages able to bypass type III-A immunity were not detected.

The unique robustness of the type III CRISPR-Cas immune response has been exploited by microorganisms that naturally encode two CRISPR loci. For example, the hyperthermophilic archaeon *S. solfataricus* carries three CRISPR arrays producing functional crRNA guides that can be used indistinguishably by three different *cas* loci: one type I and two type III (A and B). After transfection with viruses that contain a target with a proper PAM for recognition by the type I system, as well as an anti-tag region not complementary to the crRNA tag for efficient type III immunity, a low number of escapers was detected (56). Transfection with phages containing a target with a mutated PAM that are recognized only by the type III systems also resulted in a low number of escapers. However, when a mutation generating an anti-tag target that licenses only type I immunity was introduced, a significant increase of phage escapers was detected. A similar phenomenon was observed in the bacterium *M. mediterranea*, which harbors two CRISPR-Cas loci, one type I-F and another III-B, each of which uses different CRISPR arrays (with different repeat sequences). Although with low efficiency, the type III-B complex can utilize the type I-F crRNA guides to neutralize the mutant phages that escape type I-F immunity (55). Therefore, it has been proposed that type III CRISPR-Cas systems provide "immunological insurance" to hosts carrying also escapable (but more active in the acquisition of new spacers) type I systems (55), a phenomenon that provides one explanation for the frequent co-occurrence of type III and type I systems in prokaryotic genomes.

The robustness of type III systems relies on two features of the molecular mechanism of targeting: the ability to tolerate protospacer mutations and the activity of Csm6/Csx1. As opposed to Cascade-Cas3 (type I), Cas9 (type II), and Cas12 (type V), nucleases for which single point mutations in either the seed or PAM target completely abrogate target cleavage (52, 58, 59), the Cas10 complex can still trigger the degradation of the invader's DNA through the activation of the HD domain upon recognition of protospacer RNAs with a low number of mutations that generate mismatches with the crRNA (see above) (27, 49). Therefore, type III immunity can only be disrupted by (i) promoter mutations that prevent target transcription, (ii) target deletions, or (iii) the accumulation of multiple mutations within the protospacer region. These are all extremely unlikely because phages have evolved compact genomes in which most genes are essential for propagation (60); they need to be transcribed and cannot be deleted without loss of viability. And while phages have elevated rates of mutagenesis compared to prokaryotic genomes (61), the probability of the occurrence of several mutations in a region of about 30 nt is very low. As a consequence of these mechanistic differences, phages can escape more frequently type I, II, and V CRISPR-Cas systems, and type III CRISPR-Cas systems can provide a stronger immune response.

In addition to the DNA degradation activity of Cas10, refractory to protospacer point mutations present in mutant phage, the Csm6 RNase also contributes to the neutralization of escaper phage. One study found that staphylococci carrying the *S. epidermidis* type III-A system resisted infection with phages carrying up to four mutations in the protospacer; however, in the absence of Csm6, mutant phages with three or four mismatches significantly affected host growth or completely lysed the bacterial cultures, respectively (30). How the RNase activity Csm6 prevents the propagation of escapers is not fully understood; two formal possibilities exist that depend on whether the biosynthesis of cOA by the palm domain of Cas10 is affected by mismatches between protospacer and crRNA. If this is not the case, Csm6 would be triggered directly by escaper phage and will disrupt its replication through the degradation of host and phage transcripts required for propagation. However, even if mismatches abrogate Csm6 activation by the escaper phage, previous infection with wild-type phage could trigger a Csm6-mediated dormancy (see above) that would lead to the generation of inhospitable cells incapable of supporting the propagation of any phage that infects next, including escapers that would otherwise kill the host. This type of herd immunity is provided by the nonspecific RNase of type VI CRISPR-Cas systems (62) and relies on the low frequency of escapers compared to wild-type phage in viral populations, which ensures that mutant phages will most likely infect hosts that were already challenged by wild-type phage. Future studies will determine whether either of these, or a combination of both, mechanisms is in place, as well as their contribution to the neutralization of escapers relative to the degradation of escaper DNA by Cas10.

Spacer Acquisition

Spacer acquisition by type III systems has been difficult to observe under laboratory conditions. To date, the only example of type III adaptation has been described for the marine bacterium *Marinomonas mediterranea*, which carries a type III-B system with an unusual Cas1 nuclease containing an RT domain, RT-Cas1 (63, 64). Surprisingly, this is the only CRISPR-Cas system with the ability to acquire spacers both from DNA and RNA molecules (63). Spacer acquisition is still very infrequent in this bacterium and was detected only after overexpression of RT-Cas1 and Cas2 and using next-generation sequencing to analyze CRISPR loci. It was found that there is a strong correlation between the transcription and the acquisition levels of the different spacers, which was eliminated when the RT domain of Cas1 was deleted or mutated. This result suggested spacer acquisition from *M. mediterranea* transcripts, a hypothesis that was corroborated by genetics and biochemistry. First, the *M. mediterranea ssrA* and *Marme_0982* genes were interrupted by a self-splicing *td* group I intron and overexpressed from a plasmid. Next-generation sequencing of the CRISPR array revealed the presence of spacers that encompassed the splicing junction of these constructs, a finding that corroborated the acquisition of spacers directly from RNA *in vivo*, since this junction can be present only in the transcripts of these genes and not in their DNA sequences. *In vitro*, RT-Cas1 and Cas2 were purified and incubated with a labeled CRISPR locus and either short double-stranded DNA (dsDNA) or ssRNA molecules. An increase in the size of the labeled CRISPR array demonstrated that both nucleic acids were ligated into the repeat sequence. Ligation of RNA required the presence of the RT domain and was followed by the reverse transcription of the ligated sequence in the presence of deoxynucleotide triphosphates.

This work raised the possibility that type III systems could be employed to defend their hosts against RNA phages and revealed a unique mechanism for a reverse information flow from RNA to DNA during host-pathogen interactions. However, key questions about type III spacer acquisition remain unanswered. First, since these systems require transcription for immunity, only one orientation of insertion will produce an effective crRNA that is complementary to the transcript (the other orientation will generate a crRNA guide with the same sequence as the target transcript). The study described above, however, did not find an insertion bias, even from the spacers acquired from RNA, which were integrated in both orientations. Unless this result is an artifact of overexpression of RT-Cas1/Cas2, it implies that the type III CRISPR-Cas immune response is only ~50% effective, as approximately half of the adapted cells will carry a new spacer that cannot recognize the invader's transcripts to defend the host cell. Future work will determine whether this is the case. The second unanswered question regards the state of cell dormancy triggered by Csm6 (29). This activity implies that the acquisition of a new spacer has both positive and negative consequences for the host cell: it immunizes the host, but it also prevents the immune cells from growing. Therefore, upon infection, the cells in the population either die if they cannot adapt or stop replicating. While the latter is better, the impact of Csm6 activation on the host-virus coevolution dynamics is still unknown. Finally, overexpression of the *M. mediterranea* type III-B system in *Escherichia coli* did not display a correlation between spacer transcription and acquisition, suggesting that new, host-specific factors required for spacer acquisition from RNA molecules need to be found.

The Future of Type III CRISPR-Cas Research: Genetic Diversity and Accessory Effectors

Through the employment of sophisticated bioinformatic methods and even after a simple count of the number of *cas* genes, it can be inferred that type III systems possess the most complex genetic composition of all CRISPR-Cas types (65). The genetic diversity of type III *cas* operons can be classified into four subtypes, A to D (66). Only subtypes A and B have been studied in detail; future work on subtypes C and D will most likely add to the complexity of type III immunity mechanisms. For example, in subtype III-D loci, the *cas10* gene usually lacks the palm domain, which indicates that these systems do not utilize CARF domain effectors and rely exclusively on the DNase activity of the HD domain to provide immunity to the host. If so, what is the solution for the transcription-dependent weakness associated with the targeting of late-expressed viral genes, solved by Csm6 in type III-A systems (30)? On the other hand, many loci classified within the subtype III-C contain *cas10* genes without the HD domain. Do these provide immunity through a mechanism similar to abortive infection mediated by Csm6's massive RNA destruction? If so, subtype III-C systems would work similarly to type VI systems, where host and viral RNA degradation induces a perpetual cell arrest that prevents the propagation of both the host cell and its invaders (62).

Not only is there great diversity in the type III *cas* genes, but also these systems are unique due to the high abundance of accessory components, such as Csm6/Csx1 (the most common and best understood type III accessory genes). To identify these supplementary genes, two computational strategies (67, 68) relied on the "guilty by association" premise; i.e., they have to be localized in close proximity to and nonrandomly associated with the type III CRISPR-Cas locus. Using this approach, both CARF-containing and non-CARF genes were identified. After the HEPN domain present in Csm6/Csx1, one of the most common domains present in CARF proteins is the PD-D/ExK domain (69). Recently, biochemical and structural analyses determined that one such protein, *Thermus thermophilus* Can1, is activated by cA_4 binding to introduce nicks in supercoiled DNA (70). How this activity provides immunity to prokaryotes remains to be demonstrated. Other CARF domains are found linked to effector domains, such as HTH (helix-turn-helix domain, DNA binding), WYL (tryptophan-tyrosine-leucine domain, ligand binding) (71), PIL (PilT N-terminal domain, toxin) (67), and transmembrane domains fused to a Lon family protease domain (68). Among the non-CARF genes, one of the most abundant was *corA*, which encodes a divalent cation channel used for Mg^{2+} uptake (68), but also genes with homology to restriction endonucleases, rRNA maturation RNases, helicases, aspartic acid peptidases, ATPases, and primases were detected (67).

Interestingly, the CBASS (cyclic-oligonucleotide-based anti-phage signaling system) immune pathway also uses cyclic oligonucleotide second messengers to activate effector proteins (72–76). While the mechanisms of invader recognition are not completely clear yet, they activate cyclic oligonucleotide synthesis by bacterial cGAS-like enzymes. The second messenger, in turn, activates nonspecific dsDNases such as NucC (73) and Cap4 (74), which affect host viability to provide an abortive infection mode of defense (74). Many hypotheses can be made about the putative function of type III accessory genes and their functional relationship with CBASS immunity; future experiments designed to test them will undoubtedly uncover new, exciting, and complex biology behind type III CRISPR-Cas systems. Perhaps it is behind these new experiments where surprising biotechnological applications for type III systems are hiding.

Acknowledgments

L.A.M. is supported by a Burroughs Wellcome Fund PATH Award and an NIH Director's Pioneer Award (DP1GM128184). L.A.M. is an investigator of the Howard Hughes Medical Institute.

L.A.M. is a cofounder and Scientific Advisory Board member of Intellia Therapeutics and a cofounder of Eligo Biosciences.

References

1. Makarova KS, Wolf YI, Iranzo J, Shmakov SA, Alkhnbashi OS, Brouns SJJ, Charpentier E, Cheng D, Haft DH, Horvath P, Moineau S, Mojica FJM, Scott D, Shah SA, Siksnys V, Terns MP, Venclovas Č, White MF, Yakunin AF, Yan W, Zhang F, Garrett RA, Backofen R, van der Oost J, Barrangou R, Koonin EV. 2020. Evolutionary classification of CRISPR-Cas systems: a burst of class 2 and derived variants. *Nat Rev Microbiol* 18:67–83.

2. Deng L, Garrett RA, Shah SA, Peng X, She Q. 2013. A novel interference mechanism by a type IIIB CRISPR-Cmr module in *Sulfolobus*. *Mol Microbiol* 87:1088–1099.

3. Goldberg GW, Jiang W, Bikard D, Marraffini LA. 2014. Conditional tolerance of temperate phages via transcription-dependent CRISPR-Cas targeting. *Nature* 514:633–637.

4. Brüssow H, Canchaya C, Hardt WD. 2004. Phages and the evolution of bacterial pathogens: from genomic rearrangements to lysogenic conversion. *Microbiol Mol Biol Rev* 68:560–602.

5. Hale CR, Zhao P, Olson S, Duff MO, Graveley BR, Wells L, Terns RM, Terns MP. 2009. RNA-guided RNA cleavage by a CRISPR RNA-Cas protein complex. *Cell* 139:945–956.

6. Hatoum-Aslan A, Maniv I, Samai P, Marraffini LA. 2014. Genetic characterization of antiplasmid immunity through a type III-A CRISPR-Cas system. *J Bacteriol* 196:310–317.

7. Zhang J, Rouillon C, Kerou M, Reeks J, Brugger K, Graham S, Reimann J, Cannone G, Liu H, Albers SV, Naismith JH, Spagnolo L, White MF. 2012. Structure and mechanism of the CMR complex for CRISPR-mediated antiviral immunity. *Mol Cell* 45:303–313.

8. Carte J, Pfister NT, Compton MM, Terns RM, Terns MP. 2010. Binding and cleavage of CRISPR RNA by Cas6. *RNA* **16:**2181–2188.

9. Carte J, Wang R, Li H, Terns RM, Terns MP. 2008. Cas6 is an endoribonuclease that generates guide RNAs for invader defense in prokaryotes. *Genes Dev* **22:**3489–3496.

10. Hatoum-Aslan A, Maniv I, Marraffini LA. 2011. Mature clustered, regularly interspaced, short palindromic repeats RNA (crRNA) length is measured by a ruler mechanism anchored at the precursor processing site. *Proc Natl Acad Sci USA* **108:**21218–21222.

11. Sokolowski RD, Graham S, White MF. 2014. Cas6 specificity and CRISPR RNA loading in a complex CRISPR-Cas system. *Nucleic Acids Res* **42:**6532–6541.

12. Osawa T, Inanaga H, Sato C, Numata T. 2015. Crystal structure of the CRISPR-Cas RNA silencing Cmr complex bound to a target analog. *Mol Cell* **58:**418–430.

13. Rouillon C, Zhou M, Zhang J, Politis A, Beilsten-Edmands V, Cannone G, Graham S, Robinson CV, Spagnolo L, White MF. 2013. Structure of the CRISPR interference complex CSM reveals key similarities with cascade. *Mol Cell* **52:**124–134.

14. Taylor DW, Zhu Y, Staals RH, Kornfeld JE, Shinkai A, van der Oost J, Nogales E, Doudna JA. 2015. Structural biology. Structures of the CRISPR-Cmr complex reveal mode of RNA target positioning. *Science* **348:**581–585.

15. Hatoum-Aslan A, Samai P, Maniv I, Jiang W, Marraffini LA. 2013. A ruler protein in a complex for antiviral defense determines the length of small interfering CRISPR RNAs. *J Biol Chem* **288:**27888–27897.

16. Walker FC, Chou-Zheng L, Dunkle JA, Hatoum-Aslan A. 2017. Molecular determinants for CRISPR RNA maturation in the Cas10-Csm complex and roles for non-Cas nucleases. *Nucleic Acids Res* **45:**2112–2123.

17. Jia N, Mo CY, Wang C, Eng ET, Marraffini LA, Patel DJ. 2019. Type III-A CRISPR Cas Csm complexes: assembly, periodic RNA cleavage, DNase activity regulation, and autoimmunity. *Mol Cell* **73:**264–277.e5.

18. You L, Ma J, Wang J, Artamonova D, Wang M, Liu L, Xiang H, Severinov K, Zhang X, Wang Y. 2019. Structure studies of the CRISPR-Csm complex reveal mechanism of co-transcriptional interference. *Cell* **176:**239–253.e16.

19. Samai P, Pyenson N, Jiang W, Goldberg GW, Hatoum-Aslan A, Marraffini LA. 2015. Co-transcriptional DNA and RNA cleavage during type III CRISPR-Cas immunity. *Cell* **161:**1164–1174.

20. Hale CR, Cocozaki A, Li H, Terns RM, Terns MP. 2014. Target RNA capture and cleavage by the Cmr type III-B CRISPR-Cas effector complex. *Genes Dev* **28:**2432–2443.

21. Pan S, Li Q, Deng L, Jiang S, Jin X, Peng N, Liang Y, She Q, Li Y. 2019. A seed motif for target RNA capture enables efficient immune defence by a type III-B CRISPR-Cas system. *RNA Biol* **16:**1166–1178.

22. Peng W, Feng M, Feng X, Liang YX, She Q. 2015. An archaeal CRISPR type III-B system exhibiting distinctive RNA targeting features and mediating dual RNA and DNA interference. *Nucleic Acids Res* **43:**406–417.

23. Staals RHJ, Agari Y, Maki-Yonekura S, Zhu Y, Taylor DW, van Duijn E, Barendregt A, Vlot M, Koehorst JJ, Sakamoto K, Masuda A, Dohmae N, Schaap PJ, Doudna JA, Heck AJR, Yonekura K, van der Oost J, Shinkai A. 2013. Structure and activity of the RNA-targeting type III-B CRISPR-Cas complex of *Thermus thermophilus*. *Mol Cell* **52:**135–145.

24. Wang L, Mo CY, Wasserman MR, Rostøl JT, Marraffini LA, Liu S. 2019. Dynamics of Cas10 govern discrimination between self and non-self in type III CRISPR-Cas immunity. *Mol Cell* **73:**278–290.e4.

25. Elmore JR, Sheppard NF, Ramia N, Deighan T, Li H, Terns RM, Terns MP. 2016. Bipartite recognition of target RNAs activates DNA cleavage by the type III-B CRISPR-Cas system. *Genes Dev* **30:**447–459.

26. Estrella MA, Kuo FT, Bailey S. 2016. RNA-activated DNA cleavage by the type III-B CRISPR-Cas effector complex. *Genes Dev* **30:**460–470.

27. Kazlauskiene M, Tamulaitis G, Kostiuk G, Venclovas Č, Siksnys V. 2016. Spatiotemporal control of type III-A CRISPR-Cas immunity: coupling DNA degradation with the target RNA recognition. *Mol Cell* **62:**295–306.

28. Marraffini LA, Sontheimer EJ. 2008. CRISPR interference limits horizontal gene transfer in staphylococci by targeting DNA. *Science* **322:**1843–1845.

29. Rostøl JT, Marraffini LA. 2019. Non-specific degradation of transcripts promotes plasmid clearance during type III-A CRISPR-Cas immunity. *Nat Microbiol* **4:**656–662.

30. Jiang W, Samai P, Marraffini LA. 2016. Degradation of phage transcripts by CRISPR-associated RNases enables type III CRISPR-Cas immunity. *Cell* **164:**710–721.

31. Makarova KS, Aravind L, Grishin NV, Rogozin IB, Koonin EV. 2002. A DNA repair system specific for thermophilic Archaea and bacteria predicted by genomic context analysis. *Nucleic Acids Res* **30:**482–496.

32. Zhu X, Ye K. 2012. Crystal structure of Cmr2 suggests a nucleotide cyclase-related enzyme in type III CRISPR-Cas systems. *FEBS Lett* **586:**939–945.

33. Kazlauskiene M, Kostiuk G, Venclovas Č, Tamulaitis G, Siksnys V. 2017. A cyclic oligonucleotide signaling pathway in type III CRISPR-Cas systems. *Science* **357:**605–609.

34. Niewoehner O, Garcia-Doval C, Rostøl JT, Berk C, Schwede F, Bigler L, Hall J, Marraffini LA, Jinek M. 2017. Type III CRISPR-Cas systems produce cyclic oligoadenylate second messengers. *Nature* **548:**543–548.

35. Burroughs AM, Zhang D, Schäffer DE, Iyer LM, Aravind L. 2015. Comparative genomic analyses reveal a vast, novel network of nucleotide-centric systems in biological conflicts, immunity and signaling. *Nucleic Acids Res* **43:**10633–10654.

36. Lintner NG, Frankel KA, Tsutakawa SE, Alsbury DL, Copié V, Young MJ, Tainer JA, Lawrence CM. 2011. The structure of the CRISPR-associated protein Csa3 provides insight into the regulation of the CRISPR/Cas system. *J Mol Biol* **405:**939–955.

37. Han W, Stella S, Zhang Y, Guo T, Sulek K, Peng-Lundgren L, Montoya G, She Q. 2018. A type III-B Cmr effector complex catalyzes the synthesis of cyclic oligoadenylate second messengers by cooperative substrate binding. *Nucleic Acids Res* **46:**10319–10330.

38. Nasef M, Muffly MC, Beckman AB, Rowe SJ, Walker FC, Hatoum-Aslan A, Dunkle JA. 2019. Regulation of cyclic oligoadenylate synthesis by the *Staphylococcus epidermidis* Cas10-Csm complex. *RNA* **25:**948–962.

39. Rouillon C, Athukoralage JS, Graham S, Grüschow S, White MF. 2018. Control of cyclic oligoadenylate synthesis in a type III CRISPR system. *eLife* **7:**e36734.

40. Niewoehner O, Jinek M. 2016. Structural basis for the endoribonuclease activity of the type III-A CRISPR-associated protein Csm6. *RNA* **22:**318–329.

41. Foster K, Kalter J, Woodside W, Terns RM, Terns MP. 2019. The ribonuclease activity of Csm6 is required for anti-plasmid immunity by type III-A CRISPR-Cas systems. *RNA Biol* **16:**449–460.

42. Staals RH, Zhu Y, Taylor DW, Kornfeld JE, Sharma K, Barendregt A, Koehorst JJ, Vlot M, Neupane N, Varossieau K, Sakamoto K, Suzuki T, Dohmae N, Yokoyama S, Schaap PJ, Urlaub H, Heck AJ, Nogales E, Doudna JA, Shinkai A, van der Oost J. 2014. RNA targeting by the type III-A CRISPR-Cas Csm complex of *Thermus thermophilus*. *Mol Cell* **56:**518–530.

43. Tamulaitis G, Kazlauskiene M, Manakova E, Venclovas Č, Nwokeoji AO, Dickman MJ, Horvath P, Siksnys V. 2014. Programmable RNA shredding by the type III-A CRISPR-Cas system of *Streptococcus thermophilus*. *Mol Cell* **56:**506–517.

44. Jackson RN, Wiedenheft B. 2015. A conserved structural chassis for mounting versatile CRISPR RNA-guided immune responses. *Mol Cell* **58:**722–728.

45. Jia N, Jones R, Yang G, Ouerfelli O, Patel DJ. 2019. CRISPR-Cas III-A Csm6 CARF domain is a ring nuclease triggering stepwise cA_4

cleavage with ApA>p formation terminating RNase activity. *Mol Cell* 75:944–956.e6.

46. Athukoralage JS, Rouillon C, Graham S, Grüschow S, White MF. 2018. Ring nucleases deactivate type III CRISPR ribonucleases by degrading cyclic oligoadenylate. *Nature* 562:277–280.

47. Lillestøl RK, Shah SA, Brügger K, Redder P, Phan H, Christiansen J, Garrett RA. 2009. CRISPR families of the crenarchaeal genus *Sulfolobus*: bidirectional transcription and dynamic properties. *Mol Microbiol* 72:259–272.

48. Marraffini LA, Sontheimer EJ. 2010. Self versus non-self discrimination during CRISPR RNA-directed immunity. *Nature* 463:568–571.

49. Johnson K, Learn BA, Estrella MA, Bailey S. 2019. Target sequence requirements of a type III-B CRISPR-Cas immune system. *J Biol Chem* 294:10290–10299.

50. Bolotin A, Quinquis B, Sorokin A, Ehrlich SD. 2005. Clustered regularly interspaced short palindrome repeats (CRISPRs) have spacers of extrachromosomal origin. *Microbiology (Reading)* 151:2551–2561.

51. Mojica FJM, Díez-Villaseñor C, García-Martínez J, Almendros C. 2009. Short motif sequences determine the targets of the prokaryotic CRISPR defence system. *Microbiology (Reading)* 155:733–740.

52. Deveau H, Barrangou R, Garneau JE, Labonté J, Fremaux C, Boyaval P, Romero DA, Horvath P, Moineau S. 2008. Phage response to CRISPR-encoded resistance in *Streptococcus thermophilus*. *J Bacteriol* 190:1390–1400.

53. Pyenson NC, Gayvert K, Varble A, Elemento O, Marraffini LA. 2017. Broad targeting specificity during bacterial type III CRISPR-Cas immunity constrains viral escape. *Cell Host Microbe* 22:343–353.e3.

54. Majumdar S, Zhao P, Pfister NT, Compton M, Olson S, Glover CV, III, Wells L, Graveley BR, Terns RM, Terns MP. 2015. Three CRISPR-Cas immune effector complexes coexist in *Pyrococcus furiosus*. *RNA* 21:1147–1158.

55. Silas S, Lucas-Elio P, Jackson SA, Aroca-Crevillén A, Hansen LL, Fineran PC, Fire AZ, Sánchez-Amat A. 2017. Type III CRISPR-Cas systems can provide redundancy to counteract viral escape from type I systems. *eLife* 6:e27601.

56. Manica A, Zebec Z, Steinkellner J, Schleper C. 2013. Unexpectedly broad target recognition of the CRISPR-mediated virus defence system in the archaeon *Sulfolobus solfataricus*. *Nucleic Acids Res* 41:10509–10517.

57. Garrett RA, Shah SA, Erdmann S, Liu G, Mousaei M, León-Sobrino C, Peng W, Gudbergsdottir S, Deng L, Vestergaard G, Peng X, She Q. 2015. CRISPR-Cas adaptive immune systems of the Sulfolobales: unravelling their complexity and diversity. *Life (Basel)* 5:783–817.

58. Semenova E, Jore MM, Datsenko KA, Semenova A, Westra ER, Wanner B, van der Oost J, Brouns SJ, Severinov K. 2011. Interference by clustered regularly interspaced short palindromic repeat (CRISPR) RNA is governed by a seed sequence. *Proc Natl Acad Sci USA* 108:10098–10103.

59. Zetsche B, Gootenberg JS, Abudayyeh OO, Slaymaker IM, Makarova KS, Essletzbichler P, Volz SE, Joung J, van der Oost J, Regev A, Koonin EV, Zhang F. 2015. Cpf1 is a single RNA-guided endonuclease of a class 2 CRISPR-Cas system. *Cell* 163:759–771.

60. Brüssow H, Hendrix RW. 2002. Phage genomics: small is beautiful. *Cell* 108:13–16.

61. Kupczok A, Neve H, Huang KD, Hoeppner MP, Heller KJ, Franz CMAP, Dagan T. 2018. Rates of mutation and recombination in Siphoviridae phage genome evolution over three decades. *Mol Biol Evol* 35:1147–1159.

62. Meeske AJ, Nakandakari-Higa S, Marraffini LA. 2019. Cas13-induced cellular dormancy prevents the rise of CRISPR-resistant bacteriophage. *Nature* 570:241–245.

63. Silas S, Makarova KS, Shmakov S, Páez-Espino D, Mohr G, Liu Y, Davison M, Roux S, Krishnamurthy SR, Fu BXH, Hansen LL, Wang D, Sullivan MB, Millard A, Clokie MR, Bhaya D, Lambowitz AM, Kyrpides NC, Koonin EV, Fire AZ. 2017. On the origin of reverse transcriptase-using CRISPR-Cas systems and their hyperdiverse, enigmatic spacer repertoires. *mBio* 8:e00897-17.

64. Silas S, Mohr G, Sidote DJ, Markham LM, Sanchez-Amat A, Bhaya D, Lambowitz AM, Fire AZ. 2016. Direct CRISPR spacer acquisition from RNA by a natural reverse transcriptase-Cas1 fusion protein. *Science* 351:aad4234.

65. Koonin EV, Makarova KS. 2019. Origins and evolution of CRISPR-Cas systems. *Philos Trans R Soc Lond B Biol Sci* 374:20180087.

66. Koonin EV, Makarova KS, Zhang F. 2017. Diversity, classification and evolution of CRISPR-Cas systems. *Curr Opin Microbiol* 37:67–78.

67. Shah SA, Alkhnbashi OS, Behler J, Han W, She Q, Hess WR, Garrett RA, Backofen R. 2018. Comprehensive search for accessory proteins encoded with archaeal and bacterial type III CRISPR-*cas* gene cassettes reveals 39 new *cas* gene families. *RNA Biol* 16:530–542.

68. Shmakov SA, Makarova KS, Wolf YI, Severinov KV, Koonin EV. 2018. Systematic prediction of genes functionally linked to CRISPR-Cas systems by gene neighborhood analysis. *Proc Natl Acad Sci USA* 115:E5307–E5316.

69. Makarova KS, Timinskas A, Wolf YI, Gussow AB, Siksnys V, Venclovas Č, Koonin EV. 2020. Evolutionary and functional classification of the CARF domain superfamily, key sensors in prokaryotic antivirus defense. *Nucleic Acids Res* 48:8828–8847.

70. McMahon SA, Zhu W, Graham S, Rambo R, White MF, Gloster TM. 2020. Structure and mechanism of a type III CRISPR defence DNA nuclease activated by cyclic oligoadenylate. *Nat Commun* 11:500.

71. Makarova KS, Anantharaman V, Grishin NV, Koonin EV, Aravind L. 2014. CARF and WYL domains: ligand-binding regulators of prokaryotic defense systems. *Front Genet* 5:102.

72. Cohen D, Melamed S, Millman A, Shulman G, Oppenheimer-Shaanan Y, Kacen A, Doron S, Amitai G, Sorek R. 2019. Cyclic GMP-AMP signalling protects bacteria against viral infection. *Nature* 574:691–695.

73. Lau RK, Ye Q, Birkholz EA, Berg KR, Patel L, Mathews IT, Watrous JD, Ego K, Whiteley AT, Lowey B, Mekalanos JJ, Kranzusch PJ, Jain M, Pogliano J, Corbett KD. 2020. Structure and mechanism of a cyclic trinucleotide-activated bacterial endonuclease mediating bacteriophage immunity. *Mol Cell* 77:723–733.e6.

74. Lowey B, Whiteley AT, Keszei AFA, Morehouse BR, Mathews IT, Antine SP, Cabrera VJ, Kashin D, Niemann P, Jain M, Schwede F, Mekalanos JJ, Shao S, Lee ASY, Kranzusch PJ. 2020. CBASS immunity uses CARF-related effectors to sense 3′-5′- and 2′-5′-linked cyclic oligonucleotide signals and protect bacteria from phage infection. *Cell* 182:38–49.e17.

75. Whiteley AT, Eaglesham JB, de Oliveira Mann CC, Morehouse BR, Lowey B, Nieminen EA, Danilchanka O, King DS, Lee ASY, Mekalanos JJ, Kranzusch PJ. 2019. Bacterial cGAS-like enzymes synthesize diverse nucleotide signals. *Nature* 567:194–199.

76. Millman A, Melamed S, Amitai G, Sorek R. 2020. Diversity and classification of cyclic-oligonucleotide-based anti-phage signalling systems. *Nat Microbiol* 5:1608–1615.

CHAPTER 6

Type V CRISPR-Cas Systems

Morgan Quinn Beckett[1,2], Anita Ramachandran[1], and Scott Bailey[1,2]

[1]Department of Biochemistry and Molecular Biology, Bloomberg School of Public Health, Johns Hopkins University, Baltimore, MD, 21205;
[2]Department of Biophysics and Biophysical Chemistry, School of Medicine, Johns Hopkins University, Baltimore, MD, 21205

Introduction to Type V Systems

Clustered regularly interspaced short palindromic repeats (CRISPR) and CRISPR-associated (Cas) proteins are part of the bacterial and archaeal adaptive immunity that provides "memory" of past infection by foreign nucleic acids (1). CRISPR loci contain a CRISPR array which is composed of repeat sequences that are separated by spacer sequences derived from invasive genetic elements that are captured, processed, and stored by the CRISPR-Cas system (2). Following transcription and processing of the CRISPR loci into CRISPR RNA (crRNA), the crRNA is used as a guide by the CRISPR-Cas system to recognize subsequent infections through complementary binding to the invader sequence. Once bound, the target is cleaved by Cas endonucleases (3, 4). Type V systems, which comprise 10% of known CRISPR-Cas systems (5), are class 2 CRISPR-Cas systems and are defined by a signature *cas12* gene encoding a single-effector nuclease (6, 7). All studied Cas12 effectors are unified by a conserved C-terminal RuvC domain but retain little shared sequence and structural identity beyond it. The RuvC domain is related to TnpB (transposase B) proteins, a highly diverse, poorly characterized family of nucleases encoded by autonomous and nonautonomous transposons in archaea and bacteria (8). Type V systems have been found in both archaea and bacteriophages but are predominantly bacterial. Several Cas12 effectors have been developed as gene editing and nucleic acid detection tools (9–11). Studies of type V systems thus far have shed light on the molecular mechanisms of target selection (double-stranded DNA [dsDNA], single-stranded DNA [ssDNA], and RNA) and crRNA processing. This chapter expands on such topics, addressing the evolutionary origins of the type V systems and highlighting the broad similarities and differences of the molecular mechanisms of adaptive immunity among their subtypes.

Evolution and Classification of Type V Systems

Type V systems are classified based on the presence of a signature Cas12 effector that contains a characteristic RuvC nuclease domain at its C terminus (6, 12–14). Cas12 effectors from different type V subtypes are thought to have arisen from different

CRISPR: Biology and Applications, First Edition. Edited by Rodolphe Barrangou, Erik J. Sontheimer, and Luciano A. Marraffini.
© 2022 American Society for Microbiology. DOI: 10.1128/9781683673798.ch06

TnpB transposases, resulting in the extreme sequence, structural, and functional diversity we see today (8, 15, 16). Type V systems are assigned to 1 of at least 11 subtypes that include type V-A through V-K (Cas12a to -k) and a putative subtype, V-U (where "U" stands for uncharacterized) (Table 6.1) (11, 14, 17–19). It was originally thought that type V systems were exclusively bacterial. However, recently Cas12a was found in two archaeal species, Cas12f (also called Cas14) in DPANN, a superphylum of symbiotic archaea, and Cas12j (also called CasΦ) in the huge phage clade Biggiephage (8, 11, 18). The size distribution of Cas12 effectors is highly variable, ranging from 400 to 1,500 amino acids (Table 6.1), underscored by the miniature Cas12 effectors (Cas12f, Cas12j, and Cas12k) that are up to 50% smaller than other type V effectors (5, 11, 18–21). The miniature type V effectors are thought to have evolved a highly compact architecture and consolidated catalytic center due to their small host genome size, coupled with the high mutagenesis rate of prokaryotic and phage genomes to accommodate these constraints (8, 11, 18, 22).

Type V systems can also be classified based on their RNA requirements (Table 6.1) (23). Like other CRISPR systems, all type V systems use crRNA for sequence-specific binding of their targets. In addition to the crRNA, several type V systems encode an accessory noncoding RNA. To date, two distinct noncoding RNAs have been identified, a transactivating CRISPR RNA (tracrRNA) and a short-complementarity untranslated RNA (scoutRNA). When present, tracrRNA and scoutRNA are required for both the generation of mature crRNA (crRNA processing) and crRNA-guided target cleavage (13, 23, 24).

Domain Organization and Structure of Type V Effector Proteins

To date, high-resolution structures for Cas12a, Cas12b, Cas12e, and Cas12i have been reported (15, 20, 21, 25). Despite their evolutionary and sequence diversities, these Cas12 structures adopt an overall bilobed architecture consisting of the α-helical recognition (REC) lobe and the nuclease (NUC) lobe (Fig. 6.1A). The REC lobe consists of Rec1 and Rec2 domains that form stabilizing interactions with the crRNA-target strand heteroduplex. The NUC lobe contains the conserved RuvC nuclease domain as well as other domains that mediate RNA and DNA interactions (15, 20, 25–28). The only domain whose structure is conserved across all Cas12 effectors is the RuvC domain; all other domains have little or no structural similarity, although many have analogous functions (16).

In Cas12a, the NUC lobe also contains a protospacer adjacent motif (PAM)-interacting (PI) domain, an oligonucleotide binding domain (OBD; also called a wedge domain), and a target-strand loading (TSL) domain (Fig. 6.1A). The TSL domain was originally called the NUC domain because it was hypothesized to be a second DNA nuclease domain (20). However, this has proved not to be the case, and it was recently proposed that the NUC domain be renamed the TLS domain to better explain its function (15, 21). The 43-nucleotide (nt) crRNA of Cas12a consists of a 24-nt guide region and a 19-nt repeat-derived segment that forms a pseudoknot structure which is called the 5′ handle. In Cas12a from *Acidaminococcus* and *Lachnospiraceae* bacterium ND2006, the 5′ handle is positioned between the OBD and RuvC domain, where numerous contacts are formed with the phosphodiester backbone (20, 26). Similarly, Cas12a from *Francisella novicida* was shown to form extensive stabilizing contacts with the RuvC domain, OBD, and Rec2 domain (28). Cas12b has a composition similar to that of Cas12a, except that Cas12b lacks an independent PI domain. The Cas12b crRNA is composed of a 20-nt guide and 14-nt repeat segment that pairs with a 16-nt antirepeat segment of the tracrRNA. The two RNAs are recognized through extensive

Table 6.1 Descriptions of type V subtypes

Subtype	Effector	Effector length (aa[a])	PAM	Additional RNA	Pre-crRNA processing	RNA-guided target(s)	Collateral substrate	Adaptation protein(s)
V-A	Cas12a (Cpf1)	~1,300	(T)TTV	None	Auto	dsDNA and ssDNA	ssDNA	Cas1, Cas2, and Cas4
V-B	Cas12b (C2c1)	~1,300	TTN	tracrRNA	Host factor	dsDNA and ssDNA	ssDNA	Cas1, Cas2, and Cas4
V-C	Cas12c (C2c3)	~1,300	TG or TN	scoutRNA	Auto	dsDNA and ssDNA	ssDNA	Cas1
V-D	Cas12d (CasY)	~1,200	TA or TG	scoutRNA	Auto	dsDNA and ssDNA	ssDNA	Cas1
V-E	Cas12e (CasX)	~1,000	TTCN	tracrRNA	Host factor	dsDNA	ssDNA	Cas1, Cas2, and Cas4
V-F	Cas12f (Cas14)	~550	None	tracrRNA	Host factor	ssDNA	ssDNA	Cas1, Cas2, and Cas4
V-G	Cas12g	~800	None	tracrRNA	Host factor	ssDNA	ssRNA or ssDNA	None
V-H	Cas12h	~900	RTR	None	Auto	dsDNA and ssDNA	ssDNA	None
V-I	Cas12i	~1,100	TTN	None	Auto	dsDNA and ssDNA	ssDNA	None
V-J	Cas12j (Casφ)	~750	TBN	None	Auto	dsDNA and ssDNA	ssDNA	None
V-K	Cas12k (V-U5)	~650	GTN	tracrRNA	Host factor	dsDNA	None	None

[a]aa, amino acids.

interactions by the OBD and Rec2, RuvC, and TLS domains, many of which are sequence specific (Fig. 6.1A) (21).

Cas12e (also called CasX) contains the signature type V RuvC domain, but this domain has been found to share less than 16% sequence identity with the RuvC domain of Cas12a (15, 17). The cryo-electron microscopy (cryo-EM) structures of Cas12e reveal an overall effector architecture and domain organization dominated by the RNA scaffold, making up 26% of the overall single guide RNA (sgRNA)-effector complex mass (compared to 8% in *Lachnospiraceae* bacterium Cas12a and 20% in *Alicyclobacillus acidoterrestris* Cas12b). Aligned Cas12e and Cas12a structures showed that the shared structural similarities were isolated to the RuvC domain and OBD, with root mean square deviations of 2.5 Å and 3.5 Å, respectively. Cas12e is similar to Cas12a, containing various analogous domains, coupled with distinct protein and RNA folds, that include the non-target-strand binding (NTSB) domain important for target DNA unwinding, Rec1 (helical-I) domain, Rec2 (helical-II) domain, OBD, RuvC domain, and TSL domain (15).

Cas12i was found to be most structurally homologous to Cas12b, despite their low sequence identity and different RNA requirements: Cas12b requires a tracrRNA, whereas Cas12i requires only a crRNA (25). Nevertheless, significant structural and functional differences exist. The Cas12i OBD and Rec1 and PI domains are larger than those of Cas12b in order to accommodate the U-rich motifs of the crRNA stem-loop, stabilize the crRNA-target duplex, and replace the position occupied by the tracrRNA in Cas12b. However, the Cas12i RuvC and TSL domains are notably more compact. In all, Cas12i includes a REC lobe containing PI, Rec1, and Rec2 domains, while the NUC lobe is comprised of an OBD and RuvC and TSL domains. The mature 51-nt crRNA has a 28-nt guide segment and a 23-nt repeat-derived sequence which forms a stem-loop held between the Rec2 domain and OBD with the preceding, extended U-rich motif held by the OBD.

Spacer Adaptation in Type V Systems

CRISPR-Cas systems capture short viral DNA fragments and integrate them into CRISPR arrays, enabling immunity to new viruses. In most CRISPR systems, Cas1 and Cas2 form the CRISPR integrase complex and both are essential for adaptation (Fig. 6.1B) (5, 17). Viral fragments derive from processes that result in double-strand breaks and free DNA termini (29–31). Because of their inherent heterogeneity, fragments need to be trimmed to the appropriate length by exonucleases prior to integration. In several type I, II, and V systems, the Cas4 exonuclease is responsible for this trimming (32–34), while in other systems, host nucleases fulfill this role (35, 36). The CRISPR integrase is a heterohexameric complex containing two dimers of Cas1 that are bridged by a dimer of Cas2 (37, 38). Cas1 is the catalytic subunit, and Cas2 acts as a "molecule ruler" that dictates the length of the integrated fragments by spacing the Cas1 active sites (Fig. 6.1B) (39, 40). Integration occurs through nucleophilic attack of the 3′ ends of the fragment at each end of the first repeat in the CRISPR array (41, 42).

The components of type V adaptation modules vary by subtype (Table 6.1); some contain Cas1, Cas2, and Cas4 (subtypes A, B, E, and F), some contain only Cas1 (subtypes C and D), and some lack adaptation proteins entirely (subtypes G, H, I, J, K, and U) (22, 23). To date, type V acquisition has been studied only in the type V-C system (43). Type V-C and V-D are unique among all CRISPR systems not only in including a *cas1* gene and lacking a *cas2* gene but also in having notably shorter CRISPR spacers and repeats (5, 17). Purified V-C Cas1 was found to catalyze integration of DNA fragments alone and did not require Cas2 (43). Optimal integration is observed with DNA fragments that are 18 bp long, consistent with the most prevalent spacer length in native V-C CRISPR arrays. The accommodation of these short fragments likely results from the lack of Cas2, which is

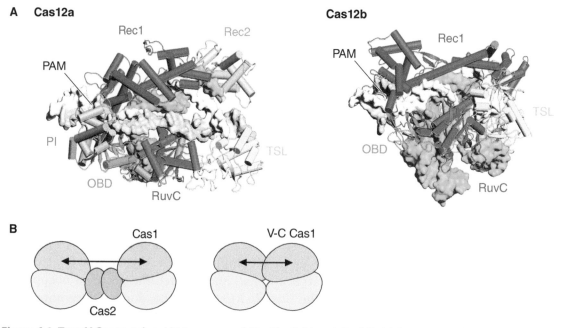

Figure 6.1 Type V Cas proteins. (A) Structures of Cas12a (left) and Cas12b (right). Protein domains are colored as follows: Rec1, blue; Rec2, light blue; OBD, orange; PI domain, pink; RuvC, red; and TSL domain, yellow. Cas12a crRNA and the Cas12b sgRNA (fusion of crRNA and tracrRNA) are green. The DNA target strand is white, and the nontarget strand is wheat. The position of the PAM is indicated. **(B)** Schematic of Cas1-Cas2 complex (left) and V-C Cas1 (right). Black arrows indicate the distance between Cas1 active sites.

proposed to effectively shorten the spacer length in a V-C Cas1 tetramer (Fig. 6.1B) (43). V-C Cas1 was also found to be promiscuous, integrating fragments at sites other than a CRISPR array, suggesting that additional host factors or mechanisms are needed to achieve specific target recognition (43).

Although several type V systems lack an adaptation module, their CRISPR arrays, particularly those of some V-U variants, show spacers of diverse sequences that can be mapped back to the genomes of bacteriophages. This suggests that spacer acquisition in these systems depends on other endogenous factors or the adaptation machinery from other CRISPR-Cas systems that coexist within the genome (8, 22). Due to the lack of Cas1-Cas2 in several type V CRISPR-Cas systems, it is speculated that the adaptation machinery might have evolved more recently and that type V subtypes lacking adaptation proteins are by-products of early CRISPR-Cas systems (7, 8).

RNA Processing in Type V Systems

CRISPR arrays, composed of alternating repeat and spacer segments, are transcribed as precursors (pre-crRNAs) that must be cleaved to produce mature crRNAs containing the sequence of a single spacer and can guide target binding. Broadly, three distinct modes of crRNA processing have been observed for type V systems, which correlate with each system's RNA requirements (Fig. 6.2 and Table 6.1).

Cas12a, Cas12i, and Cas12j directly autocleave their pre-crRNAs without the need for an additional RNA or host proteins (6, 22). Cas12a and Cas12i utilize an intrinsic endoribonuclease motif, located in the OBD of the NUC lobe, for ribose-specific cleavage (6, 18, 44). Cleavage of pre-crRNA by Cas12a and Cas12i requires magnesium, but structural studies of Cas12a reveal that bound magnesium ions are located away from the active site and therefore do not participate directly in catalysis but are required for pre-crRNA binding (25, 26, 28). Instead, Cas12a and Cas12i have been shown to cut their own pre-crRNA in a metal-independent manner, with the cut site located at the 5′ end of the upstream repeat (Fig. 6.2A). The 2′ hydroxyl upstream of the cut site is required for

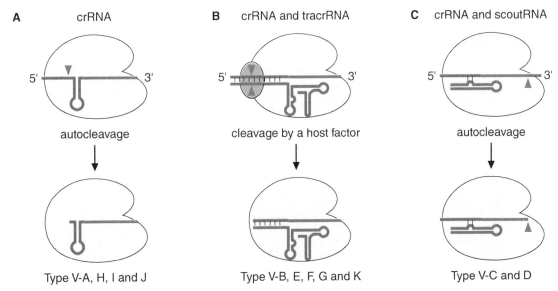

Figure 6.2 **Pre-crRNA processing by type V systems that use only crRNA (A), use crRNA and tracrRNA (B), and use crRNA and scoutRNA (C).** Green, Cas12 effectors; blue, repeat region of crRNA; pink, guide region of crRNA; brown, tracrRNA and scoutRNA; orange, host factor. Pre-crRNA cleavage sites are indicated by red triangles.

scissile phosphate cleavage, indicative of an acid-base catalyzed reaction (28). Mutation of the upstream pre-crRNA repeat, which forms a pseudoknot structure in the crRNA, disrupts sequence-specific contacts formed between the crRNA and Cas12a and ablates pre-crRNA processing (18, 44). In contrast to Cas12a and Cas12i, Cas12j autocleaves its pre-crRNA in a strictly magnesium-dependent manner (18). Mutation of the RuvC catalytic residues of Cas12j completely abolished pre-crRNA processing, without any defects in RNA binding. Furthermore, analysis of mature crRNA processed by wild-type Cas12j revealed 5′ phosphate and 2'- and 3′-hydroxyl moieties, characteristic of cleavage by the metal-dependent RuvC domain (45). Thus, Cas12j uses a single RuvC active site for both pre-crRNA processing and DNA cleavage (18). Cas12h also lacks an additional noncoding RNA, but the details of how it processes its pre-crRNA await further investigation (22).

Five type V systems (types V-B, -E, -F, -G, and -K) encode a tracrRNA, which is employed to process the pre-crRNAs. The tracrRNA hybridizes with the repeat region of the pre-crRNA due to partial sequence complementarity between the two. The resulting dsRNA structure, spanning up to 40 bp, is a substrate for host endonucleases and is the molecular basis of pre-crRNA processing (Fig. 6.2B). Cas12b and Cas12e likely utilize host RNase III to process the pre-crRNA, as first observed with Cas9 (24), but the identity of the endonuclease employed by Cas12f, Cas12k, and Cas12g is unknown.

In type V-C and V-D systems, scoutRNA is required for pre-crRNA processing and DNA cleavage (23). The secondary structures of the scoutRNA and tracrRNA are different. Unlike tracrRNA, scoutRNA bears a very short (3- to 5-nt) sequence complementary to the repeat region of the pre-crRNA, and it has not been confirmed if the two RNAs base-pair with one another. Cleavage of pre-crRNA transcripts by purified Cas12c requires scoutRNA but not an additional ribonuclease. Notably, the expected cut site in the repeat upstream of the spacer sequence, analogous to the cut site made by Cas12a, was not detected (23). Instead, the pre-crRNA was cut at the 3′ end of the spacer sequence (Fig. 6.2C). Mutation of the upstream repeat segment results in ablation of pre-crRNA processing, supporting the idea that disruption of scoutRNA binding is sufficient to abolish crRNA maturation (23).

CRISPR Interference by Cas12 Effector Proteins

Type V systems have the widest range of nucleic acid targets of any CRISPR-Cas system (Table 6.1). Individual Cas12 effectors can cleave dsDNA, ssDNA, or ssRNA in a crRNA-guided manner. Once activated by target binding, Cas12 effectors can also nonspecifically cleave ssDNA and, in one case, ssDNA or ssRNA. Although the exact *in vivo* mechanism of interference for each Cas12 is not known, several have been elucidated in extensive detail due to the presence of crystal and cryo-EM protein structures, coupled with detailed biochemical studies.

Recognition and Binding of dsDNA Targets

All Cas12s, except Cas12f, Cas12g, and Cas12k, bind and cleave complementary dsD-NAs that contain the required PAM for DNA binding (Table 6.1). Effector binding to dsDNA targets initiates at the PAM and proceeds through base-pairing between the guide sequence of the crRNA and target strand, displacing the nontarget strand to produce an R-loop (Fig. 6.3A) (15, 20, 21). Base-pairing begins at the target sequence adjacent to the PAM (called the seed region) and then proceeds over the complete target. Mutation of the PAM or mismatches within the seed have a large detrimental effect on

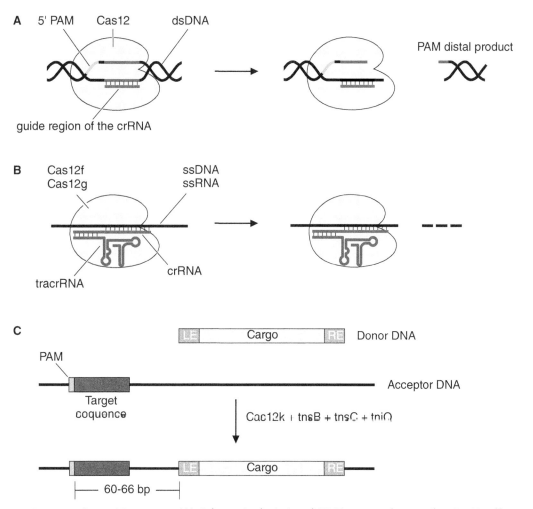

Figure 6.3 Targets of type V systems. (A) Schematic depicting dsDNA target cleavage by Cas12 effectors. For clarity, only the guide region of the crRNA is shown; i.e., the repeat region of the crRNA and additional noncoding RNAs are not shown. **(B)** Schematic depicting cleavage of ssDNA and ssRNA by Cas12f and Cas12g, respectively.

effector binding and thus target cleavage (6, 13, 28, 46). The PAM of Cas12 dsDNA targets is located at the 5′ end of the target sequence (Fig. 6.3A), the same as in type I systems and in contrast to the 3′ PAM observed in type II systems (47). Most type V PAM sequences span 2 to 4 nt and are T rich (Table 6.1), aside from the C-rich PAM for some type V-U systems (5, 6).

Biochemical and structural studies have shown that preference for a T-rich PAM is shared across Cas12a, Cas12b, Cas12e, and Cas12i effectors (15, 20, 21, 25, 26, 28). The AT-rich PAM duplex results in a narrowed minor groove, allowing for each effector to form sequence- and structure-specific contacts for accurate PAM readout. The OBD and Rec1 domain from Cas12a recognize the major groove side of the PAM duplex, while the PI domain contacts the minor groove side (Fig. 6.1A). Despite Cas12b missing the PI domain, it binds the PAM duplex in a similar location between the OBD and Rec1 domain, where the Rec1 domain forms minor groove contacts (Fig. 6.1A). Cas12i recognizes the AT-rich PAM duplex through extensive contacts formed by the PI domain, OBD, and Rec1 domain. Similar to the case with Cas12a, a Cas12i OBD loop forms hydrogen bonds with the PAM duplex major groove and the PI domain is inserted into the minor groove (25).

Following PAM recognition by Cas12a and Cas12b, the crRNA-target DNA heteroduplex is positioned in a sequence-independent manner within a central channel formed by the REC lobe, OBD, and RuvC domain. In Cas12b, the bridging helix is also part of the channel. Positively charged residues found in the OBD loops (phosphate lock loops) stabilize rotation of the PAM-proximal phosphate group of the target strand. This results in distortion and destabilization of the target strand, facilitating complementary base-pairing with the pre-ordered, A-form seed region of the guide crRNA (20, 21, 48). The backbone interactions between the crRNA-DNA heteroduplex and the cleft between the REC lobe and NUC lobe stabilize the heteroduplex for subsequent cleavage (20, 21, 26, 28). In Cas12e, the NTSB domain is thought to be responsible for initiating target dsDNA unwinding following PAM recognition and possibly stabilizing the unwound DNA duplex. Following R-loop formation, the target strand forms extensive interactions with the Rec1 and Rec2 domains and the nontarget strand is positioned in the RuvC active site.

PAM recognition by Cas12i is followed by annealing of the target strand to the pre-ordered seed region of the crRNA that is positioned by the OBD and Rec1 domain (25). The newly formed crRNA-target DNA heteroduplex is thus positioned within the Cas12i central channel, where nt 1 to 14 of the heteroduplex form extensive sequence-independent contacts with the OBD and Rec1, Rec2, and RuvC domains. Conversely, nt 15 to 26 of the heteroduplex formed sparse contacts with Cas12i, which may suggest that Cas12i can become activated with a crRNA-DNA heteroduplex less than the full-length guide segment (25).

Cas12a, Cas12b, Cas12e, and Cas12i share the common feature of an aromatic side chain at the migration point of the crRNA-target DNA heteroduplex, facilitating strand separation and subsequent downstream target strand-nontarget strand duplex reformation (15, 20, 21, 25). A second key structural feature conserved across Cas12a, Cas12b, Cas12e, and Cas12i is a lid motif loop found in the RuvC domain. Although the lid sequence is largely divergent across type V systems, structural and biochemical studies indicate its conserved function. The lid is shown to engage with the crRNA-target DNA heteroduplex via sequence-independent contacts that help stabilize target DNA binding. Heteroduplex formation results in conformational changes that lead to lid opening, revealing the RuvC active site and allowing for subsequent cleavage. Ultimately, the lid may act as a regulatory mechanism for RuvC activation (25).

RNA-Guided Cleavage of Complementary dsDNA

Cas12 effectors that target dsDNA yield staggered double-strand breaks with 5- to 12-nt 5′ overhangs (Fig. 6.3A) (6, 15, 18, 21). The nontarget strand is cleaved first at the PAM-distal end of the protospacer. Then the target strand is cleaved outside the crRNA-DNA complementary region in the PAM-distal duplex of the R-loop. This behavior is distinct from that of Cas9, which cleaves at the PAM-proximal end of the protospacer, generating blunt-end or 1-nt overhang cleavage products (12, 13). Across the structurally characterized type V subtypes, the target DNA binding and cleavage mechanism of Cas12a has been extensively studied using biophysical approaches. It has been shown that Cas12a displays robust target specificity, which rivals that of Cas9 due to its DNA target binding mechanism, which consists of two kinetically distinct steps (49–52). The first step of PAM recognition was shown to be extremely fast and reversible, while the second step of R-loop formation is largely rate limiting and mismatches at any position along the length of the R-loop contribute significantly to the rate constant. Hybridization of the crRNA-DNA heteroduplex up to 17 nt is required for formation of a hyperstable complex that undergoes cleavage, while R-loop formation prior to that point is largely reversible (49).

Following cleavage by Cas12 effectors, the PAM distal product is released from the enzyme but the PAM-proximal product remains bound (Fig. 6.3A) (49, 53). Failure to release both products results in product inhibition and makes RNA-guided dsDNA cleavage a single-turnover event. Most Cas12 effectors efficiently generate double-strand breaks. However, the efficiency with which Cas12i cleaves the target strand is much lower than the efficiency with which it cleaves the nontarget strand (22). Cleavage by Cas12i appears to be dependent on the length of the crRNA-DNA heteroduplex. There are sparse contacts found between the crRNA-DNA heteroduplex and Cas12i beyond the first 15 bp. With a minimal 14-bp heteroduplex, Cas12i displayed robust nickase activity via cleavage of the nontarget strand. Complete 28-bp heteroduplex formation, free of mismatches or deletions, was required to detect any double-strand breaks, although the predominant product was still nicked (25). This two-step activation mechanism helps clarify the requirements underpinning nicking and dsDNA cleavage by Cas12i.

Type V effectors cleave both strands of the dsDNA target with a single RuvC domain that has specificity for single-stranded nucleic acids. R-loop formation positions the nontarget strand into the RuvC active site, consistent with this being the first strand cleaved, but how RuvC cleaves the target strand was initially unclear. However, recent studies suggest that Cas12a exploits the intrinsic structural instability of the PAM-distal duplex of the R-loop to expose target strand ssDNA for cutting. The PAM proximal duplex was found not to be intrinsically unstable and retains a canonical B-form conformation. Cleavage of the target strand by the RuvC active site is therefore proposed to rely on its close proximity to the structurally unstable duplex (54). Consistently, structural studies of Cas12b and Cas12e have shown that if the PAM distal duplex is replaced with ssDNA, then that ssDNA is bent such that it is positioned into the RuvC catalytic pocket (15, 21). Structural instability and capture of the exposed target strand may be a universal mechanism by which all Cas12 effectors can cleave dsDNA with one ssDNA nuclease site.

Despite significant advances in our understanding of dsDNA cleavage by Cas12 effectors, the conformational changes that regulate cleavage of both strands are not well understood. Such insights would greatly inform our understanding of the type V interference mechanism. Interestingly, it has been proposed that the higher target specificity of type V effectors may in part be due to the increased regulation of strand cleavage, unlike Cas9, where cleavage occurs essentially in parallel and not sequentially (13, 55). Further structural and biochemical studies of the type V systems that target dsDNA will help elucidate their mechanisms, providing a more comprehensive knowledge of type V-mediated interference and a potential for the future design of highly site-specific gene editing tools.

RNA-Guided Cleavage of Complementary Single-Stranded Nucleic Acids

All purified Cas12 effectors, except Cas12e, that target and cleave dsDNA can also target and cleave complementary ssDNA (Table 6.1). This ssDNA cleavage activity does not require a PAM. However, Cas12f and Cas12g, which represent some of the smallest Cas12 effectors, cannot cleave dsDNA targets and instead cleave only ssDNA and ssRNA, respectively (Fig. 6.3B). Studies with purified Cas12f show that it cleaves ssDNA targets at sites upstream of the complementary region (11). Cleavage requires a tracrRNA and mutagenesis of the RuvC catalytic residues abolishes cleavage activity, indicating that the RuvC active site is responsible for this cleavage (11). Although cleavage of ssDNA targets by Cas12f does not require a PAM, mismatches near the middle of the targeted region strongly inhibit cleavage, revealing an internal seed region that is distinct from the PAM-proximal seed region of dsDNA targeting CRISPR systems (11). Activity of Cas12f in cells has not yet been detected. The sequences of Cas12f and its tracrRNA and crRNA were obtained from metatranscriptomic sequence analysis, and expression in a heterologous host resulted in no activity (11). Most Cas12 effectors function in heterologous systems, and therefore, further analysis will be necessary to determine the cellular requirements of Cas12f. Cas12g has no activity against ssDNA or dsDNA targets either in cells or in a purified system. Instead, it was found that Cas12g cleaves ssRNA targets; consistently, Cas12g activity is linked to transcription in cells (22). Like Cas12f, target cleavage by Cas12g requires a tracrRNA but not a PAM and is mediated by the RuvC domain (22). This is the first example of a RuvC domain that mediates cleavage of RNA substrates.

Collateral Cleavage of Single-Stranded Nucleic Acids by Target Activated Cas12

All Cas12 effectors, except Cas12k, exhibit collateral cleavage activity of non-target ssDNA or ssRNA via their RuvC active sites (Table 6.1). Collateral cleavage was originally discovered in experiments characterizing the ssDNA cleavage activity of Cas12a with a circular, single-stranded M13 DNA phage, where it was found that Cas12a induced complete degradation of the phage DNA and did not just cleave the targeted region. Subsequent experiments have shown that once Cas12a has cleaved its dsDNA targets (cis cleavage), and the PAM-distal product is released, the RuvC domain can nonspecifically cleave ssDNA substrates (trans cleavage). While cis cleavage is a single-turnover reaction, trans cleavage is a multiple-turnover reaction, which has been exploited in the development of nucleic acid detection tools (56). Purified Cas12b, Cas12c, Cas12d, Cas12e, Cas12h, Cas12i, and Cas12j, which each target dsDNA, all demonstrate collateral cleavage of nonspecific ssDNA, although Cas12e cleaves at a much lower efficiency than the others (15, 18, 22, 23, 56). The two Cas12 effectors that cleave single-stranded nucleic acids in an RNA-guided manner can also perform collateral cleavage. Cas12f exhibits robust trans cleavage of nonspecific ssDNA following binding of a specific ssDNA target (11). Cas12g, which is unique among Cas12 effectors in that it specifically targets ssRNA, is also unique in that once activated by binding specific ssRNA, it can nonspecifically cleave ssDNA or ssRNA (22).

The Transposon-Associated Type V-K Systems

Across a broad range of cyanobacteria, subtype V-K systems are found positioned within Tn7-like transposable elements (57). The type V-K locus consists of a Cas12k effector with a catalytically inactive RuvC domain, tracrRNA, CRISPR array, and upstream transposon-associated genes (tnsB, tnsC, and tniQ), all flanked by characteristic terminal inverted repeats (called LE and RE) (57, 58). No additional Cas genes are detectable, suggesting that adaptation machinery is possibly provided in trans by other active CRISPR-Cas loci in the host genome (58). When expressed in a heterologous host (Escherichia coli), the type V-K system can unidirectionally insert donor DNA, up to 10 kb in length, 60 to 66 bp downstream of a PAM (Fig. 6.3C). Transposition required all four transposon-associated genes (tnsB, tnsC, tniQ, and cas12k), an intact tracrRNA, and CRISPR array. Cas12k was found to be specifically required for transposition and could not be replaced with other Cas effectors, such as Cas9 (58). A purified and reconstituted type V-K system was also active in transposition, but consistent with its inactive RuvC domain, Cas12k was unable to cleave dsDNA targets. Thus, the V-K Tn7-like transposons utilize Cas12k to locate crRNA-guided target sites to facilitate the integration and increased mobilization of host transposons. Continued studies could dramatically enhance the potential of this system as a gene therapy tool in eukaryotic cells. Such generalizable strategies include transgene insertion into "safe harbor" loci for gene repair or therapeutic expression without potential for hot-spot preference.

Conclusion

The overview of type V systems presented herein aims to show the similarities that link type V systems while emphasizing the widespread functional and structural diversity that distinguishes each subtype. It also highlights the contrast between our in-depth understanding of the biochemical and structural properties of type V systems with our current low understanding of type V immunity *in vivo*. Growing access to genomic and metagenomic data has opened numerous avenues for increasingly accurate phylogenetic classification and discovery of novel type V systems (5, 8, 11, 22). Such insights are key to elucidating the evolutionary links underlying such extreme functional and structural diversification as well as the unique evolutionary benefits these features provide to each type V system (8). Ultimately, gaining a deeper understanding of the functional and structural features underlying each subtype's mechanisms can provide the necessary insights for potential development of Cas12 effectors as novel tools. Considering that Cas12a, Cas12b, and Cas12e have already proven to be capable tools for gene editing and/or silencing, several of the most recently studied subtypes display unique interference activities that could be used for highly specific applications (5, 6, 15, 20, 21). The dsDNA nicking activity of Cas12i has the potential to be exploited for high-fidelity gene editing, and the short 5′-TN-3′ PAM requirement for Cas12c makes it an ideal candidate for targeted cleavage of a greater number of target sites (22). More broadly, the novel C-rich PAM requirements of the previously uncharacterized V-U nuclease family expands the PAM diversity of type V systems. Finally, discovery of the highly compact Cas12j also exposes huge phage genomes as potential new sites of exploration for new CRISPR-Cas systems. Another valuable feature for potential biotechnological exploitation is type V's nearly universal *trans*-cleavage capabilities that can be used for highly precise nucleic acid detection (59). In all, continued study of these class 2 type V systems will undoubtedly be vital for the advancement and development of CRISPR-Cas systems as molecular tools.

References

1. Barrangou R, Fremaux C, Deveau H, Richards M, Boyaval P, Moineau S, Romero DA, Horvath P. 2007. CRISPR provides acquired resistance against viruses in prokaryotes. *Science* **315**:1709–1712.

2. Sternberg SH, Richter H, Charpentier E, Qimron U. 2016. Adaptation in CRISPR-Cas systems. *Mol Cell* **61**:797–808.

3. Marraffini LA. 2015. CRISPR-Cas immunity in prokaryotes. *Nature* **526**:55–61.

4. Hille F, Richter H, Wong SP, Bratovič M, Ressel S, Charpentier E. 2018. The biology of CRISPR-Cas: backward and forward. *Cell* **172**:1239–1259.

5. Shmakov S, Abudayyeh OO, Makarova KS, Wolf YI, Gootenberg JS, Semenova E, Minakhin L, Joung J, Konermann S, Severinov K, Zhang F, Koonin EV. 2015. Discovery and functional characterization of diverse class 2 CRISPR-Cas systems. *Mol Cell* **60**:385–397.

6. Zetsche B, Gootenberg JS, Abudayyeh OO, Slaymaker IM, Makarova KS, Essletzbichler P, Volz SE, Joung J, van der Oost J, Regev A, Koonin EV, Zhang F. 2015. Cpf1 is a single RNA-guided endonuclease of a class 2 CRISPR-Cas system. *Cell* **163**:759–771.

7. Koonin EV, Makarova KS. 2019. Origins and evolution of CRISPR-Cas systems. *Philos Trans R Soc Lond B Biol Sci* **374**:20180087

8. Shmakov S, Smargon A, Scott D, Cox D, Pyzocha N, Yan W, Abudayyeh OO, Gootenberg JS, Makarova KS, Wolf YI, Severinov K, Zhang F, Koonin EV. 2017. Diversity and evolution of class 2 CRISPR-Cas systems. *Nat Rev Microbiol* **15**:169–182.

9. Makarova KS, Wolf YI, Alkhnbashi OS, Costa F, Shah SA, Saunders SJ, Barrangou R, Brouns SJJ, Charpentier E, Haft DH, Horvath P, Moineau S, Mojica FJM, Terns RM, Terns MP, White MF, Yakunin AF, Garrett RA, van der Oost J, Backofen R, Koonin EV. 2015. An updated evolutionary classification of CRISPR-Cas systems. *Nat Rev Microbiol* **13**:722–736.

10. Chylinski K, Makarova KS, Charpentier E, Koonin EV. 2014. Classification and evolution of type II CRISPR-Cas systems. *Nucleic Acids Res* 42:6091–6105.

11. Harrington LB, Burstein D, Chen JS, Paez-Espino D, Ma E, Witte IP, Cofsky JC, Kyrpides NC, Banfield JF, Doudna JA. 2018. Programmed DNA destruction by miniature CRISPR-Cas14 enzymes. *Science* 362:839–842.

12. Gasiunas G, Barrangou R, Horvath P, Siksnys V. 2012. Cas9-crRNA ribonucleoprotein complex mediates specific DNA cleavage for adaptive immunity in bacteria. *Proc Natl Acad Sci USA* 109:E2579–E2586.

13. Jinek M, Chylinski K, Fonfara I, Hauer M, Doudna JA, Charpentier E. 2012. A programmable dual-RNA-guided DNA endonuclease in adaptive bacterial immunity. *Science* 337:816–821.

14. Koonin EV, Makarova KS, Zhang F. 2017. Diversity, classification and evolution of CRISPR-Cas systems. *Curr Opin Microbiol* 37:67–78.

15. Liu J-J, Orlova N, Oakes BL, Ma E, Spinner HB, Baney KLM, Chuck J, Tan D, Knott GJ, Harrington LB, Al-Shayeb B, Wagner A, Brötzmann J, Staahl BT, Taylor KL, Desmarais J, Nogales E, Doudna JA. 2019. CasX enzymes comprise a distinct family of RNA-guided genome editors. *Nature* 566:218–223.

16. Makarova KS, Koonin EV. 2015. Annotation and classification of CRISPR-Cas systems. *Methods Mol Biol* 1311:47–75.

17. Burstein D, Harrington LB, Strutt SC, Probst AJ, Anantharaman K, Thomas BC, Doudna JA, Banfield JF. 2017. New CRISPR-Cas systems from uncultivated microbes. *Nature* 542:237–241.

18. Pausch P, Al-Shayeb B, Bisom-Rapp E, Tsuchida CA, Li Z, Cress BF, Knott GJ, Jacobsen SE, Banfield JF, Doudna JA. 2020. CRISPR-CasΦ from huge phages is a hypercompact genome editor. *Science* 369:333–337.

19. Karvelis T, Bigelyte G, Young JK, Hou Z, Zedaveinyte R, Budre K, Paulraj S, Djukanovic V, Gasior S, Silanskas A, Venclovas Č, Siksnys V. 2020. PAM recognition by miniature CRISPR-Cas12f nucleases triggers programmable double-stranded DNA target cleavage. *Nucleic Acids Res* 48:5016–5023.

20. Yamano T, Nishimasu H, Zetsche B, Hirano H, Slaymaker IM, Li Y, Fedorova I, Nakane T, Makarova KS, Koonin EV, Ishitani R, Zhang F, Nureki O. 2016. Crystal structure of Cpf1 in complex with guide RNA and target DNA. *Cell* 165:949–962.

21. Yang H, Gao P, Rajashankar KR, Patel DJ. 2016. PAM-dependent target DNA recognition and cleavage by C2c1 CRISPR-Cas endonuclease. *Cell* 167:1814–1828.e12

22. Yan WX, Hunnewell P, Alfonse LE, Carte JM, Keston-Smith E, Sothiselvam S, Garrity AJ, Chong S, Makarova KS, Koonin EV, Cheng DR, Scott DA. 2019. Functionally diverse type V CRISPR-Cas systems. *Science* 363:88–91.

23. Harrington LB, Ma E, Chen JS, Witte IP, Gertz D, Paez-Espino D, Al-Shayeb B, Kyrpides NC, Burstein D, Banfield JF, Doudna JA. 2020. A scoutRNA is required for some type V CRISPR-Cas systems. *Mol Cell* 79:416–424.e5

24. Deltcheva E, Chylinski K, Sharma CM, Gonzales K, Chao Y, Pirzada ZA, Eckert MR, Vogel J, Charpentier E. 2011. CRISPR RNA maturation by trans-encoded small RNA and host factor RNase III. *Nature* 471:602–607.

25. Zhang H, Li Z, Xiao R, Chang L. 2020. Mechanisms for target recognition and cleavage by the Cas12i RNA-guided endonuclease. *Nat Struct Mol Biol* 27:1069–1076.

26. Dong D, Ren K, Qiu X, Zheng J, Guo M, Guan X, Liu H, Li N, Zhang B, Yang D, Ma C, Wang S, Wu D, Ma Y, Fan S, Wang J, Gao N, Huang Z. 2016. The crystal structure of Cpf1 in complex with CRISPR RNA. *Nature* 532:522–526.

27. Gao P, Yang H, Rajashankar KR, Huang Z, Patel DJ. 2016. Type V CRISPR-Cas Cpf1 endonuclease employs a unique mechanism for crRNA-mediated target DNA recognition. *Cell Res* 26:901–913.

28. Swarts DC, van der Oost J, Jinek M. 2017. Structural basis for guide RNA processing and seed-dependent DNA targeting by CRISPR-Cas12a. *Mol Cell* 66:221–233.e4

29. Levy A, Goren MG, Yosef I, Auster O, Manor M, Amitai G, Edgar R, Qimron U, Sorek R. 2015. CRISPR adaptation biases explain preference for acquisition of foreign DNA. *Nature* 520:505–510.

30. Modell JW, Jiang W, Marraffini LA. 2017. CRISPR-Cas systems exploit viral DNA injection to establish and maintain adaptive immunity. *Nature* 544:101–104.

31. Shiimori M, Garrett SC, Chambers DP, Glover CVC III, Graveley BR, Terns MP. 2017. Role of free DNA ends and protospacer adjacent motifs for CRISPR DNA uptake in *Pyrococcus furiosus*. *Nucleic Acids Res* 45:11281–11294.

32. Kieper SN, Almendros C, Behler J, McKenzie RE, Nobrega FL, Haagsma AC, Vink JNA, Hess WR, Brouns SJJ. 2018. Cas4 facilitates PAM-compatible spacer selection during CRISPR adaptation. *Cell Rep* 22:3377–3384.

33. Lee H, Zhou Y, Taylor DW, Sashital DG. 2018. Cas4-dependent prespacer processing ensures high-fidelity programming of CRISPR arrays. *Mol Cell* 70:48–59.e5

34. Shiimori M, Garrett SC, Graveley BR, Terns MP. 2018. Cas4 nucleases define the PAM, length, and orientation of DNA fragments integrated at CRISPR loci. *Mol Cell* 70:814–824.e6

35. Kim S, Loeff L, Colombo S, Jergic S, Brouns SJJ, Joo C. 2020. Selective loading and processing of prespacers for precise CRISPR adaptation. *Nature* 579:141–145.

36. Ramachandran A, Summerville L, Learn BA, DeBell L, Bailey S. 2020. Processing and integration of functionally oriented prespacers in the *Escherichia coli* CRISPR system depends on bacterial host exonucleases. *J Biol Chem* 295:3403–3414.

37. Yosef I, Goren MG, Qimron U. 2012. Proteins and DNA elements essential for the CRISPR adaptation process in *Escherichia coli*. *Nucleic Acids Res* 40:5569–5576.

38. Nuñez JK, Kranzusch PJ, Noeske J, Wright AV, Davies CW, Doudna JA. 2014. Cas1-Cas2 complex formation mediates spacer acquisition during CRISPR-Cas adaptive immunity. *Nat Struct Mol Biol* 21:528–534.

39. Nuñez JK, Harrington LB, Kranzusch PJ, Engelman AN, Doudna JA. 2015. Foreign DNA capture during CRISPR-Cas adaptive immunity. *Nature* 527:535–538.

40. Wang J, Li J, Zhao H, Sheng G, Wang M, Yin M, Wang Y. 2015. Structural and mechanistic basis of PAM-dependent spacer acquisition in CRISPR-Cas systems. *Cell* 163:840–853.

41. Nuñez JK, Lee ASY, Engelman A, Doudna JA. 2015. Integrase-mediated spacer acquisition during CRISPR-Cas adaptive immunity. *Nature* 519:193–198.

42. Arslan Z, Hermanns V, Wurm R, Wagner R, Pul Ü. 2014. Detection and characterization of spacer integration intermediates in type I-E CRISPR-Cas system. *Nucleic Acids Res* 42:7884–7893.

43. Wright AV, Wang JY, Burstein D, Harrington LB, Paez-Espino D, Kyrpides NC, Iavarone AT, Banfield JF, Doudna JA. 2019. A functional mini-integrase in a two-protein-type V-C CRISPR system. *Mol Cell* 73:727–737.e3

44. Fonfara I, Richter H, Bratovič M, Le Rhun A, Charpentier E. 2016. The CRISPR-associated DNA-cleaving enzyme Cpf1 also processes precursor CRISPR RNA. *Nature* 532:517–521.

45. Yang W. 2011. Nucleases: diversity of structure, *function and mechanism*. *Q Rev Biophys* 44:1–93.

46. Sternberg SH, Redding S, Jinek M, Greene EC, Doudna JA. 2014. DNA interrogation by the CRISPR RNA-guided endonuclease Cas9. *Nature* 507:62–67.

47. Wright AV, Nuñez JK, Doudna JA. 2016. Biology and applications of CRISPR systems: harnessing nature's toolbox for genome engineering. *Cell* 164:29–44.

48. Anders C, Niewoehner O, Duerst A, Jinek M. 2014. Structural basis of PAM-dependent target DNA recognition by the Cas9 endonuclease. *Nature* 513:569–573.

49. Singh D, Mallon J, Poddar A, Wang Y, Tippana R, Yang O, Bailey S, Ha T. 2018. Real-time observation of DNA target interrogation and product release by the RNA-guided endonuclease CRISPR Cpf1 (Cas12a). *Proc Natl Acad Sci USA* 115:5444–5449.

50. Kim D, Kim J, Hur JK, Been KW, Yoon S-H, Kim J-S. 2016. Genome-wide analysis reveals specificities of Cpf1 endonucleases in human cells. *Nat Biotechnol* **34**:863–868.

51. Strohkendl I, Saifuddin FA, Rybarski JR, Finkelstein IJ, Russell R. 2018. Kinetic basis for DNA target specificity of CRISPR-Cas12a. *Mol Cell* **71**:816–824.e3

52. Kleinstiver BP, Tsai SQ, Prew MS, Nguyen NT, Welch MM, Lopez JM, McCaw ZR, Aryee MJ, Joung JK. 2016. Genome-wide specificities of CRISPR-Cas Cpf1 nucleases in human cells. *Nat Biotechnol* **34**:869–874.

53. Swarts DC, Jinek M. 2019. Mechanistic insights into the cis- and trans-acting DNase activities of Cas12a. *Mol Cell* **73**:589–600.e4

54. Cofsky JC, Karandur D, Huang CJ, Witte IP, Kuriyan J, Doudna JA. 2020. CRISPR-Cas12a exploits R-loop asymmetry to form double-strand breaks. *eLife* **9**:e55143

55. Sternberg SH, LaFrance B, Kaplan M, Doudna JA. 2015. Conformational control of DNA target cleavage by CRISPR-Cas9. *Nature* **527**:110–113.

56. Chen JS, Ma E, Harrington LB, Da Costa M, Tian X, Palefsky JM, Doudna JA. 2018. CRISPR-Cas12a target binding unleashes indiscriminate single-stranded DNase activity. *Science* **360**:436–439.

57. Faure G, Shmakov SA, Yan WX, Cheng DR, Scott DA, Peters JE, Makarova KS, Koonin EV. 2019. CRISPR-Cas in mobile genetic elements: counter-defence and beyond. *Nat Rev Microbiol* **17**:513–525.

58. Strecker J, Ladha A, Gardner Z, Schmid-Burgk JL, Makarova KS, Koonin EV, Zhang F. 2019. RNA-guided DNA insertion with CRISPR-associated transposases. *Science* **365**:48–53.

59. Swarts DC. 2019. Stirring up the type V alphabet soup. *CRISPR J* **2**:14–16.

CRISPR-Cas13: Biology, Mechanism, and Applications of RNA-Guided, RNA-Targeting CRISPR Systems

Omar O. Abudayyeh and Jonathan S. Gootenberg

McGovern Institute for Brain Research at MIT, Massachusetts Institute of Technology, Cambridge, MA 02139

Discovery of CRISPR-Cas13

Clustered regularly interspaced short palindromic repeats (CRISPR) and CRISPR-associated (CRISPR-Cas) proteins have quickly become a focus of intense interest, both for their function as adaptive immune systems in their native contexts and for their numerous applications in biotechnology (1–5). Much of the initial excitement focused on leveraging the RNA-guided DNA-targeting system, CRISPR-Cas9, for genome editing applications in eukaryotes (1, 5). However, studies exploring CRISPR system diversity quickly uncovered a variety of single-effector enzymes beyond Cas9 (3, 6–9). The discovery of these new class 2 enzymes was driven by a computational pipeline that mined 25,000 bacterial genomes available in NCBI databases back in 2015 and focused on single-effector class 2 candidates (7). This pipeline relied on Cas1, a highly conserved gene involved in the acquisition of new spacers and a common marker for CRISPR systems at the time. Cas1 was used as a seed to identify all open reading frames within a set distance of the protein and determine significant associations, indicative of new CRISPR systems (7). This search identified a small family of proteins associated with Cas1/Cas2 and CRISPR arrays containing the "higher eukaryotes and prokaryotes nucleotide-binding domain" (HEPN) motif, which constitutes a metal-independent endoribonuclease active site and is present in a large superfamily of proteins (10). This family of proteins, initially named C2c2 (later renamed Cas13a), was predicted to be a programmable RNA-guided RNase due to association with Cas1 and CRISPR arrays and was the putative effector protein of the corresponding CRISPR system, type VI. Early characterization of Cas13a and the type VI system from *Leptotrichia shahii* (LshCas13a) revealed that the system was expressed in *L. shahii*, and CRISPR RNAs (crRNAs) were processed into small guide RNAs without any detectable transactivating crRNA encoded nearby (7). Subsequent investigations of the system, both biochemical and in bacteria, demonstrated that LshCas13a was functional, acting as a programmable RNase able to interfere with infection

CRISPR: Biology and Applications, First Edition. Edited by Rodolphe Barrangou, Erik J. Sontheimer, and Luciano A. Marraffini.
© 2022 American Society for Microbiology. DOI: 10.1128/9781683673798.ch07

of an RNA phage when heterologously expressed in *Escherichia coli* (8). Since the initial discovery of Cas13a, the diversity of Cas13 subfamilies and type VI systems has grown substantially, with multifarious properties and applications in biochemical, bacterial, and eukaryotic systems. In this review, we discuss the expanding set of Cas13 and type VI systems, their mechanisms of action and how they differ from other RNA targeting CRISPR systems, and the myriad applications of Cas13 for eukaryotic manipulation and nucleic acid diagnostics.

Diversity in the Cas13 Family

Four subfamilies of Cas13 have been discovered so far: Cas13a, Cas13b, Cas13c, and Cas13d (7, 9, 11–13), corresponding to the type VI-A through VI-D systems (Fig. 7.1A). These subfamilies are delineated by sequence similarity of the Cas13 effector gene sequence (orange in Fig. 7.1A). Cas13a was initially discovered by Cas1-seeded searches. However, later searches that were seeded with CRISPR arrays and thus not limited by the presence of Cas1 in the locus uncovered additional Cas13b and Cas13c families (9, 12). Cas13b and Cas13c may have lost reliance on Cas1 or Cas2 during evolution, or these activities may be provided by companion CRISPR systems in the same host. Cas13d was subsequently found by searching through additional metagenomic sequences added to databases after original searches (11, 13). Cas13d has similarity to Cas13a systems in terms of sequence identity, direct-repeat (DR) orientation, and association with Cas1 and Cas2.

Similar to the pre-crRNA maturation catalyzed by Cas12 in type V systems (14, 15), the CRISPR array is expressed in type VI systems as a pre-crRNA, which is then processed into short mature crRNAs by the RNase processing domain of the effector Cas13 protein (16). In type VI systems, the mature crRNA lengths vary between 53 and 60 nucleotides (nt), with ˜30-nt DR and spacer lengths that vary between 23 and 30 nt. The DR orientation also differs between Cas13 types: Cas13a, Cas13c, and Cas13d crRNAs have 5′ DRs, while Cas13b crRNA has a 3′ DR (Fig. 7.1B). Additionally, the DR sequences of these subfamilies are conserved. In type VI-A systems, it has been shown that the DRs stratify into two groups with interchangeable DRs and crRNAs (17). Furthermore, the Cas13a in the two groups have stereotyped base preferences for RNA cleavage, with one clade cleaving at uridine bases and the other clade cleaving at adenosine bases (17, 18).

Some Cas13 subfamilies are associated with additional accessory proteins: Cas13b systems can be subdivided into those that associate with either Csx27 (type VI-B1) or Csx28 (type VI-B2), and Cas13d is also associated with WYL domain-containing proteins such as WYL1 (Fig. 7.1C). The functions of these accessory proteins and their roles in interference are discussed in later sections.

Role of Cas13 in Bacterial Defense

The initial characterization of Cas13a as a putative single-effector system with a heterologously expressed and processed CRISPR array (7) suggested that it played a role in prokaryotic immunity through RNA-programmed phage interference. Initial evidence of RNase-based phage immunity was indeed provided in a screen against the MS2 RNA phage in *E. coli* with a heterologously expressed LshCas13a locus (8). This screen, involving a pool of every potential spacer against the MS2 genome cloned into heterologous Lsh-Cas13a arrays, determined that ˜5% of spacers could effectively provide protection against MS2. Furthermore, defense required the basic catalytic residues of the HEPN domain, establishing that interference was mediated by the RNase activity of Cas13a. Interestingly, the active spacers uncovered a 5′ sequence preference, termed the protospacer flanking sequence (PFS) (8). The PFS is conceptually similar to the sequence restriction of the

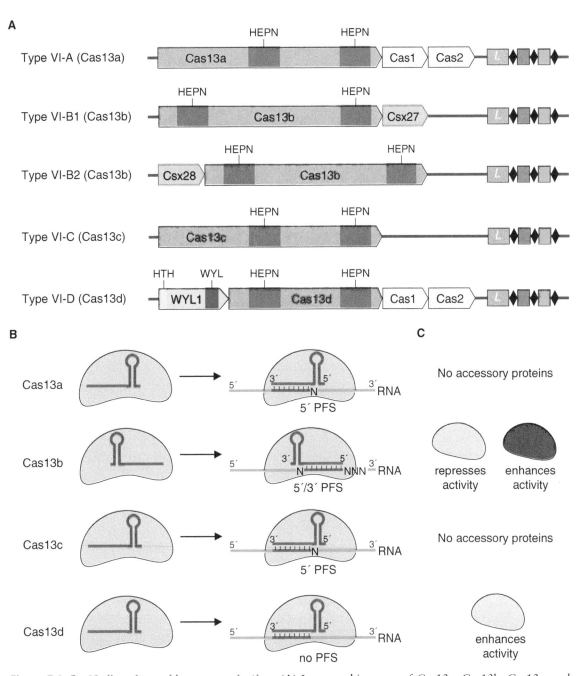

Figure 7.1 Cas13 diversity and locus organization. (A) Locus architectures of Cas13a, Cas13b, Cas13c, and Cas13d systems. The Cas13 effectors are shown in orange, with HEPN domains highlighted in red. Accessory proteins and CRISPR arrays are also shown. (B) Cas13 effector complexes shown for Cas13a, Cas13b, Cas13c, and Cas13d. The locations of the DR and PFS differ between Cas13 subtypes. (C) The effect of accessory proteins on Cas13 interference activity is shown for the subtypes indicated in panel B.

protospacer-adjacent motif in type II systems. Other Cas13 systems characterized in heterologous expression assays, such as Cas13b, have exhibited a dual PFS located at both the 3′ and 5′ ends of the protospacer (12, 19).

In vivo transcript targeting of exogenous red fluorescent protein (RFP) mRNA revealed a second aspect of Cas13 targeting, growth suppression (8). *E. coli* cultures in which the RFP transcript was successfully targeted and knocked down reached saturation at later time points, indicating either cell death or growth arrest. Additionally, this reduced growth was dependent on the expression level of the RFP mRNA, with higher levels of RFP expression resulting in later saturation in culture (8). This growth restriction suggested a form of programmed cell death by a nonspecific RNase activity, providing an additional mode of defense akin to a toxin-antitoxin or abortive infection system. The nonspecific activity was corroborated by *in vitro* observations of "collateral" cleavage of nontargeted RNA molecules (8). Further characterization of the Cas13a system of *Listeria seeligeri* in the native host demonstrated both growth suppression and nonspecific host and foreign RNA cleavage in response to specific RNA targeting (20, 21).

The combination of specific and nonspecific cleavage by type VI systems outlines a dual-action interference mechanism (8, 16) (Fig. 7.2). In bacteria harboring type VI loci, Cas13 is expressed along with the pre-crRNA transcript, which is processed by Cas13 and loaded into the effector complex (Fig. 7.2A). Upon invasion by phage, the Cas13-crRNA can target RNA phage genomes as well as transcribed RNA from DNA phage genomes for cleavage, hindering the ability of the invading phage to produce required proteins and resulting in clearance of infection (8, 16). Coincident with this targeting of phage-specific transcripts, the nonspecific RNA cleavage activity of Cas13 arrests growth of infected cells (8), sacrificing individual cells to stymie the release and spread of the phage across the bacterial population (Fig. 7.2B) (20). Additionally, nonspecific RNA cleavage increases the sensitivity of interference, as phages with escape mutations that avoid initial direct targeting by Cas13-crRNA complexes are subsequently eliminated by nonspecific cleavage (20). Aside from having the capacity to target RNA phages, an RNA-based interference mechanism circumvents potential phage mechanisms for DNA protection, such as base modifications by glucosylation, which have been demonstrated to inhibit targeting by type I and II CRISPR systems (22). In addition, certain phages may engage in strategies to physically protect their DNA, such as the construction of a nucleus-like compartment by the *Pseudomonas aeruginosa* phage ΦKZ, which assembles a protective subcellular shell that confers resistance to diverse heterologously expressed DNA-targeting defense systems but not to a heterologously expressed type VI system (23). In this approach, RNA targeting complements DNA defense.

The nonspecific cleavage of type VI systems presents a potential double-edged sword: it provides a robust mechanism for targeting mutant phages and enables defense at a population level through growth arrest, but aberrant activation in the absence of unwanted phage or foreign elements could lead to a significant fitness penalty, akin to self-targeting in DNA-targeting CRISPR systems. Although a Cas13-crRNA complex would be unable to target DNA spacers or pre-crRNA, it is possible that an activating target could be generated by antisense transcription of the CRISPR array. In many Cas13 orthologs, such as Cas13 from *Listeria seeligeri*, this scenario is avoided by inhibition of Cas13 activation by PFS restriction, which prevents activation of Cas13 when there is extended base-pairing of the DR with an anti-DR (21).

A central component of CRISPR's adaptive immunity, acquisition of sequences that guide targeting, is a peculiar circumstance in type VI systems, as there are multiple uncertainties around how RNA targets can be incorporated as DNA spacers. First, does acquisition occur upon exposure to foreign DNA or to foreign RNA? In the case of direct RNA acquisition, as in some type III systems (24), a reverse transcription step is required. While a subset of type VI-A systems have been demonstrated to be associated with Cas1-reverse transcriptase (RT) fusions (25), a majority of type VI systems lack detectable mechanisms for reverse transcription. Alternatively, spacers could be acquired into a type VI CRISPR array through a *trans*-acquisition via other RTs present in the host. In either case, the mechanism

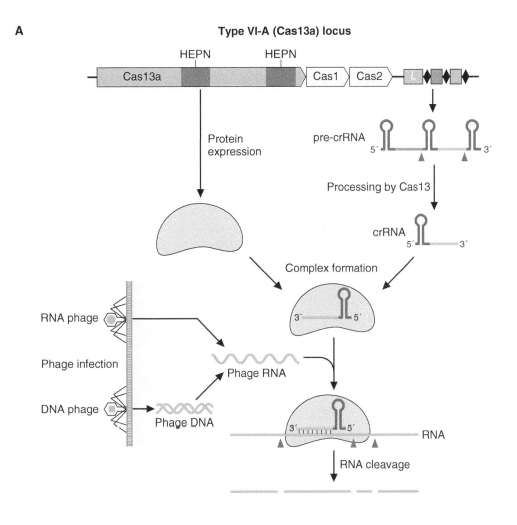

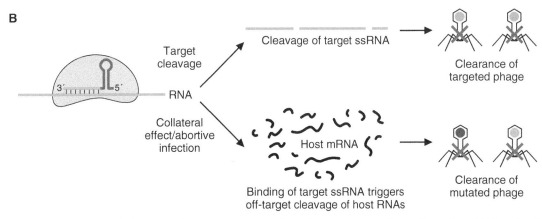

Figure 7.2 Cas13 in defense, including processing, interference, and collateral activity. (**A**) Mechanism of Cas13 interference and defense against foreign RNA, such as ssRNA phages and transcripts of DNA phages. (**B**) Dual mechanisms of Cas13 interference involving on-target cleavage and promiscuous collateral activity.

of acquired immunity for RNA phages remains unknown, as direct acquisition from an RNA phage has yet to be demonstrated. Alternatively, it is possible that type VI systems predominantly defend against DNA phages by targeting transcribed genes, a hypothesis supported by matches of type VI spacers to double-stranded DNA (dsDNA) phages (12, 13).

However, this scenario presents a challenge, as the mechanism ensuring that spacers target expressed regions in the correct orientation remains unclear; one possibility is that acquisition is independent of expression and only bacteria that incorporate spacers targeting expressed genes in the correct orientation survive infection. Furthermore, the lack of either Cas1 or Cas2 in VI-B systems suggests either the loss of acquisition entirely or potential *trans*-acquisition. Additional studies in type VI adaptation biology will hopefully shed light on these areas.

Lastly, as with DNA-targeting CRISPR systems, which have forced viral evolution of anti-CRISPR systems that can disable CRISPR immunity, type VI systems are no less susceptible to viral countermeasures. In the case of the type VI-A CRISPR system from *Listeria seeligeri*, the listeriophage (φLS46) encoding anti-CRISPR protein AcrVIA1 can inactivate the Cas13a enzyme (26). By interacting with the guide-exposed face of Cas13a, AcrVIA1 prevents the conformational changes that Cas13 requires for target cleavage. It is a highly potent inhibitor, disabling type VI-A immunity even with only a single round of infection. Further exploration of the viral evolution of anti-CRISPR systems will likely reveal even more diversity in the CRISPR arms race between microbes and phages.

Mechanisms Underlying Cas13-Mediated Target Cleavage and Pre-cRNA Processing

Initial experiments with LshCas13a demonstrated robust RNA phage interference in *E. coli* (8), but the mechanism of RNA targeting remained unclear. *In vitro* biochemical analysis of target cleavage revealed unexpected cleavage at many sites, resulting in eventual degradation of the target. This characterization determined that cleavage required the complete Cas13-crRNA complex and was programmed by the sequence of the crRNA. Cleavage was also ion and temperature dependent, with optimal cleavage achieved at 37°C with Mg^{2+} (8), an exception to the canonical, metal-independent mechanism of HEPN domains (10). Sequencing of cleavage reactions revealed that the multiple cut sites occurred at uridines in single-stranded regions. Mutation of individual uridine residues could eliminate specific cleavage sites while maintaining others, indicating that the HEPN RNase domains of Cas13 had stereotyped base preferences for cleavage, an observation confirmed by targeting of modified homopolymer sequences (8). Subsequent investigations found that other orthologs, such as the Cas13a from *Lachnospiraceae*, had different base preferences, primarily adenines, and thus different cleavage profiles than other orthologs (17). These base preferences can involve higher-order interactions, with some Cas13 orthologs preferring specific dinucleotide matches for cleavage (18). Another restriction of cleavage activity is the PFS motif, initially determined via bacterial screening. For LshCas13a, the PFS is 3′ H (i.e., not G), allowing most sites to be targeted given the short nature of the PFS (8). Upon structural analysis, the PFS seems to be dictated by the base identity of the DR sequence, as complementary target sequences will bind to the DR, affecting the binding of the Cas13 protein to the crRNA and preventing HEPN activation (21, 27, 28). In the case of Lsh-Cas13a, the first base of the DR flanking the guide is a cytosine, meaning that any targets with a 3′ guanine will bind the DR and prevent Cas13 activation. While the PFS preference seems to strongly affect activity biochemically and in bacteria, the observed PFS has not been as strict in mammalian cells with Cas13a, Cas13b, and Cas13d (11, 19, 29).

Investigation of the mechanism of target recognition by Cas13 was accomplished via introduction of mismatches. The sensitivity of Cas13 to mismatches in the target mirrors that of other class 2 enzymes. Cas13 recognition of the target sequence involves binding of the seed region of the guide to a complementary stretch in the RNA target. In comparison to other class 2 effectors with seed regions at 5′ or 3′ areas of the guide, the Cas13a and Cas13b seed regions tend to be more centered around base positions 5 to 21 from the 5′ end of the guide (8, 19, 30). The effect of mismatches disrupting cleavage in the target region helps to confirm the presence of a seed region in the crRNA as well as the exact location. Any double mismatches in the seed region completely inhibit Cas13a activation and target cleavage (8). Single mutations in the seed region can reduce activity but do not completely inhibit Cas13 activation. In addition, mutations outside of the seed regions can permit Cas13 binding while abrogating nuclease activation, revealing a decoupling of these two activities (8). Interestingly, certain mismatches between the guide and the target have been demonstrated to increase nuclease activity relative to that of wild-type sequences, at least for specific targets *in vitro* (30). Overall, the inherent specificity of Cas13 can be leveraged for specific applications for nucleic acid detection and in cells, as is discussed later.

Regarding structure, the three Cas13 subfamilies, Cas13a (Fig. 7.3A), Cas13b (Fig. 7.3B), and Cas13d (Fig. 7.3C), exhibit multiple similarities to other class 2 effectors, including a bilobed architecture consisting of recognition (REC) and nuclease (NUC) lobes (Fig. 7.3). The REC domain is responsible for binding of the crRNA, while the NUC contains the nuclease HEPN domains. In addition to the HEPN domains, all Cas13s contain two helical domains (Helical-1 and Helical-2) that, along with the HEPN domains, are predominantly alpha-helical.

Cas13a (Fig. 7.3A), the first structurally characterized Cas13, exemplifies the common properties of the Cas13 family, including (i) binding to the crRNA, (ii) formation of a HEPN RNase active site upon target recognition, and (iii) a second type of RNase activity for processing pre-crRNA transcripts (27, 28). In Cas13a complexes, the crRNA is recognized by the DR handle, which is bound by the REC lobe, a region consisting of both the Helical-1 domain and the N-terminal domain (NTD). The NUC lobe is formed by the HEPN domains and Helical-2. Upon binding, Helical-2 and HEPN1 undergo a pronounced structural rearrangement, allowing for the formation of an active site between HEPN1 and HEPN2 (27, 28). The active site requires a pair of basic residues (arginine and histidine) on each HEPN domain, and mutation of any of the four catalytic residues in either HEPN domain abolishes cleavage activity entirely (8, 16). Cas13 also processes pre-crRNA transcripts into active mature crRNAs (16), a capability shared with Cas12a. This processing initiates upon recognition of both the secondary structure and sequence of the DR in pre-crRNAs and, unlike the primary RNase activity of Cas13, is metal ion independent (27, 28). In Cas13a, the domain responsible for pre-crRNA cleavage varies between orthologs, with Helical-1 critical for processing in LshCas13a (27), HEPN2 in *Leptotrichia buccalis* Cas13a (LbuCas13a) (28), and both domains in *Lachnospiraceae* Cas13a (31). In all cases, basic residues are responsible for crRNA maturation.

Although Cas13b has the same three functions as Cas13a, there are several distinctions. In Cas13b, the 3′ region of the crRNA is bound by a DR-interacting domain, referred to as the Lid domain (32) or repeat region interaction (RRI) domain (33). These domains are also responsible for processing of the pre-crRNA array. Although there is no tertiary structure of Cas13b, HEPN1 is hypothesized to flexibly move to accept the target for crRNA binding and then rearrange to complete the catalytic pocket.

Cas13d is structurally similar to Cas13a, and it contains an NTD as part of the NUC lobe (34). Upon the formation of a crRNA:target duplex, the Helical-1 domain moves drastically, with less pronounced rearrangements of the Helical-2, HEPN1, and HEPN2 domains, resulting in the formation of the HEPN active site (34). In Cas13d, pre-crRNA processing is conferred by residues in the HEPN2 domain (35).

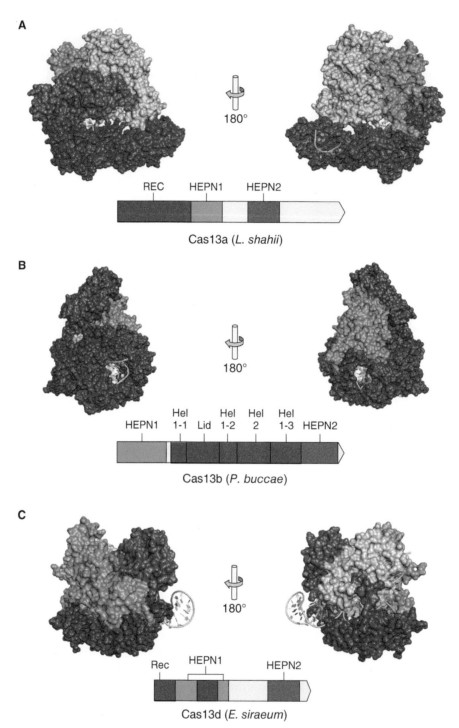

Figure 7.3 Comparison of Cas13 structures. Structures of Cas13a (**A**), Cas13b (**B**), and Cas13d (**C**) are depicted. The HEPN1 domain is shown in light blue, while HEPN2 is shown in dark blue. The REC lobes are shown in purple. Crystal structures are from the Protein Data Bank (*Leptotrichia shahii* Cas13a, PDB code 5WTK; *Prevotella buccae* Cas13b, PDB code 6DTD; *Eubacterium siraeum* Cas13d, PDB code 6E9E).

Mechanisms Underlying Nonspecific Collateral Activity

In early assays with Cas13, it was surprisingly observed that the cleavage profiles of guides at various positions across the target were roughly the same, with no variation in the size of produced fragments (8). Assuming only direct cleavage of the crRNA-bound RNA target, guides at different positions would be expected to produce unique cleavage profiles, as new cleavage sites would be accessible at different guide positions, especially as the secondary structure of the target would adjust upon Cas13 binding. Nonvarying cleavage profiles suggested that all RNA targets in the reaction were being cleaved in a defined manner regardless of the guide position. This observation became clearer when experiments were performed with labeled RNA targets with no complementarity to the guide: unexpectedly, these non-targeted nucleic acid targets were also cleaved as long as the complementary unlabeled target was present in solution (8). Importantly, these observations confirmed that Cas13a has a "collateral activity": binding of the Cas13a-crRNA to a complementary target triggers nonspecific, promiscuous cleavage of all RNAs (8, 16). This collateral activity dominates the cleavage activity in the reaction, generating stereotyped cleavage patterns visible by gel electrophoresis that are independent of the initial binding sequence (8). Indeed, this activity is the basis of stalled bacterial growth seen during active Cas13 interference in bacteria (8). Subsequent *in vitro* characterization has confirmed that many Cas13 family members possess collateral activity (8, 12, 13, 16, 18, 36).

The mechanism of the collateral cleavage activity can be understood by the exterior-facing nature of the HEPN active site. Because the HEPN domains are solvent exposed (Fig. 7.3), they not only are able to access extended regions of the target beyond the initial crRNA binding site but also can access nearby RNAs in solution (27, 28, 32, 34). Therefore, the physical separation of Cas13's crRNA binding site from the catalytic nuclease site leads to collateral cleavage of RNA triggered by target binding; this is in contrast to systems like Cas9, where the crRNA binding site is nestled within the nuclease domain, resulting in a single, internal cleavage event within the protospacer region. Confirming this proposed mechanism, inactivating mutations in the HEPN domains eliminate all nucleic acid cleavage, suggesting that specific and collateral cleavages are mediated by the same catalytic site (8, 16). Importantly, these catalytically inactive HEPN mutants retain target binding activity, enabling a host of applications for recruitment of effector domains to RNAs in cells, as is discussed in later sections.

Comparison of Cas13 to Type III CRISPR Systems and Other HEPN-Associated Systems

The RNA-targeting mechanism of type VI systems, including the reliance on HEPN domains for indiscriminate RNA cleavage, has clear parallels to the class 1 type III RNA-targeting CRISPR system. Type III systems, most of which rely on interference from a multiprotein complex including Cas7 (Csm3/5 or Cmr1/4/6) and Cas10 (Csm1 or Cmr2), have multiple mechanisms for both RNA and DNA targeting, especially through the HEPN-containing Csm6/Csx1 nucleases (37, 38). The specific, crRNA-programmed cleavage of the loaded type III effector complex occurs on both RNA and DNA. Upon guide base-pairing with a cognate RNA, Cas7 endoribonuclease activity is engaged, cleaving the target transcript at defined intervals inside the complex. In addition, targeting of a nascent transcript activates the DNA endonuclease activity of Cas10, mediated through the HD domain, to cleave the actively transcribed DNA template. Importantly, Cas10 activation also generates cyclic or linear oligoadenylate second messenger molecules through the GGDD motif of the Cas10 palm domain (39, 40). These signaling intermediates are substrates for the CARF domain of Csm6/Csx1, leading to activation of the HEPN dimer domain and indiscriminate cleavage of cellular RNA. Much like in the case of Cas13 nonspecific cleavage, Csm6/Csx1 activity aids in the antiphage response, promotes host growth arrest to prevent further replication of mobile elements, and increases tolerance and clearance of mutated transcripts (41, 42). More recently, the type III systems have been expanded to include multiple new subfamilies, including type III-D to -F (43). The type III-E family is particularly interesting because it has evolved to consolidate the multiple subunits of type III-A/B complexes into a single protein effector that has been experimentally verified to function as an RNA-guided RNase in prokaryotic (44, 45) and eukaryotic cells with no apparent collateral activity (44). Moreover, the Cas7-11 system exhibits effective RNA knockdown activity in mammalian cells without the collateral RNA cleavage or toxicity seen with Cas13 systems (44). Because the Cas7-11 protein associates with and is inhibited by its accessory protein Csx29 (a caspase-like peptidase also known as TPR-CHAT) (44, 45), an intriguing possibility is that Cas7-11 is involved in activation of a protease that can lead to general protein degradation in the cell as part of viral immunity.

The role of HEPN domains in abortive infection in both type III and VI systems is analogous with other, non-CRISPR-associated roles of HEPN in defense (10). The nondiscriminate RNase activity of HEPN domains has been characterized as a toxin in multiple roles, including the RnlAB system (46), minimal nucleotidyltransferase-HEPN system (47), and Abi abortive infection system (10). The abundance of HEPN RNase domains in various defense systems likely contributes to clustering of these domains in defense islands and the evolution of both type III and type VI systems (37).

Roles of Accessory Proteins in Cas13 Interference

While Cas13a, Cas13c, and Cas13d are associated with the adaptation proteins Cas1 and Cas2, Cas13b has no such association; instead, type VI-B systems possess either Csx27 or Csx28 proteins (12). These accessory proteins are predicted to contain one or more transmembrane domains, which may be involved in DNA uptake or degradation (48), and Csx28 also has a predicted divergent HEPN domain. When testing Cas13b interference activity against the single-stranded RNA (ssRNA) phage MS2, the expression of Csx27 was found to repress interference, while Csx28 expression enhanced interference (12). The differential regulation activities of Csx27 and Csx28 were further confirmed by essential gene screens in *E. coli*, where the RNA targeting activity of thousands of spacers could be modulated by these accessory proteins. Subsequent structural characterization of Csx28 (49) revealed that it exists as a membrane-localized octameric pore, which is activated by Cas13b viral RNA recognition, leading to metabolic suppression, cell dormancy, and the previously observed enhanced interference.

Cas13d similarly associates with a set of accessory proteins containing WYL domains (13). These WYL domain proteins also contain helix-turn-helix or ribbon-helix-helix DNA binding domains. WYL domain containing proteins are generally associated with microbial defense systems (50) and are also present in type I-D CRISPR systems, where they negatively regulate interference activity (51). In pooled guide screens in *E. coli*, Cas13d RNA cleavage activity was enhanced by WYL domain accessory proteins. Biochemical experimentation further confirmed that the WYL domain proteins could increase the RNA cleavage and collateral activity of Cas13d (13). Structural investigations into WYL1 revealed a preference to homodimerize through a C-terminal domain and strong ssRNA binding activity (52). Further exploration of these accessory proteins may reveal new roles in defense in type VI-B and VI-D systems.

Applications of Cas13

The unique RNA-programmed RNA-targeting characteristics of Cas13 enable multiple applications, both for measuring and manipulating RNA in cells (Fig. 7.4A) and for detecting nucleic acids sensitively *in vitro* (Fig. 7.4B). In cells, Cas13 has been applied to knockdown

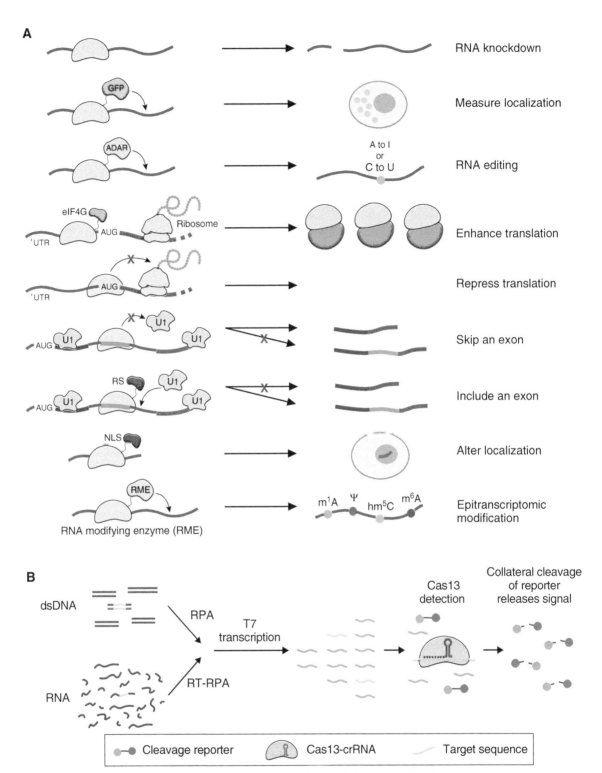

Figure 7.4 Applications of Cas13 in cells and for nucleic acid detection. (A) Applications of Cas13 in cells. Active Cas13 can be used for knockdown, while dCas13 recruitment of protein domains can be used for various applications, including RNA imaging, RNA editing, splicing, translation modulation, and epitranscriptomics. (B) Application of Cas13 for sensitive nucleic acid detection with nucleic acid preamplification and T7 transcription.

of RNA, modulation of splicing, imaging of RNA species in live cells, up- and downregulation of translational machinery, manipulation of RNA processing, and RNA editing.

Applications of Cas13 in Cells: Gene Knockdown

Cas13 knockdown was first demonstrated in bacteria with LshCas13a. The LshCas13a system was programmed with guide RNAs to target the RFP sequence, resulting in reduction of fluorescence and demonstrating the potential technological applications of Cas13 (8). As efforts to apply the system in eukaryotic cells expanded, LshCas13a was found to have weak activity in mammalian cells. Assaying multiple Cas13a orthologs showed that Cas13a from *Leptotrichia wadei* (LwaCas13a) had the most efficient knockdown across a panel of hundreds of guides tiled across a number of reporter and endogenous transcripts (29). Additionally, engineering LwaCas13a to have nuclear localization signals (NLSs) and a superfolder green fluorescent protein (GFP) tag provided the most knockdown activity (29). While some guide sequences could be found to have 80 to 90% knockdown activity, most guides achieved around 50% knockdown, and some guides had very little to no activity (29). Efforts to predict guide efficacy suggested that secondary structure and accessibility of the target might affect knockdown activity (29).

Later eukaryotic applications of Cas13b systems revealed an ortholog from *Prevotella pectinovora* strain P5-125 that achieved greater than 90% efficiency at most sites (19). Similarly, Cas13d from *Ruminococcus flavefaciens* also achieved 90 to 95% knockdown on all endogenous transcripts evaluated. Moreover, for Cas13a, Cas13b, and Cas13d, guide RNAs can be delivered within pre-crRNAs that can be cleaved into mature crRNAs by the crRNA processing domain of Cas13 (11, 19, 29). This functionality allows for delivery of multiple guides within an array for multiplexed knockdown of many transcripts at once. With Cas13a and Cas13d, it was shown that 3 or 4 transcripts could be targeted simultaneously by expressing a single pre-crRNA (11, 29).

Cas13 knockdown has several advantages over technologies such RNA interference (RNAi). Because the Cas13 system can be localized to the nucleus through fusion of NLSs, nuclear transcripts, like noncoding RNAs such as MALAT1, can be targeted for knockdown (29). Moreover, analysis of mRNA expression levels across the entire transcriptome revealed Cas13 knockdown to be substantially more specific than RNAi (29). While Cas13a, Cas13b, and Cas13d knockdown demonstrated very few to no off-target cleavage events, RNAi had hundreds to thousands of off-targets (11, 19, 29). As RNAi specificity can be a problem for screening, where off-targets can lead to false positives, and for therapeutics, where off-targets can be deleterious, the Cas13 system for knockdown is a much more attractive solution.

Given the observed collateral activity of Cas13 biochemically and in bacteria, it was surprising that Cas13 activity in mammalian cells was so specific. Collateral activity is inherently nonspecific and would lead to significant cleavage across the transcriptome, potentially leading to toxicity in cells. Attempts to measure collateral activity directly in HEK293FT cells detected no off-target cleavage by RNA sequencing or RNA integrity analysis, and there were no observed changes in cell growth (29). A subsequent study, performed in U87 glioma cells, detected cleavage of rRNA and reduction of glyceraldehyde-3-phosphate dehydrogenase transcript levels upon targeting of enhanced GFP (EGFP) reporter transcripts, confirming that collateral activity can exist in certain eukaryotic cell environments (53). Additional investigation into targeting of an epidermal growth factor receptor (EGFR) variant III (EGFRvIII) transgene demonstrated loss of rRNA integrity and suppression of cell growth (53). However, EGFRvIII knockdown itself can induce cell death in U87 cells (54), so it is difficult to separate the effect of knockdown from *bona fide* collateral action in this case. Furthermore, no changes in rRNA integrity were observed upon targeting of EGFP in 293T cells (53), consistent with previous findings that collateral activity could not be detected (29). More recent studies have also found collateral degradation in diverse cell types, including significant toxicity in mice, suggesting that cell type differences could account for the extent of collateral activity observed (55–57). The discrepancy between the reported collateral activities in different cell lines could be due to various reasons. One possibility is that the eukaryotic cytosol is densely crowded such that the activated Cas13 complexes unbind the target before they can bind to neighboring transcripts and initiate collateral cleavage. Another potential explanation is that transcripts are heavily structured or protected by RNA-binding proteins in mammalian cells, preventing significant RNA degradation. Subcellular structures and localization of condensates might also preclude access to nonspecific targets. Additionally, collateral activity may be uniform across all transcripts in a cell, causing general depletion of RNA in the cell and making it hard to detect specific collateral degradation of transcripts. This uniform collateral activity was indeed observed in HEK293T cells and could serve as another explanation for why collateral degradation is not easily observed (55). More experiments are needed to fully parse the reasons for the apparent absence of collateral activity in eukaryotic cells.

Cas13 knockdown has been applied towards developing cancer therapeutics by targeting essential genes in pancreatic cancer (58). As Cas9 can have permanent off-targets, using Cas13 offers a potentially safer approach for therapeutics without concern for permanent modifications in healthy noncancerous cells. In this study, Cas13, delivered as a ribonucleoprotein complex, was programmed to target essential oncogenic transcripts, such as KRAS, and could reduce levels of these transcripts enough to inhibit cancer growth *in vivo* and prolong mouse survival (58). Moreover, Cas13's inherent specificity could be leveraged for precise cancer cell ablation using guides against oncogenic mutations. By targeting only transcripts carrying oncogenic mutations, wild-type transcripts in healthy cells would be left intact, reducing toxicity (58). This approach could be useful in scenarios where disease-specific targeting is desirable and where off-target knockdown or genome editing in unintended cells could have dire consequences. Lastly, Cas13 has been applied for antiviral therapies (59–61), especially for knockdown of viral RNAs from viruses that have RNA genomes and no DNA intermediates in their life cycle (precluding treatment with DNA-targeting systems). Explorations of Cas13's antiviral activity have revealed potent activity against three distinct ssRNA viruses: lymphocytic choriomeningitis virus, influenza A virus, and vesicular stomatitis virus (59). Additionally, *in vivo* studies via mRNA delivery of Cas13 showed that both influenza virus and severe acute respiratory syndrome coronavirus 2 (SARS-CoV-2) infections could be controlled in the lungs in mouse models (61). Beyond therapeutics, Cas13d has been used for screening noncoding RNA function, such as circular RNAs, by leveraging the guide specificity to recognize back splicing junctions unique to circular RNAs (62). Moreover, Cas13 can be useful as a general RNA tool with applications for studying RNA function *in vivo* and in embryos, including medaka, killifish, and mouse embryos (63).

As many factors can affect guide knockdown, efforts have been dedicated towards developing design rules from large-scale Cas13 knockdown assays, akin to guide tools modeled off screening data for Cas9 and other CRISPR enzymes. One such approach screened 24,460 guide RNAs with and without mismatches to identify rules for efficient and specific knockdown and generated a design tool with high generalizability (64). Another approach screened 127,000 guide RNAs to develop a machine learning-based model for Cas13d (65).

These design tools should help with adoption of the Cas13 technology for knockdown as a scientific and therapeutic tool.

Applications of Cas13 in Cells: Splicing Modulation, RNA Imaging, Translational Regulation, and Epitranscriptomic Modifications

The programmable RNA-binding activity of Cas13 presents a rich foundation for developing molecular RNA tools beyond knockdown (Fig. 7.4A). Just as catalytic inactivation of the nuclease domains of Cas9 and fusion to effector domains has led to capabilities for gene activation, suppression, or base editing, catalytically inactivated Cas13 (dCas13) can be deployed for programmable binding, allowing targeted imaging, splicing, translational regulation, and RNA editing of specific transcripts. Initial demonstrations of dCas13 for binding established that the protein could be targeted to endogenous transcripts for pull-down and enrichment (29). This programmable binding approach extended to imaging of transcripts in live cells with dCas13-GFP fusions, although with multiple modifications: as mRNA imaging requires a mechanism to distinguish between bound and unbound dCas13-GFP, the signal-to-noise ratio was improved through two engineering approaches. First, unbound dCas13-GFP was sequestered in the nucleus via an NLS fusion and could only be translocated to the cytoplasm upon targeting of a cytoplasmic transcript, such as the *ACTB* transcript. Second, a self-repressing negative-feedback circuit was constructed by fusion to both a zinc finger and KRAB domain, leading to downregulation of dCas13a-GFP expression in the absence of binding. This negative-feedback construct, termed dCas13-NF (negative feedback), was able to image foci of *ACTB* in stress granules of live cells treated with perturbations such as sodium arsenite (61). dCas13d has also been used to image transcripts in live cells. By delivering fluorescently labeled guide RNA that targets the MS2 sequence to dCas13d-expressing cells, it was possible to image MS2 repeat-containing mRNA (66). Additional applications of the Cas13 system for imaging has allowed for single-cell RNA imaging in real time as well as dual-color imaging of two transcripts (67). Furthermore, this mRNA imaging can be combined with DNA imaging with dCas9, allowing for detection of mRNA accumulation at actively transcribing foci (67).

Programmable RNA targeting with catalytically inactive Cas13b and Cas13d has also been applied for additional perturbations of RNA processing, epitranscriptomics, and splicing. The RNA binding capacity of dCas13b has been co-opted for studying of the m^{6}A reader proteins YTHDF1 and YTHDF2, which recruit the translation and deadenylation machinery, respectively (68). Fusion of effector domains from these readers to dCas13b allowed targeting of their biological activities to reporter transcripts, enabling programmable translational activation or downregulation (68). Additional work fused a truncated METTL3 methyltransferase domain to nuclear dCas13b, or modified METTL3:METTL14 methyltransferase complexes to cytoplasmic dCas13b, to accomplish programmable m^{6}A incorporation on specific RNA transcripts (69). These tools allowed site-specific installation of methylation marks for modulating transcript abundance or alternative splicing, highlighting the potential for programmable epitranscriptomic tools. This framework enabled an exploration of the mechanism of these domains, as fusion of subfragments of YTHDF1 identified sequences sufficient to recruit the translational machinery. The dCas13b-YTHDF2 construct could also be targeted to endogenous genes with existing m^{6}A regulation, providing an alternative mechanism to knockdown activity (68). Further RNA processing tools have been engineered with dCas13b, including fusion of dCas13b to the mammalian polyadenylation factor Nudix hydrolase 21, thus generating a programmable polyadenylation and cleavage tool, *Postscriptr*, capable of directing early polyadenylation at an endogenous transcript, *SREBP1* (70). These applications showcase the value of RNA tool development both for RNA manipulation and to understand fundamental RNA biology.

Manipulation of splicing via dCas13 can be achieved by both the steric block of splicing sites mediating exon inclusion or exclusion or through the recruitment of factors that normally mediate exclusion or inclusion. Targeting of dCas13d to the splice acceptor of an exon was able to significantly bias splicing towards targeted exon exclusion in a fluorescent protein reporter system, and this effect could be somewhat enhanced via additional recruitment of the C-terminal domain of the hnRNPa1 splicing factor (65). Targeting of other sites, such as the branch point, intra-exon region, or splice donor site, was less effective, although multiplexed targeting of all four sites demonstrated the highest level of exon exclusion. This approach was able to modulate splicing of the *MAPT* gene in neurons in a model of frontotemporal dementia with Parkinsonism linked to chromosome 17 (66). Exon inclusion can also be manipulated by adding an RNA aptamer to the guide RNA and fusing the corresponding aptamer binding protein to splicing factors such as RBFOX1 and RBM38. dCas13d targeting then recruits these factors and modulates exon inclusion; this approach has been demonstrated on reporter assays (71). Further exploration of splicing machinery via recruitment with dCas13 will provide a rich toolkit for understanding splicing biology and developing novel therapeutics.

Applications of Cas13 in Cells: RNA Base Editing

Applications of Cas9 and Cas12 for genome editing have been transformative for understanding cellular biology, mechanisms of disease, and gene therapy. While homology-directed repair can be used for gene editing, precise base editing in dividing or nondividing cells can be achieved with catalytically inactive Cas9 fused to deaminases, such as cytidine deaminases (e.g., APOBEC) (72) and evolved adenine deaminases (e.g., TadA) (73). Analogous to DNA base editing, RNA base editing can be performed with catalytically inactive Cas13 fused to adenosine deaminase acting on RNA 2 (ADAR2) for programmable and precise RNA editing (19, 74).

Existing, non-CRISPR-based programmable RNA editing tools rely on RNA base-pairing interactions between a guide RNA and target RNA (75–80), as opposed to protein-mediated RNA binding. The guides are modified to have hairpins that can recruit either entire ADAR2 or ADAR1 proteins or just the ADAR1/2 deaminase domains, which catalytically deaminate adenosines to inosines (functionally interpreted as guanines in cells). Although editing can be detected with these approaches, RNA:RNA hybridization is not very efficient and may lead to nonspecific interactions. Cas13-based target recognition can be more effective, as the Cas13 protein facilitates specific target recognition, tolerating only a single mismatch between the guide and target. To enable programmable editing with Cas13, the catalytically inactive version of Cas13b from *Prevotella* was fused to the deaminase domain of ADAR2, generating the RNA editing for programmable A to I replacement (REPAIR) system (75, 76). This technology was capable of efficient editing of both reporter and endogenous transcripts with high precision. Unlike DNA base editors, which can edit multiple bases within a given window, the desired RNA base to edit can be dictated via a programmed mismatch in the Cas13 guide sequence, as ADAR2 prefers to deaminate within a mismatch bubble in an RNA duplex. While an exact base to deaminate can be programmed with high efficiency, REPAIR treatment resulted in thousands of off-targets elsewhere in the transcriptome, due to overexpression of the ADAR2 deaminase domain and subsequent deamination of natural substrates. To address this issue, a more specific version of the REPAIR system was created by mutating the ADAR2 enzyme such that the catalytic activity was reduced, disfavoring deamination of most off-targets. REPAIR was demonstrated to edit more than 40 disease-relevant mutations, and truncated versions could fit in adeno-associated vectors (AAV), which would enable *in vivo* RNA base editing (75). Subsequent studies have extended the RNA editing toolbox with additional smaller orthologs (81, 82), demonstrating AAV delivery (81).

While A-to-G conversion is useful for many diseases, C-to-U conversion would allow for additional disease mutations to be addressed, as well as for more posttranslational modification sites to be modulated, including serines and tyrosines. While APOBEC enzymes are capable of C-to-U editing on RNA, they usually operate on single-stranded substrates and cannot easily be directed to a specific base. Moreover, there are no known enzymes that deaminate cytidines within dsRNA mismatch bubbles in dsRNA. To enable programmable Cas13 editing of cytidines to uridines, the ADAR2 deaminase domain was engineered via directed evolution to adjust the catalytic pocket to fit a cytidine and allow for cytidine deamination (74). After 16 rounds of evolution, the Cas13-cytidine deaminase fusion, termed RNA editing for specific C to U exchange (RESCUE), was capable of C-to-U editing rates up to 70% on reporter transcripts and 40% on endogenous transcripts, a >1,000-fold improvement in cytidine deamination activity (74). RESCUE was shown to be useful for activating cell growth via modulation of phosphorylation sites on beta-catenin, and a high-specificity version (RESCUE-S) was generated with very few off-targets. RESCUE has also been extended to smaller Cas13 ortholog families (81, 82), facilitating AAV packaging and delivery.

While RESCUE and REPAIR have been successfully deployed for RNA editing *in vitro*, their activity has yet to be tested *in vivo*. Truncated forms of these systems will allow for packaging in AAV, and future work will show whether these systems can achieve high enough efficiency for modulating disease phenotypes. Many disease-related alleles are potential targets for REPAIR, including Rett syndrome, which sometimes has common pretermination mutations that can be corrected through A-to-G conversions. Similarly, for RESCUE, many diseases can be addressed, such as Niemann-Pick disease. Furthermore, RNA editing can be applied to the transient modulation of protective variants that are not necessarily driving disease. For many protective variants, not all potential biological effects of introducing the variant are thoroughly studied, so permanent introduction may be undesirable. As an example, the protective ApoE2 variant for Alzheimer's disease could be generated from the high-risk variant ApoE4 using RESCUE; if for any reason the variant must be reverted, the RNA editing systems can be withdrawn. REPAIR and RESCUE can also be useful for affecting other biological effects reversibly, such as modulating growth in the context of regeneration (e.g., after acute liver failure) or for temporary modulation of the immune system or pain receptors. Reversible editing of RNA has numerous applications for treating diseases and improving health and serves as a complementary toolset to genome editing with DNA targeting CRISPR tools.

Applications of Cas13 for Nucleic Acid Detection

While most applications of Cas13 rely on either on-target RNA cleavage or programmable binding with inactivated Cas13, the collateral activity can be very useful for sensing nucleic acids outside of cells for molecular diagnostics in health care, agriculture, and environmental applications (Fig. 7.4B) (83, 84). Target detection with Cas13 was initially performed by fluorescently labeling a nontargeted nucleic acid and visualizing its degradation by gel electrophoresis in the presence of a targeted nucleic acid (8). By adding a quencher to the sensor nucleic acid, fluorescence could be monitored using fluorimeters by the collateral-mediated liberation of fluorophores (79). These sensors enabled Cas13 detection with sensitivity in the picomolar to mid-femtomolar range with LwaCas13a or Lbu-Cas13a. While these levels of detection are useful for certain applications, they are limited to only sensing millions of molecules per microliter, which is not sensitive enough for many clinical applications that require limits of detection as low as 1 molecule per microliter.

The sensitivity of the collateral activity can be significantly improved through the addition of a preamplification step (36), most commonly an isothermal amplification, such as recombinase polymerase amplification (RPA) (85). RPA enables exponential amplification of very few nucleic acid molecules at 37°C, removing the need for thermocycling. By using primers with T7 promoter handles, the amplified nucleic acid from RPA can be transcribed and sensed by Cas13. This technology, called specific high-sensitivity enzymatic reporter unlocking (SHERLOCK), is capable of single-molecule detection (~2 aM with 1-μl sample volume) and is highly specific, enabling single-nucleotide distinction of mutations involved in cancer or human genotyping (80, 81). While SHERLOCK is modular (83), allowing any preamplification technology to be used in the first step, in practice RPA and loop-mediated isothermal amplification work the best (86, 87). In the case of RPA, all reaction steps are compatible, allowing for single-step detection, and versions of the technology can be read out by lateral flow strips or colorimetrically. Second-generation developments of SHER-LOCK (18), and other technologies created later, incorporated detection of amplified nucleic acid by Cas12 collateral without T7 transcription (88). Combining Cas12 detection along with other Cas13 orthologs with orthogonal base preferences enabled up to four targets to be simultaneously detected in separate fluorescent channels (18). Higher-order multiplexing with Cas13 was accomplished via a scalable platform involving nanoliter droplets that contain either specific crRNAs or sample reagents and can be crossed for screening through many target-sample pairings (89). This platform was able to differentiate all 169 human-associated viruses using 4,500 crRNA-target pairs in a single assay (89). Further mechanistic and structural understanding of how Cas13 recognizes its target in the HEPN domain will shed light on the distinct RNA base preferences of different orthologs, allowing rational engineering of more orthogonal base preferences for higher-order multiplexing in nucleic acid detection reactions.

The SHERLOCK technology has been adapted for detection from crude samples, including saliva, blood, serum, and urine, without nucleic acid purification, enabling rapid detection of Zika and dengue virus from patient samples (90). Other SHERLOCK applications include detection of cancer mutations from cell-free DNA (82), bacterial genomic DNA (84–87), human papillomavirus DNA from cervical swab samples (88), and opportunistic infections, such as BK polyomavirus DNA and cytomegalovirus DNA, from blood and urine samples of patients experiencing acute kidney transplant rejection (91). Another attractive feature of SHERLOCK is that a single reaction costs only $0.60, enabling rapid and portable deployment of molecular diagnostics at low cost. For tracking epidemics in the field or other point-of-care applications, low cost and portability are key technology features of SHERLOCK. With the pandemic spread of SARS-CoV-2, many Cas13-based diagnostic assays were developed for viral detection for both lab-based and point-of-care settings (92–95), demonstrating the flexibility of SHERLOCK and CRISPR diagnostics to respond to emerging health care needs.

Beyond health care, SHERLOCK has also been deployed for rapid detection of plant genes (in under 20 min) by lateral flow from crushed soybean seeds without requiring sample purification (96). This rapid detection is useful for crop breeding, enabling easy tracking of engineered traits, and can also be useful for tracking diseases in a field and preventing widespread crop damage. CRISPR diagnostics can also be used to detect environmental DNA for tracking species in conservation efforts and ecological monitoring (97). SHERLOCK has significant applications in both health care and agriculture, and CRISPR diagnostics help to address limitations of current molecular diagnostic systems (83).

Future of Cas13

The type VI CRISPR systems are unique among the class 2 systems because of their RNA-guided RNA targeting interference activity. Cas13, discovered through searches for additional class 2 effectors beyond Cas9, stood out from other class 2 effectors because it lacked identifiable domains other than the HEPN domains. These domains, in combination with the nearby CRISPR array, suggested programmable RNase activity. Extensive biochemical and functional genomic characterization has revealed that Cas13 is capable of robust phage defense in bacteria against RNA phages and DNA phages. Direct RNA targeting is only part of the mechanism, however, as Cas13 utilizes a promiscuous cleavage activity, termed the collateral effect, to indiscriminately cleave host and phage RNAs upon target recognition. This cleavage leads to growth inhibition, allowing the Cas13 time to clear phage transcripts, and adds an additional benefit of being able to clear mutated phage transcripts as well, preventing phage escape. If phage replication is not overcome, the collateral effect eventually results in host death as part of a programmed cell death pathway, preventing spread of the phage. The dual nature of Cas13 interference, including target degradation and general collateral activity, allows for robust defense against any foreign RNA threats. While the mechanism of Cas13 is largely elucidated, some questions about the natural targets of Cas13 systems remain, as most spacers do not align to any known targets. This is perhaps a consequence of there not being many RNA phages known or perhaps an issue of poor sampling of the natural viruses of Cas13-containing hosts.

The programmable RNA targeting activity of Cas13 has been harnessed for numerous cellular applications, including RNA knockdown, RNA editing, RNA imaging, RNA splicing, and RNA stability via poly(A) modifications. Moreover, many applications utilizing the collateral effect have been realized for nucleic acid detection *in vitro*. As many diseases require better diagnostics that are cheaper, more sensitive, portable, and easy to use, CRISPR diagnostics can offer a significant advantage over other molecular diagnostic technologies. Overall, Cas13 is an excellent example of evolution producing a unique enzyme for host defense in the viral arms race and how the unexpected properties of the system can broaden our understanding of bacterial gene systems but also contribute to a myriad tools and technologies that can offer great benefit to human society.

Acknowledgments

We thank former colleagues of the Feng Zhang lab and our mentors for helpful discussions and inspiration. We thank the McGovern Institute for supporting and funding our respective labs as McGovern Fellows. We also thank Sabbi Lall for helpful comments and editing of the review and Ian Slaymaker for feedback on structural aspects of Cas13.

O.O.A. and J.S.G. are coinventors on a number of issued patents and patent applications relating to CRISPR-based technologies. O.O.A. and J.S.G. are cofounders of Sherlock Biosciences, Proof Diagnostics, Tome Biosciences, and Moment Biosciences. O.O.A. and J.S.G. were also advisors for Beam Therapeutics.

References

1. Hsu PD, Lander ES, Zhang F. 2014. Development and applications of CRISPR-Cas9 for genome engineering. *Cell* **157**:1262–1278.

2. Doudna JA, Charpentier E. 2014. Genome editing. The new frontier of genome engineering with CRISPR-Cas9. *Science* **346**:1258096.

3. Koonin EV, Makarova KS. 2017. Mobile genetic elements and evolution of CRISPR-Cas systems: all the way there and back. *Genome Biol Evol* **9**:2812–2825.

4. Cox DBT, Platt RJ, Zhang F. 2015. Therapeutic genome editing: prospects and challenges. *Nat Med* **21**:121–131.

5. Rees HA, Liu DR. 2018. Base editing: precision chemistry on the genome and transcriptome of living cells. *Nat Rev Genet* **19**:770–788.

6. Makarova KS, Haft DH, Barrangou R, Brouns SJJ, Charpentier E, Horvath P, Moineau S, Mojica FJM, Wolf YI, Yakunin AF, van der Oost J, Koonin EV. 2011. Evolution and classification of the CRISPR-Cas systems. *Nat Rev Microbiol* **9**:467–477.

7. Shmakov S, Abudayyeh OO, Makarova KS, Wolf YI, Gootenberg JS, Semenova E, Minakhin L, Joung J, Konermann S, Severinov K, Zhang F, Koonin EV. 2015. Discovery and functional characterization of diverse class 2 CRISPR-Cas systems. *Mol Cell* **60**:385–397.

8. Abudayyeh OO, Gootenberg JS, Konermann S, Joung J, Slaymaker IM, Cox DBT, Shmakov S, Makarova KS, Semenova E, Minakhin L, Severinov K, Regev A, Lander ES, Koonin EV, Zhang F. 2016. C2c2 is a single-component programmable RNA-guided RNA-targeting CRISPR effector. *Science* **353**:aaf5573.

9. Shmakov S, Smargon A, Scott D, Cox D, Pyzocha N, Yan W, Abudayyeh OO, Gootenberg JS, Makarova KS, Wolf YI, Severinov K, Zhang F, Koonin EV. 2017. Diversity and evolution of class 2 CRISPR-Cas systems. *Nat Rev Microbiol* **15**:169–182.

10. Anantharaman V, Makarova KS, Burroughs AM, Koonin EV, Aravind L. 2013. Comprehensive analysis of the HEPN superfamily: identification of novel roles in intra-genomic conflicts, defense, pathogenesis and RNA processing. *Biol Direct* **8**:15.

11. Konermann S, Lotfy P, Brideau NJ, Oki J, Shokhirev MN, Hsu PD. 2018. Transcriptome engineering with RNA-targeting type VI-D CRISPR effectors. *Cell* **173**:665–676.e14.

12. Smargon AA, Cox DBT, Pyzocha NK, Zheng K, Slaymaker IM, Gootenberg JS, Abudayyeh OA, Essletzbichler P, Shmakov S, Makarova KS, Koonin EV, Zhang F. 2017. Cas13b is a type VI-B CRISPR-associated RNA-guided RNase differentially regulated by accessory proteins Csx27 and Csx28. *Mol Cell* **65**:618–630.e7.

13. Yan WX, Chong S, Zhang H, Makarova KS, Koonin EV, Cheng DR, Scott DA. 2018. Cas13d is a compact RNA-targeting type VI CRISPR effector positively modulated by a WYL-domain-containing accessory protein. *Mol Cell* **70**:327–339.e5.

14. Fonfara I, Richter H, Bratovič M, Le Rhun A, Charpentier E. 2016. The CRISPR-associated DNA-cleaving enzyme Cpf1 also processes precursor CRISPR RNA. *Nature* **532**:517–521.

15. Zetsche B, Heidenreich M, Mohanraju P, Fedorova I, Kneppers J, DeGennaro EM, Winblad N, Choudhury SR, Abudayyeh OO, Gootenberg JS, Wu WY, Scott DA, Severinov K, van der Oost J, Zhang F. 2017. Multiplex gene editing by CRISPR-Cpf1 using a single crRNA array. *Nat Biotechnol* **35**:31–34.

16. East-Seletsky A, O'Connell MR, Knight SC, Burstein D, Cate JHD, Tjian R, Doudna JA. 2016. Two distinct RNase activities of CRISPR-C2c2 enable guide-RNA processing and RNA detection. *Nature* **538**:270–273.

17. East-Seletsky A, O'Connell MR, Burstein D, Knott GJ, Doudna JA. 2017. RNA targeting by functionally orthogonal type VI-A CRISPR-Cas enzymes. *Mol Cell* **66**:373–383.e3.

18. Gootenberg JS, Abudayyeh OO, Kellner MJ, Joung J, Collins JJ, Zhang F. 2018. Multiplexed and portable nucleic acid detection platform with Cas13, Cas12a, and Csm6. *Science* **360**:439–444.

19. Cox DBT, Gootenberg JS, Abudayyeh OO, Franklin B, Kellner MJ, Joung J, Zhang F. 2017. RNA editing with CRISPR-Cas13. *Science* **358**:1019–1027.

20. Meeske AJ, Nakandakari-Higa S, Marraffini LA. 2019. Cas13-induced cellular dormancy prevents the rise of CRISPR-resistant bacteriophage. *Nature* **570**:241–245.

21. Meeske AJ, Marraffini LA. 2018. RNA guide complementarity prevents self-targeting in type VI CRISPR systems. *Mol Cell* **71**:791–801.e3.

22. Vlot M, Houkes J, Lochs SJA, Swarts DC, Zheng P, Kunne T, Mohanraju P, Anders C, Jinek M, van der Oost J, Dickman MJ, Brouns SJJ. 2018. Bacteriophage DNA glucosylation impairs target DNA binding by type I and II but not by type V CRISPR-Cas effector complexes. *Nucleic Acids Res* **46**:873–885.

23. Mendoza SD, Nieweglowska ES, Govindarajan S, Leon LM, Berry JD, Tiwari A, Chaikeeratisak V, Pogliano J, Agard DA, Bondy-Denomy J. 2020. A bacteriophage nucleus-like compartment shields DNA from CRISPR nucleases. *Nature* **577**:244–248.

24. Silas S, Mohr G, Sidote DJ, Markham LM, Sanchez-Amat A, Bhaya D, Lambowitz AM, Fire AZ. 2016. Direct CRISPR spacer acquisition from RNA by a natural reverse transcriptase-Cas1 fusion protein. *Science* **351**:aad4234.

25. Toro N, Mestre MR, Martínez-Abarca F, González-Delgado A. 2019. Recruitment of reverse transcriptase-Cas1 fusion proteins by type VI-A CRISPR-Cas systems. *Front Microbiol* **10**:2160.

26. Meeske AJ, Jia N, Cassel AK, Kozlova A, Liao J, Wiedmann M, Patel DJ, Marraffini LA. 2020. A phage-encoded anti-CRISPR enables complete evasion of type VI-A CRISPR-Cas immunity. *Science* **369**:54–59.

27. Liu L, Li X, Wang J, Wang M, Chen P, Yin M, Li J, Sheng G, Wang Y. 2017. Two distant catalytic sites are responsible for C2c2 RNase activities. *Cell* **168**:121–134.e12.

28. Liu L, Li X, Ma J, Li Z, You L, Wang J, Wang M, Zhang X, Wang Y. 2017. The molecular architecture for RNA-guided RNA cleavage by Cas13a. *Cell* **170**:714–726.e10.

29. Abudayyeh OO, Gootenberg JS, Essletzbichler P, Han S, Joung J, Belanto JJ, Verdine V, Cox DBT, Kellner MJ, Regev A, Lander ES, Voytas DF, Ting AY, Zhang F. 2017. RNA targeting with CRISPR-Cas13. *Nature* **550**:280–284.

30. Tambe A, East-Seletsky A, Knott GJ, Doudna JA, O'Connell MR. 2018. RNA binding and HEPN-nuclease activation are decoupled in CRISPR-Cas13a. *Cell Rep* **24**:1025–1036.

31. Knott GJ, East-Seletsky A, Cofsky JC, Holton JM, Charles E, O'Connell MR, Doudna JA. 2017. Guide-bound structures of an RNA-targeting A-cleaving CRISPR-Cas13a enzyme. *Nat Struct Mol Biol* **24**:825–833.

32. Slaymaker IM, Mesa P, Kellner MJ, Kannan S, Brignole E, Koob J, Feliciano PR, Stella S, Abudayyeh OO, Gootenberg JS, Strecker J, Montoya G, Zhang F. 2019. High-resolution structure of Cas13b and biochemical characterization of RNA targeting and cleavage. *Cell Rep* **26**:3741–3751.e5.

33. Zhang B, Ye W, Ye Y, Zhou H, Saeed AFUH, Chen J, Lin J, Perčulija V, Chen Q, Chen C-J, Chang M-X, Choudhary MI, Ouyang S. 2018. Structural insights into Cas13b-guided CRISPR RNA maturation and recognition. *Cell Res* **28**:1198–1201.

34. Zhang C, Konermann S, Brideau NJ, Lotfy P, Wu X, Novick SJ, Strutzenberg T, Griffin PR, Hsu PD, Lyumkis D. 2018. Structural basis for the RNA-guided ribonuclease activity of CRISPR-Cas13d. *Cell* **175**:212–223.e17.

35. Zhang B, Ye Y, Ye W, Perčulija V, Jiang H, Chen Y, Li Y, Chen J, Lin J, Wang S, Chen Q, Han Y-S, Ouyang S. 2019. Two HEPN domains dictate CRISPR RNA maturation and target cleavage in Cas13d. *Nat Commun* **10**:2544.

36. Gootenberg JS, Abudayyeh OO, Lee JW, Essletzbichler P, Dy AJ, Joung J, Verdine V, Donghia N, Daringer NM, Freije CA, Myhrvold C, Bhattacharyya RP, Livny J, Regev A, Koonin EV, Hung DT, Sabeti PC, Collins JJ, Zhang F. 2017. Nucleic acid detection with CRISPR-Cas13a/C2c2. *Science* **356**:438–442.

37. Koonin EV, Makarova KS. 2019. Origins and evolution of CRISPR-Cas systems. *Philos Trans R Soc Lond B Biol Sci* 374:20180087.

38. Tamulaitis G, Venclovas Č, Siksnys V. 2017. Type III CRISPR-Cas immunity: major differences brushed aside. *Trends Microbiol* 25:49–61.

39. Niewoehner O, Garcia-Doval C, Rostøl JT, Berk C, Schwede F, Bigler L, Hall J, Marraffini LA, Jinek M. 2017. Type III CRISPR-Cas systems produce cyclic oligoadenylate second messengers. *Nature* 548:543–548.

40. Kazlauskiene M, Kostiuk G, Venclovas Č, Tamulaitis G, Siksnys V. 2017. A cyclic oligonucleotide signaling pathway in type III CRISPR-Cas systems. *Science* 357:605–609.

41. Jiang W, Samai P, Marraffini LA. 2016. Degradation of phage transcripts by CRISPR-associated RNases enables type III CRISPR-Cas immunity. *Cell* 164:710–721.

42. Rostøl JT, Marraffini LA. 2019. Non-specific degradation of transcripts promotes plasmid clearance during type III-A CRISPR-Cas immunity. *Nat Microbiol* 4:656–662.

43. Makarova KS, Wolf YI, Iranzo J, Shmakov SA, Alkhnbashi OS, Brouns SJJ, Charpentier E, Cheng D, Haft DH, Horvath P, Moineau S, Mojica FJM, Scott D, Shah SA, Siksnys V, Terns MP, Venclovas Č, White MF, Yakunin AF, Yan W, Zhang F, Garrett RA, Backofen R, van der Oost J, Barrangou R, Koonin EV. 2020. Evolutionary classification of CRISPR-Cas systems: a burst of class 2 and derived variants. *Nat Rev Microbiol* 18:67–83.

44. Özcan A, Krajeski R, Ioannidi E, Lee B, Gardner A, Makarova KS, Koonin EV, Abudayyeh OO, Gootenberg JS. 2021. Programmable RNA targeting with the single-protein CRISPR effector Cas7-11. *Nature* 597:720–725.

45. van Beljouw SPB, Haagsma AC, Rodríguez-Molina A, van den Berg DF, Vink JNA, Brouns SJJ. 2021. The gRAMP CRISPR-Cas effector is an RNA endonuclease complexed with a caspase-like peptidase. *Science* 373:1349–1353.

46. Koga M, Otsuka Y, Lemire S, Yonesaki T. 2011. *Escherichia coli* rnlA and rnlB compose a novel toxin-antitoxin system. *Genetics* 187:123–130.

47. Yao J, Guo Y, Zeng Z, Liu X, Shi F, Wang X. 2015. Identification and characterization of a HEPN-MNT family type II toxin-antitoxin in *Shewanella oneidensis*. *Microb Biotechnol* 8:961–973.

48. Makarova KS, Gao L, Zhang F, Koonin EV. 2019. Unexpected connections between type VI-B CRISPR-Cas systems, bacterial natural competence, ubiquitin signaling network and DNA modification through a distinct family of membrane proteins. *FEMS Microbiol Lett* 366:fnz088.

49. VanderWal AR, Park J-U, Polevoda B, Kellogg EH, O'Connell MR. 2021. CRISPR-Csx28 forms a Cas13b-activated membrane pore required for robust CRISPR-Cas adaptive immunity. *bioRxiv*.

50. Makarova KS, Anantharaman V, Grishin NV, Koonin EV, Aravind L. 2014. CARF and WYL domains: ligand-binding regulators of prokaryotic defense systems. *Front Genet* 5:102.

51. Hein S, Scholz I, Voß B, Hess WR. 2013. Adaptation and modification of three CRISPR loci in two closely related cyanobacteria. *RNA Biol* 10:852–864.

52. Zhang H, Dong C, Li L, Wasney GA, Min J. 2019. Structural insights into the modulatory role of the accessory protein WYL1 in the type VI-D CRISPR-Cas system. *Nucleic Acids Res* 47:5420–5428.

53. Wang Q, Liu X, Zhou J, Yang C, Wang G, Tan Y, Wu Y, Zhang S, Yi K, Kang C. 2019. The CRISPR-Cas13a gene-editing system induces collateral cleavage of RNA in glioma cells. *Adv Sci (Weinh)* 6:1901299.

54. Yamoutpour F, Bodempudi V, Park SE, Pan W, Mauzy MJ, Kratzke RA, Dudek A, Potter DA, Woo RA, O'Rourke DM, Tindall DJ, Farassati F. 2008. Gene silencing for epidermal growth factor receptor variant III induces cell-specific cytotoxicity. *Mol Cancer Ther* 7:3586–3597.

55. Shi P, Murphy MR, Aparicio AO, Kesner JS, Fang Z, Chen Z, Trehan A, Wu X. 2021. RNA-guided cell targeting with CRISPR/RfxCas13d collateral activity in human cells. *bioRxiv*.

56. Tong H, Huang J, Xiao Q, He B, Dong X, Liu Y, Yang X, Han D, Wang Z, Ying W, Zhang R, Wei Y, Wang X, Xu C, Zhou Y, Li Y, Cai M, Wang Q, Xue M, Li G, Fang K, Zhang H, Yang H. 2021. High-fidelity Cas13 variants for targeted RNA degradation with minimal collateral effect. *bioRxiv*.

57. Li Y, Xu J, Guo X, Li Z, Cao L, Liu S, Guo Y, Wang G, Luo Y, Zhang Z, Wei X, Zhao Y, Liu T, Wang X, Xia H, Kuang M, Guo Q, Li J, Chen L, Wang Y, Li Q, Wang F, Liu Q, You F. 2022. Collateral cleavage of 28s rRNA by RfxCas13d causes death of mice. *bioRxiv*.

58. Zhao X, Liu L, Lang J, Cheng K, Wang Y, Li X, Shi J, Wang Y, Nie G. 2018. A CRISPR-Cas13a system for efficient and specific therapeutic targeting of mutant KRAS for pancreatic cancer treatment. *Cancer Lett* 431:171–181.

59. Freije CA, Myhrvold C, Boehm CK, Lin AE, Welch NL, Carter A, Metsky HC, Luo CY, Abudayyeh OO, Gootenberg JS, Yozwiak NL, Zhang F, Sabeti PC. 2019. Programmable inhibition and detection of RNA viruses using Cas13. *Mol Cell* 76:826–837.e11.

60. Abbott TR, Dhamdhere G, Liu Y, Lin X, Goudy L, Zeng L, Chemparathy A, Chmura S, Heaton NS, Debs R, Pande T, Endy D, La Russa MF, Lewis DB, Qi LS. 2020. Development of CRISPR as an antiviral strategy to combat SARS-CoV-2 and influenza. *Cell* 181:865–876.e12.

61. Blanchard EL, Vanover D, Bawage SS, Tiwari PM, Rotolo L, Beyersdorf J, Peck HE, Bruno NC, Hincapie R, Michel F, Murray J, Sadhwani H, Vanderheyden B, Finn MG, Brinton MA, Lafontaine ER, Hogan RJ, Zurla C, Santangelo PJ. 2021. Treatment of influenza and SARS-CoV-2 infections via mRNA-encoded Cas13a in rodents. *Nat Biotechnol* 39:717–726.

62. Li S, Li X, Xue W, Zhang L, Yang L-Z, Cao S-M, Lei Y-N, Liu C-X, Guo S-K, Shan L, Wu M, Tao X, Zhang J-L, Gao X, Zhang J, Wei J, Li J, Yang L, Chen L-L. 2021. Screening for functional circular RNAs using the CRISPR-Cas13 system. *Nat Methods* 18:51–59.

63. Kushawah G, Hernandez-Huertas L, Abugattas-Nuñez Del Prado J, Martinez-Morales JR, DeVore ML, Hassan H, Moreno-Sanchez I, Tomas-Gallardo L, Diaz-Moscoso A, Monges DE, Guelfo JR, Theune WC, Brannan EO, Wang W, Corbin TJ, Moran AM, Sánchez Alvarado A, Málaga-Trillo E, Takacs CM, Bazzini AA, Moreno-Mateos MA. 2020. CRISPR-Cas13d induces efficient mRNA knockdown in animal embryos. *Dev Cell* 54:805–817.e7.

64. Wessels H-H, Méndez-Mancilla A, Guo X, Legut M, Daniloski Z, Sanjana NE. 2020. Massively parallel Cas13 screens reveal principles for guide RNA design. *Nat Biotechnol* 38:722–727.

65. Wei J, Lotfy P, Faizi K, Kitano H, Hsu PD, Konermann S. 2021. Deep learning of Cas13 guide activity from high-throughput gene essentiality screening. *bioRxiv*.

66. Wang H, Nakamura M, Abbott TR, Zhao D, Luo K, Yu C, Nguyen CM, Lo A, Daley TP, La Russa M, Liu Y,

Qi LS. 2019. CRISPR-mediated live imaging of genome editing and transcription. *Science* 365:1301–1305.

67. Yang L-Z, Wang Y, Li S-Q, Yao R-W, Luan P-F, Wu H, Carmichael GG, Chen L-L. 2019. Dynamic imaging of RNA in living cells by CRISPR-Cas13 systems. *Mol Cell* 76:981–997.e7.

68. Rauch S, Dickinson BC. 2019. Targeted m6A reader proteins to study the epitranscriptome. *Methods Enzymol* 621:1–16.

69. Wilson C, Chen PJ, Miao Z, Liu DR. 2020. Programmable m6A modification of cellular RNAs with a Cas13-directed methyltransferase. *Nat Biotechnol* 38:1431–1440.

70. Anderson KM, Poosala P, Lindley SR, Anderson DM. 2019. Targeted cleavage and polyadenylation of RNA by CRISPR-Cas13. *bioRxiv*.

71. Du M, Jillette N, Zhu JJ, Li S, Cheng AW. 2020. CRISPR artificial splicing factors. *Nat Commun* 11:2973.

72. Komor AC, Kim YB, Packer MS, Zuris JA, Liu DR. 2016. Programmable editing of a target base in genomic DNA without double-stranded DNA cleavage. *Nature* 533:420–424.

73. Gaudelli NM, Komor AC, Rees HA, Packer MS, Badran AH, Bryson DI, Liu DR. 2017. Programmable base editing of A•T to G•C in genomic DNA without DNA cleavage. *Nature* 551:464–471.

74. Abudayyeh OO, Gootenberg JS, Franklin B, Koob J, Kellner MJ, Ladha A, Joung J, Kirchgatterer P, Cox DBT, Zhang F. 2019. A cytosine deaminase for programmable single-base RNA editing. *Science* 365:382–386.

75. Fukuda M, Umeno H, Nose K, Nishitarumizu A, Noguchi R, Nakagawa H. 2017. Construction of a guide-RNA for site-directed RNA mutagenesis utilising intracellular A-to-I RNA editing. *Sci Rep* 7:41478.

76. Katrekar D, Chen G, Meluzzi D, Ganesh A, Worlikar A, Shih Y-R, Varghese S, Mali P. 2019. In vivo RNA editing of point mutations via RNA-guided adenosine deaminases. *Nat Methods* 16:239–242.

77. Merkle T, Merz S, Reautschnig P, Blaha A, Li Q, Vogel P, Wettengel J, Li JB, Stafforst T. 2019. Precise RNA editing by recruiting endogenous ADARs with antisense oligonucleotides. *Nat Biotechnol* 37:133–138.

78. Montiel-González MF, Vallecillo-Viejo IC, Rosenthal JJC. 2016. An efficient system for selectively altering genetic information within mRNAs. *Nucleic Acids Res* 44:e157.

79. Vogel P, Moschref M, Li Q, Merkle T, Selvasaravanan KD, Li JB, Stafforst T. 2018. Efficient and precise editing of endogenous transcripts with SNAP-tagged ADARs. *Nat Methods* 15:535–538.

80. Wettengel J, Reautschnig P, Geisler S, Kahle PJ, Stafforst T. 2017. Harnessing human ADAR2 for RNA repair—recoding a PINK1 mutation rescues mitophagy. *Nucleic Acids Res* 45:2797–2808.

81. Kannan S, Altae-Tran H, Jin X, Madigan VJ, Oshiro R, Makarova KS, Koonin EV, Zhang F. 2021. Compact RNA editors with small Cas13 proteins. *Nat Biotechnol* 40:194–197.

82. Xu C, Zhou Y, Xiao Q, He B, Geng G, Wang Z, Cao B, Dong X, Bai W, Wang Y, Wang X, Zhou D, Yuan T, Huo X, Lai J, Yang H. 2021. Programmable RNA editing with compact CRISPR-Cas13 systems from uncultivated microbes. *Nat Methods* 18:499–506.

83. Kaminski MM, Abudayyeh OO, Gootenberg JS, Zhang F, Collins JJ. 2021. CRISPR-based diagnostics. *Nat Biomed Eng* 5:643–656.

84. Abudayyeh OO, Gootenberg JS. 2021. CRISPR diagnostics. *Science* 372:914–915.

85. Piepenburg O, Williams CH, Stemple DL, Armes NA. 2006. DNA detection using recombination proteins. *PLoS Biol* 4:e204.

86. Li S-Y, Cheng Q-X, Wang J-M, Li X-Y, Zhang Z-L, Gao S, Cao R-B, Zhao G-P, Wang J. 2018. CRISPR-Cas12a-assisted nucleic acid detection. *Cell Discov* 4:20.

87. Li L, Li S, Wu N, Wu J, Wang G, Zhao G, Wang J. 2019. HOLMESv2: a CRISPR-Cas12b-assisted platform for nucleic acid detection and DNA methylation quantitation. *ACS Synth Biol* 8:2228–2237.

88. Chen JS, Ma E, Harrington LB, Da Costa M, Tian X, Palefsky JM, Doudna JA. 2018. CRISPR-Cas12a target binding unleashes indiscriminate single-stranded DNase activity. *Science* 360:436–439.

89. Ackerman CM, Myhrvold C, Thakku SG, Freije CA, Metsky HC, Yang DK, Ye SH, Boehm CK, Kosoko-Thoroddsen TF, Kehe J, Nguyen TG, Carter A, Kulesa A, Barnes JR, Dugan VG, Hung DT, Blainey PC, Sabeti PC. 2020. Massively multiplexed nucleic acid detection with Cas13. *Nature* 582:277–282.

90. Myhrvold C, Freije CA, Gootenberg JS, Abudayyeh OO, Metsky HC, Durbin AF, Kellner MJ, Tan AL, Paul LM, Parham LA, Garcia KF, Barnes KG, Chak B, Mondini A, Nogueira ML, Isern S, Michael SF, Lorenzana I, Yozwiak NL, MacInnis BL, Bosch I, Gehrke L, Zhang F, Sabeti PC. 2018. Field-deployable viral diagnostics using CRISPR-Cas13. *Science* 360:444–448.

91. Kaminski MM, Alcantar MA, Lape IT, Greensmith R, Huske AC, Valeri JA, Marty FM, Klämbt V, Azzi J, Akalin E, Riella LV, Collins JJ. 2020. A CRISPR-based assay for the detection of opportunistic infections post-transplantation and for the monitoring of transplant rejection. *Nat Biomed Eng* 4:601–609.

92. Patchsung M, Jantarug K, Pattama A, Aphicho K, Suraritdechachai S, Meesawat P, Sappakhaw K, Leelahakorn N, Ruenkam T, Wongsatit T, Athipanyasilp N, Eiamthong B, Lakkanasirorat B, Phoodokmai T, Niljianskul N, Pakotiprapha D, Chanarat S, Homchan A, Tinikul R, Kamutira P, Phiwkaow K, Soithongcharoen S, Kantiwiriyawanitch C, Pongsupasa V, Trisrivirat D, Jaroensuk J, Wongnate T, Maenpuen S, Chaiyen P, Kamnerdnakta S, Swangsri J, Chuthapisith S, Sirivatanauksorn Y, Chaimayo C, Sutthent R, Kantakamalakul W, Joung J, Ladha A, Jin X, Gootenberg JS, Abudayyeh OO, Zhang F, Horthongkham N, Uttamapinant C. 2020. Clinical validation of a Cas13-based assay for the detection of SARS-CoV-2 RNA. *Nat Biomed Eng* 4:1140–1149.

93. Fozouni P, Son S, Díaz de León Derby M, Knott GJ, Gray CN, D'Ambrosio MV, Zhao C, Switz NA, Kumar GR, Stephens SI, Boehm D, Tsou C-L, Shu J, Bhuiya A, Armstrong M, Harris AR, Chen P-Y, Osterloh JM, Meyer-Franke A, Joehnk B, Walcott K, Sil A, Langelier C, Pollard KS, Crawford ED, Puschnik AS, Phelps M, Kistler A, DeRisi JL, Doudna JA, Fletcher DA, Ott M. 2021. Amplification-free detection of SARS-CoV-2 with CRISPR-Cas13a and mobile phone microscopy. *Cell* 184:323–333.e9.

94. Arizti-Sanz J, Freije CA, Stanton AC, Petros BA, Boehm CK, Siddiqui S, Shaw BM, Adams G, Kosoko-Thoroddsen TF, Kemball ME, Uwanibe JN, Ajogbasile FV, Eromon PE, Gross R, Wronka L, Caviness K, Hensley LE, Bergman NH, MacInnis BL, Happi CT, Lemieux JE, Sabeti PC, Myhrvold C. 2020. Streamlined inactivation, amplification, and Cas13-based detection of SARS-CoV-2. *Nat Commun* 11:5921.

95. Liu TY, Knott GJ, Smock DCJ, Desmarais JJ, Son S, Bhuiya A, Jakhanwal S, Prywes N, Agrawal S, Díaz de

León Derby M, Switz NA, Armstrong M, Harris AR, Charles EJ, Thornton BW, Fozouni P, Shu J, Stephens SI, Kumar GR, Zhao C, Mok A, Iavarone AT, Escajeda AM, McIntosh R, Kim S, Dugan EJ, Pollard KS, Tan MX, Ott M, Fletcher DA, Lareau LF, Hsu PD, Savage DF, Doudna JA, IGI Testing Consortium. 2021. Accelerated RNA detection using tandem CRISPR nucleases. *Nat Chem Biol* 17:982–988.

96. Abudayyeh OO, Gootenberg JS, Kellner MJ, Zhang F. 2019. Nucleic acid detection of plant genes using CRISPR-Cas13. *CRISPR J* 2:165–171.

97. Baerwald MR, Goodbla AM, Nagarajan RP, Gootenberg JS, Abudayyeh OO, Zhang F, Schreier AD. 2020. Rapid and accurate species identification for ecological studies and monitoring using CRISPR-based SHERLOCK. *Mol Ecol Resour* 20:961–970.

SECTION II

CRISPR-Cas Biology

CRISPR-Cas, Horizontal Gene Transfer, and the Flexible (Pan) Genome

Uri Gophna

The Shmunis School of Biomedicine and Cancer Research, Tel Aviv University, Tel Aviv, Israel, 69978

Introduction

Most known CRISPR-Cas systems primarily provide acquired, heritable immunity to bacteria and archaea against invasion by selfish DNA elements. Consequently, the main activity of these systems is to destroy incoming DNA that matches their immune memory—spacers within the CRISPR array. However, not all incoming DNA is as purely selfish as lytic viruses (the term viruses will often be used to refer more broadly to both bacteriophages and archaeal viruses). Indeed, plasmids and temperate viruses, while selfish, can often provide substantial benefits for their respective hosts, which under specific conditions outweigh the costs of their maintenance. Since CRISPR immunity against such elements is often acquired, and since CRISPR-Cas systems can then efficiently destroy these elements, having CRISPR-Cas systems can compromise the potential of bacteria and archaea for genomic innovation. In the following chapter, the evidence for and against such a trade-off (1) between CRISPR-Cas and the ability to maintain high genomic diversity (and the benefits that come with it [Fig. 8.1]) via horizontal gene transfer (HGT) will be discussed.

Proviruses/Prophages

Most viruses that infect bacteria and archaea and are capable of a temperate (nonlytic) lifestyle can integrate into the genomes of their hosts. Integrated proviruses represent an important part of microbial pangenomes, can provide a benefit to the host under a variety of stresses and environmental conditions (2, 3), and can contribute to bacterial virulence (4, 5). CRISPR-Cas systems have been shown to prevent provirus integration, also known as lysogenization (6), and could therefore hinder the acquisition of potentially beneficial proviruses and provirus-encoded functions. Indeed, CRISPR-negative *Streptococcus pyogenes* strains have been shown to be more prophage rich, and there was an inverse association between the number of spacers and the number of prophages in strains of that species (7). If CRISPR-Cas systems truly represent a barrier for provirus integration, one would hypothesize that CRISPR-negative strains of a species would have more integrated proviruses than those that are CRISPR positive. This would be especially notable in species in which most variation in genome size across strains is

CRISPR: Biology and Applications, First Edition. Edited by Rodolphe Barrangou, Erik J. Sontheimer, and Luciano A. Marraffini.
© 2022 American Society for Microbiology. DOI: 10.1128/9781683673798.ch08

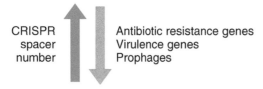

Figure 8.1 A "CRISPR trade-off." Continuously acquiring CRISPR spacers can have major short-term consequences for strain evolution, specifically a lower capacity to acquire virulence and antibiotic resistance genes.

due to provirus integration rather than the presence of plasmids (8). One would therefore expect the pattern observed in *S. pyogenes* to be observed across multiple species. However, a comprehensive analysis of 2,246 prophages in complete bacterial genomes showed unexpectedly complicated associations between lysogeny and CRISPR content (9). On the one hand, lysogens (bacteria that carry lysogenic prophages in their genomes) were more likely to encode CRISPR-Cas systems than nonlysogens, although the number of prophages in the genomes of the lysogens was not associated with the presence or absence of CRISPR-Cas systems. In contrast, in the CRISPR-positive genomes, the number of spacers (which corresponds to the system's activity in terms of acquiring and maintaining immune memory) was moderately but significantly negatively associated with the number of prophages in the genome. Lysogens also had fewer spacers than nonlysogens (Table 8.1), again indicating that a large immune repertoire and a large prophage repertoire are incompatible.

It may at first appear to be contradictory that lysogens are more likely to have CRISPR-Cas systems but have smaller immune memories once those CRISPR-Cas systems are established. However, there are ecological as well as CRISPR-related explanations for observations such as these. First, there are many bacteria (such as intracellular symbionts) that have low phage exposure, and those typically have neither prophages nor CRISPR-Cas. Additionally, many bacteria have small and static CRISPR arrays, some of which have not acquired new spacers for thousands of years (10). These systems may now play alternative roles that promote conservation in most or all strains of the species, yet they no longer accumulate new spacers or limit the integration of new prophages into the genome, as has been predicted for *Escherichia coli* (11).

Another intriguing observation was that different CRISPR-Cas system types were associated with different prophage contents: genomes with type I CRISPR-Cas systems had the highest mean number of prophages, while those encoding type III systems were less likely to have any lysogenic phages, and when they did, carried fewer prophages (9). The latter finding is counterintuitive, since type III systems require transcription of the target gene and therefore allow lysogeny of CRISPR-targeted phages (12) and so should, in principle, be more phage tolerant. To resolve this apparent contradiction, the authors suggested that phages that infect type III-encoding bacteria have "evolved away" from lysogeny, rather than being continuously enslaved by the host, serving as a beneficial gene reservoir while

Table 8.1 CRISPR spacer-related properties of genomes of lysogens versus nonlysogens

Lysogens	Nonlysogens (no prophages)
More likely to have CRISPR-Cas systems	Less likely to have CRISPR-Cas systems
Fewer spacers	More spacers
Spacer number is negatively associated with prophage number	NA[a]
Type I systems, more prophages; type III systems, fewer prophages	NA

[a]NA, not applicable.

being unable to multiply and transfer horizontally between hosts. Alternatively, the lack of prophages in genomes with type III systems may reflect the fact that these systems tend to co-occur with additional CRISPR-Cas systems in the same genome (only 53/177 type III systems in that study were the exclusive CRISPR-Cas systems in their respective genomes), and their genomes therefore may be "overprotected."

In agreement with the moderate effects of CRISPR-Cas on lysogeny previously observed (10), our work has shown little dependence between CRISPR-Cas activity (number of spacers) and recently horizontally acquired genes, many of which are prophage borne (8, 9), in the genomes of bacteria and archaea (13). In addition to the ability of type III systems to maintain lysogens, as long as they do not progress to the lytic phase (see above [12]), there is a key viral factor that may explain why CRISPR-Cas systems do not keep the borders of the genome tightly closed: anti-CRISPR (Acr) proteins (14). These virus-encoded proteins usually act by binding to components of the interference machinery, thereby preventing the degradation of nucleic acids, protecting the virus. When such an Acr is constitutively expressed by an integrated prophage, it will protect not just the prophage that encodes it but also all DNA elements that may be targeted by the spacers of the inhibited CRISPR-Cas system. Acr proteins have also been identified in integrated elements other than prophages (15, 16). Importantly, anti-CRISPR activity does not disable the system completely (17), and since the levels of the interference complex are regulated (18–20) and sometimes induced upon infection (21), the system can still function as an antiviral defense (albeit an attenuated one) even in the presence of anti-CRISPR genes in the host genome. Thus, a CRISPR-*cas* locus can be maintained in a genome encoding one or more anti-CRISPR proteins and provide a benefit not just to the host microbe but also to all its integrated prophages by protecting from coinfection with competing phages and host death. Since diverse Acr proteins have been shown to inhibit multiple different systems using distinct mechanisms (22–26), and many still await discovery, it is reasonable to assume that many genomes that have CRISPR-Cas and multiple prophages will also encode Acr proteins.

In addition to their contribution to the genome as integrated elements, phages can also enhance gene and allele exchange across bacteria by performing generalized transduction. Generalized transduction occurs when phages that have mistakenly packaged plasmid or chromosomal DNA of their host and are therefore defective, known as transducing particles, deliver DNA into a new host. Such transducing particles represent a small minority of the otherwise fully infective phage particle population but may constitute a very important form of HGT. Generalized transduction will be successful if the transferred DNA is self-replicating (such as a plasmid) or similar enough to the chromosomal DNA of the host to be successfully integrated via homologous recombination. It can also rely on site-specific recombination, as in the case of integrative conjugative elements (ICEs). These elements, formerly known as conjugative transposons, typically encode both recombinases for excision from and later integration into a host replicon and conjugation machinery that facilitates their horizontal transmission (27). Thus, if the transduced DNA is an ICE, it can successfully integrate into the new host genome via site-specific recombination. Work with the plant-pathogenic bacterium *Pectobacterium atrosepticum* has shown that its native type I-F CRISPR-Cas system successfully inhibited generalized transduction of chromosomal, plasmid, and ICE DNA (28). However, in a more natural scenario in which the transducing particles were a small fraction of a large lytic phage population, the transduction efficiency of bacteria that had a spacer targeting the phage DNA was actually increased. This is because a larger fraction of transduced cells survived when they were immune to the lytic phage (28). Thus, CRISPR-Cas systems can have a positive effect on HGT that is mediated by generalized transduction.

Natural Transformation

The process of natural transformation (also known as competence), the uptake of naked DNA from the environment, has been known to be common in many bacterial lineages, including both Gram-negative (e.g., *Acinetobacter* [29], *Haemophilus* [30], and *Neisseria* [31]) and Gram-positive (e.g., *Streptococcus* [32] and *Bacillus* [33]) bacteria. Unlike other forms of HGT that are mediated by selfish mobile DNA entities, natural transformation primarily involves uptake of chromosomal genome fragments, although naturally some of that DNA may also encode transposable elements. Nevertheless, CRISPR-Cas has been shown to be a barrier for transformation in *Streptococcus pneumoniae* expressing the CRISPR-Cas system from *Streptococcus pyogenes* (1) and in *Neisseria meningitidis* (34). Notably, many spacers of *N. meningitidis* matched the genomes of other *Neisseria* species, often matching surface modification and other virulence-related genes that could have originated in integrated mobile elements. These matches indicate that it is very likely that DNA fragments taken up by natural transformation have become substrates for spacer acquisition and could thereafter limit gene flow between different species of *Neisseria* (34). Similar observations of spacers that match other strains or species of the same genus have also been made in another naturally competent human pathogen, *Porphyromonas gingivalis* (35, 36). An interesting observation is that strains of *Aggregatibacter actinomycetemcomitans* that have lost their capacity for natural transformation have also lost parts of their CRISPR-Cas systems (37). This could imply that these systems are more beneficial, and therefore more frequently retained, in lineages that are naturally transformable, perhaps because they protect the genome from overexposure to transposons and insertion sequences that could disrupt bacterial genes (35).

Antibiotic Resistance

Possibly the greatest pressure to increase genome innovation for bacteria is antibiotics. Emerging pathogens in the past decades have been shown time and time again to have acquired antibiotic resistance genes by HGT from other lineages (38–40). Most antibiotic resistance genes are encoded on mobile elements, and thus, CRISPR-Cas-positive strains may be less likely to acquire antibiotic resistance than their CRISPR-Cas-negative counterparts. The best known example for such an association is *Enterococcus faecalis*, in which the emerging multidrug-resistant lineages, especially those with vancomycin resistance, did not have either of the two CRISPR loci of that species; a similar pattern was observed for the single CRISPR-Cas locus in *Enterococcus faecium* genomes (41). The authors speculated that under the selective pressures of antibiotic treatment, the acquisition/protection balance may have shifted and that it is more beneficial for human pathogens to acquire new genetic material than to be protected from selfish elements. The same research group later showed that plasmid recipients (known as transconjugants) can be obtained transiently in the presence of active CRISPR defense and that if plasmid presence is selected for, spacer loss will occur. A combination of simulations and experiments with *Staphylococcus epidermidis* has shown the existence of a mutant subpopulation that has lost CRISPR-Cas activity by different mechanisms, from loss of spacers to inactivating mutations in *cas* genes, and that the frequency of such mutants can be as high as 10^{-4} (42). This mutant subpopulation enables the acquisition of a targeted plasmid by a substantial fraction of the population. Taken together, these studies imply that having CRISPR-Cas systems will not prevent the emergence of plasmid-encoded antibiotic resistance but rather that antibiotic pressure may select for lineages that either have lost specific spacers or have had their CRISPR immunity inactivated by various mechanisms. Many *cas* operons are actually flanked from both ends by their CRISPR arrays in bacteria (43) and archaea (44–46) and consequently can be lost due to break repair that involves recombination or annealing between distant repeats from both ends. Such events can be more common than point mutations in *cas* genes (46, 47). Complete or near-complete loss has been observed in many lineages whose ancestors are inferred to have had an intact system, including those that have a single CRISPR array, such as *Mycobacterium tuberculosis* (48). Loss of CRISPR-Cas can then potentially expose bacterial genomes to expansion via acquisition of novel mobile genetic elements such as plasmids or phages.

Mobility of CRISPR-Cas Systems

If CRISPR-Cas systems are so frequently lost (42), why are they still so often present in many or even most strains of CRISPR-positive species? The best explanation is that these systems, or parts of them (e.g., CRISPR arrays), can be regained via horizontal reacquisition. Indeed, CRISPR-Cas systems were shown in the lab to be easily mobilized by generalized transduction (28), and because some of them (especially class II systems that use a single protein for interference) are relatively small clusters, their transferability as a functional module is quite high. Packaging by viruses in a generalized transduction-related process could have contributed to the emergence of virus-encoded CRISPR-Cas systems (49). CRISPR-Cas systems are often also encoded by integrated mobile genetic elements (50) and plasmids (47, 51). Phylogenetic analysis has long supported frequent horizontal transmission of CRISPR-Cas systems (52, 53), which appears to be a recurring theme in microbial evolution.

CRISPR-Cas Systems as Pangenome Editors

The pangenome of a bacterial lineage is defined as the set of all genes present in the genomes of that lineage (54, 55). In genomes of certain bacteria, such as *Streptococcus agalactiae* (group B streptococcus), which can cause human newborn infections and bovine mastitis, the major contributors to genome diversity and the pangenome of the species are ICEs (56). Type II-A CRISPR-Cas systems are ubiquitous in *S. agalactiae* genomes, and a large and diverse collection of strains (351 strains) showed a broad variety of spacers, collectively amounting to 949 distinct sequences (57). As expected, these CRISPR loci contain multiple spacers that are derived from ICEs, and when present, they can drastically reduce successful invasion of the strain by the respective ICE via conjugation (57). Since self-targeting of the chromosome is toxic, a strain with an intact CRISPR-Cas system can have an ICE or a spacer that matches it, but usually not both (as long as an effective protospacer-adjacent motif is present). The authors concluded that in *S. agalactiae*, the reason that type II-A CRISPR-Cas systems are ubiquitous is that they are beneficial at the population level, since for every ICE or prophage, they allow carrier and noncarrier strains to coexist, preserving the diversity of mobile elements (sometimes known as the mobilome [58]) that those bacteria can carry (Fig. 8.2).

Along similar lines, the presence of CRISPR-Cas systems may help prevent an overaccumulation of integrated mobile elements, allowing the maintenance of an extremely large pangenome with a genome size that does not exceed a certain range. In the halophilic archaeal genus *Haloferax*, our lab has observed that different strains have highly variable genome contents yet a very narrow range of genome sizes (usually not exceeding 4 Mbp [including all plasmids]). All *Haloferax* strains that have had their genomes sequenced to

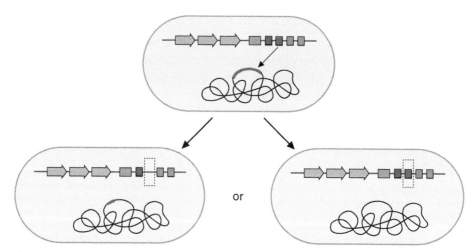

Figure 8.2 CRISPR-Cas systems as pangenome editors. By acquiring spacers that target existing proviruses and genomic islands, CRISPR-Cas systems modulate the pangenome and generate populations where strains have either a prophage- or island-targeting spacer or an island, but not both.

date contain CRISPR-Cas systems. Our lab has shown that those systems are preferentially acquired from integrated mobile elements during their replication (59), so it is likely that those systems help keep the *Haloferax* genome size in check and may also do likewise for other microbial species. The observation that most spacers that have database matches target integrated genetic elements rather than lytic viruses (60, 61) is also in agreement with the notion that CRISPR-Cas systems can serve as natural "pangenome" editors, so that no individual genome will amass too many auxiliary functions that are present in mobile elements (such as antibiotic resistance or virulence genes), yet at the population level many such functions will be retained. There should therefore be no true trade-off between CRISPR-Cas and genome diversity at the population level, as long as conditions are not taken to extremes, as in the specific case of exaggerated antibiotic selection pressure caused by human activity. Nonetheless, there could be lineages in which frequent ecological shifts may be able to naturally generate such extreme selective pressures. Indeed, a recent analysis of 1,871 *Bacillus cereus* genomes has shown that strains with intact CRISPR-Cas systems occupy fewer ecological niches and have fewer plasmids and prophages than those in which CRISPR-Cas systems are incomplete or absent (62).

Acknowledgments

This work was supported by European Research Council grant CRISPR-EVOl AdG 787514. U.G. was also supported by BSF-NSF grant 2016671 and ISF grant 535/15.

I thank Philippe Glaser for helpful discussions and Israela Turgeman-Grott, Leah Reshef, and Netta Shemesh for their constructive comments about this chapter.

References

1. Bikard D, Hatoum-Aslan A, Mucida D, Marraffini LA. 2012. CRISPR interference can prevent natural transformation and virulence acquisition during in vivo bacterial infection. *Cell Host Microbe* **12**:177–186.

2. Wang X, Kim Y, Ma Q, Hong SH, Pokusaeva K, Sturino JM, Wood TK. 2010. Cryptic prophages help bacteria cope with adverse environments. *Nat Commun* **1**:147.

3. Bobay LM, Touchon M, Rocha EP. 2014. Pervasive domestication of defective prophages by bacteria. *Proc Natl Acad Sci U S A* **111**:12127–12132.

4. Wagner PL, Waldor MK. 2002. Bacteriophage control of bacterial virulence. *Infect Immun* **70**:3985–3993.

5. Fortier LC, Sekulovic O. 2013. Importance of prophages to evolution and virulence of bacterial pathogens. *Virulence* **4**:354–365.

6. Edgar R, Qimron U. 2010. The *Escherichia coli* CRISPR system protects from λ lysogenization, lysogens, and prophage induction. *J Bacteriol* **192**:6291–6294.

7. Nozawa T, Furukawa N, Aikawa C, Watanabe T, Haobam B, Kurokawa K, Maruyama F, Nakagawa I. 2011.

CRISPR inhibition of prophage acquisition in *Streptococcus pyogenes*. *PLoS One* 6:e19543.

8. Zeng H, Zhang J, Li C, Xie T, Ling N, Wu Q, Ye Y. 2017. The driving force of prophages and CRISPR-Cas system in the evolution of *Cronobacter sakazakii*. *Sci Rep* 7:40206.

9. Touchon M, Bernheim A, Rocha EP. 2016. Genetic and life-history traits associated with the distribution of prophages in bacteria. *ISME J* 10:2744–2754.

10. Touchon M, Rocha EP. 2010. The small, slow and specialized CRISPR and anti-CRISPR of *Escherichia* and *Salmonella*. *PLoS One* 5:e11126.

11. Bozic B, Repac J, Djordjevic M. 2019. Endogenous gene regulation as a predicted main function of type I-E CRISPR/Cas system in *E. coli*. *Molecules* 24:784.

12. Goldberg GW, Jiang W, Bikard D, Marraffini LA. 2014. Conditional tolerance of temperate phages via transcription-dependent CRISPR-Cas targeting. *Nature* 514:633–637.

13. Gophna U, Kristensen DM, Wolf YI, Popa O, Drevet C, Koonin EV. 2015. No evidence of inhibition of horizontal gene transfer by CRISPR-Cas on evolutionary timescales. *ISME J* 9:2021–2027.

14. Pawluk A, Davidson AR, Maxwell KL. 2018. Anti-CRISPR: discovery, mechanism and function. *Nat Rev Microbiol* 16:12–17.

15. Bondy-Denomy J, Pawluk A, Maxwell KL, Davidson AR. 2013. Bacteriophage genes that inactivate the CRISPR/Cas bacterial immune system. *Nature* 493:429–432.

16. Marino ND, Zhang JY, Borges AL, Sousa AA, Leon LM, Rauch BJ, Walton RT, Berry JD, Joung JK, Kleinstiver BP, Bondy-Denomy J. 2018. Discovery of widespread type I and type V CRISPR-Cas inhibitors. *Science* 362:240–242.

17. Landsberger M, Gandon S, Meaden S, Rollie C, Chevallereau A, Chabas H, Buckling A, Westra ER, van Houte S. 2018. Anti-CRISPR phages cooperate to overcome CRISPR-Cas immunity. *Cell* 174:908–916.e12.

18. Patterson AG, Jackson SA, Taylor C, Evans GB, Salmond GPC, Przybilski R, Staals RHJ, Fineran PC. 2016. Quorum sensing controls adaptive immunity through the regulation of multiple CRISPR-Cas systems. *Mol Cell* 64:1102–1108.

19. Patterson AG, Chang JT, Taylor C, Fineran PC. 2015. Regulation of the type I-F CRISPR-Cas system by CRP-cAMP and GalM controls spacer acquisition and interference. *Nucleic Acids Res* 43:6038–6048.

20. Høyland-Kroghsbo NM, Muñoz KA, Bassler BL. 2018. Temperature, by controlling growth rate, regulates CRISPR-Cas activity in *Pseudomonas aeruginosa*. *mBio* 9:e02184-18.

21. Agari Y, Sakamoto K, Tamakoshi M, Oshima T, Kuramitsu S, Shinkai A. 2010. Transcription profile of *Thermus thermophilus* CRISPR systems after phage infection. *J Mol Biol* 395:270–281.

22. Bondy-Denomy J, Garcia B, Strum S, Du M, Rollins MF, Hidalgo-Reyes Y, Wiedenheft B, Maxwell KL, Davidson AR. 2015. Multiple mechanisms for CRISPR-Cas inhibition by anti-CRISPR proteins. *Nature* 526:136–139.

23. Dong L, Guan X, Li N, Zhang F, Zhu Y, Ren K, Yu L, Zhou F, Han Z, Gao N, Huang Z. 2019. An anti-CRISPR protein disables type V Cas12a by acetylation. *Nat Struct Mol Biol* 26:308–314.

24. He F, Bhoobalan-Chitty Y, Van LB, Kjeldsen AL, Dedola M, Makarova KS, Koonin EV, Brodersen DE, Peng X. 2018. Anti-CRISPR proteins encoded by archaeal lytic viruses inhibit subtype I-D immunity. *Nat Microbiol* 3:461–469.

25. Hynes AP, Rousseau GM, Agudelo D, Goulet A, Amigues B, Loehr J, Romero DA, Fremaux C, Horvath P, Doyon Y, Cambillau C, Moineau S. 2018. Widespread anti-CRISPR proteins in virulent bacteriophages inhibit a range of Cas9 proteins. *Nat Commun* 9:2919.

26. Watters KE, Fellmann C, Bai HB, Ren SM, Doudna JA. 2018. Systematic discovery of natural CRISPR-Cas12a inhibitors. *Science* 362:236–239.

27. Wozniak RA, Waldor MK. 2010. Integrative and conjugative elements: mosaic mobile genetic elements enabling dynamic lateral gene flow. *Nat Rev Microbiol* 8:552–563.

28. Watson BNJ, Staals RHJ, Fineran PC. 2018. CRISPR-Cas-mediated phage resistance enhances horizontal gene transfer by transduction. *mBio* 9:e02406-17.

29. Juni E. 1972. Interspecies transformation of *Acinetobacter*: genetic evidence for a ubiquitous genus. *J Bacteriol* 112:917–931.

30. Alexander HE, Leidy G. 1951. Determination of inherited traits of *H. influenzae* by desoxyribonucleic acid fractions isolated from type-specific cells. *J Exp Med* 93:345–359.

31. Alexander HE, Redman W. 1953. Transformation of type specificity of meningococci; change in heritable type induced by type-specific extracts containing desoxyribonucleic acid. *J Exp Med* 97:797–806.

32. Avery OT, Macleod CM, McCarty M. 1944. Studies on the chemical nature of the substance inducing transformation of pneumococcal types: induction of transformation by a desoxyribonucleic acid fraction isolated from pneumococcus type III. *J Exp Med* 79:137–158.

33. Spizizen J. 1958. Transformation of biochemically deficient strains of *Bacillus subtilis* by deoxyribonucleate. *Proc Natl Acad Sci U S A* 44:1072–1078.

34. Zhang Y, Heidrich N, Ampattu BJ, Gunderson CW, Seifert HS, Schoen C, Vogel J, Sontheimer EJ. 2013. Processing-independent CRISPR RNAs limit natural transformation in *Neisseria meningitidis*. *Mol Cell* 50:488–503.

35. Watanabe T, Nozawa T, Aikawa C, Amano A, Maruyama F, Nakagawa I. 2013. CRISPR regulation of intraspecies diversification by limiting IS transposition and intercellular recombination. *Genome Biol Evol* 5:1099–1114.

36. Watanabe T, Shibasaki M, Maruyama F, Sekizaki T, Nakagawa I. 2017. Investigation of potential targets of *Porphyromonas* CRISPRs among the genomes of *Porphyromonas* species. *PLoS One* 12:e0183752.

37. Jorth P, Whiteley M. 2012. An evolutionary link between natural transformation and CRISPR adaptive immunity. *mBio* 3:e00309-12.

38. Liu YY, Wang Y, Walsh TR, Yi LX, Zhang R, Spencer J, Doi Y, Tian G, Dong B, Huang X, Yu LF, Gu D, Ren H, Chen X, Lv L, He D, Zhou H, Liang Z, Liu JH, Shen J. 2016. Emergence of plasmid-mediated colistin resistance mechanism MCR-1 in animals and human beings in China: a microbiological and molecular biological study. *Lancet Infect Dis* 16:161–168.

39. Sommer MOA, Munck C, Toft-Kehler RV, Andersson DI. 2017. Prediction of antibiotic resistance: time for a new preclinical paradigm? *Nat Rev Microbiol* 15:689–696.

40. Courvalin P. 2006. Vancomycin resistance in gram-positive cocci. *Clin Infect Dis* 42(Suppl 1):S25–S34.

41. Palmer KL, Gilmore MS. 2010. Multidrug-resistant enterococci lack CRISPR-Cas. *mBio* 1:e00227-10.

42. Jiang W, Maniv I, Arain F, Wang Y, Levin BR, Marraffini LA. 2013. Dealing with the evolutionary downside of CRISPR immunity: bacteria and beneficial plasmids. *PLoS Genet* 9:e1003844.

43. Staals RH, Zhu Y, Taylor DW, Kornfeld JE, Sharma K, Barendregt A, Koehorst JJ, Vlot M, Neupane N, Varossieau K, Sakamoto K, Suzuki T, Dohmae N, Yokoyama S, Schaap PJ, Urlaub H, Heck AJ, Nogales E, Doudna JA, Shinkai A, van der Oost J. 2014. RNA targeting by the type III-A CRISPR-Cas Csm complex of *Thermus thermophilus*. *Mol Cell* 56:518–530.

44. Terns RM, Terns MP. 2013. The RNA- and DNA-targeting CRISPR-Cas immune systems of *Pyrococcus furiosus*. *Biochem Soc Trans* 41:1416–1421.

45. Jaubert C, Danioux C, Oberto J, Cortez D, Bize A, Krupovic M, She Q, Forterre P, Prangishvili D, Sezonov G. 2013. Genomics and genetics of *Sulfolobus islandicus* LAL14/1, a model hyperthermophilic archaeon. *Open Biol* 3:130010.

46. Fischer S, Maier LK, Stoll B, Brendel J, Fischer E, Pfeiffer F, Dyall-Smith M, Marchfelder A. 2012. An archaeal immune system can detect multiple protospacer adjacent motifs (PAMs) to target invader DNA. *J Biol Chem* 287:33351–33363.

47. Stachler AE, Turgeman-Grott I, Shtifman-Segal E, Allers T, Marchfelder A, Gophna U. 2017. High tolerance to self-targeting of the genome by the endogenous CRISPR-Cas system in an archaeon. *Nucleic Acids Res* 45:5208–5216.

48. Wei W, Zhang S, Fleming J, Chen Y, Li Z, Fan S, Liu Y, Wang W, Wang T, Liu Y, Ren B, Wang M, Jiao J, Chen Y, Zhou Y, Zhou Y, Gu S, Zhang X, Wan L, Chen T, Zhou L, Chen Y, Zhang X-E, Li C, Zhang H, Bi L. 2019. *Mycobacterium tuberculosis* type III-A CRISPR/Cas system crRNA and its maturation have atypical features. *FASEB J* 33:1496–1509.

49. Seed KD, Lazinski DW, Calderwood SB, Camilli A. 2013. A bacteriophage encodes its own CRISPR/Cas adaptive response to evade host innate immunity. *Nature* 494:489–491.

50. Krupovic M, Makarova KS, Wolf YI, Medvedeva S, Prangishvili D, Forterre P, Koonin EV. 2019. Integrated mobile genetic elements in *Thaumarchaeota*. *Environ Microbiol* 21:2056–2078.

51. Millen AM, Horvath P, Boyaval P, Romero DA. 2012. Mobile CRISPR/Cas-mediated bacteriophage resistance in *Lactococcus lactis*. *PLoS One* 7:e51663.

52. Godde JS, Bickerton A. 2006. The repetitive DNA elements called CRISPRs and their associated genes: evidence of horizontal transfer among prokaryotes. *J Mol Evol* 62:718–729.

53. Chakraborty S, Snijders AP, Chakravorty R, Ahmed M, Tarek AM, Hossain MA. 2010. Comparative network clustering of direct repeats (DRs) and cas genes confirms the possibility of the horizontal transfer of CRISPR locus among bacteria. *Mol Phylogenet Evol* 56:878–887.

54. Tettelin H, Masignani V, Cieslewicz MJ, Donati C, Medini D, Ward NL, Angiuoli SV, Crabtree J, Jones AL, Durkin AS, Deboy RT, Davidsen TM, Mora M, Scarselli M, Margarit y Ros I, Peterson JD, Hauser CR, Sundaram JP, Nelson WC, Madupu R, Brinkac LM, Dodson RJ, Rosovitz MJ, Sullivan SA, Daugherty SC, Haft DH, Selengut J, Gwinn ML, Zhou L, Zafar N, Khouri H, Radune D, Dimitrov G, Watkins K, O'Connor KJ, Smith S, Utterback TR, White O, Rubens CE, Grandi G, Madoff LC, Kasper DL, Telford JL, Wessels MR, Rappuoli R, Fraser CM. 2005. Genome analysis of multiple pathogenic isolates of *Streptococcus agalactiae*: implications for the microbial "pan-genome." *Proc Natl Acad Sci U S A* 102:13950–13955.

55. Lapierre P, Gogarten JP. 2009. Estimating the size of the bacterial pan-genome. *Trends Genet* 25:107–110.

56. Brochet M, Couvé E, Glaser P, Guédon G, Payot S. 2008. Integrative conjugative elements and related elements are major contributors to the genome diversity of *Streptococcus agalactiae*. *J Bacteriol* 190:6913–6917.

57. Lopez-Sanchez MJ, Sauvage E, Da Cunha V, Clermont D, Ratsima Hariniaina E, Gonzalez-Zorn B, Poyart C, Rosinski-Chupin I, Glaser P. 2012. The highly dynamic CRISPR1 system of *Streptococcus agalactiae* controls the diversity of its mobilome. *Mol Microbiol* 85:1057–1071.

58. Frost LS, Leplae R, Summers AO, Toussaint A. 2005. Mobile genetic elements: the agents of open source evolution. *Nat Rev Microbiol* 3:722–732.

59. Turgeman-Grott I, Joseph S, Marton S, Eizenshtein K, Naor A, Soucy SM, Stachler AE, Shalev Y, Zarkor M, Reshef L, Altman-Price N, Marchfelder A, Gophna U. 2019. Pervasive acquisition of CRISPR memory driven by inter-species mating of archaea can limit gene transfer and influence speciation. *Nat Microbiol* 4:177–186.

60. Shmakov SA, Sitnik V, Makarova KS, Wolf YI, Severinov KV, Koonin EV. 2017. The CRISPR spacer space is dominated by sequences from species-specific mobilomes. *mBio* 8:e01397-17.

61. Cady KC, White AS, Hammond JH, Abendroth MD, Karthikeyan RSG, Lalitha P, Zegans ME, O'Toole GA. 2011. Prevalence, conservation and functional analysis of *Yersinia* and *Escherichia* CRISPR regions in clinical *Pseudomonas aeruginosa* isolates. *Microbiology* 157:430–437.

62. Zheng Z, Zhang Y, Liu Z, Dong Z, Xie C, Bravo A, Soberón M, Mahillon J, Sun M, Peng D. 2020. The CRISPR-Cas systems were selectively inactivated during evolution of *Bacillus cereus* group for adaptation to diverse environments. *ISME J* 14:1479–1493.

CHAPTER 9

Evasion Tactics Manifested by Bacteriophages against Bacterial Immunity

Jenny Y. Zhang, Sutharsan Govindarajan, and Joseph Bondy-Denomy

Department of Microbiology & Immunology, University of California, San Francisco, San Francisco, CA 94143

Introduction

Wherever bacteria are found, so are the viruses that infect them. These viruses, called bacteriophages (phages), are the most abundant biological entities on Earth (1, 2). To initiate an infection, phages bind to specific bacterial surface proteins or cell wall components to inject their genetic material into the cell. Upon injection of DNA or RNA, the phage can adopt one of two replication strategies—lytic or lysogenic infection (Fig. 9.1). During a lytic infection, phages utilize proteins of phage origin and hijacked host components to achieve transcription, translation, DNA replication, protein production, and virion assembly. Newly made phage components then assemble into functional phage particles, which are released through the lysing of the bacterial host. In lysogenic replication, phage genomes integrate into the bacterial chromosome or are maintained as nonintegrated episomes. These phage genomes, known as prophages, repress lytic genes to maintain lysogeny, replicating together with the bacterial genome to be passed onto future generations. Prophages can be induced to replicate in the lytic cycle once again, often in response to environmental conditions (3).

To prevent infection by bacteriophages, bacteria have developed an arsenal of defense mechanisms. CRISPR-Cas, the subject of this volume, is one of the best-known systems of bacterial defense against phage. CRISPR-Cas systems are divided into six distinct types based on their evolution, protein constituents, and structures. Aside from CRISPR-Cas systems, restriction-modification (R-M) is another major form of nuclease-dependent antiphage immunity. R-M systems are ubiquitous amongst bacteria and archaea and are extremely diverse (4). They can be broadly classified into four major types based on mode of action as well as subunit composition. In considering the ways that phages evade bacterial immunity, many phage evasion tactics were discovered for R-M systems many years ago and form the conceptual basis of this field. With this in mind, we start with a focus on R-M.

CRISPR: Biology and Applications, First Edition. Edited by Rodolphe Barrangou, Erik J. Sontheimer, and Luciano A. Marraffini.
© 2022 American Society for Microbiology. DOI: 10.1128/9781683673798.ch09

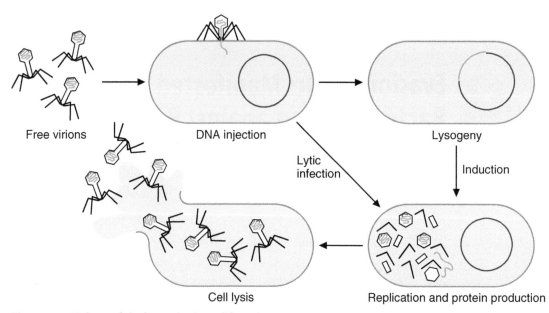

Figure 9.1 Biology of the bacteriophage life cycle. Phages can adopt two different life cycles upon infection of a host cell. A lysogenic life cycle requires the integration of phage DNA into the bacterial chromosome or the maintenance of phage genomes as a nonintegrated episome. The alternative life cycle is lytic infection, in which the phage replicates its genome and assembles new phage particles, eventually lysing the host cell and releasing new virions. Phages in the lysogenic cycle, named prophages, can be induced to undergo the lytic life cycle.

Restriction-Modification Systems

R-M systems contain two major components: a restriction endonuclease and a methyltransferase. The restriction endonuclease recognizes short DNA motifs for cleavage, sequences that exist in both host and pathogen, while the cognate methyltransferase provides protection from cleavage through the methylation of adenine and cytosine bases. Bacterial DNA is modified by the methyltransferase shortly after replication, while phage DNA that is not appropriately methylated is vulnerable to cleavage. With this combination, the R-M system is able to discriminate between self- and non-self-DNA and confer protection through the destruction of foreign DNA elements (5).

Phages have evolved strategies to evade targeting by R-M systems (Fig. 9.2). Studies of phages that mainly infect the model bacterium *Escherichia coli*, in particular, have identified a great variety of different antirestriction mechanisms. Due to the sizes of sequence motifs recognized by R-M systems (typically 4 to 8 bp), they can occur many times in a phage genome. One common strategy is to prevent recognition of the phage DNA by an R-M system. As a result, some phages have purged from their genomes the recognition motifs of the endogenous restriction enzyme in its host. R-M components often interact with both strands of the DNA double helix in a symmetrical manner to adopt the specific conformation required to achieve cleavage. This manifests as often palindromic or bipartite restriction motifs (5). Phages can escape restriction targeting by disrupting this interaction. For example, *E. coli* phages T3 and T7 both contain restriction sites for the type II restriction endonuclease EcoRII but resist cleavage. EcoRII needs to bind to two or more copies of its target sequence for activity. It is likely that these phages evolved longer distances between target sites to escape EcoRII. In fact, when supplemented with DNA containing an abundance of EcoRII recognition sequences, the phage genomes became sensitive to endonuclease activity (6). Some phages also employ unusual bases or apply their own

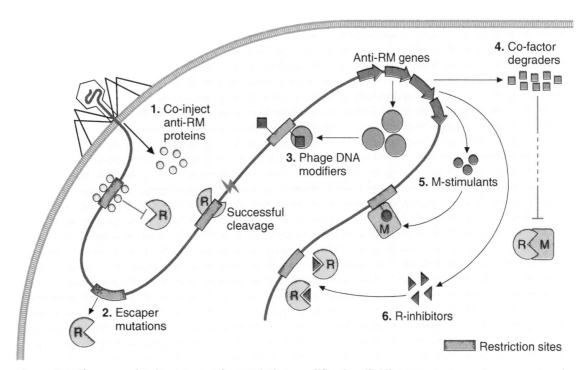

Figure 9.2 Phage mechanisms to evade restriction-modification (R-M) systems. Six major strategies that phages use to escape R-M targeting include coinjection of accessory proteins with the phage genome to prevent genome cleavage by occlusion of restriction sites (1), R-M escaper mutations that include both single base pair mutations in the restriction sequences and the disruption of entire restriction sites (2), DNA modification proteins that prevent genome binding/cleavage by restriction enzymes (3), production of "anti-RM" proteins that degrade or promote the destruction of R-M cofactors (4), stimulation of host methyltransferases to promote host methylation of phage DNA before cleavage by the corresponding restriction endonucleases (5), and direct inhibitors of restriction enzymes that block R-M targeting (6).

DNA modifications to prevent binding by restriction endonucleases. The *E. coli* phage T4, for example, has adopted this strategy. In place of cytosines in its DNA, T4 has glucosyl-hydroxymethylcytosine (ghmC) residues (7), which protect T4 against all type I to III R-M systems with recognition sequences that involve cytosines (8). Glucosylation facilitates evasion of the *E. coli* McrBC system, an R-M system that specifically cleaves hmC-containing DNA (9–12). Additionally, phages can physically occlude restriction sites directly, with phage-encoded accessory proteins that protect the phage DNA. One such example involves the enterobacterial phage P1, whose accessory proteins, DarA and DarB, are coinjected alongside the P1 genome and prevent DNA cleavage by the type IA and IB R-M systems through occluding their restriction sites (13).

Each strategy above describes a *cis*-acting mechanism, broadly characterized as "self" protection. Another common protective strategy adopted by phages is to act in *trans*, directly interacting with proteins of R-M systems. For example, phage λ uses the Ral protein to hyperstimulate the host methyltransferases. Consequently, the type IA methyltransferases modify phage DNA before recognition and cleavage by the corresponding restriction enzymes (14). Another method is to destroy certain cofactors required for R-M function. For example, *S*-adenosylmethionine (SAM) is the designated substrate for methyltransferases of the type I and type III R-M systems. The *E. coli* phage T3 encodes a SAM hydrolase that hydrolyzes intracellular SAM and reduces the overall concentration of this important cofactor, which weakens the cell's R-M defense (15). Lastly, mirroring the discussion of anti-CRISPRs later in this chapter, phages can encode inhibitors of R-M systems. Overcome classical restriction protein (Ocr) is the most extensively studied antirestriction protein. Ocr is the first expression product of the *E. coli* phage T7 upon infection and inhibits type I R-M enzymes (16, 17). The crystal structure of Ocr reveals that it functions as a dimer and blocks the DNA binding site of type I R-M enzymes through mimicking the shape and surface charge distribution of B-form DNA. As a result, Ocr is able to successfully inhibit both the restriction and modification abilities of the type I enzymes (18).

R-M systems and CRISPR-Cas systems are both heavily studied modes of bacterial defense, but more and more immune systems are being unveiled to us all the time. Recent bioinformatics and experimental efforts, for example, have introduced nine new antiphage systems in the microbial pangenome (19). As for phage-developed strategies to thwart these systems, we can only suspect that they exist in great varieties but remain unidentified.

CRISPR-Cas

CRISPR-Cas systems are comprised of a DNA-based array of clustered regularly interspaced short palindromic repeats and CRISPR-associated genes that enable antiphage function. Cas proteins are responsible for the integration of phage or plasmid DNA into the CRISPR array, for processing the transcript from the CRISPR array into small CRISPR RNAs (crRNAs), and for mediating effector cleavage function of target nucleic acids (Fig. 9.3). The spacer sequences between repeats in the CRISPR array specify the sequence identity of the foreign target RNA or DNA. This immune system is widely disseminated in bacteria and archaea, presenting phages and viruses with yet an additional hurdle to overcome.

Mutation

CRISPR-Cas systems can target DNA, RNA, or both. In all cases, the primary mechanism by which the CRISPR-Cas system executes its function is via base-pairing between the target and the crRNA (20). Given the single recognition site, mutations can disrupt the base-pairing necessity to escape from the action of CRISPR-Cas systems. Early studies on phages that escaped different CRISPR-Cas systems have shown that escapers arise by various mutational outcomes, including single base changes. The nature of mutations that lead to escape provided information about the target requirements of various CRISPR-Cas systems as well as the plasticity of the phage genome. The first example of mutation-mediated escape from CRISPR-Cas activity was reported for phages infecting *Streptococcus thermophilus*, an

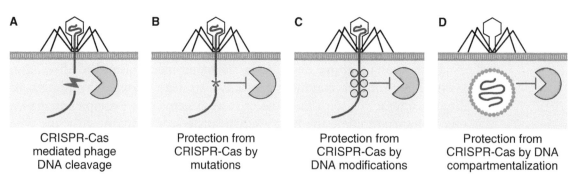

A	B	C	D
CRISPR-Cas mediated phage DNA cleavage	Protection from CRISPR-Cas by mutations	Protection from CRISPR-Cas by DNA modifications	Protection from CRISPR-Cas by DNA compartmentalization

Figure 9.3 CRISPR-Cas evasion mechanisms. A schematic summarizing phage-host coevolution mechanisms during CRISPR-Cas selection. (**A**) Phage DNA being cleaved by a CRISPR nuclease (red lightning bolt); (**B**) point mutation(s) (*) in the phage genome that weaken or abolish Cas:crRNA binding to the target sequence and prevent cleavage; (**C**) DNA base modifications (yellow circles) that prevent DNA binding or cleavage; (**D**) a proteinaceous structure that protects the phage DNA from cleavage by Cas and restriction nucleases.

industrially important lactic acid bacterium (21). *S. thermophilus* strains that had acquired type II-A spacers enabling CRISPR-based resistance against a phage were still infected by a subpopulation of the same phage. Sequencing analysis revealed that these "escaper" phages contained mutations in their protospacer regions that prevented crRNA base-pairing and enabled avoidance of CRISPR-Cas targeting (Fig. 9.4). Analysis of the escapers revealed that a range of mutations were present, including point mutations, deletions, or duplications in the protospacer itself or point mutations within the protospacer-adjacent motif (PAM; a 2- to 6-bp DNA sequence that flanks the protospacer region) (21). A detailed analysis of CRISPR-Cas evasion via mutation also came from another study that analyzed the nature of phages that escape type I-E CRISPR-Cas targeting (22). This work systematically introduced point mutations into the crRNA in addition to isolating spontaneous phage mutants that avoided CRISPR-Cas targeting. In addition to mutations in the PAM, the authors also identified a PAM-proximal protospacer region called the "seed" where crucial mutations were often found. This work also demonstrated a biochemical basis for this effect where single point mutations in the PAM and seed regions of the protospacer were sufficient to weaken Cascade-DNA interactions. In the type I-F system, spontaneous phage mutations that evaded CRISPR-Cas targeting were also found in the PAM and seed regions, as well as spread throughout the protospacer (23). However, not all mutations in the target protospacer region were found to evade CRISPR-Cas activity and certain mutations in the protospacer allowed interference, albeit at a reduced efficiency, while other mutations completely blocked it. This work demonstrated that a perfect match between the crRNA and the

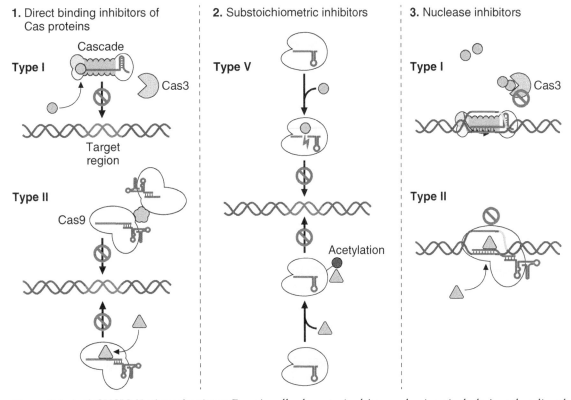

Figure 9.4 Anti-CRISPR (Acr) mechanisms. Functionally characterized Acr mechanisms include Acrs that directly bind to Cas proteins and inhibit CRISPR-Cas function (1), substoichiometric inhibitors such as enzymes (2), and Acrs that inhibit CRISPR nuclease domains or subunits (3). Type I-F CRISPR-Cas3, type II-A CRISPR-Cas9, and type V-A CRISPR-Cas12a are shown as representative type I, type II, and type V systems, respectively.

target region is not an absolute requirement for CRISPR-Cas activity. Similar observations were also made for natural escapers of type I-E systems of *E. coli* and *Vibrio cholerae* (24, 25). However, in addition to point mutations in the seed region and PAM motif, deletions were also observed in some cases.

Interestingly, in contrast to DNA-targeting CRISPR systems, single nucleotide mutations are not an effective strategy in escaping from RNA-targeting type III CRISPR-Cas systems. Rather, large DNA deletions or multiple mutations are commonly observed as escaping mechanisms for these systems. The type III CRISPR-Cas system employs transcription-dependent DNA interference to target both RNA and DNA (26). When type III crRNA base-pairs with a target RNA, it activates two of its nuclease activities—RNA cleavage as well as single-stranded DNA cleavage (27). Interestingly, when type III CRISPR-Cas systems target a phage, escape mutants arise very infrequently compared to the case with type II CRISPR-Cas9. Sequence analysis indicated that type III escapers contained large deletions that range between 700 and 2,500 bp encompassing the protospacer region or point mutations in the promoter that inactivate transcription. The strategies likely limit escape potential when an essential gene is targeted. The relaxed specificity and lack or reliance of seed-like sequence for type III CRISPR-Cas targeting appears to be the reason for escaper formation by large deletion instead of point mutations (27). In addition to its relaxed specificity, the type III system was recently shown to co-opt crRNAs from the type I-F system present in the same bacteria. Under such conditions, phages that escaped type I-F targeting by single nucleotide mutations were still targeted by the type III system, making it a remarkably effective bacterial adaptive immune system (28).

DNA Modification

Bacteriophage genomes are modified by a wide variety of chemical modifications. These modifications, which include amino acids, polyamines, monosaccharides, and disaccharides, are added to the nucleotide in a manner that does not affect its base-pairing ability (29). Bioinformatics studies have identified the presence of a great diversity of DNA modification enzymes exclusively in phage genomes (30, 31), suggesting that phages may contain the widest range of the DNA base modifications found in nature. Phage DNA modifications are known to be important for protection from host defense systems. This phenomenon was first described for R-M evasion, in which it was observed that phages that acquire methylation marks by a host methyltransferase cannot be targeted by a cognate restriction enzyme and hence are capable of subsequent infection. Several phage DNA modifications are known to block the activities of restriction endonucleases. Modification of thymine to 5-hydroxymethyluracil by some *Bacillus subtilis* phages protects them from host restriction enzymes (5). Similarly, hmC and ghmC modifications present in T-even phages provide protection against host restriction enzymes (29), as discussed above. More recently, two different thymidine hypermodifications, namely, 5-(2-aminoethyl) uridine and 5-(2-aminoethoxy) methyluridine, were shown to inhibit cleavage by several restriction endonucleases (31).

In contrast to our understanding of DNA modification-dependent antirestriction, the role of phage DNA modification in protection from CRISPR-Cas is only beginning to be explored. *E. coli* phage T4, which contains ghmC modification, is so far the best-characterized modification that protects the genome from specific CRISPR-Cas systems. Initially, Yaung and colleagues tested whether T4 phages are sensitive to heterologous Cas9 nuclease activity (32). They found that *Streptococcus pyogenes* Cas9 is active against methylation modifications, including N^6-methyladenine and 5-methylcytosine, and can target the DNA of T4 phage modified with hmC or ghmC. The presence of a partially matching natural spacer against phages T2 and CC31, which also contain ghmC modification, suggests that the native *E. coli* type I-E system may be capable of targeting phages possessing large DNA modifications like ghmC (32). However, two later studies suggested that the bulkier ghmC modification of T4 is indeed sufficient to protect the phage DNA from Cas9 nuclease activity (33, 34) (Fig. 9.3C). It was found that Cas9 sensitivity of ghmC-modified T4 observed by Yaung and colleagues is due to the use of synthetic single guide RNAs (sgRNAs) that contain crRNAs and transactivating crRNAs (tracrRNAs) fused together. In contrast, when *S. pyogenes* Cas9 is loaded with natural guide RNAs, which include crRNAs and tracrRNAs as two independent units, the system is incapable of degrading ghmC-modified T4. Similarly, ghmC modification also protected T4 from the native type I-E Cascade of *E. coli* (33, 35). Moreover, when the phage DNA contained a high density of less bulky hmC modifications, it still protected T4 from Cas9 nuclease activity (33). Interestingly, the same ghmC modification of T4 DNA is insufficient to provide protection against type V-A nuclease Cas12a (Cpf1). It was suggested that an intrinsically more open architecture of Cas12a allowed its binding to the protospacer region irrespective of its modification state (34). In contrast, Cas9 and Cascade seem to lack such an open architecture, resulting in imperfect target recognition due to modification, enabling evasion.

Another example of DNA modification-mediated CRISPR evasion comes from the study of bacteriophage φR1-37 (35). φR1-37 is a *Yersinia* phage that contains deoxyuridine (dU) instead of thymidine in its DNA (35). φR1-37 is capable of propagating in *E. coli* that expresses the appropriate lipopolysaccharide receptor. A comprehensive study that tested the efficacy of type I-E targeting against various *E. coli* phages identified T4 and φR1-37 as type I-E resistant phages (35). It was suggested that ghmC modification of T4 and dU modification of φR1-37 provided protection against type I Cascade. While ghmC is confirmed as a DNA modification that protects phages from certain Cas proteins, due to the testing of isogenic phage mutants lacking the modification, the role of dU DNA modification in CRISPR-Cas evasion has not been formally proven. This is due to the possibility that φR1-37, which is a jumbophage with a 262-kb genome, could evade CRISPR-Cas targeting through other mechanisms, including anti-CRISPR proteins or by physically occluding its DNA, similar to a family of jumbophages infecting *Pseudomonas aeruginosa* (see below) (36).

DNA Protection

An unexplored area pertaining to CRISPR immunity is the ability of phages to inject or produce proteins that protect phage DNA from cleavage. Previous work has uncovered multiple examples of restriction enzyme inhibitors being injected by phages. A family of "internal proteins" is encoded and packaged into phage particles by T-even phages like T4. The model protein IPI (internalized protein 1) has been shown to inhibit the prophage-encoded GmrS/GmrD restriction enzyme in *E. coli* (37). The GmrS/GmrD enzymes can cleave the ghmC-modified DNA (described above) possessed by T-even phages when IPI is knocked out, suggesting that this is an additional R-M protection beyond the already protective ghmC modification. IPI is also coencoded with two other internalized proteins (IPII and IPIII) of unknown function. A T4 mutant lacking IPI to IPIII maintained resistance to Cas9 immunity when expressed in *E. coli*; however, this mutant phage still had ghmC modifications (33). More work is needed in this area to determine whether mechanisms previously attributed to restriction enzyme evasion (38) or new mechanisms involving injected proteins impart resistance to CRISPR-Cas immunity.

Recent work has identified a family of *Pseudomonas aeruginosa* and *Serratia* phages that assemble a proteinaceous "nucleus-like" structure surrounding the phage genome during infection (39, 40). The large compartment houses the phage genome, where DNA replication and transcription occur, while phage mRNAs leave the nucleus-like compartment and are translated in the cytoplasm. Phage particles are built in the cytoplasm and dock directly at the shell to receive DNA (41). This elaborate phage life cycle correlates with a phenotype of CRISPR and restriction enzyme resistance in both *P. aeruginosa* and *Serratia* (36, 40). The phage displays complete resistance to targeting by two endogenous immune systems in *Pseudomonas*, the type I-C CRISPR-Cas system and the type I R-M system (Fig. 9.3D). Moreover, this phage is also recalcitrant to targeting by the type II-A (Cas9) and type V-A (Cas12a) systems, which led the authors to consider that the nucleus-like shell protects the DNA from immunity, as opposed to mechanisms to specifically inactivate immune systems to which this phage has likely not been exposed. Moreover, the sensitivity of these phages to RNA-targeting type III-A and type VI-A, but complete resistance to each DNA targeting system, bolsters these conclusions. These are the first examples of a clear mechanism to broadly evade DNA-targeting immune systems. The most potent phage protection mechanisms reported to date, such as DNA hypermodification, have not provided resistance as broad as this, since even ghmC modifications can be subject to partial targeting in cytidine-depleted regions of the genome, by modification-dependent restriction systems, and by type V-A CRISPR-Cas immunity (33, 34).

CRISPR-Cas Inhibition by Anti-CRISPR Proteins

Anti-CRISPR (Acr) proteins (42) are specific CRISPR-Cas inhibitors that are highly diverse in sequence, structure, and mechanism of action (Fig. 9.4). Currently, ~90 distinct families of Acr proteins have been identified that inhibit a variety of different CRISPR-Cas systems, both class I and class II (43, 44), and in both bacteria and archaea (45). Anti-CRISPR proteins are often encoded by mobile genetic elements such as phages and prophages and are often small hypothetical proteins that receive the "anti-CRISPR" label when they are seen to inhibit at least one CRISPR-Cas system *in vivo* or *in vitro*. Besides this phenotype, Acr proteins often lack known domains or a predictable biochemical function. Experimental studies have revealed that Acr proteins inhibit different steps in the CRISPR-Cas immune pathway, including protein biogenesis and stability, crRNA loading, DNA target binding, nuclease activation, and second messenger signaling, or act on the DNA substrate to relieve torsion. Acr proteins that target the same stage of CRISPR-Cas function can do so through specific and often unique binding sites, stoichiometries, and modes of action (46).

A subset of Acr proteins have characterized mechanisms and often function through inhibition of the target binding or nuclease activation processes, usually by direct binding of Cas proteins and stoichiometric or enzymatic inactivation. The majority of anti-CRISPR proteins fall under the category of target binding inhibitors. AcrIF1, AcrIF2, and AcrIF10 are all inhibitors of DNA binding by the type I-F CRISPR-Cas system. DNA binding of this system is carried out by the cascade effector complex that is made up of nine subunits of four Cas proteins where a Cas8f:Cas5f (Csy1:Csy2) heterodimer and a Cas6f (Csy4) monomer bind the 5′ and 3′ ends of the crRNA respectively, with six copies of Cas7fs (Csy3) making up the backbone (47). A separate nuclease (Cas3) is recruited subsequent to target recognition by the complex (48, 49). Early size-exclusion chromatography studies immediately differentiated the abilities of AcrIF1 and AcrIF2 to bind distinct components of the type I-F surveillance complex. Two or three molecules of AcrIF1 were proposed to be bound to the Cas7f backbone, while one copy of AcrIF2 interacted with the Cas8f:Cas5f heterodimer (50). Cryo-electron microscopy analysis further elucidated their mechanisms, suggesting that the binding of two AcrIF1 proteins induces a conformational change in the complex backbone that restricts access by the target DNA to the guide RNA, directly interfering with the hybridization of guide to target. In contrast, AcrIF2 takes advantage of a nook, or "lysine-rich vise," in the complex important for DNA binding, interacting with several positively charged residues on Cas8f and Cas7f at the tail end of the complex. A pseudo-helical display of acidic charges on the surface of AcrIF2 is suspected to mimic the negative charge distribution on a DNA duplex backbone, allowing AcrIF2 to compete directly for the DNA binding site (51–53). Interestingly, similar studies on AcrIF10 described this anti-CRISPR as a DNA mimic, but with several differences from AcrIF2. While AcrIF10 also seems to interact with Cas8f:Cas5f and Cas7f at the DNA binding site, it does not contain a surface charge distribution that mirrors that of double-stranded DNA (dsDNA). Despite this difference, AcrIF10 binding appears to cause the complex to adopt a stable DNA-bound conformation by binding in the exact location DNA would occupy (52).

Anti-CRISPRs against the type II CRISPR-Cas systems have also been shown to prevent target binding. Type II systems are single-effector CRISPR systems utilizing Cas9. Twenty-three inhibitors of type II-A Cas9 (AcrIIA1 to AcrIIA23) have been identified, along with six inhibitors of type II-C Cas9 (AcrIIC1 to AcrIIC6). Cryo-electron microscopy revealed that AcrIIC3 binding to Cas9 induces the dimerization of type II-C Cas9 and limits DNA binding (54), though DNA-bound but nuclease-inhibited structural states have been defined (55). While the exact mechanisms are still unknown, AcrIIC4 and AcrIIC5 have been shown in electrophoretic mobility shift assays to inhibit target DNA binding by Nme-Cas9, but neither anti-CRISPR interferes with the loading of the crRNA (56). AcrIIC2, however, has been shown to inhibit sgRNA loading through an interaction with the bridge helix (57, 58). AcrIIA4 inhibits the type II-A Cas9, has been extensively studied, and has been shown through X-ray crystallography to bind to the PAM-binding pocket of Cas9, preventing PAM sequence binding (59–61). AcrIIA2, another type II-A Cas9 inhibitor, was shown through cryo-electron microscopy and X-ray crystallography to be another anti-CRISPR likely functioning as a DNA mimic. Similar to AcrIIA4, AcrIIA2 contains a surface distribution of negative charges. It also occupies the PAM-binding pocket of Cas9 by forming a stable complex with crRNA-bound Cas9. Unlike target DNA, the binding of AcrIIA2 does not induce Cas9 to undergo a conformational change (62, 63).

Three of the first five identified inhibitors of type V-A CRISPR-Cas12a systems (another single-effector machinery) were also revealed as DNA binding inhibitors (64, 65). AcrVA4 is the first Acr shown to inhibit CRISPR-Cas at several stages. AcrVA4 binds both the LbCas12a-crRNA and the LbCas12a-crRNA-dsDNA complexes, where it prevents DNA recognition and promotes release of bound target DNA, respectively. AcrVA4 was also found to bind Cas12a proteins postcleavage, and this mechanism was proposed to block recycling of the enzyme (66, 67). Two other DNA binding inhibitors of Cas12a were shown to function through substoichiometric enzymatic activities. The first, AcrVA1, contains RNase activity and induces the cleavage of crRNA. Cryo-electron microscopy has shown that AcrVA1 binds to Cas12a by mimicking the DNA substrate and cleaves crRNA in a Cas12a-dependent manner (66, 67). The other inhibitor, AcrVA5, acetylates Cas12a directly in *Moraxella bovoculi* and *Lachnospiraceae* bacterium orthologs. Structural and biochemical assays have revealed that AcrVA5 acetylates specific lysine residues on Cas12a that are critical for PAM recognition, preventing DNA binding through steric hindrance. On the other arm of the host-pathogen warfare, the authors of this study also observed that certain *M. bovoculi* strains contained single lysine-to-arginine substitutions that rendered AcrVA5 inactive and restored CRISPR activity (68).

Nuclease activation is the other major target for known anti-CRISPR mechanisms. Some anti-CRISPR proteins have been characterized to inhibit the binding (type I) and cleavage (type I and II) activities of the nuclease domain. Size-exclusion chromatography identified AcrIF3 to be an inhibitor of the type I nuclease Cas3 (50). X-ray crystallography and cryo-electron microscopy studies showed that AcrIF3 dimers form a complex with Cas3, with each AcrIF3 monomer interacting with different domains of the Cas3 protein. This coverage of the Cas3 protein shields the entrance of the DNA-binding cleft of the helicase domain, blocking it (69). X-ray crystallography and copurification experiments also showed that AcrIE1, which inhibits the type I-E CRISPR-Cas system, functions as a dimer and inhibits the type I nuclease Cas3 (70).

Of the type II systems, AcrIIC1 is the only anti-CRISPR demonstrated to block nuclease activation specifically. From cryo-electron microscopy structures, AcrIIC1 interacts with residues in the active interface of the HNH nuclease domain of Cas9, whose proper activation by crRNA and target DNA pairing is crucial to eventual cleavage of the target. By directly blocking the nuclease domain of Cas9, AcrIIC1 can inhibit not only NmeCas9 activity but also a broad range of Cas9 orthologs, as the HNH domain is highly conserved and contains essential residues that are difficult to mutate without compromising Cas9 activity (54).

In 2020 to 2021, a further explosion in anti-CRISPR discovery and mechanistic characterization occurred. Over 90 diverse families of Acr proteins have been uncovered, as tracked and updated in real time in a single resource (www.tinyurl.com/anti-crispr) (71). Acr proteins have also been found that inhibit RNA-targeting type VI-A systems (72, 73), and for type III-A and type III-B systems (74, 75). Notably, a protein called AcrIII-1 inhibits both type III subtypes because it enzymatically cleaves a second messenger molecule required for nonspecific ribonuclease activity, as opposed to directly binding to any Cas proteins (75).

We have examined known examples of anti-CRISPR mechanisms and can immediately deduce that there is a great deal of diversity in both the stage of CRISPR function that is targeted and the specific mode of anti-CRISPR inhibition at each stage. It is interesting to contemplate how such diversity came into being and whether anti-CRISPRs inhibiting the same process or through a similar mechanism (i.e., DNA mimicry) have common ancestry or were simply the outcomes of convergent evolutionary events. We can also contemplate each strategy that we have observed and whether some inhibition mechanisms are "better" than the others. Anti-CRISPRs inhibiting a broad spectrum of CRISPR orthologs (i.e., AcrIIC1) can perceivably protect against a wide spectrum of hosts, while an anti-CRISPR evolved to block more than one domain of its target (i.e., AcrIF3) would provide extremely efficient inhibition. Notably, despite its strong binding affinity and broad spectrum, AcrIIC1 is a fairly weak inhibitor of its cognate NmeCas9 (56), perhaps because Cas9 is still allowed to bind DNA; any dissociation of AcrIIC1 will result in cleavage immediately.

With respect to anti-CRISPR evolution, two recent studies reported the discovery of anti-CRISPRs with broad phylogenetic distributions—AcrIF11, AcrVA2 (65), and AcrIIA7-IIA10 (76). We imagine that the distinct contrast between the widespread nature of these anti-CRISPRs and previously known anti-CRISPRs suggests that their mechanisms allow them to be successful in a broad range of cellular environments and CRISPR-Cas orthologs. Notably, AcrVA3 is a dual inhibitor of the type V-A and type I-C CRISPR-Cas systems but is quite weak. While instances of dual inhibitors have been observed before (i.e., AcrIF6 and AcrIE4-F7), this is the first instance of an anti-CRISPR that blocks completely different CRISPR types, as the type V-A and type I-C systems do not share any protein components. While the exact mechanism of this anti-CRISPR has not yet been determined, broad-spectrum anti-CRISPRs such as this are likely advantageous when hosts have multiple CRISPR-Cas systems of completely distinct classes, as *Moraxella* does.

Anti-CRISPR Phage Biology

Phage-encoded *acr* genes are typically clustered and located upstream of an a̲c̲r̲-a̲ssociated (*aca*) gene. *aca* genes encode a helix-turn-helix DNA-binding motif and were thus hypothesized to be transcriptional regulators. To date, seven distinct families of *aca* genes have been characterized. Like *acr* genes, *aca* genes span a wide domain of bacterial species, encoded by phages, prophages, plasmids, and conjugative elements (42, 65, 77, 78). These gene families have been defined as encoding transcriptional repressors of the *acr* promoter. One study of the Mu-like phage JBD30, containing the *aca* gene *aca1*, showed that Acr proteins are rapidly produced shortly upon injection of the phage DNA, likely due to the strength of the *acr* promoter. Transcription of the *acr* locus precedes that of even well-described phage early genes, providing a burst of Acr protein. Aca1 is required after this initial burst of transcription to repress the promoter by binding to two inverted-repeat (IR) sequences and stop *acr* production. This is essential for phage replication; an *aca1* deletion or a point mutation that abolished Aca1 DNA binding prevents phage replication in a CRISPR-Cas-independent manner. This phenotype was attributed to the strong *acr* promoter that, if not repressed, leads to readthrough transcription that disrupts the production of downstream essential genes involved in capsid morphogenesis. This results in loss of phage viability during lytic growth but, interestingly, does not impair lysogen-forming ability, as lysogeny does not require particle formation. The authors also validated two other *aca* genes, *aca2* and *aca3*, as transcriptional repressors of their respective *acr* promoters (79).

Another study also characterized the function of *aca2*, working in *Pectobacterium carotovorum*. Similar to the case with *aca1*, two IR sequences were identified in the *acr* promoter, and Aca2 was shown to dimerize and bind both sequences. Aca2 has a higher affinity for the first inverted repeat, IR1, than for the second, IR2, and binding to only IR1 was required for transcriptional repression. The authors speculate that IR2 might be used for lytic phage infections when high levels of Aca2 bind both IR1 and IR2 to quickly shut down Acr production. The study also showed that *aca2* has a weaker ribosomal binding site than that of the *acr* gene, which contributes to creating the burst of Acr proteins immediately postinfection, after which Aca proteins accumulate at a slower pace and eventually shut off *acr* transcription. Moreover, the study showed low levels of Aca2 production during lysogeny, which suggests a balancing between prophage Acr production and maintaining some host CRISPR activity to prevent superinfection (80).

Despite this *acr* locus architecture, phages must synthesize Acrs *de novo* upon infection, while bacteria can be armed with preexpressed and crRNA-loaded CRISPR-Cas complexes that actively patrol the cell for foreign DNA. When infecting a population with a diverse collection of spacers, in fact, phages will only survive and replicate if they encode anti-CRISPRs (81). However, the intrinsic challenge of neutralizing preexpressed CRISPR-Cas with anti-CRISPRs (82) lacked a mechanistic explanation until recently. To overcome the challenge associated with this problem, phages can take advantage of the tight binding affinities of Acr proteins for their target Cas protein, ensuring that this protein-protein interaction outlives the phage genome if it gets degraded. In this manner, phages deploying anti-CRISPR proteins rely on multiple infections of the same cell to neutralize CRISPR-Cas. This manifests as a multiplicity-of-infection-dependent outcome, where phage infection can succeed or fail based on the number of infections and the anti-CRISPR binding strength (83, 84). This deposition of an anti-CRISPR protein to the cell could be susceptible to "cheaters," though phages that do not encode their own anti-CRISPR receive only a very subtle benefit from this public good (83). Once lysogens have formed, however, the continued expression of anti-CRISPR proteins from the prophage does fully neutralize CRISPR-Cas immunity (42, 85), which is interestingly countered by other prophage-expressed genes that prevent superinfection through other mechanisms, such as receptor modification (86).

Future Directions

Other steps of CRISPR function remain vulnerable to inhibition, but no specific proteins have been identified yet. Anti-CRISPRs could block the acquisition of new spacers, for instance, preventing its parent nucleic acids from being recognized by CRISPR. Anti-CRISPRs could inhibit CRISPR array or *cas* gene transcription, Cas protein translation, or crRNA/tracrRNA processing. Additionally, nucleic acid-based anti-CRISPRs that mimic the crRNA or a fragment of it could potentially outcompete the endogenous molecule (80). Inhibition of CRISPR effector biogenesis, or any mechanism to prevent functional CRISPR-Cas complexes from assembling, would likely also be an adaptive long-term strategy for the many integrative elements that encode anti-CRISPRs. The possibilities of anti-CRISPR mechanisms and distribution are endless. Based on the discussion of restriction enzyme inhibitors above, we also put forth that entirely new classes of CRISPR evasion mechanisms may remain to be discovered, such as the installation of novel DNA modifications or the direct injection of proteins that protect the genome. We also consider whether host modifications exist that protect the bacterial genome from CRISPR-Cas immunity and whether those could be co-opted or hyperactivated by phages. Furthermore, mirroring the classic example of SAM destruction to inactivate R-M discussed above, recent discoveries have revealed an anti-CRISPR ring nuclease that degrades a cyclic nucleotide molecule required for CRISPR-Cas function (75). This suggests that the enzymatic degradation of other small-molecule cofactors required for antiphage immune function may await discovery.

The constant competitive coevolution of CRISPR-Cas and anti-CRISPRs occurring in the immense numbers of microorganisms has produced a trove of diverse mechanisms that will be uncovered by ongoing scientific investigation.

References

1. Hendrix RW, Smith MCM, Burns RN, Ford ME, Hatfull GF. 1999. Evolutionary relationships among diverse bacteriophages and prophages: all the world's a phage. *Proc Natl Acad Sci U S A* 96:2192–2197.

2. Cobián Güemes AG, Youle M, Cantú VA, Felts B, Nulton J, Rohwer F. 2016. Viruses as winners in the game of life. *Annu Rev Virol* 3:197–214.

3. Bossi L, Fuentes JA, Mora G, Figueroa-Bossi N. 2003. Prophage contribution to bacterial population dynamics. *J Bacteriol* 185:6467–6471.

4. Roberts RJ, Vincze T, Posfai J, Macelis D. 2005. REBASE—restriction enzymes and DNA methyltransferases. *Nucleic Acids Res* 33:D230–D232.

5. Tock MR, Dryden DTF. 2005. The biology of restriction and antirestriction. *Curr Opin Microbiol* 8:466–472.

6. Krüger DH, Barcak GJ, Reuter M, Smith HO. 1988. EcoRII can be activated to cleave refractory DNA recognition sites. *Nucleic Acids Res* 16:3997–4008.

7. Wyatt GR, Cohen SS. 1953. The bases of the nucleic acids of some bacterial and animal viruses: the occurrence of 5-hydroxymethylcytosine. *Biochem J* 55:774–782.

8. Bickle TA, Krüger DH. 1993. Biology of DNA restriction. *Microbiol Rev* 57:434–450.

9. Raleigh EA, Trimarchi R, Revel H. 1989. Genetic and physical mapping of the mcrA (rglA) and mcrB (rglB) loci of Escherichia coli K-12. *Genetics* 122:279–296.

10. Raleigh EA. 1992. Organization and function of the mcrBC genes of *Escherichia coli* K-12. *Mol Microbiol* 6:1079–1086.

11. Hattman S. 1964. The functioning of T-even phages with unglucosylated DNA in restricting *Escherichia coli* host cells. *Virology* 24:333–348.

12. Fukasawa T. 1964. The course of infection with abnormal bacteriophage T4 containing non-glucosylated DNA on *Escherichia coli* strains. *J Mol Biol* 9:525–536.

13. Iida S, Streiff MB, Bickle TA, Arber W. 1987. Two DNA antirestriction systems of bacteriophage P1, *darA*, and *darB*: characterization of *darA⁻* phages. *Virology* 157:156–166.

14. Zabeau M, Friedman S, Van Montagu M, Schell J. 1980. The ral gene of phage λ. I. Identification of a non-essential gene that modulates restriction and modification in *E. coli*. *Mol Gen Genet* 179:63–73.

15. Studier FW, Movva NR. 1976. SAMase gene of bacteriophage T3 is responsible for overcoming host restriction. *J Virol* 19:136–145.

16. Studier FW. 1973. Analysis of bacteriophage T7 early RNAs and proteins on slab gels. *J Mol Biol* 79:237–248.

17. Bandyopadhyay PK, Studier FW, Hamilton DL, Yuan R. 1985. Inhibition of the type I restriction-modification enzymes EcoB and EcoK by the gene 0.3 protein of bacteriophage T7. *J Mol Biol* 182:567–578.

18. Walkinshaw MD, Taylor P, Sturrock SS, Atanasiu C, Berge T, Henderson RM, Edwardson JM, Dryden DTF. 2002. Structure of Ocr from bacteriophage T7, a protein that mimics B-form DNA. *Mol Cell* 9:187–194.

19. Doron S, Melamed S, Ofir G, Leavitt A, Lopatina A, Keren M, Amitai G, Sorek R. 2018. Systematic discovery of antiphage defense systems in the microbial pangenome. *Science* 359:eaar4120.

20. Klompe SE, Sternberg SH. 2018. Harnessing "a billion years of experimentation": the ongoing exploration and exploitation of CRISPR-Cas immune systems. *CRISPR J* 1:141–158.

21. Deveau H, Barrangou R, Garneau JE, Labonté J, Fremaux C, Boyaval P, Romero DA, Horvath P, Moineau S. 2008. Phage response to CRISPR-encoded resistance in *Streptococcus thermophilus*. *J Bacteriol* 190:1390–1400.

22. Semenova E, Jore MM, Datsenko KA, Semenova A, Westra ER, Wanner B, van der Oost J, Brouns SJJ, Severinov K. 2011. Interference by clustered regularly interspaced short palindromic repeat (CRISPR) RNA is governed by a seed sequence. *Proc Natl Acad Sci U S A* 108:10098–10103.

23. Cady KC, Bondy-Denomy J, Heussler GE, Davidson AR, O'Toole GA. 2012. The CRISPR/Cas adaptive immune system of *Pseudomonas aeruginosa* mediates resistance to naturally occurring and engineered phages. *J Bacteriol* 194:5728–5738.

24. Fineran PC, Gerritzen MJH, Suárez Diez M, Künne T, Boekhorst J, van Hijum SAFT, Staals RHJ, Brouns SJJ. 2014. Degenerate target sites mediate rapid primed CRISPR adaptation. *Proc Natl Acad Sci U S A* 111:E1629–E1638.

25. Box AM, McGuffie MJ, O'Hara BJ, Seed KD. 2015. Functional analysis of bacteriophage immunity through a type I-E CRISPR-Cas system in *Vibrio cholerae* and its application in bacteriophage genome engineering. *J Bacteriol* 198:578–590.

26. Goldberg GW, Jiang W, Bikard D, Marraffini LA. 2014. Conditional tolerance of temperate phages via transcription-dependent CRISPR-Cas targeting. *Nature* 514:633–637.

27. Pyenson NC, Gayvert K, Varble A, Elemento O, Marraffini LA. 2017. Broad targeting specificity during bacterial type III CRISPR-Cas immunity constrains viral escape. *Cell Host Microbe* 22:343–353.e3.

28. Silas S, Makarova KS, Shmakov S, Páez-Espino D, Mohr G, Liu Y, Davison M, Roux S, Krishnamurthy SR, Fu BXH, Hansen LL, Wang D, Sullivan MB, Millard A, Clokie MR, Bhaya D, Lambowitz AM, Kyrpides NC, Koonin EV, Fire AZ. 2017. On the origin of reverse transcriptase-using CRISPR-Cas systems and their hyperdiverse, enigmatic spacer repertoires. *mBio* 8:e00897-17.

29. Weigele P, Raleigh EA. 2016. Biosynthesis and function of modified bases in bacteria and their viruses. *Chem Rev* 116:12655–12687.

30. Iyer LM, Zhang D, Burroughs AM, Aravind L. 2013. Computational identification of novel biochemical systems involved in oxidation, glycosylation and other complex modifications of bases in DNA. *Nucleic Acids Res* 41:7635–7655.

31. Lee Y-J, Dai N, Walsh SE, Müller S, Fraser ME, Kauffman KM, Guan C, Corrêa IR, Jr, Weigele PR. 2018. Identification and biosynthesis of thymidine hypermodifications in the genomic DNA of widespread bacterial viruses. *Proc Natl Acad Sci U S A* 115:E3116–E3125.

32. Yaung SJ, Esvelt KM, Church GM. 2014. CRISPR/Cas9-mediated phage resistance is not impeded by the DNA modifications of phage T4. *PLoS One* 9:e98811.

33. Bryson AL, Hwang Y, Sherrill-Mix S, Wu GD, Lewis JD, Black L, Clark TA, Bushman FD. 2015. Covalent modification of bacteriophage T4 DNA inhibits CRISPR-Cas9. *mBio* 6:e00648-15.

34. Vlot M, Houkes J, Lochs SJA, Swarts DC, Zheng P, Kunne T, Mohanraju P, Anders C, Jinek M, van der Oost J, Dickman MJ, Brouns SJJ. 2018. Bacteriophage DNA glucosylation impairs target DNA binding by type I and II but not by type V CRISPR-Cas effector complexes. *Nucleic Acids Res* 46:873–885.

35. Strotskaya A, Savitskaya E, Metlitskaya A, Morozova N, Datsenko KA, Semenova E, Severinov K. 2017. The action of *Escherichia coli* CRISPR-Cas system on lytic bacteriophages with different lifestyles and development strategies. *Nucleic Acids Res* 45:1946–1957.

36. Mendoza SD, Nieweglowska ES, Govindarajan S, Leon LM, Berry JD, Tiwari A, Chaikeeratisak V, Pogliano J, Agard DA, Bondy-Denomy J. 2020. A bacteriophage nucleus-like compartment shields DNA from CRISPR nucleases. *Nature* 577:244–248.

37. Bair CL, Black LW. 2007. A type IV modification dependent restriction nuclease that targets glucosylated hydroxymethyl cytosine modified DNAs. *J Mol Biol* 366:768–778.

38. Labrie SJ, Samson JE, Moineau S. 2010. Bacteriophage resistance mechanisms. *Nat Rev Microbiol* 8:317–327.

39. Chaikeeratisak V, Nguyen K, Khanna K, Brilot AF, Erb ML, Coker JKC, Vavilina A, Newton GL, Buschauer R, Pogliano K, Villa E, Agard DA, Pogliano J. 2017. Assembly of a nucleus-like structure during viral replication in bacteria. *Science* 355:194–197.

40. Malone LM, Warring SL, Jackson SA, Warnecke C, Gardner PP, Gumy LF, Fineran PC. 2020. A jumbo phage that forms a nucleus-like structure evades CRISPR-Cas DNA targeting but is vulnerable to type III RNA-based immunity. *Nat Microbiol* 5:48–55.

41. Chaikeeratisak V, Khanna K, Nguyen KT, Sugie J, Egan ME, Erb ML, Vavilina A, Nonejuie P, Nieweglowska E, Pogliano K, Agard DA, Villa E, Pogliano J. 2019. Viral capsid trafficking along treadmilling tubulin filaments in bacteria. *Cell* 177:1771–1780.e12.

42. Bondy-Denomy J, Pawluk A, Maxwell KL, Davidson AR. 2013. Bacteriophage genes that inactivate the CRISPR/Cas bacterial immune system. *Nature* 493:429–432.

43. Borges AL, Davidson AR, Bondy-Denomy J. 2017. The discovery, mechanisms, and evolutionary impact of anti-CRISPRs.

44. Pawluk A, Davidson AR, Maxwell KL. 2018. Anti-CRISPR: discovery, mechanism and function. *Nat Rev Microbiol* 16:12–17.

45. He F, Bhoobalan-Chitty Y, Van LB, Kjeldsen AL, Dedola M, Makarova KS, Koonin EV, Brodersen DE, Peng X. 2018. Anti-CRISPR proteins encoded by archaeal lytic viruses inhibit subtype I-D immunity. *Nat Microbiol* 3:461–469.

46. Hwang S, Maxwell KL. 2019. Meet the anti-CRISPRs: widespread protein inhibitors of CRISPR-Cas systems. *CRISPR J* 2:23–30.

47. Wiedenheft B, van Duijn E, Bultema JB, Waghmare SP, Zhou K, Barendregt A, Westphal W, Heck AJR, Boekema EJ, Dickman MJ, Doudna JA. 2011. RNA-guided complex from a bacterial immune system enhances target recognition through seed sequence interactions. *Proc Natl Acad Sci U S A* 108:10092–10097.

48. Westra ER, van Erp PBG, Künne T, Wong SP, Staals RHJ, Seegers CLC, Bollen S, Jore MM, Semenova E, Severinov K, de Vos WM, Dame RT, de Vries R, Brouns SJJ, van der Oost J. 2012. CRISPR immunity relies on the consecutive binding and degradation of negatively supercoiled invader DNA by Cascade and Cas3. *Mol Cell* 46:595–605.

49. Huo Y, Nam KH, Ding F, Lee H, Wu L, Xiao Y, Farchione MD Jr, Zhou S, Rajashankar K, Kurinov I, Zhang R, Ke A. 2014. Structures of CRISPR Cas3 offer mechanistic insights into Cascade-activated DNA unwinding and degradation. *Nat Struct Mol Biol* 21:771–777.

50. Bondy-Denomy J, Garcia B, Strum S, Du M, Rollins MF, Hidalgo-Reyes Y, Wiedenheft B, Maxwell KL, Davidson AR. 2015. Multiple mechanisms for CRISPR-Cas inhibition by anti-CRISPR proteins. *Nature* 526:136–139.

51. Chowdhury S, Carter J, Rollins MF, Golden SM, Jackson RN, Hoffmann C, Nosaka L, Bondy-Denomy J, Maxwell KL, Davidson AR, Fischer ER, Lander GC, Wiedenheft B. 2017. Structure reveals mechanisms of viral suppressors that intercept a CRISPR RNA-guided surveillance complex. *Cell* 169:47–57.e11.

52. Guo TW, Bartesaghi A, Yang H, Falconieri V, Rao P, Merk A, Eng ET, Raczkowski AM, Fox T, Earl LA, Patel DJ, Subramaniam S. 2017. Cryo-EM structures reveal mechanism and inhibition of DNA targeting by a CRISPR-Cas surveillance complex. *Cell* 171:414–426.e12.

53. Peng R, Xu Y, Zhu T, Li N, Qi J, Chai Y, Wu M, Zhang X, Shi Y, Wang P, Wang J, Gao N, Gao GF. 2017. Alternate binding modes of anti-CRISPR viral suppressors AcrF1/2 to Csy surveillance complex revealed by cryo-EM structures. *Cell Res* 27:853–864.

54. Harrington LB, Doxzen KW, Ma E, Liu J-J, Knott GJ, Edraki A, Garcia B, Amrani N, Chen JS, Cofsky JC, Kranzusch PJ, Sontheimer EJ, Davidson AR, Maxwell KL, Doudna JA. 2017. A broad-spectrum inhibitor of CRISPR-Cas9. *Cell* 170:1224–1233.e15.

55. Sun W, Yang J, Cheng Z, Amrani N, Liu C, Wang K, Ibraheim R, Edraki A, Huang X, Wang M, Wang J, Liu L, Sheng G, Yang Y, Lou J, Sontheimer EJ, Wang Y. 2019. Structures of *Neisseria meningitidis* Cas9 complexes in catalytically poised and anti-CRISPR-inhibited states. *Mol Cell* 76:938–952.e5.

56. Lee J, Mir A, Edraki A, Garcia B, Amrani N, Lou HE, Gainetdinov I, Pawluk A, Ibraheim R, Gao XD, Liu P, Davidson AR, Maxwell KL, Sontheimer EJ. 2018. Potent Cas9 inhibition in bacterial and human cells by AcrIIC4 and AcrIIC5 anti-CRISPR proteins. *mBio* 9:e02321-18.

57. Thavalingam A, Cheng Z, Garcia B, Huang X, Shah M, Sun W, Wang M, Harrington L, Hwang S, Hidalgo-Reyes Y, Sontheimer EJ, Doudna J, Davidson AR, Moraes TF, Wang Y, Maxwell KL. 2019. Inhibition of CRISPR-Cas9 ribonucleoprotein complex assembly by anti-CRISPR AcrIIC2. *Nat Commun* 10:2806.

58. Zhu Y, Gao A, Zhan Q, Wang Y, Feng H, Liu S, Gao G, Serganov A, Gao P. 2019. Diverse mechanisms of CRISPR-Cas9 inhibition by type IIC anti-CRISPR proteins. *Mol Cell* 74:296–309.e7.

59. Dong D, Guo M, Wang S, Zhu Y, Wang S, Xiong Z, Yang J, Xu Z, Huang Z. 2017. Structural basis of CRISPR-SpyCas9 inhibition by an anti-CRISPR protein. *Nature* 546:436–439.

60. Yang H, Patel DJ. 2017. Inhibition mechanism of an anti-CRISPR suppressor AcrIIA4 targeting SpyCas9. *Mol Cell* 67:117–127.e5.

61. Shin J, Jiang F, Liu J-J, Bray NL, Rauch BJ, Baik SH, Nogales E, Bondy-Denomy J, Corn JE, Doudna JA. 2017. Disabling Cas9 by an anti-CRISPR DNA mimic. *Sci Adv* 3:e1701620.

62. Jiang F, Liu J-J, Osuna BA, Xu M, Berry JD, Rauch BJ, Nogales E, Bondy-Denomy J, Doudna JA. 2019. Temperature-responsive competitive inhibition of CRISPR-Cas9. *Mol Cell* 73:601–610.e5.

63. Liu L, Yin M, Wang M, Wang Y. 2019. Phage AcrIIA2 DNA mimicry: structural basis of the CRISPR and anti-CRISPR arms race. *Mol Cell* 73:611–620.e3.

64. Watters KE, Fellmann C, Bai HB, Ren SM, Doudna JA. 2018. Systematic discovery of natural CRISPR-Cas12a inhibitors. *Science* 362:236–239.

65. Marino ND, Zhang JY, Borges AL, Sousa AA, Leon LM, Rauch BJ, Walton RT, Berry JD, Joung JK, Kleinstiver BP, Bondy-Denomy J. 2018. Discovery of widespread type I and type V CRISPR-Cas inhibitors. *Science* 362:240–242.

66. Knott GJ, Thornton BW, Lobba MJ, Liu J-J, Al-Shayeb B, Watters KE, Doudna JA. 2019. Broad-spectrum enzymatic inhibition of CRISPR-Cas12a. *Nat Struct Mol Biol* 26:315–321.

67. Zhang H, Li Z, Daczkowski CM, Gabel C, Mesecar AD, Chang L. 2019. Structural basis for the inhibition of CRISPR-Cas12a by anti-CRISPR proteins. *Cell Host Microbe* 25:815–826.e4.

68. Dong L, Guan X, Li N, Zhang F, Zhu Y, Ren K, Yu L, Zhou F, Han Z, Gao N, Huang Z. 2019. An anti-CRISPR protein disables type V Cas12a by acetylation. *Nat Struct Mol Biol* 26:308–314.

69. Wang J, Ma J, Cheng Z, Meng X, You L, Wang M, Zhang X, Wang Y. 2016. A CRISPR evolutionary arms race: structural insights into viral anti-CRISPR/Cas responses. *Cell Res* 26:1165–1168.

70. Pawluk A, Shah M, Mejdani M, Calmettes C, Moraes TF, Davidson AR, Maxwell KL. 2017. Disabling a type I-E CRISPR-Cas nuclease with a bacteriophage-encoded anti-CRISPR protein. *mBio* 8:e01751-17.

71. Bondy-Denomy J, Davidson AR, Doudna JA, Fineran PC, Maxwell KL, Moineau S, Peng X, Sontheimer EJ, Wiedenheft B. 2018. A unified resource for tracking anti-CRISPR names. *CRISPR J* 1:304–305.

72. Meeske AJ, Jia N, Cassel AK, Kozlova A, Liao J, Wiedmann M, Patel DJ, Marraffini LA. 2020. A phage-encoded anti-CRISPR enables complete evasion of type VI-A CRISPR-Cas immunity. *Science* 369:54–59.

73. Lin P, Qin S, Pu Q, Wang Z, Wu Q, Gao P, Schettler J, Guo K, Li R, Li G, Huang C, Wei Y, Gao GF, Jiang J, Wu M. 2020. CRISPR-Cas13 inhibitors block RNA editing in bacteria and mammalian cells. *Mol Cell* 78:850–861.e5.

74. Bhoobalan-Chitty Y, Johansen TB, Di Cianni N, Peng X. 2019. Inhibition of type III CRISPR-Cas immunity by an archaeal virus-encoded anti-CRISPR protein. *Cell* 179:448–458.e11 http://dx.doi.org/10.1016/j.cell.2019.09.003.

75. Athukoralage JS, McMahon SA, Zhang C, Grüschow S, Graham S, Krupovic M, Whitaker RJ, Gloster TM, White MF. 2020. An anti-CRISPR viral ring nuclease subverts type III CRISPR immunity. *Nature* 577:572–575.

76. Uribe RV, van der Helm E, Misiakou M-A, Lee S-W, Kol S, Sommer MOA. 2019. Discovery and characterization of Cas9 inhibitors disseminated across seven bacterial phyla. *Cell Host Microbe* 25:233–241.e5.

77. Pawluk A, Staals RHJ, Taylor C, Watson BNJ, Saha S, Fineran PC, Maxwell KL, Davidson AR. 2016. Inactivation of CRISPR-Cas systems by anti-CRISPR proteins in diverse bacterial species. *Nat Microbiol* 1:16085.

78. Pawluk A, Amrani N, Zhang Y, Garcia B, Hidalgo-Reyes Y, Lee J, Edraki A, Shah M, Sontheimer EJ, Maxwell KL, Davidson AR. 2016. Naturally occurring off-switches for CRISPR-Cas9. *Cell* 167:1829–1838.e9.

79. Stanley SY, Borges AL, Chen K-H, Swaney DL, Krogan NJ, Bondy-Denomy J, Davidson AR. 2019. Anti-CRISPR-associated proteins are crucial repressors of anti-CRISPR transcription. *Cell* 178:1452–1464.e13.

80. Birkholz N, Fagerlund RD, Smith LM, Jackson SA, Fineran PC. 2019. The autoregulator Aca2 mediates anti-CRISPR repression. *Nucleic Acids Res* 47:9658–9665.

81. van Houte S, Ekroth AKE, Broniewski JM, Chabas H, Ashby B, Bondy-Denomy J, Gandon S, Boots M, Paterson S, Buckling A, Westra ER. 2016. The diversity-generating benefits of a prokaryotic adaptive immune system. *Nature* 532:385–388.

82. Li Y, Bondy-Denomy J. 2021. Anti-CRISPRs go viral: the infection biology of CRISPR-Cas inhibitors. *Cell Host Microbe* 29:704–714.

83. Borges AL, Zhang JY, Rollins MF, Osuna BA, Wiedenheft B, Bondy-Denomy J. 2018. Bacteriophage cooperation suppresses CRISPR-Cas3 and Cas9 immunity. *Cell* 174:917–925.e10.

84. Landsberger M, Gandon S, Meaden S, Rollie C, Chevallereau A, Chabas H, Buckling A, Westra ER, van Houte S. 2018. Anti-CRISPR phages cooperate to overcome CRISPR-Cas immunity. *Cell* 174:908–916.e12.

85. Rauch BJ, Silvis MR, Hultquist JF, Waters CS, McGregor MJ, Krogan NJ, Bondy-Denomy J. 2017. Inhibition of CRISPR-Cas9 with bacteriophage proteins. *Cell* 168:150–158.e10.

86. Bondy-Denomy J, Qian J, Westra ER, Buckling A, Guttman DS, Davidson AR, Maxwell KL. 2016. Prophages mediate defense against phage infection through diverse mechanisms. *ISME J* 10:2854–2866.

Regulation of CRISPR-Cas Expression and Function

Leah M. Smith[1*], Aroa Rey Campa[1,2*], and Peter C. Fineran[1,2]

[1]Department of Microbiology and Immunology, University of Otago, Dunedin 9054, New Zealand
[2]Bio-Protection Research Centre, University of Otago, Dunedin 9054, New Zealand

Introduction

Bacteria are continually exposed to bacteriophages (phages) and other mobile genetic elements (MGEs). Phages are highly abundant, present in a vast variety of environments, and outnumber bacteria approximately 10-fold (1). Phages, together with MGEs, can pose a threat to bacteria; therefore, bacteria have evolved different defense systems to overcome these dangers. The different defenses can be broadly classified into two types: innate and adaptive immunity (2). Examples of bacterial innate immunity include restriction-modification systems, surface modification, and toxin-antitoxin and abortive infection mechanisms. Adaptive immunity is characterized by the generation of memories of previous infections. So far, CRISPR-Cas (clustered regularly interspaced short palindromic repeats and CRISPR associated) systems are the only type of adaptive immunity described for bacteria (3).

CRISPR-Cas loci consist of two components: *cas* genes and CRISPR arrays. The *cas* genes encode a number of proteins fundamental for the immune response, including helicases and endonucleases (4). The presence of different *cas* genes forms the basis of grouping systems into different classes, types, and subtypes (5). CRISPR arrays act as memory banks of previous infections, containing short sequences of DNA (spacers) that are derived from, and hence match, the invader nucleic acid (6). Spacers are separated by short repeat sequences of a similar length.

CRISPR defense consists of three steps. During the first step, adaptation, a new spacer is incorporated into the CRISPR array (7, 8). In the second step, expression of the *cas* genes and transcription of the CRISPR arrays occurs, forming the pre-CRISPR RNA (pre-crRNA). Cas proteins and other components then process the pre-crRNA into mature crRNAs (9). Finally, during the third step, interference, the invader DNA is recognized and cleaved by the ribonucleoprotein complex (10).

CRISPR-Cas systems are found in about half of sequenced bacteria and in nearly all archaea (11). The immunity they provide can be beneficial through protection from

---

*Equal contributions

CRISPR: Biology and Applications, First Edition. Edited by Rodolphe Barrangou, Erik J. Sontheimer, and Luciano A. Marraffini.

phages. However, there are also disadvantages associated with the presence of these systems (12). Horizontal gene transfer (HGT) plays an important role in the evolution of bacterial communities, contributing to environmental adaptation (13). The presence of CRISPR-Cas immunity can limit the acquisition of horizontally acquired elements, potentially influencing fitness in certain environments. It has been observed that bacteria can lose CRISPR-Cas function under conditions in which horizontal transfer is important for survival (14). Indeed, in some bacterial pathogens, the presence of antibiotic resistance genes negatively correlates with CRISPR-Cas systems (15, 16). Therefore, there appears to be a requirement for a balance between protection and acquisition of new elements to enable rapid evolutionary adaptation.

It is not uncommon for spacers in CRISPR arrays to match chromosomal sequences. In fact, a study on lactic acid bacteria showed that 23% of spacers match chromosomal genes (17). It was suggested that self-targeting spacers could have a regulatory role (18); however, further studies analyzing the spacer origins indicated that it is more likely a case of autoimmunity (19). Sequencing of CRISPR arrays showed that chromosomal spacers are almost always the last spacer acquired (20), providing evidence in support of the autoimmunity hypothesis. The presence of self-targeting spacers is cytotoxic and can lead to genome reshaping, causing mutations or deletions of the *cas* locus or the targeted regions, including the loss of entire genomic islands (21, 22).

With the need to balance cell defense, horizontally acquired elements, and autoimmune effects, it is clear that CRISPR-Cas systems need regulation to provide defense when required but also to limit the fitness costs associated with their expression. While the mechanisms, classification, and biotechnological applications of CRISPR-Cas systems have been well studied and exploited, their regulation is not so well understood. In this chapter we examine the effects that different regulators have on CRISPR immunity.

QS Influences CRISPR-Cas Immunity

Quorum sensing (QS) is a cell-cell communication system in which chemical signals are produced by bacteria, accumulate as cell density increases, and elicit changes in gene expression in bacterial populations (Fig. 10.1). In this manner, QS coordinately regulates diverse physiological processes in bacterial groups, including biofilm formation, virulence, antibiotic production, and motility (23). In Gram-negative bacteria, the most prevalent signaling molecules are *N*-acyl-homoserine lactones (AHLs), which are predominantly produced by LuxI family AHL synthase enzymes (24). AHLs, often known as autoinducers since some activate their own synthesis (Fig. 10.1; e.g., LasIR), are released by cells and accumulate as cell density rises under conditions of limited diffusion (25). AHLs are typically recognized and bound by specific DNA-binding transcriptional regulators of the LuxR family. When a certain threshold is reached, AHLs bind to their LuxR receptors, causing a conformational change that alters the ability of LuxR to bind DNA and alter gene expression (24, 26). Frequently, LuxR proteins activate gene expression when bound by their cognate AHLs, but they can also act as repressors that become derepressed in the presence of QS signals.

The risk of a viral epidemic increases as cell density rises, as bacteriophages require a certain level of bacterial cell density to proliferate, known as the replication threshold (27). This suggests that CRISPR immunity would be more important during times of high cell density, when the threat of a phage epidemic or spread of MGEs by conjugation or transformation is highest. QS regulates CRISPR-Cas activity in a cell density-dependent manner in both *Serratia* sp. strain ATCC 39006 and *Pseudomonas aeruginosa* PA14 (Fig. 10.1) (28, 29). In *Serratia*, the SmaIR QS system regulates immunity of type I-E, I-F,

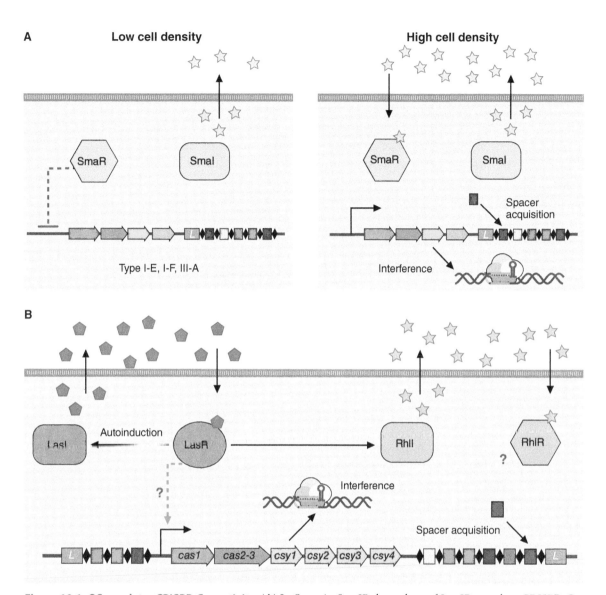

Figure 10.1 QS regulates CRISPR-Cas activity. (**A**) In *Serratia*, SmaIR, homologs of LuxIR, regulate CRISPR-Cas activity in a cell density-dependent manner. When AHL signals are low, repression of *cas* genes and CRISPR arrays is caused by SmaR. When a certain threshold is reached, AHL molecules bind to SmaR, resulting in derepression of *cas* expression, increased interference, and spacer acquisition. (**B**) In *P. aeruginosa*, both LasI and RhlI are involved in CRISPR-Cas regulation. These systems are interconnected, as LasR activates *rhlI*. When the signals accumulate, *cas* expression increases, leading to stronger interference and increased spacer acquisition. Solid arrows, direct regulation; dashed arrows, unknown; *L*, leader; ?, role of protein in CRISPR-Cas regulation is unknown.

and III-A CRISPR-Cas systems by controlling expression of *cas* genes and CRISPR arrays, as well as influencing spacer acquisition and interference (28). During low cell density, when the signaling molecules (AHLs) are less abundant, CRISPR immunity is repressed by SmaR. When cell density increases, AHL signals accumulate, alleviating SmaR-dependent repression; consequently, CRISPR immunity increases (Fig. 10.1A) (28). *P. aeruginosa* PA14 harbors a type I-F CRISPR-Cas system that is controlled by its two primary QS systems, LasIR and RhlIR. In the absence of QS signals at early growth phases, expression of *cas* genes and CRISPR-Cas immunity is reduced. During later growth stages, AHLs accumulate, enhancing plasmid loss and spacer acquisition (Fig. 10.1B) (29); however, the role, if any, of the LasR and RhlR signal receptors is unknown. The effect of QS on CRISPR-Cas is also observed in other bacteria.

In *Burkholderia glumae* PG1, decreased type I-F *cas* expression was seen by RNA sequencing in mutants unable to produce AHLs (30). Furthermore, AHL-dependent regulation of a type I-F system was detectable in a microarray study of *Pectobacterium atrosepticum* (28, 31).

A link between CRISPR-Cas immunity, QS, and temperature has been described for *P. aeruginosa* (32). At low temperature, CRISPR adaptation was enhanced. This effect was both QS independent (i.e., it occurred when *lasIR* and *rhlIR* were deleted) and QS dependent, since addition of AHLs increased spacer acquisition at intermediate temperatures. The low-temperature effect may be due to lowered growth rates, since slower growth led to enhanced plasmid interference. It is hypothesized that a lower growth rate provided the cells more time to adapt to clear the infection.

More recently, a regulator of QS with a role in CRISPR-Cas regulation was identified in *P. aeruginosa* PA14 (33). CdpR (ClpAP degradation and pathogenicity regulator), an AraC family regulator, decreases QS signals by reducing transcripts of *lasI* and *rhlI*. The lower QS signal levels caused Vfr (virulence factor regulator) to decline, and Vfr can bind to a *cis* response element in the *cas1* promoter to activate gene expression. Therefore, CdpR, by repressing QS regulators, leads to a reduction of Vfr, which causes decreased *cas* expression and CRISPR-Cas adaptation and interference. Interestingly, in *P. aeruginosa* PA14, the type I-F CRISPR-Cas system is reported to promote cleavage of some host mRNAs—a possible case of "autoimmunity"—and CdpR-mediated CRISPR-Cas repression reduces endogenous mRNA cleavage, potentially mitigating detrimental targeting (33).

CRISPR-Cas Repression through H-NS

Nucleoid-associated proteins (NAPs) are DNA-binding proteins that alter the global and local topology of the chromosome, and they are responsible for a basal level of transcriptional repression in bacteria (34). The best-studied NAP is the heat-stable nucleoid structuring protein (H-NS). H-NS does not recognize a specific target sequence; however, it has an affinity for curved DNA *in vitro* (35, 36) and AT-rich sequences (37, 38)—features often found in promoter regions. The binding and oligomerization of H-NS near promoters are responsible for widespread transcriptional repression in bacteria (39).

In *Escherichia coli* strain K-12, the type I-E CRISPR-Cas system appears to be inactive under laboratory conditions (40, 41). Introduction of a spacer targeting phage λ, either on an artificial plasmid-encoded CRISPR array (42) or in the chromosome (40), offers protection against phage infection. However, this effect was dependent on, or enhanced by, overexpression of *cas* genes, indicating that wild-type levels of *cas* expression are insufficient to stimulate CRISPR immunity (40, 42). Subsequent mutation of *hns* in a strain containing a chromosomally introduced λ phage spacer abolishes infection, indicating that the CRISPR-Cas system is normally silenced by H-NS (40). Furthermore, H-NS directly represses CRISPR-Cas expression by binding to the *cas8e* (*casA*) and CRISPR1 promoters, likely inhibiting RNA polymerase (RNAP) binding and transcriptional initiation (43). Expression of *cas3* is also controlled by H-NS during stationary phase (44).

An *hns* mutant of an *E. coli* Stx2 (Shiga toxin-converting bacteriophage) lysogen showed elevated *cas* expression, as well as decreased transformation efficiency of plasmids carrying protospacers with matches in either the CRISPR1 or CRISPR2 array (45). Phage replication and lysogenization were also reduced in the *hns* mutant background when strains were engineered to harbor anti-phage protospacers. Interestingly, an *hns* mutation in the wild-type Stx2 lysogen background exhibited decreased swarming motility and biofilm formation, which is reminiscent of a reduction in these phenotypes observed in a *P. aeruginosa* strain PA14 DMS3 lysogen (46). In *P. aeruginosa*, the effect of lysogeny on these group behaviors is due to overexpression of SOS response-related genes (including pyocin genes), which are stimulated by Cas3-mediated DNA damage following partial target recognition at an integrated DMS3 prophage (47). However, the CRISPR-Cas dependency of these phenotypes in the *E. coli* Stx2 lysogen has not been established.

In *Klebsiella pneumoniae*, deletion of *hns* led to increased type I-E CRISPR-Cas activity, as measured by transformation efficiency, spacer acquisition, and plasmid loss (48). The addition of the carbapenem antibiotic imipenem resulted in elevated *hns* and decreased *cas3* mRNA levels, suggesting that increased H-NS further represses *cas* genes. In agreement, overexpression of *hns*, through the addition of imipenem, stabilized a CRISPR-Cas-targeted plasmid (48). The role of imipenem as an inducer of *hns* expression is unclear but may be related generally to cellular stress responses. This model of CRISPR-Cas silencing by H-NS fits with the assumption that these defense systems may be tightly regulated and activated only in response to specific threats or stress signals. CRISPR-Cas silencing by H-NS may provide an advantage under situations where the acquisition of horizontally acquired elements (i.e., antibiotic resistance genes) may be advantageous.

How does the cell "know" when to alleviate H-NS repression in favor of the transcription of CRISPR-Cas loci? Derepression of H-NS can be achieved via several mechanisms, one of which includes interaction with, or antagonism by, other DNA-binding proteins, such as the LysR family transcriptional factor LeuO (Fig. 10.2) (49). LeuO, a leucine biosynthesis regulator, binds to the promoter region of *cas8e* (*casA/cse1*), derepressing the

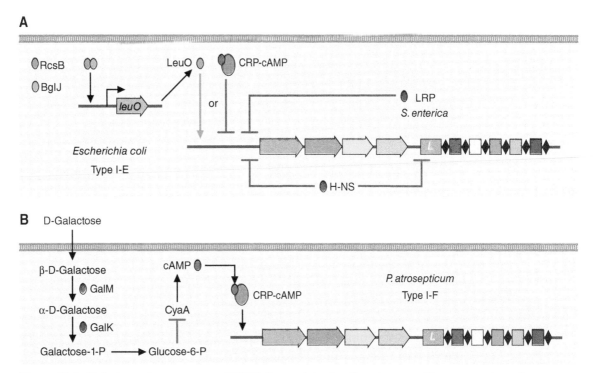

Figure 10.2 Global regulators govern CRISPR-Cas activity by directly controlling expression of *cas* genes and CRISPR arrays. (**A**) In *E. coli*, H-NS represses CRISPR-Cas by inhibiting expression of both *cas* genes and CRISPR arrays. Similarly, LRP in *S. enterica* also represses *cas* expression. LeuO, which can be activated by the RcsB/BglJ complex, causes increased CRISPR-Cas activity by activation of *cas* expression, likely through derepression of H-NS. Antagonism between regulators is also observed, as CRP-cAMP competes with LeuO for binding, repressing *cas* expression. (**B**) In *Pectobacterium atrosepticum*, galactose metabolism plays a role in CRISPR-Cas regulation through glucose-mediated repression of CyaA, which produces cAMP. CRP complexed with cAMP activates *cas* expression.

H-NS-silenced *cas* operon in *E. coli* K-12. Overexpression of *leuO* increases protection against λ phage in a strain containing a chromosomally introduced spacer (41). Furthermore, deletion of *leuO* leads to an increase in P1 phage titer in a strain containing endogenous spacers with partial complementarity to P1 (50). Together, these findings demonstrate that LeuO is required for CRISPR-Cas activation. Expression of *leuO* is, in turn, activated by the heterodimeric RcsB/BglJ complex (51, 52). RcsB, a response regulator involved in sensing membrane perturbation (53), complexes with the BglJ transcription factor to induce *leuO* expression (Fig. 10.2A) (51). Although *cas* transcription levels were increased in a BglJ overexpression strain, the effect on CRISPR-Cas activity has not been assessed. Notably, there was a decrease in the accumulation of mature crRNAs, which was proposed to be a result of destabilization of the Cascade complex, but the mechanism remains unknown. It is likely that a LeuO-independent target of BglJ regulation could influence the stability of the Cascade complex in *E. coli* (52).

In addition to regulation of the type I-E CRISPR-Cas system by H-NS and LeuO (54, 55), the leucine-responsive regulatory protein (LRP) also silences CRISPR-Cas expression in *Salmonella enterica* serovar Typhimurium by binding upstream and downstream of the *cas8e* (*casA*) promoter (Fig. 10.2) (55). LRP is a global regulator involved in amino acid transport (56) and other cellular processes, including virulence and pilus formation (57). In fact, a recent experiment using both chromatin immunoprecipitation sequencing and RNA sequencing (RNA-seq) has revealed that the LRP regulon constitutes about one-third of the *E. coli* genome, where the majority of changes are due to indirect effects (58). Interestingly, unlike H-NS, LRP does not bind to the *cas* promoter of *E. coli* or influence *cas* expression (43, 58). However, LRP has been shown to directly bind the *leuO* promoter region in *E. coli* (58, 59). While silencing of CRISPR-Cas by H-NS occurs in divergent bacterial lineages, negative regulation by LRP appears to be restricted to certain lineages, such as *Salmonella*. The helix-turn-helix (HTH) domain of LRP recognizes a palindromic sequence in the DNA (60, 61), and perhaps the architecture of the *E. coli* promoter precludes direct LRP regulation. Taken together, the data on LeuO and LRP suggest a link between cellular stresses, like membrane perturbation and amino acid availability, and CRISPR-Cas activation. However, it is important to note that these studies have not directly tested the role of such stresses in CRISPR-Cas activity.

Interestingly, the main determinant of H-NS localization appears to be base composition, as exhibited in *S.* Typhimurium (37, 38), rather than DNA topology, as was previously hypothesized (35, 36). The correlation of H-NS associating with AT-rich regions was 20 times stronger than its correlation with GC-rich regions, while the association of H-NS with curved DNA regions is only double that of noncurved DNA (37). These findings pose the interesting theory that H-NS can selectively silence foreign DNA, as AT-rich content is often associated with horizontally acquired elements. H-NS was shown experimentally to localize to and repress expression of foreign DNA introduced from *Helicobacter pylori* into *S. enterica* (38). H-NS also controls the expression of many genes introduced by lateral gene transfer, and deletion of *hns* leads to uncontrolled expression of pathogenicity islands in *S.* Typhimurium (37).

The ability of H-NS to discriminate self from non-self, via GC content differences, may represent a high-order level of protection to ensure that possibly deleterious, horizontally acquired genes are not freely expressed within the new recipient host. This theory, proposed by the "xenogenic silencing" model, suggests that H-NS restricts expression of foreign DNA until it can be integrated into host regulatory circuits (62). It has been suggested that incomplete xenogenic silencing may permit the replication of foreign DNA, which would, in turn, titrate H-NS away from CRISPR-Cas promoters and thus activate expression and subsequent interference (63). Unsurprisingly, in the constant arms race between bacteria and phage, there are examples of phages and mobile genetic elements overcoming xenogenic silencing induced by H-NS. Antagonists of H-NS, such as Ler and H-NST, are often associated with pathogenicity/genomic islands, and evidence suggests that such antagonists have emerged via HGT (62). Additionally, the T7 phage protein gp5.5 directly binds and inhibits H-NS, enabling expression of H-NS-repressed loci (64). It has also been postulated that phage-encoded H-NS may function to suppress host gene expression, possibly providing a fitness advantage to the phage if CRISPR-Cas or other defenses against MGEs are repressed (65).

CRP Modulates CRISPR-Cas Immunity

The cyclic AMP (cAMP) receptor protein (CRP) and its effector molecule cAMP are global regulators in bacteria and are best known for transcriptional regulation in response to carbon source availability (66, 67). In *E. coli*, the CyaA adenylate cyclase binds the glucose transporter EIIA^{Glc} when glucose levels are low and EIIA^{Glc} is predominantly phosphorylated, activating cAMP synthesis and thus leading to increased cAMP (68). When cAMP binds to CRP, a conformational change occurs, leading to CRP binding to conserved palindromic sequences present in the promoters of target genes. The position of CRP binding sites can determine whether the result is transcriptional activation or repression (69). Although the main role of CRP-cAMP resides with metabolic regulation, it is also involved in regulation of several other diverse processes, including CRISPR-Cas defense mechanisms (Fig. 10.2). The *cas* genes were initially identified as CRP targets during studies to determine the role of a CRP homolog in the thermophile *Thermus thermophilus* (70), and now several studies have demonstrated the effects of CRP homologs on defense regulation in different organisms (50, 71, 72).

In *E. coli* K-12, CRP-cAMP abundance, associated with low glucose availability, leads to a reduction of *cas* gene expression from the type I-E system (50). LeuO, a transcriptional activator of *cas* genes (41), partially shares its binding site with the cAMP-CRP complex on the *E. coli cas* operon promoter. This overlap results in antagonism between these two proteins, whereby LeuO binding, in the absence of cAMP, displaces CRP and allows for expression of the *cas* genes (Fig. 10.2A).

Based on these findings, it could be hypothesized that when resources are scarce, cells limit their immune responses to lower fitness costs and save energy. However, opposite effects on CRISPR regulation through CRP are observed in *Thermus thermophilus* HB8 and *Pectobacterium atrosepticum*, where CRP-cAMP activates *cas* expression in both bacteria (70, 71). In *T. thermophilus* two *cas* operons regulated by CRP have been identified, the *csm* (RAMP domain type III-A) and the *cse* (Cascade complex type I-E) operons (70). Expression of CRP-regulated genes, including these CRISPR-Cas systems, increased upon ΦYS4 phage infection, potentially due to elevated cAMP levels (72). However, some *cas* genes were still stimulated by phage infection in a Δ*crp* strain, suggesting that unknown additional regulatory elements induced upon phage infection might have been present (72).

P. atrosepticum possesses a type I-F CRISPR-Cas system with three CRISPR arrays. Similar to *T. thermophilus*, CRP also activates the *cas* operon *P. atrosepticum*, but in response to carbon source availability (71). Transposon mutagenesis was performed to explore novel *cas* regulators, and two mutants displaying strong expression phenotypes were identified: an insertion in the *crp* promoter, which led to lower *cas* expression, and a *galM* insertion that increased *cas* expression (71). The presence of a CRP binding site in the *cas1* promoter suggested direct *cas* operon regulation. In agreement, mutations of the CRP binding site led to reduced *cas* expression. *P. atrosepticum* possesses two operons involved in galactose metabolism: *galME* and *galKT*. The *galM* gene encodes a galactose mutarotase that catalyzes the conversion of β-D-galactose to α-D-galactose (73). A more recent study also revealed that *galK* is involved in *cas* regulation (74). The increase in *cas* expression in the *gal* mutants is likely caused by altered carbon metabolism. Indeed, both GalM and GalK are involved in a pathway that converts galactose to glucose, ultimately causing a decrease in cAMP and the CRP-dependent activation of the *cas* operon (Fig. 10.2B). Interestingly, the Gal-cAMP-CRP pathway has no effect on CRISPR expression (71, 74). Taken together, the studies show that CRP directly regulates CRISPR-Cas activity at all three levels—expression, adaptation, and interference—by binding the *cas* promoter, and this regulation is modulated by carbon metabolism.

Specialized Sigma Factors Influence CRISPR-Cas Expression

Another high-order form of transcriptional control comes from the RNAP holoenzyme itself, in the form of sigma (σ) factors. Sigma factors orient RNAP to specific regions of the DNA where transcription is initiated. Changes in environmental conditions lead to the transition from housekeeping σ factor RpoD (σ^{70}) to specialized factors such as RpoS (σ^{38}) and RpoN (σ^{54}), which, in turn, induce widespread transcriptional changes within the cell (75, 76).

The σ^{54} sigma factor directs transcription of seemingly disparate genes, such as those involved in nitrogen assimilation, flagellar biosynthesis, virulence, and carbon source utilization (77). While σ^{70} primarily recognizes the −35 and −10 regions upstream of the transcription start site, σ^{54} alternatively recognizes conserved motifs at −24 and −12 (78). Additionally, transcription from the σ^{54}-containing RNAP requires the presence of bacterial enhancer binding proteins (bEBPs), which recognize an upstream activation sequence and facilitate the transition of RNAP from a closed to an open complex, via ATP hydrolysis (79). Many bEBPs are part of signal transduction pathways, adding an extra layer of regulation to genes under σ^{54} control (80). To define the σ^{54} regulon in *S.* Typhimurium LT2 (81), researchers overexpressed the previously characterized constitutively active bEBP DctD from *Sinorhizobium meliloti* (82). They subsequently identified a σ^{54}-dependent promoter upstream of *cas1*, but transcriptional control was not experimentally confirmed. Interestingly, σ^{54} also regulates expression of QS components in *P. aeruginosa* PAO1 (83–85). Mutation of σ^{54} leads to repression of the *pqs* (*Pseudomonas* quinoline signal) QS system, likely due to the absence, or inactivation, of PqsR (85). In a σ^{54} mutant, the *lasRI* and *rhlRI* QS systems were derepressed and production of QS molecules (OdDHL and BHL) was increased in nutrient yeast broth (83), while in minimal media, *rhlRI* expression and BHL production were reduced (84). This highlights how QS systems, and genes comprising their regulons, may be modulated in response to nutrient availability. While *P. aeruginosa* PAO1 lacks CRISPR-Cas, it is tempting to speculate that σ^{54} may have a role in regulating *cas* expression, possibly in response to nutrient availability, in other strains containing CRISPR-Cas systems.

Extracytoplasmic function (ECF) σ factors are a divergent group of σ factors that are activated under various cellular conditions, usually in response to extracytoplasmic signals (86, 87). Activation of ECF σ factors is often controlled by a transmembrane protein (anti-σ), making the σ/anti-σ system analogous to two-component signal transduction systems (88). The ECF σ/anti-σ pair DdvS/A, along with the proteins CarD and CarG, regulate expression of a type III-B CRISPR-Cas system in *Myxococcus xanthus* (Fig. 10.3) (89). The membrane-associated protein (anti-σ) DdvA sequesters DdvS until an unknown signal induces the release of DdvS. DdvS can then complex with RNAP, where it is directed to

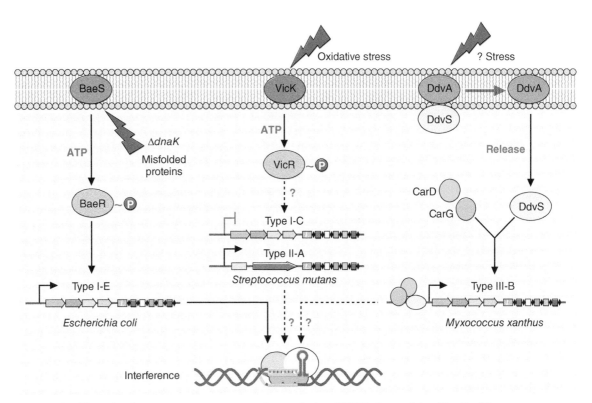

Figure 10.3 The role of stress response systems in regulating CRISPR-Cas activity. The BaeSR two-component membrane stress response system activates type I-E *cas* expression in *E. coli*, likely in response to misfolded protein aggregation in the absence of chaperone protein DnaK (left). The VicRK two-component oxidative stress response system represses type I-C *cas* expression while promoting type II-A *cas* expression in *S. mutans* (center). Upon an unknown stress signal(s), the extracytoplasmic sigma factor DdvS is released from anti-sigma factor DdvA and complexes with CarD and CarG to promote transcription of the type III-B system in *M. xanthus*. Solid arrows, direct regulation; dashed arrows, unknown; ?, role of protein is unknown.

certain promoters to enhance expression. It is unclear how CarD/CarG facilitate this inter-action, but it is hypothesized that the complex may guide RNAP to the promoter via simul-taneous interactions with AT-rich DNA and RNAP itself (90, 91). DdvS, CarD, and CarG directly bind to the *cas* promoter of the type III-B system *in vivo*, resulting in transcription of the *cas* operon through to the end of the CRISPR array, which is located directly down-stream. Because this polycistronic RNA includes *cas6*, which encodes the enzyme respon-sible for crRNA maturation, as well as the CRISPR array, crRNA biogenesis may also, in theory, fall under DdvS/DdvA control.

While most ECF σ factors are activated via signaling through the anti-σ, such as the redox stress-sensing RsrA in *Streptomyces coelicolor* (92, 93), some can be directly activated without this "signal transduction." In *Bacillus subtilis*, transcription of the σ factor genes *sigX* and *sigM*, which are regulated by their corresponding anti-σ proteins (σ^X and σ^M) (94), can also be induced by glucose, independent of the anti-σ proteins (95). Likewise, σ^E in *E. coli* can be activated by (p)ppGpp, a global signaling molecule which responds to nutrient limitation, independently of the anti-σ protein RseA (96, 97). Elevated (p)ppGpp levels, under amino acid or phosphate starvation conditions or artificial induction, activate σ^E activity, whereas a deficiency in (p)ppGpp production lowers σ^E activity (98). While only one example of CRISPR-Cas regulation via an ECF σ/anti-σ pair has been described to date, the ability of these systems to respond to various cellular stresses (physical and nutrient) leads us to predict that they are likely to play a role in the modulation of bacterial defense systems.

Stress Response Systems Implicated in CRISPR-Cas Regulation

The cellular envelope represents the first barrier to foreign genetic elements. Therefore, sensing envelope stress may be an important trigger for modulating bacterial adaptive immunity (99). Bacteria have evolved several systems to sense and respond to extracytoplasmic (envelope) perturbations, including multiple two-component signal transduction systems (100).

The BaeSR (bacterial adaptive response) two-component system, which responds to envelope stress and regulates certain efflux pumps in *E. coli* (101, 102), is also implicated in the activation of the type I-E CRISPR-Cas system under conditions in which misfolded proteins may accumulate (Fig. 10.3) (103). Overexpression of a twin arginine transporter (Tat) signal peptide (ssTorA) fused to green fluorescent protein (GFP) activated CRISPR-Cas expression and interference in a Δ*dnaK* background. The Tat secretory pathway is responsible for translocating proteins across the cytoplasmic membrane (104), while the DnaK (heat shock protein 70) chaperone facilitates proper protein folding (105). It was proposed that in the absence of DnaK, an accumulation of misfolded ssTorA-GFP could localize at the membrane and potentially activate BaeSR. Expression of *cas* was increased in the Δ*dnaK* background upon ssTorA-GFP expression, which was abrogated by deletion of *baeR* or *baeS*, indicating that induction was BaeSR dependent. Electrophoretic mobility shift assays confirmed that purified BaeR binds within the *cas8e* (*casA*) gene, perhaps alleviating repression caused by H-NS binding upstream (43), but how this occurs is unknown. Although BaeR overexpression upregulated *cas8e* (formerly *ygcL*) expression in a previous study in *E. coli* (102), the mechanism of this pathway and its implications for CRISPR-Cas under normal conditions remain unknown.

In addition to DnaK, whose absence stimulated the BaeSR response pathway, another heat stress-induced chaperone is involved in the modulation of CRISPR-Cas activity. High-temperature protein G (HtpG), a homolog of eukaryotic heat shock protein 90 (106), is required for CRISPR-Cas function in *E. coli* K-12, likely through the stabilization of Cas3 levels (107). Deletion of *htpG* led to an increase in transformation efficiency and reduced protection from lysogenization at 32°C in strains carrying anti-phage spacers. In contrast, *cas3* overexpression in the Δ*htpG* background restored CRISPR-Cas activity. A similar role was also observed for HtpG in an *hns* mutant, where *cas3* expression levels likely represent an important limiting factor in protection from phage infection at 37°C (44).

Finally, VicRK, a two-component system that is activated in response to hydrogen peroxide or oxygen stress in *Streptococcus mutans* (108), differentially regulates CRISPR-Cas expression (Fig. 10.3) (109). *S. mutans* UA159 contains both type II-A and I-C CRISPR-Cas systems, with putative VicR binding sites just downstream of the transcription start site of the *cas9* promoter (type II-A) and just upstream of the −35 position of the *cas3* promoter (type I-C). Accordingly, expression of these two systems is differentially modulated by VicRK. While direct regulation or effects on CRISPR-Cas activity have not been shown, reverse transcription-quantitative PCR in the *vicK* histidine kinase mutant indicates that type II-A expression is repressed while type I-C is activated by this two-component system.

Viral Infection Stimulates CRISPR-Cas Response

Metabolic, proteomic, and gene expression studies have been carried out with different bacteria and archaea to elucidate response mechanisms initiated upon viral infection (72, 110–115). While many global changes occur in host metabolism in response to infection, some CRISPR-Cas-specific changes are also evident (72, 112, 113). Microarray analysis revealed that expression of *T. thermophilus* HB8 type I-E and type III-A CRISPR-Cas systems was elevated in response to the lytic phage YS40 (72). In the archaeon *Sulfolobus islandicus* LAL14/1, RNA-seq showed that infection by the lytic virus SIRV2 induced the expression of type I-A, I-D, and III-B *cas* operons, as well as CRISPR arrays (113). Finally, in *Streptococcus thermophilus* DGCC7710, a shotgun proteomics experiment revealed that while some Cas proteins were constitutively expressed (i.e., in the absence of phage infection), levels of other Cas proteins, such as Cas9, were elevated in response to phage 2927 (112). In summary, these studies show that phage infection can trigger increases in gene expression and protein levels of CRISPR-Cas components, possibly bolstering the defense capability of the cells in response to specific threats.

Autoregulators and Exclusive CRISPR Regulators

Most of the proteins and regulators of CRISPR-Cas described above are pleiotropic regulators involved in several different cellular processes and widely conserved among bacteria. Other genes have been identified, often located near *cas* operons, whose products have roles as direct CRISPR-Cas regulators. The first examples of these regulators are the DNA-binding proteins Cbp1, Csa3a, and Csa3b from the *Sulfolobus* genus (Fig. 10.4).

Cbp1, a homolog of the protein formerly called SRSR repeat-binding protein (116), binds directly to the CRISPR repeats. Cbp1 modulates transcription of the CRISPR locus by allowing the formation of longer and more regular transcripts generated from the leader. Cpb1 is proposed to suppress spurious transcriptional activity derived from spacers or repeats within the CRISPR array. It is also speculated that Cpb1 binding could be involved in protecting against spacer loss via recombination within repeat regions (117).

Csa3b, together with the Cascade complex, represses the activity of the interference module of the type I-A system in *Sulfolobus* (118). The Csa3b protein binds to two tandem repeats in the promoter of the interference genes (P*cas*) and represses its activity. In the absence of phage infection, the Cascade complex (and a subcomplex) and Csa3b bind the promoter region, causing transcriptional repression. Upon infection, the Cascade components are recruited to the matching invader DNA, alleviating the repression of P*cas* and consequently leading to transcriptional activation of the interference cassette (118).

Csa3a regulates the activity of the adaptation module (termed a*cas*) in *Sulfolobus*. The *csa3a* gene, also called *csx1*, is commonly associated with *cmr* gene cassettes and occasionally

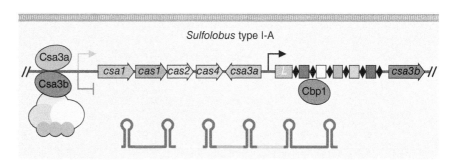

Figure 10.4 Archaeal CRISPR-Cas regulation in *Sulfolobus*. Several regulators have been identified in the *Sulfolobus* genus. Csa3a, by binding to the *cas* promoter, leads to increased *cas* expression, while Csa3b, together with Cascade, inhibits *cas* expression. Cbp1 binds to repeats and promotes longer CRISPR transcripts.

with type I systems. In *S. islandicus*, *csa3a* is located downstream of the adaptation *cas* genes *csa1*, *cas1*, *cas2*, and *cas4*. Csa3a binds to imperfect palindromes in both the *csa1* and *cas1* promoters, increasing transcription from the a*cas* operon. Overexpression of Csa3a also causes increased *de novo* spacer acquisition (119). Additionally, Csa3a transcriptionally upregulates DNA repair genes, including the nuclease *nurA* and the helicase *herA* genes, by binding to their promoters. CRISPR spacer acquisition requires DNA repair genes (120). Therefore, it has been hypothesized that Csa3a may act as a global CRISPR regulator during incorporation of new spacers (121). The structure of the closely related protein Csa3 from *Sulfolobus solfataricus* revealed a C-terminal MarR-like winged HTH domain, suggesting its function as a transcriptional regulator. The crystal structure showed a cleft in the N-terminal domain of this protein, suggesting a conserved ligand binding site for a small molecule that could allosterically modulate the effect of this protein (122). Cyclic oligoadenylate is a key second messenger produced by Cas10 of type III CRISPR-Cas systems upon invader recognition, and it is feasible that this signal might bind to this transcriptional factor and alter its activity (123).

Another example of an exclusive regulator is a gene initially named *devS*. As part of the *dev* operon, together with *devR* and *devT*, these genes are involved in spore differentiation inside *M. xanthus* fruiting bodies (124). These genes were subsequently identified as part of a type I-C *cas* operon including *cas7* (*devR*), *cas5* (*devS*) (125), and *cas8c* (*devT*) (126, 127). Interestingly, Cas5 negatively autoregulates the *cas* operon, and this effect requires regulatory elements in the *dev* promoter region (124). Exactly how this observed Cas5-mediated regulation occurs is unclear, but there might be commonalities with the Cascade-dependent regulation in *Sulfolobus* described earlier (118). *M. xanthus* also harbors a type III-B CRISPR-Cas system with one CRISPR array. Transposon mutagenesis showed that the type III-B *cas* genes, together with the CRISPR array, were involved in the development of fruiting bodies and production of exopolysaccharide (125), again exhibiting their role in sporulation in *M. xanthus*. These roles in sporulation suggest that CRISPR-Cas systems can influence activities other than immunity (128).

Potential regulators are also located near *cas* operons in the cyanobacterium *Synechocystis* (129). *Synechocystis* strain PCC6803 contains three CRISPR-Cas systems: two type III systems and a type I-D system. Mutation of *sll7009*, associated with the type I-D system, resulted in increased CRISPR1 transcripts, suggesting its role as a repressor. However, the effect on CRISPR-Cas immunity was not examined. This protein is predicted to have both HTH DNA-binding and WYL domains and therefore may alter its DNA binding in response to a modified nucleotide (130).

Potential Cooperation between Defense Systems

Eukaryotic Argonaute proteins are involved in RNA silencing in eukaryotes (131), and homologs exist in prokaryotes (pAgos). However, pAgos are not involved in RNA silencing but instead bind RNA or DNA guides and preferentially degrade DNA targets (132). A study of *T. thermophilus* Argonaute (TtAgo) showed that TtAgo did not greatly affect chromosomal gene expression (133). However, TtAgo-dependent plasmid targeting led to increased *cas1*, *cas2*, and *csx1* levels, as detected by RNA-seq, suggesting that TtAgo stimulates CRISPR adaptation from plasmid DNA (133). Furthermore, *Natronobacterium gregoryi* Argonaute interacts with RecA, which is involved in homologous recombination in bacteria (134). Although speculative, since CRISPR adaptation involves DNA repair genes (135), the interaction between TtAgo and RecA might assist in the acquisition of new spacers.

Future Directions

It is clear that bacteria employ multiple tiers of regulation over CRISPR-Cas systems, fitting with the concept that CRISPR-Cas immunity must be fine-tuned to maximize defense while minimizing fitness costs. The highest order form of control is exerted by "global regulators"; however, it is often unclear whether these directly or indirectly exercise transcriptional control. Further work defining precise binding sites of regulators and their positions within hierarchies is needed to decipher mechanisms controlling CRISPR-Cas expression and tease apart direct and indirect regulatory effects. Conversely, some organisms exploit dedicated CRISPR-Cas regulators located near, or even as part of, the *cas* operons. These regulators modulate CRISPR immunity, but occasionally, they regulate other activities particularly important for the life cycle of the organisms (124, 125). This highlights the importance of characterizing other genes associated with *cas* loci, most of which are of yet-unknown function (136).

The importance of regulation in terms of fitness and autoimmune costs, as well as HGT, is an area in need of further exploration. Differential responses to nutrient availability may represent the fine-tuning of regulatory pathways of bacteria occupying diverse niches (2, 137), as it is clear that catabolite repression is important determinant in CRISPR-Cas regulation (50, 70, 71, 74). The ecological niche inhabited also appears to influence HGT (13), as well as the types of immune systems present in bacteria (2). CRISPR-Cas interference has been shown to limit HGT (138), and there is a negative association of CRISPR-Cas systems with antibiotic resistance genes (15, 16). Perhaps the absence, or fine-tuning via regulation, of CRISPR-Cas systems may facilitate the acquisition of potentially beneficial elements for certain organisms (139). Further exploration of how CRISPR-Cas systems or components are horizontally acquired (140, 141) and subsequently integrated into host regulatory circuits is also crucial to understanding the evolution of adaptive immunity.

Membrane perturbation and extracellular cues also appear to be determinants of CRISPR-Cas regulation. Several signal transduction systems have been implicated in the control of CRISPR-Cas activity (89, 103, 109). However, a major limitation of these studies is that such stressors have not been directly shown to trigger CRISPR-Cas defense. To elucidate the role of adaptive immunity in response to membrane perturbation and extracellular signals, it is pertinent to perform experiments which directly compromise membrane integrity or stimulate extracellular stress response pathways—such as might be expected from phage infection or conjugation events. Additionally, the mechanisms of CRISPR-Cas modulation in response to phage infection are unclear. While it may be presumed that a host can change expression levels in response to phage-induced cues, whether these cues are sensed extracellularly (i.e., upon adsorption) or intercellularly (i.e., upon initiation of phage replication) needs further investigation. This brings into question the time frame in which cells sense a perturbation and alter regulation and whether this could feasibly result in increased defense capability once an infection has commenced.

Much of our knowledge of CRISPR-Cas regulation has focused superficially on transcriptional regulation of selected model organisms, for example, *E. coli*. It is essential to dive deeper into proposed pathways, as no complete regulation profile for a single bacterium is currently known. Additionally, studies should move to include diverse organisms, especially those whose CRISPR-Cas systems, unlike that of *E. coli*, are active under normal laboratory conditions. Unraveling CRISPR-Cas modulation in other organisms will inform us on the prevalence of conserved versus strain-specific regulatory circuits. Also, it is appropriate to think of what happens beyond transcription initiation, such as posttranscriptional and posttranslational control (142). While many studies show transcriptional changes in response to some genetic change, few go beyond to indicate how these changes translate into effects on CRISPR-Cas activity. Potential regulators should also be tested for roles in interference and/or adaptation to get a complete view on their involvement in immunity.

Another limitation in our current view of CRISPR-Cas regulation is due to how regulators are discovered. Most information we have about regulation comes from classical genetic screens or from targeted mutagenesis of alleles implicated in CRISPR-Cas modulation. Other regulators are discovered by chance, in studies not directly tasked with understanding regulation. While such discoveries are very useful, there is a need to move toward systematic unbiased (genetic and phenotypic) approaches to build a comprehensive understanding of CRISPR-Cas regulation. This will enable a greater understanding of the natural role of CRISPR-Cas systems and will open up avenues to manipulate their activity for effective phage therapy where CRISPR resistance is undesirable, or to reduce the spread of antibiotic resistance genes by increasing CRISPR immunity.

Acknowledgments

Research in the Fineran laboratory on CRISPR-Cas systems is supported by the Marsden Fund, RSNZ, the Bio-Protection Centre of Research Excellence, and the University of Otago. L.M.S. was supported by a University of Otago Doctoral Scholarship, and A.R.C. was financed by a scholarship from the Bio-Protection Centre of Research Excellence (Tertiary Education Commission, NZ).

References

1. Weinbauer MG. 2004. Ecology of prokaryotic viruses. *FEMS Microbiol Rev* **28**:127–181.

2. van Houte S, Buckling A, Westra ER. 2016. Evolutionary ecology of prokaryotic immune mechanisms. *Microbiol Mol Biol Rev* **80**:745–763.

3. Rath D, Amlinger L, Rath A, Lundgren M. 2015. The CRISPR-Cas immune system: biology, mechanisms and applications. *Biochimie* **117**:119–128.

4. Makarova KS, Grishin NV, Shabalina SA, Wolf YI, Koonin EV. 2006. A putative RNA-interference-based immune system in prokaryotes: computational analysis of the predicted enzymatic machinery, functional analogies with eukaryotic RNAi, and hypothetical mechanisms of action. *Biol Direct* **1**:7.

5. Koonin EV, Makarova KS, Zhang F. 2017. Diversity, classification and evolution of CRISPR-Cas systems. *Curr Opin Microbiol* **37**:67–78.

6. Mojica FJ, Díez-Villaseñor C, García-Martínez J, Soria E. 2005. Intervening sequences of regularly spaced prokaryotic repeats derive from foreign genetic elements. *J Mol Evol* **60**:174–182.

7. Amitai G, Sorek R. 2016. CRISPR-Cas adaptation: insights into the mechanism of action. *Nat Rev Microbiol* **14**:67–76.

8. Jackson SA, McKenzie RE, Fagerlund RD, Kieper SN, Fineran PC, Brouns SJ. 2017. CRISPR-Cas: adapting to change. *Science* **356**:356.

9. Wiedenheft B, Lander GC, Zhou K, Jore MM, Brouns SJJ, van der Oost J, Doudna JA, Nogales E. 2011. Structures of the RNA-guided surveillance complex from a bacterial immune system. *Nature* **477**:486–489.

10. Garneau JE, Dupuis ME, Villion M, Romero DA, Barrangou R, Boyaval P, Fremaux C, Horvath P, Magadán AH, Moineau S. 2010. The CRISPR/Cas bacterial immune system cleaves bacteriophage and plasmid DNA. *Nature* **468**:67–71.

11. Grissa I, Vergnaud G, Pourcel C. 2007. The CRISPRdb database and tools to display CRISPRs and to generate dictionaries of spacers and repeats. *BMC Bioinformatics* **8**:172.

12. Westra ER, van Houte S, Oyesiku-Blakemore S, Makin B, Broniewski JM, Best A, Bondy-Denomy J, Davidson A, Boots M, Buckling A. 2015. Parasite exposure drives selective evolution of constitutive versus inducible defense. *Curr Biol* **25**:1043–1049.

13. van Elsas JD, Bailey MJ. 2002. The ecology of transfer of mobile genetic elements. *FEMS Microbiol Ecol* **42**:187–197.

14. Hatoum-Aslan A, Marraffini LA. 2014. Impact of CRISPR immunity on the emergence and virulence of bacterial pathogens. *Curr Opin Microbiol* **17**:82–90.

15. Palmer KL, Gilmore MS. 2010. Multidrug-resistant enterococci lack CRISPR-cas. *mBio* **1**:e00227-10.

16. Shehreen S, Chyou TY, Fineran PC, Brown CM. 2019. Genome-wide correlation analysis suggests different roles of CRISPR-Cas systems in the acquisition of antibiotic resistance genes in diverse species. *Philos Trans R Soc Lond B Biol Sci* **374**:20180384.

17. Horvath P, Coûté-Monvoisin AC, Romero DA, Boyaval P, Fremaux C, Barrangou R. 2009. Comparative analysis of CRISPR loci in lactic acid bacteria genomes. *Int J Food Microbiol* **131**:62–70.

18. Sorek R, Kunin V, Hugenholtz P. 2008. CRISPR—a widespread system that provides acquired resistance against phages in bacteria and archaea. *Nat Rev Microbiol* **6**:181–186.

19. Stern A, Keren L, Wurtzel O, Amitai G, Sorek R. 2010. Self-targeting by CRISPR: gene regulation or autoimmunity? *Trends Genet* **26**:335–340.

20. Staals RH, Jackson SA, Biswas A, Brouns SJ, Brown CM, Fineran PC. 2016. Interference-driven spacer acquisition is dominant over naive and primed adaptation in a native CRISPR-Cas system. *Nat Commun* **7**:12853.

21. Vercoe RB, Chang JT, Dy RL, Taylor C, Gristwood T, Clulow JS, Richter C, Przybilski R, Pitman AR, Fineran PC. 2013. Cytotoxic chromosomal targeting by CRISPR/Cas systems can reshape bacterial genomes and expel or remodel pathogenicity islands. *PLoS Genet* **9**:e1003454.

22. Hampton HG, McNeil MB, Paterson TJ, Ney B, Williamson NR, Easingwood RA, Bostina M, Salmond GPC, Fineran PC. 2016. CRISPR-Cas gene-editing reveals RsmA and RsmC act through FlhDC to repress the SdhE flavinylation factor and control motility

and prodigiosin production in Serratia. *Microbiology (Reading)* **162**:1047–1058.

23. Miller MB, Bassler BL. 2001. Quorum sensing in bacteria. *Annu Rev Microbiol* **55**:165–199.

24. Papenfort K, Bassler BL. 2016. Quorum sensing signal-response systems in Gram-negative bacteria. *Nat Rev Microbiol* **14**:576–588.

25. Redfield RJ. 2002. Is quorum sensing a side effect of diffusion sensing? *Trends Microbiol* **10**:365–370.

26. Rajput A, Kaur K, Kumar M. 2016. SigMol: repertoire of quorum sensing signaling molecules in prokaryotes. *Nucleic Acids Res* **44**(D1):D634–D639.

27. Kasman LM, Kasman A, Westwater C, Dolan J, Schmidt MG, Norris JS. 2002. Overcoming the phage replication threshold: a mathematical model with implications for phage therapy. *J Virol* **76**:5557–5564.

28. Patterson AG, Jackson SA, Taylor C, Evans GB, Salmond GPC, Przybilski R, Staals RHJ, Fineran PC. 2016. Quorum sensing controls adaptive immunity through the regulation of multiple CRISPR-Cas systems. *Mol Cell* **64**:1102–1108.

29. Høyland-Kroghsbo NM, Paczkowski J, Mukherjee S, Broniewski J, Westra E, Bondy-Denomy J, Bassler BL. 2017. Quorum sensing controls the Pseudomonas aeruginosa CRISPR-Cas adaptive immune system. *Proc Natl Acad Sci USA* **114**:131–135.

30. Gao R, Krysciak D, Petersen K, Utpatel C, Knapp A, Schmeisser C, Daniel R, Voget S, Jaeger KE, Streit WR. 2015. Genome-wide RNA sequencing analysis of quorum sensing-controlled regulons in the plant-associated Burkholderia glumae PG1 strain. *Appl Environ Microbiol* **81**:7993–8007.

31. Bowden SD, Eyres A, Chung JC, Monson RE, Thompson A, Salmond GP, Spring DR, Welch M. 2013. Virulence in Pectobacterium atrosepticum is regulated by a coincidence circuit involving quorum sensing and the stress alarmone, (p)ppGpp. *Mol Microbiol* **90**:457–471.

32. Høyland-Kroghsbo NM, Muñoz KA, Bassler BL. 2018. Temperature, by controlling growth rate, regulates CRISPR-Cas activity in Pseudomonas aeruginosa. *mBio* **9**:e02184-18.

33. Lin P, Pu Q, Shen G, Li R, Guo K, Zhou C, Liang H, Jiang J, Wu M. 2019. CdpR inhibits CRISPR-cas adaptive immunity to lower anti-viral defense while avoiding self-reactivity. *iScience* **13**:55–68.

34. Dillon SC, Dorman CJ. 2010. Bacterial nucleoid-associated proteins, nucleoid structure and gene expression. *Nat Rev Microbiol* **8**:185–195.

35. Owen-Hughes TA, Pavitt GD, Santos DS, Sidebotham JM, Hulton CS, Hinton JC, Higgins CF. 1992. The chromatin-associated protein H-NS interacts with curved DNA to influence DNA topology and gene expression. *Cell* **71**:255–265.

36. Spurio R, Falconi M, Brandi A, Pon CL, Gualerzi CO. 1997. The oligomeric structure of nucleoid protein H-NS is necessary for recognition of intrinsically curved DNA and for DNA bending. *EMBO J* **16**:1795–1805.

37. Lucchini S, Rowley G, Goldberg MD, Hurd D, Harrison M, Hinton JC. 2006. H-NS mediates the silencing of laterally acquired genes in bacteria. *PLoS Pathog* **2**:e81.

38. Navarre WW, Porwollik S, Wang Y, McClelland M, Rosen H, Libby SJ, Fang FC. 2006. Selective silencing of foreign DNA with low GC content by the H-NS protein in Salmonella. *Science* **313**:236–238.

39. Dorman CJ. 2004. H-NS: a universal regulator for a dynamic genome. *Nat Rev Microbiol* **2**:391–400.

40. Pougach K, Semenova E, Bogdanova E, Datsenko KA, Djordjevic M, Wanner BL, Severinov K. 2010. Transcription, processing and function of CRISPR cassettes in Escherichia coli. *Mol Microbiol* **77**:1367–1379.

41. Westra ER, Pul U, Heidrich N, Jore MM, Lundgren M, Stratmann T, Wurm R, Raine A, Mescher M, Van Heereveld L, Mastop M, Wagner EG, Schnetz K, Van Der Oost J, Wagner R, Brouns SJ. 2010. H-NS-mediated repression of CRISPR-based immunity in Escherichia coli K12 can be relieved by the transcription activator LeuO. *Mol Microbiol* **77**:1380–1393.

42. Brouns SJ, Jore MM, Lundgren M, Westra ER, Slijkhuis RJ, Snijders AP, Dickman MJ, Makarova KS, Koonin EV, van der Oost J. 2008. Small CRISPR RNAs guide antiviral defense in prokaryotes. *Science* **321**:960–964.

43. Pul U, Wurm R, Arslan Z, Geissen R, Hofmann N, Wagner R. 2010. Identification and characterization of E. coli CRISPR-cas promoters and their silencing by H-NS. *Mol Microbiol* **75**:1495–1512.

44. Majsec K, Bolt EL, Ivančić-Baće I. 2016. Cas3 is a limiting factor for CRISPR-Cas immunity in Escherichia coli cells lacking H-NS. *BMC Microbiol* **16**:28.

45. Fu Q, Li S, Wang Z, Shan W, Ma J, Cheng Y, Wang H, Yan Y, Sun J. 2017. H-NS mutation-mediated CRISPR-Cas activation inhibits phage release and toxin production of Escherichia coli Stx2 phage lysogen. *Front Microbiol* **8**:652.

46. Zegans ME, Wagner JC, Cady KC, Murphy DM, Hammond JH, O'Toole GA. 2009. Interaction between bacteriophage DMS3 and host CRISPR region inhibits group behaviors of Pseudomonas aeruginosa. *J Bacteriol* **191**:210–219.

47. Heussler GE, Cady KC, Koeppen K, Bhuju S, Stanton BA, O'Toole GA. 2015. Clustered regularly interspaced short palindromic repeat-dependent, biofilm-specific death of Pseudomonas aeruginosa mediated by increased expression of phage-related genes. *mBio* **6**:e00129-15.

48. Lin TL, Pan YJ, Hsieh PF, Hsu CR, Wu MC, Wang JT. 2016. Imipenem represses CRISPR-Cas interference of DNA acquisition through H-NS stimulation in Klebsiella pneumoniae. *Sci Rep* **6**:31644.

49. Stoebel DM, Free A, Dorman CJ. 2008. Anti-silencing: overcoming H-NS-mediated repression of transcription in Gram-negative enteric bacteria. *Microbiology (Reading)* **154**:2533–2545.

50. Yang CD, Chen YH, Huang HY, Huang HD, Tseng CP. 2014. CRP represses the CRISPR/Cas system in Escherichia coli: evidence that endogenous CRISPR spacers impede phage P1 replication. *Mol Microbiol* **92**:1072–1091.

51. Stratmann T, Pul Ü, Wurm R, Wagner R, Schnetz K. 2012. RcsB-BglJ activates the Escherichia coli leuO gene, encoding an H-NS antagonist and pleiotropic regulator of virulence determinants. *Mol Microbiol* **83**:1109–1123.

52. Arslan Z, Stratmann T, Wurm R, Wagner R, Schnetz K, Pul Ü. 2013. RcsB-BglJ-mediated activation of Cascade operon does not induce the maturation of CRISPR RNAs in E. coli K12. *RNA Biol* **10**:708–715.

53. Majdalani N, Gottesman S. 2005. The Rcs phosphorelay: a complex signal transduction system. *Annu Rev Microbiol* **59**:379–405.

54. Hernández-Lucas I, Gallego-Hernández AL, Encarnación S, Fernández-Mora M, Martínez-Batallar AG, Salgado H, Oropeza R, Calva E. 2008. The LysR-type transcriptional regulator LeuO controls expression of several genes in Salmonella enterica serovar Typhi. *J Bacteriol* **190**:1658–1670.

55. Medina-Aparicio L, Rebollar-Flores JE, Gallego-Hernández AL, Vázquez A, Olvera L, Gutiérrez-Ríos RM, Calva E, Hernández-Lucas I. 2011. The CRISPR/Cas immune system is an operon regulated by LeuO, H-NS, and leucine-responsive regulatory protein in Salmonella enterica serovar Typhi. *J Bacteriol* **193**:2396–2407.

56. Haney SA, Platko JV, Oxender DL, Calvo JM. 1992. Lrp, a leucine-responsive protein, regulates branched-chain amino acid transport genes in Escherichia coli. *J Bacteriol* **174**:108–115.

57. Newman EB, Lin R. 1995. Leucine-responsive regulatory protein: a global regulator of gene expression in E. coli. *Annu Rev Microbiol* **49**:747–775.

58. Kroner GM, Wolfe MB, Freddolino PL. 2019. Escherichia coli Lrp regulates one-third of the genome via direct, cooperative, and indirect routes. *J Bacteriol* **201**:e00411-18.

59. Shimada T, Saito N, Maeda M, Tanaka K, Ishihama A. 2015. Expanded roles of leucine-responsive regulatory protein in transcription regulation of the Escherichia coli genome: genomic SELEX screening of the regulation targets. *Microb Genom* **1**:e000001.

60. Leonard PM, Smits SH, Sedelnikova SE, Brinkman AB, de Vos WM, van der Oost J, Rice DW, Rafferty JB. 2001. Crystal structure of the Lrp-like transcriptional regulator from the archaeon Pyrococcus furiosus. *EMBO J* **20**:990–997.

61. de los Rios S, Perona JJ. 2007. Structure of the Escherichia coli leucine-responsive regulatory protein Lrp reveals a novel octameric assembly. *J Mol Biol* **366**:1589–1602.

62. Navarre WW, McClelland M, Libby SJ, Fang FC. 2007. Silencing of xenogeneic DNA by H-NS-facilitation of lateral gene transfer in bacteria by a defense system that recognizes foreign DNA. *Genes Dev* **21**:1456–1471.

63. Mojica FJM, Díez-Villaseñor C. 2010. The on-off switch of CRISPR immunity against phages in Escherichia coli. *Mol Microbiol* **77**:1341–1345.

64. Liu Q, Richardson CC. 1993. Gene 5.5 protein of bacteriophage T7 inhibits the nucleoid protein H-NS of Escherichia coli. *Proc Natl Acad Sci USA* **90**:1761–1765.

65. Skennerton CT, Angly FE, Breitbart M, Bragg L, He S, McMahon KD, Hugenholtz P, Tyson GW. 2011. Phage encoded H-NS: a potential Achilles heel in the bacterial defence system. *PLoS One* **6**:e20095.

66. Botsford JL, Harman JG. 1992. Cyclic AMP in prokaryotes. *Microbiol Rev* **56**:100–122.

67. You C, Okano H, Hui S, Zhang Z, Kim M, Gunderson CW, Wang YP, Lenz P, Yan D, Hwa T. 2013. Coordination of bacterial proteome with metabolism by cyclic AMP signalling. *Nature* **500**:301–306.

68. Deutscher J, Aké FM, Derkaoui M, Zébré AC, Cao TN, Bouraoui H, Kentache T, Mokhtari A, Milohanic E, Joyet P. 2014. The bacterial phosphoenolpyruvate:carbohydrate phosphotransferase system: regulation by protein phosphorylation and phosphorylation-dependent protein-protein interactions. *Microbiol Mol Biol Rev* **78**:231–256.

69. Busby S, Ebright RH. 1999. Transcription activation by catabolite activator protein (CAP). *J Mol Biol* **293**:199–213.

70. Shinkai A, Kira S, Nakagawa N, Kashihara A, Kuramitsu S, Yokoyama S. 2007. Transcription activation mediated by a cyclic AMP receptor protein from Thermus thermophilus HB8. *J Bacteriol* **189**:3891–3901.

71. Patterson AG, Chang JT, Taylor C, Fineran PC. 2015. Regulation of the type I-F CRISPR-Cas system by CRP-cAMP and GalM controls spacer acquisition and interference. *Nucleic Acids Res* **43**:6038–6048.

72. Agari Y, Sakamoto K, Tamakoshi M, Oshima T, Kuramitsu S, Shinkai A. 2010. Transcription profile of Thermus thermophilus CRISPR systems after phage infection. *J Mol Biol* **395**:270–281.

73. Thoden JB, Kim J, Raushel FM, Holden HM. 2003. The catalytic mechanism of galactose mutarotase. *Protein Sci* **12**:1051–1059.

74. Hampton HG, Patterson AG, Chang JT, Taylor C, Fineran PC. 2019. GalK limits type I-F CRISPR-Cas expression in a CRP-dependent manner. *FEMS Microbiol Lett* **366**:fnz137.

75. Browning DF, Busby SJ. 2016. Local and global regulation of transcription initiation in bacteria. *Nat Rev Microbiol* **14**:638–650.

76. Gruber TM, Gross CA. 2003. Multiple sigma subunits and the partitioning of bacterial transcription space. *Annu Rev Microbiol* **57**:441–466.

77. Reitzer L, Schneider BL. 2001. Metabolic context and possible physiological themes of sigma(54)-dependent genes in Escherichia coli. *Microbiol Mol Biol Rev* **65**:422–444.

78. Barrios H, Valderrama B, Morett E. 1999. Compilation and analysis of sigma(54)-dependent promoter sequences. *Nucleic Acids Res* **27**:4305–4313.

79. Bush M, Dixon R. 2012. The role of bacterial enhancer binding proteins as specialized activators of σ54-dependent transcription. *Microbiol Mol Biol Rev* **76**:497–529.

80. Xu H, Hoover TR. 2001. Transcriptional regulation at a distance in bacteria. *Curr Opin Microbiol* **4**:138–144.

81. Samuels DJ, Frye JG, Porwollik S, McClelland M, Mrázek J, Hoover TR, Karls AC. 2013. Use of a promiscuous, constitutively-active bacterial enhancer-binding protein to define the σ54 (RpoN) regulon of Salmonella Typhimurium LT2. *BMC Genomics* **14**:602.

82. Xu H, Gu B, Nixon BT, Hoover TR. 2004. Purification and characterization of the AAA+ domain of Sinorhizobium meliloti DctD, a sigma54-dependent transcriptional activator. *J Bacteriol* **186**:3499–3507.

83. Heurlier K, Dénervaud V, Pessi G, Reimmann C, Haas D. 2003. Negative control of quorum sensing by RpoN (sigma54) in Pseudomonas aeruginosa PAO1. *J Bacteriol* **185**:2227–2235.

84. Thompson LS, Webb JS, Rice SA, Kjelleberg S. 2003. The alternative sigma factor RpoN regulates the quorum sensing gene rhlI in Pseudomonas aeruginosa. *FEMS Microbiol Lett* **220**:187–195.

85. Cai Z, Liu Y, Chen Y, Yam JK, Chew SC, Chua SL, Wang K, Givskov M, Yang L. 2015. RpoN regulates virulence factors of Pseudomonas aeruginosa via modulating the PqsR quorum sensing regulator. *Int J Mol Sci* **16**:28311–28319.

86. Sineva E, Savkina M, Ades SE. 2017. Themes and variations in gene regulation by extracytoplasmic function (ECF) sigma factors. *Curr Opin Microbiol* **36**:128–137.

87. Lonetto MA, Brown KL, Rudd KE, Buttner MJ. 1994. Analysis of the Streptomyces coelicolor sigE gene reveals the existence of a subfamily of eubacterial RNA polymerase sigma factors involved in the regulation of extracytoplasmic functions. *Proc Natl Acad Sci USA* **91**:7573–7577.

88. Helmann JD. 2002. The extracytoplasmic function (ECF) sigma factors. *Adv Microb Physiol* **46**:47–110.

89. Bernal-Bernal D, Abellón-Ruiz J, Iniesta ΛΛ, Pajares-Martínez E, Bastida-Martínez E, Fontes M, Padmanabhan S, Elías-Arnanz M. 2018. Multifactorial control of the expression of a CRISPR-Cas system by an extracytoplasmic function σ/anti-σ pair and a global regulatory complex. *Nucleic Acids Res* **46**:6726–6745.

90. Abellón-Ruiz J, Bernal-Bernal D, Abellán M, Fontes M, Padmanabhan S, Murillo FJ, Elías-Arnanz M. 2014. The CarD/CarG regulatory complex is required for the action of several members of the large set of Myxococcus xanthus extracytoplasmic function σ factors. *Environ Microbiol* **16**:2475–2490.

91. Peñalver-Mellado M, García-Heras F, Padmanabhan S, García-Moreno D, Murillo FJ, Elías-Arnanz M. 2006. Recruitment of a novel zinc-bound transcriptional factor by a bacterial HMGA-type protein is required for regulating multiple processes in Myxococcus xanthus. *Mol Microbiol* **61**:910–926.

92. Paget MSB, Kang JG, Roe JH, Buttner MJ. 1998. σR, an RNA polymerase sigma factor that modulates expression of the thioredoxin system in response to oxidative stress in Streptomyces coelicolor A3(2). *EMBO J* **17**:5776–5782.

93. Lee KL, Yoo JS, Oh GS, Singh AK, Roe JH. 2017. Simultaneous activation of iron- and thiol-based sensor-regulator systems by redox-active compounds. *Front Microbiol* **8**:139.

94. Asai K, Yamaguchi H, Kang C-M, Yoshida K, Fujita Y, Sadaie Y. 2003. DNA microarray analysis of Bacillus subtilis sigma factors of extracytoplasmic function family. *FEMS Microbiol Lett* **220**:155–160.

95. Ogura M, Asai K. 2016. Glucose induces ECF sigma factor genes, sigX and sigM, independent of cognate anti-sigma factors through acetylation of CshA in Bacillus subtilis. *Front Microbiol* **7**:1918.

96. Costanzo A, Ades SE. 2006. Growth phase-dependent regulation of the extracytoplasmic stress factor, σE, by guanosine 3′,5′-bispyrophosphate (ppGpp). *J Bacteriol* **188**:4627–4634.

97. Costanzo A, Nicoloff H, Barchinger SE, Banta AB, Gourse RL, Ades SE. 2008. ppGpp and DksA likely regulate the activity of the extracytoplasmic stress factor sigmaE in Escherichia coli by both direct and indirect mechanisms. *Mol Microbiol* **67**:619–632.

98. Gopalkrishnan S, Nicoloff H, Ades SE. 2014. Co-ordinated regulation of the extracytoplasmic stress factor, sigmaE, with other Escherichia coli sigma factors by (p)ppGpp and DksA may be achieved by specific regulation of individual holoenzymes. *Mol Microbiol* **93**:479–493.

99. Ratner HK, Sampson TR, Weiss DS. 2015. I can see CRISPR now, even when phage are gone: a view on alternative CRISPR-Cas functions from the prokaryotic envelope. *Curr Opin Infect Dis* **28**:267–274.

100. MacRitchie DM, Buelow DR, Price NL, Raivio TL. 2008. Two-component signaling and gram negative envelope stress response systems. *Adv Exp Med Biol* **631**:80–110.

101. Raffa RG, Raivio TL. 2002. A third envelope stress signal transduction pathway in Escherichia coli. *Mol Microbiol* **45**:1599–1611.

102. Baranova N, Nikaido H. 2002. The BaeSR two-component regulatory system activates transcription of the yegMNOB (mdtA-BCD) transporter gene cluster in Escherichia coli and increases its resistance to novobiocin and deoxycholate. *J Bacteriol* **184**:4168–4176.

103. Perez-Rodriguez R, Haitjema C, Huang Q, Nam KH, Bernardis S, Ke A, DeLisa MP. 2011. Envelope stress is a trigger of CRISPR RNA-mediated DNA silencing in Escherichia coli. *Mol Microbiol* **79**:584–599.

104. Berks BC, Sargent F, Palmer T. 2000. The Tat protein export pathway. *Mol Microbiol* **35**:260–274.

105. Bukau B, Horwich AL. 1998. The Hsp70 and Hsp60 chaperone machines. *Cell* **92**:351–366.

106. Bardwell JCA, Craig EA. 1987. Eukaryotic Mr 83,000 heat shock protein has a homologue in Escherichia coli. *Proc Natl Acad Sci USA* **84**:5177–5181.

107. Yosef I, Goren MG, Kiro R, Edgar R, Qimron U. 2011. High-temperature protein G is essential for activity of the Escherichia coli clustered regularly interspaced short palindromic repeats (CRISPR)/Cas system. *Proc Natl Acad Sci USA* **108**:20136–20141.

108. Deng DM, Liu MJ, ten Cate JM, Crielaard W. 2007. The VicRK system of Streptococcus mutans responds to oxidative stress. *J Dent Res* **86**:606–610.

109. Serbanescu MA, Cordova M, Krastel K, Flick R, Beloglazova N, Latos A, Yakunin AF, Senadheera DB, Cvitkovitch DG. 2015. Role of the Streptococcus mutans CRISPR-Cas systems in immunity and cell physiology. *J Bacteriol* **197**:749–761.

110. Poranen MM, Ravantti JJ, Grahn AM, Gupta R, Auvinen P, Bamford DH. 2006. Global changes in cellular gene expression during bacteriophage PRD1 infection. *J Virol* **80**:8081–8088.

111. De Smet J, Zimmermann M, Kogadeeva M, Ceyssens PJ, Vermaelen W, Blasdel B, Bin Jang H, Sauer U, Lavigne R. 2016. High coverage metabolomics analysis reveals phage-specific alterations to Pseudomonas aeruginosa physiology during infection. *ISME J* **10**:1823–1835.

112. Young JC, Dill BD, Pan C, Hettich RL, Banfield JF, Shah M, Fremaux C, Horvath P, Barrangou R, Verberkmoes NC. 2012. Phage-induced expression of CRISPR-associated proteins is revealed by shotgun proteomics in Streptococcus thermophilus. *PLoS One* **7**:e38077.

113. Quax TE, Voet M, Sismeiro O, Dillies MA, Jagla B, Coppée JY, Sezonov G, Forterre P, van der Oost J, Lavigne R, Prangishvili D. 2013. Massive activation of archaeal defense genes during viral infection. *J Virol* **87**:8419–8428.

114. Leskinen K, Blasdel BG, Lavigne R, Skurnik M. 2016. RNA-sequencing reveals the progression of phage-host interactions between φR1-37 and Yersinia enterocolitica. *Viruses* **8**:111.

115. Sacher JC, Flint A, Butcher J, Blasdel B, Reynolds HM, Lavigne R, Stintzi A, Szymanski CM. 2018. Transcriptomic analysis of the Campylobacter jejuni response to T4-like phage NCTC 12673 infection. *Viruses* **10**:332.

116. Peng X, Brügger K, Shen B, Chen L, She Q, Garrett RA. 2003. Genus-specific protein binding to the large clusters of DNA repeats (short regularly spaced repeats) present in Sulfolobus genomes. *J Bacteriol* **185**:2410–2417.

117. Deng L, Kenchappa CS, Peng X, She Q, Garrett RA. 2012. Modulation of CRISPR locus transcription by the repeat-binding protein Cbp1 in Sulfolobus. *Nucleic Acids Res* **40**:2470–2480.

118. He F, Vestergaard G, Peng W, She Q, Peng X. 2017. CRISPR-Cas type I-A Cascade complex couples viral infection surveillance to host transcriptional regulation in the dependence of Csa3b. *Nucleic Acids Res* **45**:1902–1913.

119. Liu T, Li Y, Wang X, Ye Q, Li H, Liang Y, She Q, Peng N. 2015. Transcriptional regulator-mediated activation of adaptation genes triggers CRISPR de novo spacer acquisition. *Nucleic Acids Res* **43**:1044–1055.

120. Levy A, Goren MG, Yosef I, Auster O, Manor M, Amitai G, Edgar R, Qimron U, Sorek R. 2015. CRISPR adaptation biases explain preference for acquisition of foreign DNA. *Nature* **520**:505–510.

121. Liu T, Liu Z, Ye Q, Pan S, Wang X, Li Y, Peng W, Liang Y, She Q, Peng N. 2017. Coupling transcriptional activation of CRISPR-Cas system and DNA repair genes by Csa3a in Sulfolobus islandicus. *Nucleic Acids Res* **45**:8978–8992.

122. Lintner NG, Frankel KA, Tsutakawa SE, Alsbury DL, Copié V, Young MJ, Tainer JA, Lawrence CM. 2011. The structure of the CRISPR-associated protein Csa3 provides insight into the regulation of the CRISPR/Cas system. *J Mol Biol* **405**:939–955.

123. Rouillon C, Athukoralage JS, Graham S, Grüschow S, White MF. 2019. Investigation of the cyclic oligoadenylate signaling pathway of type III CRISPR systems. *Methods Enzymol* **616**:191–218.

124. Viswanathan P, Murphy K, Julien B, Garza AG, Kroos L. 2007. Regulation of dev, an operon that includes genes essential for Myxococcus xanthus development and CRISPR-associated genes and repeats. *J Bacteriol* **189**:3738–3750.

125. Wallace RA, Black WP, Yang X, Yang Z. 2014. A CRISPR with roles in Myxococcus xanthus development and exopolysaccharide production. *J Bacteriol* **196**:4036–4043.

126. Hochstrasser ML, Taylor DW, Kornfeld JE, Nogales E, Doudna JA. 2016. DNA targeting by a minimal CRISPR RNA-guided cascade. *Mol Cell* **63**:840–851.

127. Haft DH, Selengut J, Mongodin EF, Nelson KE. 2005. A guild of 45 CRISPR-associated (Cas) protein families and multiple CRISPR/Cas subtypes exist in prokaryotic genomes. *PLOS Comput Biol* **1**:e60.

128. Westra ER, Buckling A, Fineran PC. 2014. CRISPR-Cas systems: beyond adaptive immunity. *Nat Rev Microbiol* **12**:317–326.

129. Hein S, Scholz I, Voß B, Hess WR. 2013. Adaptation and modification of three CRISPR loci in two closely related cyanobacteria. *RNA Biol* **10**:852–864.

130. Makarova KS, Anantharaman V, Grishin NV, Koonin EV, Aravind L. 2014. CARF and WYL domains: ligand-binding regulators of prokaryotic defense systems. *Front Genet* **5**:102.

131. Hutvagner G, Simard MJ. 2008. Argonaute proteins: key players in RNA silencing. *Nat Rev Mol Cell Biol* **9**:22–32.

132. Lisitskaya L, Aravin AA, Kulbachinskiy A. 2018. DNA interference and beyond: structure and functions of prokaryotic Argonaute proteins. *Nat Commun* **9**:5165.

133. Swarts DC, Koehorst JJ, Westra ER, Schaap PJ, van der Oost J. 2015. Effects of argonaute on gene expression in Thermus thermophilus. *PLoS One* **10**:e0124880.

134. Fu L, Xie C, Jin Z, Tu H, Han L, Jin M, Xiang Y, Zhang A. 2019. The prokaryotic Argonaute proteins enhance homology sequence-directed recombination in bacteria. *Nucleic Acids Res* **47**:3568–3579.

135. Cubbon A, Ivancic-Bace I, Bolt EL. 2018. CRISPR-Cas immunity, DNA repair and genome stability. *Biosci Rep* **38**:BSR20180457.

136. Shmakov SA, Makarova KS, Wolf YI, Severinov KV, Koonin EV. 2018. Systematic prediction of genes functionally linked to CRISPR-Cas systems by gene neighborhood analysis. *Proc Natl Acad Sci USA* **115**:E5307–E5316.

137. Green J, Stapleton MR, Smith LJ, Artymiuk PJ, Kahramanoglou C, Hunt DM, Buxton RS. 2014. Cyclic-AMP and bacterial cyclic-AMP receptor proteins revisited: adaptation for different ecological niches. *Curr Opin Microbiol* **18**:1–7.

138. Marraffini LA, Sontheimer EJ. 2008. CRISPR interference limits horizontal gene transfer in staphylococci by targeting DNA. *Science* **322**:1843–1845.

139. García-Martínez J, Maldonado RD, Guzmán NM, Mojica FJM. 2018. The CRISPR conundrum: evolve and maybe die, or survive and risk stagnation. *Microb Cell* **5:**262–268.

140. Watson BNJ, Staals RHJ, Fineran PC. 2018. CRISPR-Cas-mediated phage resistance enhances horizontal gene transfer by transduction. *mBio* **9:**e02406-17.

141. Varble A, Meaden S, Barrangou R, Westra ER, Marraffini LA. 2019. Recombination between phages and CRISPR-cas loci facilitates horizontal gene transfer in staphylococci. *Nat Microbiol* **4:**956–963.

142. Leon LM, Mendoza SD, Bondy-Denomy J. 2018. How bacteria control the CRISPR-Cas arsenal. *Curr Opin Microbiol* **42:**87–95.

CRISPR-Based Technologies and Applications

CHAPTER

Genome Editing with CRISPR-Cas Systems

Peter Lotfy[1,2,3] and Patrick D. Hsu[4,5,6]

[1]Biological and Biomedical Sciences PhD Program, Harvard Medical School, Boston, MA 02115
[2]Division of Gastroenterology, Hepatology, and Nutrition, Boston Children's Hospital, Boston, MA 02115
[3]Broad Institute of MIT and Harvard, Cambridge, MA 02142
[4]Department of Bioengineering, University of California, Berkeley, Berkeley, CA 94720
[5]Innovative Genomics Institute, University of California, Berkeley, Berkeley, CA 94704
[6]Center for Computational Biology, University of California, Berkeley, Berkeley, CA 94720

Introduction to Genome Editing

Since the advent of modern molecular biology, biologists have sought approaches to precisely manipulate DNA sequences to dissect the genetic underpinnings of cell biology and organismal physiology. Collectively known as "genome editing," such methodologies facilitate the generation of gene disruptions or conversions at any target site in diverse genomes. "Gene targeting"—the first generation of genome editing—was born from the observation that exogenous DNA flanked by sequences that are homologous to endogenous genes could spontaneously integrate into chromosomal DNA at very low frequencies (1, 2) (Fig. 11.1A). This method could mediate allelic replacement in 1 in approximately 10^4 to 10^7 electroporated mouse embryonic stem cells. This discovery transformed the biological sciences, spurring the generation of genetically engineered mice (3) and plants (4) with broad applications across basic research and biotechnology.

Modern genome editing utilizes a site-specific nuclease to induce DNA damage, followed by exploitation of cellular DNA repair mechanisms to introduce desired changes into the genome. The first examples of modern genome editing came from Haber and Jasin and colleagues, who built on traditional gene targeting methods by deploying meganucleases, which cleave DNA at specific 18-nucleotide (nt) recognition sites, to introduce DNA double-stranded breaks (DSBs) and promote the introduction of exogenous sequences into a targeted locus (5–9). However, meganucleases are limited to their specific recognition sequence and are difficult to retarget to a site of interest. Designer nucleases such as transcription activator-like effector nucleases (TALENs) and zinc finger nucleases (ZFNs), were engineered in the early 2000s by fusing programmable DNA-binding modules to the catalytic domain of the dimerization-dependent endonuclease FokI, generating DSBs in a more programmable fashion (10, 11). The DNA binding domain of each subunit of a zinc finger protein recognizes a specific nucleotide triplet, while each subunit of a TALEN recognizes a single nucleotide, facilitating modular assembly of an effector that can recognize nearly any sequence (12). Once two TALENs or ZFNs have bound to adjacent target sites, the fused FokI nuclease domains can dimerize and nonspecifically cleave the intervening DNA sequence (Fig. 11.1A).

CRISPR: Biology and Applications, First Edition. Edited by Rodolphe Barrangou, Erik J. Sontheimer, and Luciano A. Marraffini.
© 2022 American Society for Microbiology. DOI: 10.1128/9781683673798.ch11

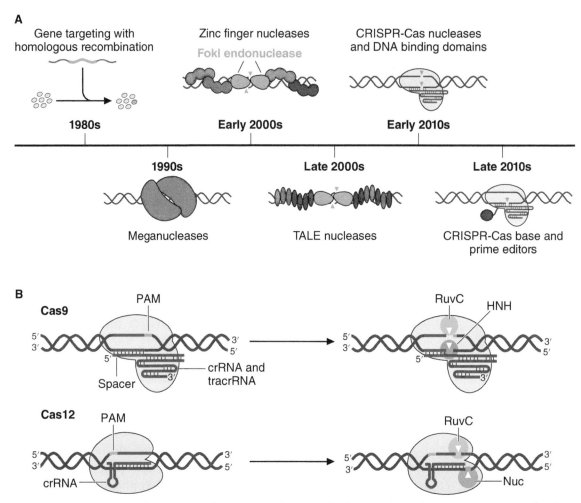

Figure 11.1 Overview of genome editing technologies. (**A**) Progression of genome editing technology development. TALE, transcription activator-like effector; Cas, CRISPR associated. (**B**) Mechanisms of DNA binding and cleavage by Cas9 and Cas12 nucleases. Cas9 recognizes a protospacer-adjacent motif (PAM) and unwinds DNA toward the PAM-distal end of the protospacer. The guide RNA (gRNA) contains a spacer sequence that base-pairs with the target sequence. Cas9 undergoes progressive conformational changes, activating the RuvC and HNH nuclease domains to generate blunt DSBs in target DNA. Upon gRNA binding, Cas12 activates its Nuc and RuvC nuclease domains, generating a staggered DSB in the PAM-distal protospacer sequence.

The subsequent DSBs stimulate endogenous repair machinery to mediate gene knock-in via homology-directed repair (HDR) or gene knockout via nonhomologous end-joining (NHEJ). Although these designer nucleases possess tremendous value to basic research (13, 14) and human therapeutics (15, 16), they require extensive protein engineering to generate desired effector enzymes (17) and can exhibit context-dependent activity or variable specificity (18), limiting their utility in high-throughput contexts.

Typical of many advances in biotechnology (19), a solution to these pitfalls can be found at the heart of an arms race between prokaryotic life and foreign genetic elements. To defend themselves against predatory phage, many bacteria can summon enzyme complexes that encode their genetic target with antisense oligonucleotide guides, including Argonaute and clustered regularly interspaced short palindromic repeats (CRISPR)-Cas systems. In recent years, a series of studies have unraveled the basic mechanisms of the CRISPR-associated nuclease Cas9, which uses a cognate guide RNA to cleave target DNA (20–26). Demonstrations that Cas9 can be programmed to cleave desired DNA by specifying a 20-nt spacer within a CRISPR RNA (crRNA) or single guide RNA (sgRNA) (27, 28) and engineered for multiplexed genome editing in human cells (29, 30) have culminated in a growing suite of molecular tools for nucleic acid manipulation that has revolutionized molecular biology (Fig. 11.1B). In this chapter, we discuss the basic mechanisms of CRISPR-Cas genome editors, molecular engineering efforts to improve their efficacy and specificity, the interplay between DNA repair and genome editing outcomes, and applications of genome editing in basic research and human therapeutics.

DNA Targeting with CRISPR-Cas Nucleases

Overview

CRISPR-Cas systems confer adaptive immunity to bacteria and archaea by mediating sequence-dependent recognition and destruction of invading genomes from bacteriophage or mobile genetic elements (31). A "CRISPR locus," stored as a functional operon within the bacterial genome, encodes many or all of the components required for the three phases of CRISPR-mediated adaptive immunity: adaptation, expression, and interference. Adaptation is the process of spacer acquisition, in which foreign DNA elements are incorporated into a CRISPR array. During the expression phase, the array is transcribed and processed into individual crRNAs, which are used to guide an effector module to cleave an invading target sequence during interference.

CRISPR effectors are classified into two primary classes and, at present, six types, each with several subtypes. Class 1 CRISPR-Cas systems, including type I, III, and IV effectors, utilize multisubunit complexes for target interference and comprise ~90% of CRISPR loci found in microbial genomes (32). Conversely, class 2 CRISPR-Cas systems, including type II and type V, utilize a single RNA-guided protein for target interference (32, 33), facilitating simple reprogramming in heterologous systems for genome editing (27, 29, 30). The best-characterized and most broadly applied CRISPR system is the class 2, type II CRISPR-Cas9 from *Streptococcus pyogenes* (SpCas9). This section focuses on the key concepts of target search, recognition, and interference by SpCas9 in eukaryotic cells while highlighting mechanistic differences from other type II and type V systems and describing engineering efforts to enhance the utility of these enzymes for genome editing.

Mechanisms of Target Search and Recognition

A fundamental element of every class 2 DNA-targeting CRISPR effector identified to date is the protospacer adjacent motif (PAM), a DNA element immediately flanking the target sequence that facilitates target binding (34, 35) by the effector enzyme. In their native bacteria, PAMs serve as a primary go/no-go signal for CRISPR immunity, conferring specificity for non-self-DNA by preventing self-targeting of the originating CRISPR locus (31, 36). In eukaryotic cells, Cas enzymes probe euchromatin for PAMs (37), likely via 3-dimensional lateral diffusion (38, 39) and 1-dimensional diffusion (40). Upon recognition of a PAM, Cas9 begins to unwind adjacent DNA to check for spacer complementarity (38, 41). In the absence of a complementary target sequence, Cas9 dissociates from its initial binding site to search for a correct target (38, 42). If the spacer is complementary to the PAM-proximal target, Cas9 then undergoes a series of conformational changes to enable stable binding of target DNA (38, 43, 44) prior to its transformation into a cleavage-competent state (45–48).

SpCas9 has a primary 5′-NGG and secondary 5′-NAG PAM (29, 49), which occur approximately every 8 bp (on average), thereby making most mammalian genomic regions accessible for gene targeting. However, the stringent requirements of the SpCas9 PAM may limit the targeting range required for applications in which sequence context is restricted. Therefore, multiple groups have sought to expand the targeting range of CRISPR nucleases by relaxing their PAM specificities. These efforts can be grouped into 3 major categories: identification and characterization of new class 2 enzymes with unique PAMs, rational engineering of previously characterized nucleases to recognize noncanonical PAMs, and directed evolution of nucleases with broader PAM requirements.

Class 2 enzymes, including Cas9 and Cas12, have many naturally occurring orthologs with divergent properties, including size, PAM specificity, and recognition of unique crRNA scaffolds. These diverse characteristics can be leveraged for different gene editing applications. For example, *Staphylococcus aureus* Cas9 (SaCas9) recognizes a 5′-NNGRRT PAM, where R can be A or G (50), *Neisseria meningitidis* Cas9 (NmeCas9) recognizes a 5′-NNNNGATT PAM (51, 52), and *Lachnospiraceae* bacterium Cas12a (LbCas12a) and *Acidaminococcus* sp. Cas12a (AsCas12a) require an upstream 5′-TTTV PAM (53–55). The diversity of PAMs in highly active genome editors confers tremendous flexibility for targeting applications, and the stringency of each of those PAMs can be exploited for novel applications of multiplexed genome editing (56).

In a structural study of the biochemical basis of PAM recognition, Anders et al. (47) found that the R1333 and R1335 residues in the PAM-interacting domain of SpCas9 make base-specific contacts with the NGG PAM and that mutating either of these residues hampers cleavage efficiency. Another structural study by Nishimasu et al. (46) showed that St3Cas9, which recognizes a 5′-NGGNG PAM, can be engineered to recognize 5′-NGG PAMs by swapping out its PAM-interacting domain (amino acids [aa] 1102 to 1394) with the PAM-interacting domain of SpCas9 (aa 1099 to 1368). These data suggested that the PAM specificities of CRISPR enzymes can be reprogrammed. In follow-up work, structure-guided rational engineering efforts have successfully produced several highly active genome editors with novel PAMs (57–61) (Table 11.1). One such variant is NG-Cas9, which includes an R1335V mutation in SpCas9, disrupting a base-specific contact with the G in position 3 of the NGG PAM, and incorporates several compensatory mutations that stabilize the PAM duplex, facilitating targeting at 5′-NG PAMs (62).

Another parallel approach for generating genome editors with expanded PAMs is directed evolution. Two studies by Kleinstiver et al. generated plasmid libraries encoding SpCas9 (63) or SaCas9 (64) with randomly mutagenized PAM-interacting domains. By applying several rounds of selective pressure, they derived variants, including VQR SpCas9, VRER SpCas9, and KKH SaCas9 (Table 11.1), that could successfully perturb the CcdB toxin at target sequences flanked by noncanonical PAMs in *Escherichia coli*. Building on this work, Hu et al. (65) engineered xCas9, which recognizes 5′-NG, 5′-GAA, or 5′-GAT PAMs, using phage-assisted continuous evolution (66). Altogether, advances in CRISPR-Cas system discovery, characterization, and protein engineering have offered synergistic approaches to building a flexible ensemble of custom enzymes for a range of genome editing applications.

Determinants of Target Cleavage Efficiency and Specificity

For all class 2 CRISPR-Cas nucleases, targeting specificity is conferred by base-pairing of a 20-to 30-nt spacer sequence encoded within a crRNA to a complementary target sequence. These spacers can be programmed to target any sequence and are constrained only by the PAM requirement (27, 67). However, gene targeting efficiency varies widely between spacers, indicating that sequence-intrinsic factors affect target recognition by the

Table 11.1 CRISPR-Cas enzyme variants for genome editing in mammalian cells

Enzyme	Classification	Mutation(s)	Size (aa)	PAM(s)	Reference(s)
SpCas9	Naturally occurring	NA[a]	1,368	NGG	27, 29, 30
eSpCas9(1.1)	High-fidelity SpCas9	K848A/K1003A/R1060A	1,368	NGG	88
Cas9-HF1	High-fidelity SpCas9	N497A/R661A/Q695A/Q926A	1,368	NGG	89
HypaCas9	High-fidelity SpCas9	N692A/M694A/Q695A/H698A	1,368	NGG	90
evoCas9	High-fidelity SpCas9	M495V/Y515N/K526E/R661Q	1,368	NGG	92
HiFi-Cas9	High-fidelity SpCas9	R691A	1,368	NGG	91
VQR-SpCas9	SpCas9 PAM variant	D1135V/R1335Q/T1337R	1,368	NGAN, NGNG	63
EQR-SpCas9	SpCas9 PAM variant	D1135E/R1335Q/T1337R	1,368	NGAG	63
VRER-SpCas9	SpCas9 PAM variant	D1135V/G1218R/R1335E/T1337R	1,368	NGCG	63
xCas9(3.7)	SpCas9 PAM variant	A262T/R324L/S409I/E480K/E543D/ M694I/E1219V	1,368	NG, GAA, GAT	65
NG-Cas9	SpCas9 PAM variant	L1111R/D1135V/G1218R/E1219F/ A1322R/R1335V/T1337R	1,368	NG	62
SaCas9	Naturally occurring	NA	1,053	NNGRRT	50
KKH-SaCas9	SaCas9 PAM variant	E782K/N968K/R1015H	1,053	NNNRRT	64
NmeCas9	Naturally occurring	NA	1,082	NNNNGATT	324, 325
FnCas9	Naturally occurring	NA	1,629	NGG	57
CjCas9	Naturally occurring	NA	984	NNNNRYAC	290
LbCas12a	Naturally occurring	NA	1,228	TTTV	53, 55
AsCas12a	Naturally occurring	NA	1,307	TTTV	53, 55
AsCas12a RR	AsCas12a PAM variant	S542R/K607R	1,307	TYCV	58
AsCas12a RVR	AsCas12a PAM variant	S542R/K548V/N552R	1,307	TATV	50
enAsCas12a	AsCas12a PAM variant	E174R/S542R/K548R	1,307	TTYN, VTTV, TRTV	59
enAsCas12a-HF1	High fidelity enCas12a	E174R/N282A/S542R/K548R	1,307	TTYN, VTTV, TRTV	59
AaCas12b	Naturally occurring	NA	1,129	TTN	326
DpbCas12e (CasX)	Naturally occurring	NA	986	TTCN	327

[a] NA, not applicable.

interference complex (29). Further investigation of these features has revealed that base composition (68–70) and duplex stability (71, 72) can predictably modulate activity. Thus, several computational tools have emerged to algorithmically design highly active sgRNAs (73, 74). However, the accuracy and reliability of these *in silico* predictions are often variable across diverse genome editing contexts, perhaps because they typically do not account for the effect of local chromatin accessibility (43), a cell-type- and cell-state-specific property that can affect Cas9 targeting efficiency (75, 76). Further elucidation of these effects will facilitate the development of more precise predictive algorithms for highly active sgRNAs, which will be critical for pooled genetic screens and therapeutic applications.

Early studies of CRISPR-Cas systems, including Cas9, showed that R-loop formation tolerates spacer-protospacer mismatches in the PAM-distal region and that Cas enzymes can cleave target sequences with only partial complementarity (27, 49, 77–79). Mismatch tolerance can affect the specificity of targeting in genome editing, and therefore, several studies have interrogated the critical factors determining off-target binding (80) and cleavage by Cas9 (49, 81–83) in mammalian cells. Overall, these studies showed that Cas9 sgRNAs exhibit mismatch tolerance profiles where PAM-proximal mismatches generally confer a higher penalty than PAM distal mismatches (49), confirming that spacer-protospacer binding has a sensitive "seed" region and informing the development of computational algorithms that predict sgRNA specificity in cells (49, 74). Moreover, high-throughput methods for assessing sgRNA cleavage fidelity (84, 85) have enabled the unbiased assessment of indels at off-target sites, facilitating finer algorithm optimization and sgRNA development.

To improve the fidelity of target cleavage by Cas9, paired Cas9 nickases (86) or FokI fusions (87) can be employed to generate DNA DSBs with extended base-pairing requirements. These approaches exhibit lower cleavage efficiency but mediate targeting with up to 1,000-fold-improved fidelity as assessed by indel formation at predicted off-target sites. Moreover, several groups have sought to decrease the affinity of Cas9 for mismatched DNA targets by introducing mutations that affect stable interaction of Cas9 with target DNA or increase the energetic threshold for conformational activation (Table 11.1) (88–92). In practice, however, many high-specificity variants can be less efficient. Moving forward, improving cleavage fidelity while maintaining on-target efficiency will continue to be a major focus of the development of CRISPR-Cas tools for basic research and human therapeutics.

DNA Repair and Genome Editing Outcomes

Overview

DSBs are detrimental lesions that compromise genome integrity. As a result, eukaryotic cells have evolved multiple mechanisms for repairing DSBs with little to no change in the original sequence. These mechanisms are broadly categorized as end joining or homology-directed repair (HDR). As the names suggest, end-joining mechanisms do not use a homologous template to guide repair, while HDR utilizes substantial sequence homology—from either a sister chromatid or an exogenous template—for more precise repair (Fig. 11.2A). In this section, we briefly discuss the mechanisms of each of these pathways and the current approaches to exploit them for genome editing.

Figure 11.2 Genome editing strategies and outcomes exploit endogenous DNA repair pathways. (A) DSBs generated by programmable nucleases can be repaired by nonhomologous end joining (NHEJ), alternative end joining (a-EJ), or homology-directed repair (HDR). NHEJ is initiated by the binding of KU70/80 to DNA ends, followed by processing by nucleases and DNA ligase IV, resulting in high-fidelity repair or indel formation. If cuts are faithfully repaired, Cas9 can repeatedly cleave the locus until an inactivating indel is generated. a-EJ is initiated by end resection by CtIP-MRN, followed by processing by DNA polymerase theta (Pol θ) and DNA ligase IIIα. a-EJ-mediated repair generates indels, which can cause gene knockouts. HDR also proceeds through end resection, followed by binding of replication protein A (RPA) and RAD51, promoting strand invasion of homologous DNA, which can be exogenously supplied for gene knock-in. (B) Cytidine deaminases and adenine deaminases chemically modify cytosine to uracil or adenine to inosine, respectively. Base editors mediate modification of target bases in Cas9 target sites without generating DSBs. BER, base-excision repair; UGI, uracil DNA glycosylase inhibitor. (C) Base editors can be limited by restricted targeting space, as defined by the positioning of a target base relative to the PAM and the editing window, as well as the potential for undesired editing events within the editing window and at distal loci. However, improved variants of CRISPR-Cas base editors are being generated to expand their targeting range with novel PAMs, narrow the editing window, and decrease off-target editing. Moving forward, C-to-T and A-to-G editing has recently been augmented by C-to-G editing systems, but additional sequence changes via base editing appear unlikely. (D) Prime editing utilizes a Cas9(H840A) nickase fused to a variant of the M-MLV reverse transcriptase (RT) and a prime editing guide RNA (pegRNA), which includes an sgRNA fused to a primer binding site and a donor sequence. After introducing a nick into the non-target strand, the pegRNA primer binding site hybridizes and initiates templated synthesis of the donor by the reverse transcriptase. The flap containing the unedited sequence can be excised by endogenous flap exonucleases, resulting in the formation of a heteroduplex with partial noncomplementarity. To resolve this heteroduplex, an additional sgRNA targeting the edited strand can be used to guide the Cas9 nickase to generate a nick in the unedited strand, directing mismatch repair to use the edited strand as a template.

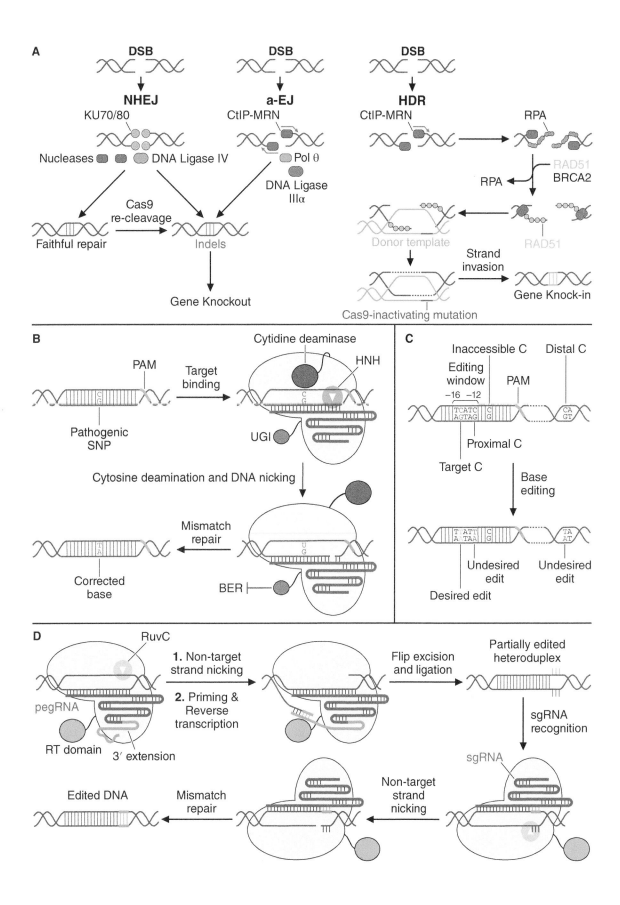

Gene Knockout

Nonhomologous end joining (NHEJ) is a high-fidelity repair process that religates staggered DSB ends with minimal end processing (93, 94). Classical NHEJ is initiated by the recruitment of 53BP1 and binding of KU70-KU80 heterodimers to DNA ends, preventing end resection. Upon KU-mediated stabilization, DNA ligase IV uses 0 to 4 nt of terminal sequence homology (microhomology) to stitch the ends back together. If DSB ends are partially processed and are thus incompatible for ligation, accessory factors, including nucleases and polymerases, are recruited by KU to process the ends and expose microhomology that can facilitate ligation (94, 95). Consequently, NHEJ-mediated repair can introduce insertions or deletions (referred to as indels) that can disrupt a coding sequence and is therefore considered "error prone" in a genome editing context (Fig. 11.2A).

Alternative end joining (a-EJ), also known as microhomology-mediated end joining or polymerase θ-mediated end joining, is an infrequent and error-prone end-joining mechanism that differs from NHEJ by the absence of KU-mediated stabilization of DSB ends. a-EJ is initiated by DSB end resection, which proceeds until 5 to 25 bp of microhomology is revealed and RPA is displaced by DNA polymerase θ (96). DNA ligase IIIα then mediates ligation with the coordinated action of the endonuclease FEN1 (97) (Fig. 11.2A). If end resection reveals longer stretches of homology due to the presence of repeat elements flanking a DSB, then a repair process called single-strand annealing (SSA) can be initiated. In SSA, extended stretches of single-stranded DNA (ssDNA) become coated with the ssDNA-binding protein RPA, which recruits RAD52 to mediate annealing of long (>25-bp) complementary sequences in the repetitive resected ends. Similarly to a-EJ, SSA-mediated repair results in the deletion of the region between the annealed ends of the microhomology sequence, thereby repairing the genomic lesion while compromising the original sequence (93).

In CRISPR-Cas9 gene editing, Cas9 is guided to a target site and cleaves double-stranded DNA (dsDNA) to generate a blunt DSB (27). These DSBs can be reliably religated by NHEJ (98), but faithfully repaired target sites can be repeatedly cleaved by Cas9 until an indel disrupts the PAM or crRNA recognition sequence, completing the editing reaction. Because DNA repair by NHEJ, a-EJ, or SSA enriches for the introduction of indels, Cas9-induced DSBs have been immediately useful for generating genetic knockouts in cell lines and animal models (99). Typically, Cas9-sgRNA complexes are targeted to constitutive exons near the portion of a gene corresponding to the N terminus to generate early frameshifts that can result in premature termination codons, inducing nonsense-mediated decay (74, 100, 101). Although this gene knockout strategy is typically robust, emerging studies have shown that cellular compensatory mechanisms can sometimes produce truncated or alternatively spliced proteins that maintain partial functionality (102, 103). Accordingly, it is important to carefully characterize edited cell lines (104). Some protocols suggest the use of paired sgRNAs to delete a genomic region of interest or sgRNAs that specifically perturb relevant functional domains within a coding sequence (105). Interestingly, because NHEJ, a-EJ, and SSA-mediated repair are dependent on cut site microhomology, the indel landscape of individual sgRNAs can be computationally predicted based on sequence context. This enables the deployment of sgRNAs that are highly likely to introduce indels that generate functional frameshifts and promote nonsense-mediated decay (106, 107).

Gene Knock-In

Ideally, genome editing outcomes would be more deterministic than semirandom NHEJ repair. Gene insertion or correction therefore requires the ability to more precisely program DNA repair outcomes after induced DSBs (99). As a result, most applications of gene editing have focused on exploiting the HDR pathway due to its higher fidelity and presumed higher programmability than NHEJ. Unlike NHEJ, classical HDR proceeds without KU-mediated protection of DSB ends. Instead, CtIP, MRN (MRE11, RAD50, and NBS1), and other nucleases (108) initiate 5′-to-3′ end resection, and the resulting ssDNA overhangs recruit RPA to prevent self-binding of the 3′ overhangs. BRCA1 and BRCA2 then mediate displacement of RPA by RAD51, which promotes strand invasion into complementary sequences on a homologous dsDNA donor. This facilitates template-guided repair of the DSB by HDR (109) (Fig. 11.2A). In normal cell biology, HDR uses the sister chromatid as a template, but in genome editing experiments, users can introduce an exogenous DNA template including a transgene with homology arms to facilitate HDR. These dsDNA templates on a plasmid or viral genome can deliver payloads up to ~11 kb in size to knock in gene corrections or large cassettes (110). HDR donors also typically introduce silent mutations to the PAM sequence or to the protospacer seed sequence to prevent cleavage of the donor upon integration at the targeted locus (111).

In addition to dsDNA donors, single-stranded oligodeoxyribonucleotides (ssODNs) with shorter homology tracts (<100 nt) can be efficiently inserted into Cas9-induced DSBs, offering a more straightforward approach for genome editing when small changes are desired (111–113). ssODNs integrate into DNA breaks via a RAD51-independent mechanism that is dependent on the Fanconi anemia pathway, a DNA repair pathway that is known to coordinate with HDR machinery to repair DSBs at stalled replication forks (114). ssODNs can be designed to have complementarity to either strand but can integrate more efficiently when annealed to the nontarget strand, which is not bound to the sgRNA and thus dissociates from Cas9 more quickly than the target strand (112). These donors can be programmed to encode a mutation of interest or a cassette of 0.2 to 1.5 kb flanked by homology arms of less than 100 nt. ssODNs are typically of higher efficiency than dsDNA donors and can mediate HDR in 30 to 60% and sometimes up to 100% of alleles (110).

Altogether, HDR offers a precise repair pathway for introducing mutations or large cassettes for gene correction (99) or knock-in (115, 116). However, HDR-dependent approaches exhibit significant limitations, including overall poor efficiency and restricted cell cycle availability (117). Accordingly, the frequency of cells with an NHEJ-mediated outcome typically dominates the frequency of those with HDR, and HDR in postmitotic tissues can be of extremely low frequency or inactive. This has motivated the development of methods that promote HDR at the expense of NHEJ, including small-molecule inhibition of NHEJ (118, 119), fusion of Cas9 to HDR-enhancing factors (120–122), and timed delivery of Cas9 into cell cycle-arrested cells (123). Still, HDR is not accessible to nondividing cells, prompting the use of knock-in methods that are compatible with NHEJ (124, 125). Such techniques introduce Cas9, sgRNA, and a vehicle containing a donor sequence flanked by sgRNA target sites into cells. Upon cleavage of chromosomal DNA and excision of the donor from the vehicle, the three DNA fragments can undergo NHEJ-mediated stitching (125, 126). Further development of these methodologies holds great promise for improving the efficiency and scalability of CRISPR-mediated knock-ins for both *in vitro* and *in vivo* applications in postmitotic cell types.

Gene Editing without DSBs

Due to concerns over off-target cleavage by Cas9, the potential toxicity of Cas9-induced DSBs (127), and the range of possible outcomes following repair, several strategies have emerged to edit DNA without generating DSBs. The first idea was to use a Cas9 "nickase," which has a point mutation that renders one of two nuclease domains, HNH and RuvC, catalytically inactive. These nickases create single-stranded breaks that are subsequently repaired primarily through the base excision repair pathway (128) or homologous recombination at much lower levels (29, 86, 129, 130). Early attempts to induce nickase-mediated HDR showed that low levels of donor integration occur without indel formation (29). Although these strategies improve targeting specificity and may reduce DSB-induced toxicity, they exhibit low efficiency and are not feasible for high-throughput purposes.

In recent years, an efficient approach for precise gene modification has emerged called base editing, in which specified DNA nucleobases are chemically modified to introduce point mutations while eliminating the need for a DSB or template-guided repair (131, 132). "Base editors" consist of the fusion of a catalytically inactive nuclease to a cytidine deaminase enzyme (e.g., rAPOBEC1, which converts cytosine to uracil) or an adenine deaminase enzyme (e.g., TadA, which converts adenine to inosine). After target DNA unwinding and R-loop formation, the catalytic domain deaminates accessible bases within a specified window in the unpaired ssDNA (Fig. 11.2B). The earliest variant of base editors, termed BE1, exhibited poor deamination efficiency due to the recognition of the U/G mismatch in the edited sequence by endogenous uracil N-glycosylase and reversion of the edited pair to C/G via base excision repair (133). Thus, in the second variant, BE2, a phage-derived uracil DNA glycosylase inhibitor, was fused to BE1 to inhibit this proofreading mechanism, successfully improving editing efficiencies in bacteria (134) and human cells (133). Newer iterations such as BE3, BE4, and ABE have further improved the editing efficiency by instead using the D10A Cas9 nickase to introduce a single-stranded nick at the target locus. This nick stimulates DNA mismatch repair to modify the base complementary to the deaminated nucleotide, resulting in the successful conversion of both strands (135, 136).

Overall, base editing is significantly more efficient than NHEJ- or HDR-mediated conversion of single nucleotide polymorphisms and avoids the generation of indel by-products. Accordingly, it has quickly become a promising genome editing tool for a range of applications, including therapeutic gene correction in postmitotic cells (137), generation of transgenic animal models (138, 139), cellular recordings (134, 140) and a molecular tool for directed evolution (141). However, base editing currently exhibits several limitations that restrict its utility in many practical settings. First, the targeting space of base editors is limited not only by PAM accessibility but also by the distance of the PAM from the target base (133) (Fig. 11.2C). Deaminases also have intrinsic biases that limit their targeting scope, such as non-GC for rAPOBEC1 (131, 133). Moreover, base editors frequently deaminate target-proximal nucleotides within a 5- to 10-nt "editing window" and are therefore further limited by sequence context (Fig. 11.2C). Cytidine and adenine base editors can even mediate unexpected (C-to-A, C-to-G, or A-C) conversions (141–143) as well as off-target deamination of DNA (144–147) and mRNA (148, 149). To address these limitations, systematic protein engineering efforts have focused on improving the efficiency of base editors at therapeutically relevant target sites (150), tightening or shifting the editing window (151, 152), developing other CRISPR-Cas base editors with a wide range of PAMs (59, 65, 151), and engineering deaminases to expand their targeting scope (152–157) or abrogate off-target DNA and RNA editing (148, 158–160). Key directions moving forward will be engineering base editor variants that can more precisely edit specific target nucleotides. Such enhancements will pave the way for the development of base editors that are suitable for highly efficient genome editing for diverse research and clinical applications (131) (Fig. 11.2C).

While base editors aim to convert single nucleotides, a recently reported method called prime editing enables small designer insertion, deletion, or substitution edits while eliminating the need for codelivered donor DNA. Prime editors exploit a Moloney murine leukaemia virus (M-MLV) reverse transcriptase fused to Cas9 nickase to incorporate an sgRNA-encoded donor template into nicked DNA (161), introducing single nucleotide changes and insertions (up to 31 nt) or deletions (up to 80 nt). Briefly, in prime editing, a Cas9 nickase uses a "prime editing guide RNA" (pegRNA) that includes an sgRNA fused to a 3′ donor sequence and an RNA primer that binds to the nicked nontarget strand to initiate synthesis by the reverse transcriptase (Fig. 11.2D). Following synthesis of the donor into the nicked strand, endogenous flap nucleases such as EXO1 or FEN1 can excise the unedited flap, generating a DNA heteroduplex with one edited strand and one unedited strand. To improve prime editing efficiency, an additional sgRNA can guide a Cas9 nickase to nick the unedited strand and stimulate DNA mismatch repair using the edited strand as a template. In this initial study, prime editors did not exhibit overt toxicity in immortalized cancer cell lines or primary neurons, suggesting that prime editing may be a viable method for therapeutic gene editing applications. Although further characterization of editing outcomes *in vivo* is required (162–164), prime editors have the potential to enable precision genome engineering beyond the capability of base editors. Moving forward, a major focus will be improving the efficiency of prime editors, which will likely include pegRNA optimization, reverse transcriptase engineering and evolution, and the development of strategies that facilitate long gene insertions and deletions both *in vivo* and *ex vivo*.

Applications

Genetic Studies

A central challenge in modern genetics is understanding the contribution of single genetic variants to complex traits. Key methodological advances such as genome-wide association studies (GWAS) and expression quantitative trait locus studies have identified single nucleotide variants (SNVs) associated with specific phenotypes (165). This has generated a wealth of genomic data that provide insight into the genetic contribution to disease susceptibility (166–168), behavioral traits (169–171), and even socioeconomic outcomes (172, 173). However, these associations are often unsuccessful at directly identifying causal mechanisms, requiring substantial experimental validation to understand the exact roles of these variants in phenotypic outputs (174, 175).

To complement these efforts, methodologies that can directly perturb genes and gene networks would enable direct linkages between these trait- or disease-associated variants and their molecular mechanisms. To this end, genome editing technologies including ZFNs, TALENs, and CRISPR-Cas nucleases have been widely used to perturb and interrogate genetic variants in animal cells (11, 13). To interrogate disease biology, Cas9 has been widely exploited to generate isogenic stem cell-based models that can be differentiated into appropriate cell types to investigate the causal role of single variants in disease-associated phenotypes in Alzheimer's disease (176–179), Parkinson's disease (180), type 2 diabetes (181), and many other disorders (182–184) (Fig. 11.3A).

In late-onset Alzheimer's disease (LOAD), the *APOE* gene, encoding apolipoprotein E, is polymorphic, with three predominant haplotypes, *APOE*-ε2, *APOE*-ε3, and *APOE*-ε4, conferring differential risk for LOAD based on GWAS (185). To characterize the *APOE* allele-specific contributions to Alzheimer's disease, one group (177, 186) used CRISPR-based gene editing to generate isogenic stem cell lines with different *APOE* alleles. To do this, they delivered SpCas9 and *APOE*-targeting sgRNA into homozygous ε3 stem cells alongside an ssODN encoding a silent PAM mutation and the C112R mutation that

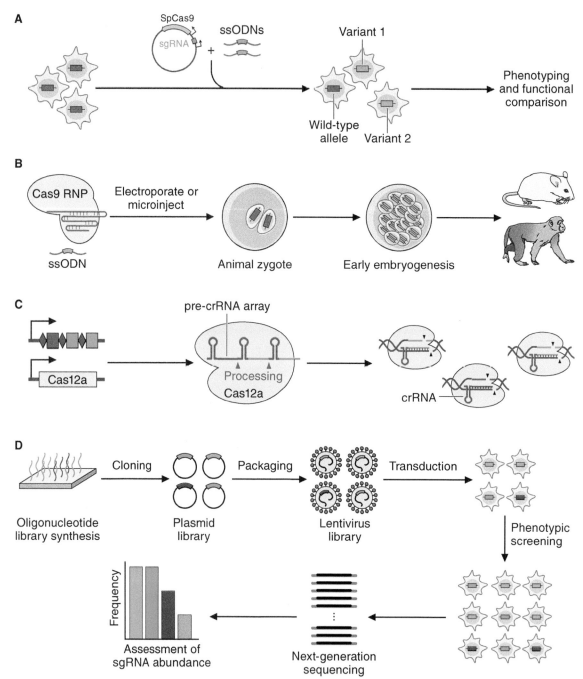

Figure 11.3 Applications of CRISPR-Cas genome editing in fundamental research. (**A**) Modeling of genetic variants in human cells. Plasmids encoding Cas9 and sgRNA can be cotransfected or electroporated into cells with ssODNs encoding the desired mutations for introduction. The generation of isogenic cell lines is typically followed by phenotypic characterization to interrogate the role of a genetic variant in a desired biological process. (**B**) Cas9 ribonucleoproteins (RNP) can be introduced into animal zygotes to generate transgenic knockout or knock-in animals more rapidly than previous approaches. (**C**) Cas12a (Cpf1) is capable of processing its pre-crRNA array into individual crRNAs. This property enables simple multiplexed targeting of multiple loci on a single-cell basis. (**D**) Genetic screens with CRISPR-Cas genome editing. An oligonucleotide library encoding different sgRNAs is synthesized and then cloned into a lentiviral plasmid backbone. Lentiviral particles are packaged in a pool and then used to transduce cells at a low multiplicity of infection. The cells undergo selection to enrich for single lentiviral integration events. After phenotypic screening, the enrichment or depletion of sgRNAs is assessed by next-generation sequencing.

constitutes the ε4 risk allele. After fluorescence-activated cell sorting-mediated enrichment, edited stem cells were screened for homozygous ε3-to-ε4 conversion and then differentiated into human brain cell types, including neurons, astrocytes, and microglia-like cells. Transcriptomic and phenotypic profiling of these cells revealed genotype- and cell-type-specific differences such as increased cholesterol accumulation in ε4 astrocytes, reduced amyloid-beta uptake by ε4 microglia-like cells, and early maturation of ε4 neurons, all of which are associated with neurodegeneration (177, 186). These studies thereby bridge the gap between risk gene identification and disease mechanism for the most predominant risk allele for LOAD, paving the way for therapeutic progress.

Transgenic Animal Models

Since their adoption almost 100 years ago, mouse models have served as a fundamental platform for genetic studies interrogating complex mammalian biological processes such as development and metabolism. As a result, the genome engineering field had an early focus on simplifying the generation of diverse genetically engineered mouse models. Traditional gene targeting methods have relied on spontaneous homologous recombination to introduce transgenes into the genome of a mouse embryonic stem cell (mESC) after microinjection, with limited efficiency (187). In such workflows, edited mESCs are injected into a mouse blastocyst and implanted into a pseudopregnant female, generating chimeric mice that are then capable of passing down the transgene through their germ line. Over time, more efficient methods have emerged for integrating transgenes at specific loci in mESCs using homology-containing DNA donors (188). Although these methods have been widely adopted, they are laborious, expensive, limited to a few mouse strains, and challenging to translate to other model organisms (189).

The emergence of CRISPR-Cas9 has improved the efficiency, speed, and flexibility of gene targeting in animals by eliminating the need for ESC intermediates and thereby allowing for one-step generation of nonchimeric mutant embryos (190–193). In these workflows, Cas9 is commonly microinjected or electroporated into zygotes as a ribonucleoprotein (RNP) complex with or without a donor template. In the cell, Cas9 can immediately cleave its target locus before cell division can occur, thereby limiting mosaic editing outcomes (194). Newer methods even bypass *ex vivo* microinjection of embryos by delivering Cas9 components into the oviducts of day 0.7 pregnant females, followed by *in vivo* electroporation (195). To date, CRISPR-Cas9 has been used in a wide variety of animal germ lines, including *Caenorhabditis elegans* (196), *Drosophila melanogaster* (197), zebrafish (198), mice (190), rats (199), pigs (200), cynomolgus monkeys (191), the common marmoset (201), and even humans (202) (Fig. 11.3B).

Multiplexed Genome Editing

CRISPR-based genome editing typically involves a class 2 CRISPR nuclease and an sgRNA to target a single genomic locus in a cell. However, many biological networks are sustained by genetic redundancy, in which gene paralogs can compensate for single-gene loss of function, or orchestrated by the coordinated interaction of multiple genes. To dissect these networks, researchers have engineered CRISPR-based methods to efficiently perturb multiple genes simultaneously. In native bacteria, a pre-crRNA array is transcribed from a Cas9-associated CRISPR locus and then processed by RNase III into individual crRNAs (26). These mature crRNAs then hybridize with an endogenous transactivating crRNA (tracrRNA) to form a guide RNA that is recognized by Cas9 and used to target invading sequences (26). CRISPR guide arrays have been leveraged to facilitate multiplexed gene targeting in mammalian cells through several orthogonal strategies.

Cong et al. expressed a pre-crRNA array containing spacers targeting two endogenous genes alongside a tracrRNA and SpCas9 in human cells (29). They found that the array was processed into individual crRNAs, presumably by endogenous RNases, and could induce deletions at low frequencies. Single-vector systems can also contain multiple U6 promoter-driven sgRNA cassettes, facilitating simple delivery of multiple mature sgRNAs into difficult-to-transfect cell lines (203). One facile solution to multiplexing comes from other class 2 CRISPR systems. The type V CRISPR-Cas12a enzyme can process pre-crRNA arrays into individual crRNAs without the need for endogenous RNases or a tracrRNA, greatly simplifying CRISPR-based multiplexing efforts (53, 204). This intrinsic property of Cas12a has been leveraged to mediate targeting of several genes simultaneously in bacteria (205), mammalian cell culture (206), mouse brain (207), and plants (208) (Fig. 11.3C). The continued development and optimization of Cas12a multiplexing technology (209) promise to expand the scalability of highly efficient pairwise genetic perturbations in the dissection of gene networks.

Genome-Scale Screens

In addition to studying single genetic variants, CRISPR-Cas genome editing can be scaled to systematically perturb genes in reverse genetic screens that introduce a wide assortment of targeted genetic variants in a single pool of cells. This approach leverages advances in DNA synthesis, which can generate a pool of up to a few hundred thousand short oligonucleotides. In a "CRISPR screen," these oligonucleotides are synthesized to contain spacers targeting thousands of different genes and then cloned into Cas9 or Cas12a expression vectors and introduced into mammalian cells at single copy to perturb one gene per cell. After application of a selective pressure, cell populations are sequenced to assess the depletion or enrichment of sgRNAs relative to the initial population (Fig. 11.3D). Thus far, such studies have provided insights into gene essentiality in cancer cells (68), mechanisms of drug resistance in melanoma (100, 210), neuron-specific essential genes (211), and host factors mediating toxin susceptibility (212).

Similar approaches with short hairpin RNA (shRNA)- and small interfering RNA-based platforms have elucidated gene function in basic cellular processes such as apoptosis and cell division (213, 214) and revealed synthetic lethal gene interactions in cancer cell lines (215, 216). However, RNA interference (RNAi) has been limited by confounding off-target effects due to intrinsic properties of shRNA processing and seed specificity (217). Because it is just as simple to reprogram while also generating full knockouts rather than partial knockdowns and conferring greater specificity of targeting, CRISPR-based genetic screening has usurped RNAi-based approaches in recent years.

One recent study conducted parallel genome-wide CRISPR-Cas9 screens in 324 human cancer cell lines representing 30 cancer types, generating a comprehensive cancer dependency map that identified genotype- and lineage-specific dependencies (218). For example, disruption of WRN, a RecQ DNA helicase, was specifically deleterious for the proliferation of cancer cells with microsatellite instability (MSI), which is common in gastric, colon, ovarian, and endometrial tumors. Further investigation of the mechanisms of WRN dependency has the potential to identify opportunities for therapeutic intervention in MSI-associated cancers (219, 220).

CRISPR screens can be extended beyond protein-coding genes to manipulate noncoding genetic elements such as long noncoding RNAs (lncRNAs), transcription factor binding sites, enhancers, and microRNAs. One group targeted Cas9 to candidate splice acceptor and splice donor sites to disrupt alternative splicing of lncRNAs and interrogate their role in cancer cell proliferation (221). However, because these sequences are noncoding, it is difficult to predict whether indels will generate functional nulls for these regulatory elements. This limitation can be partially addressed using paired sgRNAs to excise noncoding loci (222), CRISPR interference to repress transcription by inducing local chromatin methylation (223), or CRISPR activation to activate transcription (224, 225).

CRISPR-Cas genetic screens can also be used to study the role of SNVs in gene function at scale. Rather than characterizing single genetic variants in clonal isogenic lines, an orthogonal approach known as saturation genome editing utilizes CRISPR-based perturbations in conjunction with HDR-mediated donor integration. This strategy has been used to assess the effect of every possible SNV within specific critical protein domains (226) or on cellular fitness across short coding (227) and noncoding (228) regions. In this approach, ssODN libraries encoding an exon containing every variation of SNV at each locus, as well as synonymous codon mutations in the Cas9 protospacer to inhibit recleavage, are synthesized and introduced into cells alongside Cas9 and sgRNA plasmids. Upon Cas9 cleavage at the targeted exon, HDR mediates the integration of a donor template, introducing an SNV into the target gene. The consequences of each SNV on cellular fitness can then be assessed by measuring the depletion or enrichment of cells carrying that SNV over time. A BRCA1 saturation mutagenesis screen in haploid HAP1 cells identified hundreds of SNVs that disrupted BRCA1 function and were highly concordant with clinical outcomes in breast cancer patients (227). Because of the limitations of HDR-based approaches to saturation mutagenesis, recent work has pioneered the application of high-throughput base editing screens to probe the effects of human genetic variants on cancer cell fitness following drug administration (229, 230). The ability to systematically characterize the functional consequences of SNVs in clinically relevant genes offers an exciting path to study the effects of SNVs in primary human cells and develop patient-specific therapeutic strategies.

Gene Drives

Gene drives exploit homology-directed repair to promote "super-Mendelian" inheritance of a gene that can rapidly spread throughout a population (231). In CRISPR-based gene drives, a cassette containing a constitutively expressed Cas9, sgRNA, and flanking homology arms is knocked into the genome of a target organism. The encoded sgRNA targets the unmodified allele of that gene, and the homology arms promote duplication of the Cas9-containing allele via HDR. When that organism reproduces with an unedited mate, its progeny will inherit one copy of the drive cassette, which will express Cas9 and sgRNA to then target and convert the second allele via HDR. As a result, the gene drive cassette can be passed on to all progeny at a rate that exceeds normal Mendelian inheritance.

Gene drives have been demonstrated to prevent malaria transmission by vector mosquitoes (232) and suppress mosquito population levels by rendering females sterile (233, 234), demonstrating promising routes to control pest populations and eliminate malaria. Although gene drives are traditionally discussed in an insect engineering context, Grunwald et al. recently used gene drives in the germ lines of female mice to mediate up to 72% inheritance of converted genes (235), raising the possibility that gene drives could succeed in mammalian populations (236, 237).

Thus far, gene drives have been confined to the laboratory due to concerns over unanticipated effects on wild animal populations (238, 239). To address this, multiple groups are developing molecular safeguarding modalities such as split drive systems (240) and molecular "brakes" that prevent further transmission (241). Moreover, the gene drive community has discussed ethical standards for the application of certain types of drives (242), offering recommendations for the enactment of federally enforced regulatory safeguards (243) to ensure safe application of this technology.

Therapeutic Genome Editing

The technological innovations outlined in this chapter have paved the way for the development of efficient and safe CRISPR-based modalities for therapeutic genetic editing of human somatic and germ cells. CRISPR-based therapies can extend beyond the reach of other therapeutic strategies by mediating permanent correction of harmful mutations or stable integration of advantageous transgenes into the genomes of affected patient cells. In this section, we discuss potential therapeutic applications of CRISPR-Cas genome editing and the advantages and limitations of different therapeutic modalities and delivery vehicles, and we offer a perspective on the outlook of genome editing therapies.

Therapeutic Outcomes

Genome editing via gene disruption, gene insertion, or gene conversion holds therapeutic promise in a variety of disease contexts (Fig. 11.4A). Gene disruption via Cas9-mediated DSBs and subsequent repair by NHEJ will be an important application of CRISPR-based therapeutics to perturb deleterious genes. For example, β-thalassemia is an inherited blood disorder characterized by loss-of-function mutations in the *HBB* gene that affect expression of normal adult hemoglobin in erythrocytes. Because persistent fetal hemoglobin (HbF) expression is associated with improved clinical outcomes (244), therapeutic gene editing strategies are currently geared toward activating HbF expression by disrupting an enhancer element in the *BCL11A* that negatively regulates HbF (245, 246). To this end, Cas9 can be introduced into hematopoietic stem cells (HSCs) with an sgRNA targeting the enhancer locus to induce NHEJ-mediated generation of disruptive indels.

Moreover, CRISPR-based therapeutics can facilitate HDR-mediated insertion of beneficial transgenes into target cells, such as chimeric-antigen receptor (CAR) genes into T cells. At present, most CAR-T cells are generated by transducing human T cells with a randomly integrating virus that introduces a CAR with enhanced tumor-targeting ability (Fig. 11.4A). This approach has the potential to result in unpredictable, deleterious disruptions due to random integration of the delivery vehicle. In contrast, the precision and efficacy of CRISPR-Cas enable HDR-mediated insertion of large transgenes into specified loci such as the AAVS1 "safe harbor locus," which can tolerate insertions without adverse side effects (247). Furthermore, Eyquem et al. demonstrated that CRISPR-mediated insertion of a CAR into the endogenous T cell receptor alpha constant locus (*TRAC*) results in enhanced antitumor targeting compared to traditionally engineered CARs in mouse models (248), highlighting the unique potential of programmable nucleases for advancing the next generation of immunotherapies.

A third important mode of therapeutic gene editing—gene conversion—offers an exciting approach for treating monogenic diseases by directly modifying disease-causing mutations in affected cells or their progenitors. CRISPR-Cas base editors and HDR-based approaches are capable of mediating single nucleotide changes at almost any locus in the genome. They have been applied in several preclinical studies (249) to correct mutations in cell-based or animal models of monogenic diseases such as sickle cell disease (SCD) (91), Fanconi anemia (250, 251), and cystic fibrosis (252). For example, SCD is caused by a recessive missense mutation in the *HBB* gene that causes hypoxia-induced polymerization of hemoglobin, resulting in a wide range of clinical presentations that affect the quality of life of affected individuals (253). CRISPR-mediated gene conversion of one mutant *HBB* allele

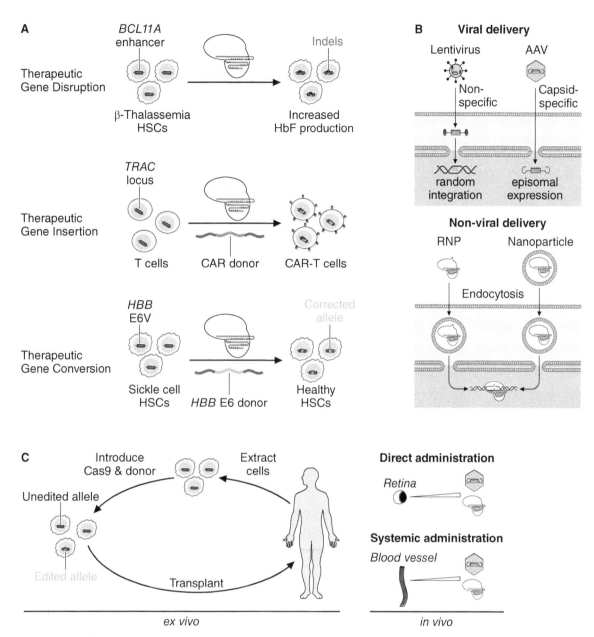

Figure 11.4 Therapeutic genome editing with CRISPR-Cas systems. (A) Genome editing can be used to mediate therapeutic outcomes through different modes of DNA repair, including gene insertion and conversion via HDR-mediated knock-ins or gene disruption via NHEJ-induced indels. (B) CRISPR-Cas therapeutics can be delivered to target cells via viral delivery, including lentivirus and adeno-associated virus (AAV), or nonviral delivery, including RNPs and lipid or gold-conjugated nanoparticles. Lentivirus delivers its payload nonspecifically into cells, while AAV capsids exhibit tissue-specific tropism that can be leveraged for tissue-specific delivery. Lentiviral cassettes are integrated into the host genome and AAV is episomally maintained in host cells, facilitating stable expression of Cas9 and sgRNA. RNPs and nanoparticles can be internalized via receptor-mediated or receptor-independent endocytosis and can mediate gene editing upon endosomal escape. (C) Therapeutic genome editing can occur both *in vivo* and *ex vivo*. For *ex vivo* therapies, target cells are extracted from a patient, treated with genome editing agents, and then transplanted autologously into the donor. For *in vivo* genome editing, gene editing agents such as AAV-encoded Cas9 and sgRNA or cell-penetrating peptide-fused Cas9 RNP can be injected locally into target tissue (e.g., retina) or introduced intravenously into systemic circulation.

in HSCs is sufficient to revert the phenotypes associated with SCD in differentiated erythroblasts (254). These proof-of-concept studies have established that genome editing is an effective therapeutic platform for treating diverse types of disorders. Further development will focus on effective methods for delivering desired gene perturbation machinery to target cells and tissue *ex vivo* or *in vivo*.

Ex Vivo Genome Editing

Ex vivo genome editing is the process of extracting cells from an individual, introducing genetic modifications via gene knock-in (255–257) or knockout (258–262) (Fig. 11.4A), and then conducting autologous or allogeneic transplantation of the edited cells into a patient. These approaches can be achieved through clinical-scale electroporation of Cas9 RNP into primary cell types that are amenable to *ex vivo* handling and can be reintroduced into target tissue, such as T cells and HSCs (Fig. 11.4B). *Ex vivo* editing with programmable nucleases presents an opportunity to treat a range of disorders by engineering immune cells to target cancer (263, 264) or evade human immunodeficiency virus (HIV) (15), repopulating stem cell niches with gene-edited HSCs to treat blood disorders (265–268), or correcting hereditary monogenic disease-causing mutations in germ cells or zygotes (269).

However, the broad application of *ex vivo* genome editing in therapeutics is still limited by the low efficiency of editing, the difficulty of screening for cells that have undergone successful and specific editing, and the high cost of generating sufficient numbers of patient-specific cells for transplantation. Therefore, one key focus moving forward will be developing strategies for manufacturing universal donor cells that contain desired gene edits (including CAR-T cells) and are capable of evading immune rejection in allogeneic hosts (270). Indeed, T cells engineered to express a tumor-specific T cell receptor with targeted indels in the *TRAC*, *TRBC1*, *TRBC2*, and *PDCD1* genes have been demonstrated to provide effective and long-lasting (up to 9 months) antitumor immunity in 3 human patients (271, 272). Moreover, several studies have found that DSB-dependent gene editing is toxic to human pluripotent stem cells (hPSCs) and is inhibited by the tumor suppressor p53 (127), also known as "the guardian of the genome." Therefore, successful editing of hPSCs may select for p53 mutant cells that are predisposed for cancerous transformation (273). Although these complications might be circumvented through the use of base editors or transient inhibition of p53 response during genome editing (274), such studies highlight the importance of investigating the long-term effects of gene editing in pluripotent stem cells (e.g., induced pluripotent stem cells) or multipotent stem cells (e.g., HSCs) to ensure therapeutic safety.

In Vivo Genome Editing

Similar to traditional somatic cell gene therapy approaches in which exogenous DNA sequences are introduced into nonreproductive cells of an individual to treat disease, CRISPR-Cas components are capable of being delivered to target tissue *in vivo*. A fundamental requirement for *in vivo* gene therapy is the availability of robust and safe vehicles for targeted delivery of therapeutic cargo, including transgenes or genome editors, to affected tissue. The two primary modes of *in vivo* delivery are broadly characterized as viral and nonviral, each conferring specific benefits and limitations that affect their therapeutic efficacy (Fig. 11.4B). Common viral modalities include retrovirus (including lentivirus) and adeno-associated virus (AAV). Lentivirus, derived from the HIV, efficiently delivers gene therapy cargo by nonspecifically penetrating cell membranes and integrating cargo into the host genome. Lentiviral vectors have been used extensively for clinical *ex vivo* engineering of CAR-T cells and gene therapy for HSCs, but the immunogenicity, lack of tissue-specific tropism, and potential for unwanted gene perturbations due to semirandom genome integration have limited the use of lentiviral vectors for *in vivo* gene therapy of long-lived cells (275).

These pitfalls have stimulated the development of methods that instead utilize AAV, which is a nonintegrating ssDNA virus that requires a helper virus such as adenovirus to replicate (276). In recent years, AAV has served as a successful gene therapy vector in humans (277, 278), owing to its long-term episomal stability in human cells (279), low immunogenicity (280), and tissue-specific capsid tropism (281). Many studies have demonstrated the delivery of AAV-packaged CRISPR-Cas agents into animal tissues via intravenous (50, 282–284) or local (207, 285, 286) injection. However, due to the large size of SpCas9 (4.2 kb) and limited packaging capacity of the AAV genome (4.7 to 5.2 kb) (287), it is challenging to generate all-in-one AAV vectors that encode both SpCas9 and sgRNA. Therefore, most strategies for *in vivo* delivery of SpCas9 require the simultaneous use of two AAV vectors that separately encode the enzyme and sgRNA (285, 288), but these may not cotransduce efficiently *in vivo* (289). The discovery of smaller CRISPR-Cas orthologs such as SaCas9 (3.16 kb) (50) and *Campylobacter jejuni* Cas9 (CjCas9) (2.95 kb) (290, 291) has facilitated the development of all-in-one viruses (292), paving the way for ongoing clinical trials of AAV-Cas9 gene therapies to treat congenital blindness (293, 294).

Although the long-term stability of AAV in nondividing cells is crucial for traditional gene therapy efforts, gene editing outcomes are permanent and therefore do not require persistent expression of CRISPR-Cas9, which is associated with increased off-target cutting (49, 295) and immune stimulation (296–298). These shortcomings have motivated the application of nonviral delivery modalities to facilitate transient expression of CRISPR-Cas components, including administration of "naked" RNPs (299), lipid nanoparticle-encapsulated RNPs (300), and gold nanoparticle-conjugated RNPs (257, 301, 302) (Fig. 11.4B). For example, Gao et al. demonstrated that cationic lipid nanoparticles can deliver Cas9 RNPs into mouse sensory hair cells to mediate allele-specific disruption of a mutant gene to ameliorate autosomal-dominant hearing loss (303). Other studies have demonstrated nonviral CRISPR-Cas approaches to treat disease in mouse models of Duchenne muscular dystrophy (301), fragile X syndrome (302), and liver disease (304). Moving forward, a primary focus will be to address major challenges with nonviral approaches including *in vivo* RNP stability and efficacy (305, 306), cell-type- and cell-state-dependent delivery and activation of RNPs (307–310), and immunogenicity (311, 312). These improvements present exciting opportunities for the development of safe and effective CRISPR-based biopharmaceutical drugs compatible with both *ex vivo* and *in vivo* delivery.

Conclusion and Perspectives

In the brief period since the discovery that class 2 CRISPR-Cas enzymes can be programmed to target any sequence of DNA (27) in mammalian systems (29, 30), the scientific community has envisaged a number of ambitious applications of these DNA-targeting modalities, ranging from interrogating genetic variants in cell culture to developing genome editing-based human therapeutics. As the community has been captivated by these technological breakthroughs, the success of CRISPR-Cas tools illustrates the transformative impact of basic research as well as the symbiotic relationship between basic discovery and translational application. Because of the continued focus on sampling of bacterial and viral genomes and metagenomes, the CRISPR-Cas toolbox is ever expanding, revealing new systems such as CRISPR-associated transposases (313–316) and RNA-targeting CRISPR families (317–323) that may provide completely new modes of CRISPR-based genome editing. Moving forward, we expect that further exploration of the tremendous diversity in CRISPR-Cas systems will pave the way for new breakthroughs in molecular tool development for genome editing and beyond.

References

1. Smithies O, Gregg RG, Boggs SS, Koralewski MA, Kucherlapati RS. 1985. Insertion of DNA sequences into the human chromosomal beta-globin locus by homologous recombination. *Nature* **317**:230–234.

2. Thomas KR, Folger KR, Capecchi MR. 1986. High frequency targeting of genes to specific sites in the mammalian genome. *Cell* **44**:419–428.

3. Capecchi MR. 1989. Altering the genome by homologous recombination. *Science* **244**:1288–1292.

4. Paszkowski J, Baur M, Bogucki A, Potrykus I. 1988. Gene targeting in plants. *EMBO J* **7**:4021–4026.

5. Puchta H, Dujon B, Hohn B. 1993. Homologous recombination in plant cells is enhanced by in vivo induction of double strand breaks into DNA by a site-specific endonuclease. *Nucleic Acids Res* **21**:5034–5040.

6. Sargent RG, Brenneman MA, Wilson JH. 1997. Repair of site-specific double-strand breaks in a mammalian chromosome by homologous and illegitimate recombination. *Mol Cell Biol* **17**:267–277.

7. Rouet P, Smih F, Jasin M. 1994. Introduction of double-strand breaks into the genome of mouse cells by expression of a rare-cutting endonuclease. *Mol Cell Biol* **14**:8096–8106.

8. Plessis A, Perrin A, Haber JE, Dujon B. 1992. Site-specific recombination determined by I-SceI, a mitochondrial group I intron-encoded endonuclease expressed in the yeast nucleus. *Genetics* **130**:451–460.

9. Rudin N, Sugarman E, Haber JE. 1989. Genetic and physical analysis of double-strand break repair and recombination in *Saccharomyces cerevisiae*. *Genetics* **122**:519–534.

10. Bibikova M, Golic M, Golic KG, Carroll D. 2002. Targeted chromosomal cleavage and mutagenesis in Drosophila using zinc-finger nucleases. *Genetics* **161**:1169–1175.

11. Miller JC, Tan S, Qiao G, Barlow KA, Wang J, Xia DF, Meng X, Paschon DE, Leung E, Hinkley SJ, Dulay GP, Hua KL, Ankoudinova I, Cost GJ, Urnov FD, Zhang HS, Holmes MC, Zhang L, Gregory PD, Rebar EJ. 2011. A TALE nuclease architecture for efficient genome editing. *Nat Biotechnol* **29**:143–148.

12. Kim HJ, Lee HJ, Kim H, Cho SW, Kim JS. 2009. Targeted genome editing in human cells with zinc finger nucleases constructed via modular assembly. *Genome Res* **19**:1279–1288.

13. Urnov FD, Rebar EJ, Holmes MC, Zhang HS, Gregory PD. 2010. Genome editing with engineered zinc finger nucleases. *Nat Rev Genet* **11**:636–646.

14. Joung JK, Sander JD. 2013. TALENs: a widely applicable technology for targeted genome editing. *Nat Rev Mol Cell Biol* **14**:49–55.

15. Tebas P, Stein D, Tang WW, Frank I, Wang SQ, Lee G, Spratt SK, Surosky RT, Giedlin MA, Nichol G, Holmes MC, Gregory PD, Ando DG, Kalos M, Collman RG, Binder-Scholl G, Plesa G, Hwang WT, Levine BL, June CH. 2014. Gene editing of CCR5 in autologous CD4 T cells of persons infected with HIV. *N Engl J Med* **370**:901–910.

16. Urnov FD, Miller JC, Lee YL, Beausejour CM, Rock JM, Augustus S, Jamieson AC, Porteus MH, Gregory PD, Holmes MC. 2005. Highly efficient endogenous human gene correction using designed zinc-finger nucleases. *Nature* **435**:646–651.

17. Maeder ML, Thibodeau-Beganny S, Osiak A, Wright DA, Anthony RM, Eichtinger M, Jiang T, Foley JE, Winfrey RJ, Townsend JA, Unger-Wallace E, Sander JD, Müller-Lerch F, Fu F, Pearlberg J, Göbel C, Dassie JP, Pruett-Miller SM, Porteus MH, Sgroi DC, Iafrate AJ, Dobbs D, McCray PB, Jr, Cathomen T, Voytas DF, Joung JK. 2008. Rapid "open-source" engineering of customized zinc-finger nucleases for highly efficient gene modification. *Mol Cell* **31**:294–301.

18. Juillerat A, Dubois G, Valton J, Thomas S, Stella S, Maréchal A, Langevin S, Benomari N, Bertonati C, Silva GH, Daboussi F, Epinat JC, Montoya G, Duclert A, Duchateau P. 2014. Comprehensive analysis of the specificity of transcription activator-like effector nucleases. *Nucleic Acids Res* **42**:5390–5402.

19. Pingoud A, Wilson GG, Wende W. 2016. Type II restriction endonucleases—a historical perspective and more. *Nucleic Acids Res* **44**:8011.

20. Mojica FJ, Díez-Villaseñor C, García-Martínez J, Soria E. 2005. Intervening sequences of regularly spaced prokaryotic repeats derive from foreign genetic elements. *J Mol Evol* **60**:174–182.

21. Barrangou R, Fremaux C, Deveau H, Richards M, Boyaval P, Moineau S, Romero DA, Horvath P. 2007. CRISPR provides acquired resistance against viruses in prokaryotes. *Science* **315**:1709–1712.

22. Marraffini LA, Sontheimer EJ. 2010. CRISPR interference: RNA-directed adaptive immunity in bacteria and archaea. *Nat Rev Genet* **11**:181–190.

23. Marraffini LA, Sontheimer EJ. 2008. CRISPR interference limits horizontal gene transfer in staphylococci by targeting DNA. *Science* **322**:1843–1845.

24. Brouns SJ, Jore MM, Lundgren M, Westra ER, Slijkhuis RJ, Snijders AP, Dickman MJ, Makarova KS, Koonin EV, van der Oost J. 2008. Small CRISPR RNAs guide antiviral defense in prokaryotes. *Science* **321**:960–964.

25. Garneau JE, Dupuis ME, Villion M, Romero DA, Barrangou R, Boyaval P, Fremaux C, Horvath P, Magadán AH, Moineau S. 2010. The CRISPR/Cas bacterial immune system cleaves bacteriophage and plasmid DNA. *Nature* **468**:67–71.

26. Deltcheva E, Chylinski K, Sharma CM, Gonzales K, Chao Y, Pirzada ZA, Eckert MR, Vogel J, Charpentier E. 2011. CRISPR RNA maturation by trans-encoded small RNA and host factor RNase III. *Nature* **471**:602–607.

27. Jinek M, Chylinski K, Fonfara I, Hauer M, Doudna JA, Charpentier E. 2012. A programmable dual-RNA-guided DNA endonuclease in adaptive bacterial immunity. *Science* **337**:816–821.

28. Gasiunas G, Barrangou R, Horvath P, Siksnys V. 2012. Cas9-crRNA ribonucleoprotein complex mediates specific DNA cleavage for adaptive immunity in bacteria. *Proc Natl Acad Sci USA* **109**:E2579–E2586.

29. Cong L, Ran FA, Cox D, Lin S, Barretto R, Habib N, Hsu PD, Wu X, Jiang W, Marraffini LA, Zhang F. 2013. Multiplex genome engineering using CRISPR/Cas systems. *Science* **339**:819–823.

30. Mali P, Yang L, Esvelt KM, Aach J, Guell M, DiCarlo JE, Norville JE, Church GM. 2013. RNA-guided human genome engineering via Cas9. *Science* **339**:823–826.

31. Marraffini LA. 2015. CRISPR-Cas immunity in prokaryotes. *Nature* **526**:55–61.

32. Makarova KS, Wolf YI, Alkhnbashi OS, Costa F, Shah SA, Saunders SJ, Barrangou R, Brouns SJ, Charpentier E, Haft DH, Horvath P, Moineau S, Mojica FJ, Terns RM, Terns MP, White MF, Yakunin AF, Garrett RA, van der Oost J, Backofen R, Koonin EV. 2015. An updated evolutionary classification of CRISPR-Cas systems. *Nat Rev Microbiol* **13**:722–736.

33. Shmakov S, Smargon A, Scott D, Cox D, Pyzocha N, Yan W, Abudayyeh OO, Gootenberg JS, Makarova KS, Wolf YI, Severinov K, Zhang F, Koonin EV. 2017. Diversity and evolution of class 2 CRISPR-Cas systems. *Nat Rev Microbiol* **15**:169–182.

34. Horvath P, Romero DA, Coûté-Monvoisin AC, Richards M, Deveau H, Moineau S, Boyaval P, Fremaux C, Barrangou R. 2008. Diversity, activity, and evolution of CRISPR loci in *Streptococcus thermophilus*. *J Bacteriol* **190**:1401–1412.

35. Shah SA, Erdmann S, Mojica FJ, Garrett RA. 2013. Protospacer recognition motifs: mixed identities and functional diversity. *RNA Biol* **10**:891–899.

36. Marraffini LA, Sontheimer EJ. 2010. Self versus non-self discrimination during CRISPR RNA-directed immunity. *Nature* **463**:568–571.

37. Kallimasioti-Pazi EM, Thelakkad Chathoth K, Taylor GC, Meynert A, Ballinger T, Kelder MJE, Lalevée S, Sanli I, Feil R, Wood AJ. 2018. Heterochromatin delays CRISPR-Cas9 mutagenesis but does

not influence the outcome of mutagenic DNA repair. *PLoS Biol* **16**:e2005595.

38. Sternberg SH, Redding S, Jinek M, Greene EC, Doudna JA. 2014. DNA interrogation by the CRISPR RNA-guided endonuclease Cas9. *Nature* **507**:62–67.

39. Globyte V, Lee SH, Bae T, Kim JS, Joo C. 2019. CRISPR/Cas9 searches for a protospacer adjacent motif by lateral diffusion. *EMBO J* **38**:e99466.

40. Jeon Y, Choi YH, Jang Y, Yu J, Goo J, Lee G, Jeong YK, Lee SH, Kim IS, Kim JS, Jeong C, Lee S, Bae S. 2018. Direct observation of DNA target searching and cleavage by CRISPR-Cas12a. *Nat Commun* **9**:2777.

41. Mekler V, Minakhin L, Severinov K. 2017. Mechanism of duplex DNA destabilization by RNA-guided Cas9 nuclease during target interrogation. *Proc Natl Acad Sci USA* **114**:5443–5448.

42. Martens KJA, van Beljouw SPB, van der Els S, Vink JNA, Baas S, Vogelaar GA, Brouns SJJ, van Baarlen P, Kleerebezem M, Hohlbein J. 2019. Visualisation of dCas9 target search in vivo using an open-microscopy framework. *Nat Commun* **10**:3552.

43. Singh D, Sternberg SH, Fei J, Doudna JA, Ha T. 2016. Real-time observation of DNA recognition and rejection by the RNA-guided endonuclease Cas9. *Nat Commun* **7**:12778.

44. Dahlman JE, Abudayyeh OO, Joung J, Gootenberg JS, Zhang F, Konermann S. 2015. Orthogonal gene knockout and activation with a catalytically active Cas9 nuclease. *Nat Biotechnol* **33**:1159–1161.

45. Jiang F, Taylor DW, Chen JS, Kornfeld JE, Zhou K, Thompson AJ, Nogales E, Doudna JA. 2016. Structures of a CRISPR-Cas9 R-loop complex primed for DNA cleavage. *Science* **351**:867–871.

46. Nishimasu H, Ran FA, Hsu PD, Konermann S, Shehata SI, Dohmae N, Ishitani R, Zhang F, Nureki O. 2014. Crystal structure of Cas9 in complex with guide RNA and target DNA. *Cell* **156**:935–949.

47. Anders C, Niewoehner O, Duerst A, Jinek M. 2014. Structural basis of PAM-dependent target DNA recognition by the Cas9 endonuclease. *Nature* **513**:569–573.

48. Sternberg SH, LaFrance B, Kaplan M, Doudna JA. 2015. Conformational control of DNA target cleavage by CRISPR-Cas9. *Nature* **527**:110–113.

49. Hsu PD, Scott DA, Weinstein JA, Ran FA, Konermann S, Agarwala V, Li Y, Fine EJ, Wu X, Shalem O, Cradick TJ, Marraffini LA, Bao G, Zhang F. 2013. DNA targeting specificity of RNA-guided Cas9 nucleases. *Nat Biotechnol* **31**:827–832.

50. Ran FA, Cong L, Yan WX, Scott DA, Gootenberg JS, Kriz AJ, Zetsche B, Shalem O, Wu X, Makarova KS, Koonin EV, Sharp PA, Zhang F. 2015. In vivo genome editing using *Staphylococcus aureus* Cas9. *Nature* **520**:186–191.

51. Zhang Y, Heidrich N, Ampattu BJ, Gunderson CW, Seifert HS, Schoen C, Vogel J, Sontheimer EJ. 2013. Processing-independent CRISPR RNAs limit natural transformation in *Neisseria meningitidis*. *Mol Cell* **50**:488–503.

52. Hou Z, Zhang Y, Propson NE, Howden SE, Chu LF, Sontheimer EJ, Thomson JA. 2013. Efficient genome engineering in human pluripotent stem cells using Cas9 from *Neisseria meningitidis*. *Proc Natl Acad Sci USA* **110**:15644–15649.

53. Zetsche B, Gootenberg JS, Abudayyeh OO, Slaymaker IM, Makarova KS, Essletzbichler P, Volz SE, Joung J, van der Oost J, Regev A, Koonin EV, Zhang F. 2015. Cpf1 is a single RNA-guided endonuclease of a class 2 CRISPR-Cas system. *Cell* **163**:759–771.

54. Kim HK, Song M, Lee J, Menon AV, Jung S, Kang YM, Choi JW, Woo E, Koh HC, Nam JW, Kim H. 2017. In vivo high-throughput profiling of CRISPR-Cpf1 activity. *Nat Methods* **14**:153–159.

55. Yamano T, Zetsche B, Ishitani R, Zhang F, Nishimasu H, Nureki O. 2017. Structural basis for the canonical and non-canonical PAM recognition by CRISPR-Cpf1. *Mol Cell* **67**:633–645.e3.

56. Najm FJ, Strand C, Donovan KF, Hegde M, Sanson KR, Vaimberg EW, Sullender ME, Hartenian E, Kalani Z, Fusi N, Listgarten J, Younger ST, Bernstein BE, Root DE, Doench JG. 2018. Orthogonal CRISPR-Cas9 enzymes for combinatorial genetic screens. *Nat Biotechnol* **36**:179–189.

57. Hirano H, Gootenberg JS, Horii T, Abudayyeh OO, Kimura M, Hsu PD, Nakane T, Ishitani R, Hatada I, Zhang F, Nishimasu H, Nureki O. 2016. Structure and engineering of *Francisella novicida* Cas9. *Cell* **164**:950–961.

58. Gao L, Cox DBT, Yan WX, Manteiga JC, Schneider MW, Yamano T, Nishimasu H, Nureki O, Crosetto N, Zhang F. 2017. Engineered Cpf1 variants with altered PAM specificities. *Nat Biotechnol* **35**:789–792.

59. Kleinstiver BP, Sousa AA, Walton RT, Tak YE, Hsu JY, Clement K, Welch MM, Horng JE, Malagon-Lopez J, Scarfò I, Maus MV, Pinello L, Aryee MJ, Joung JK. 2019. Engineered CRISPR-Cas12a variants with increased activities and improved targeting ranges for gene, epigenetic and base editing. *Nat Biotechnol* **37**:276–282.

60. Ma D, Xu Z, Zhang Z, Chen X, Zeng X, Zhang Y, Deng T, Ren M, Sun Z, Jiang R, Xie Z. 2019. Engineer chimeric Cas9 to expand PAM recognition based on evolutionary information. *Nat Commun* **10**:560.

61. Walton RT, Christie KA, Whittaker MN, Kleinstiver BP. 2020. Unconstrained genome targeting with near-PAMless engineered CRISPR-Cas9 variants. *Science* **368**:290–296.

62. Nishimasu H, Shi X, Ishiguro S, Gao L, Hirano S, Okazaki S, Noda T, Abudayyeh OO, Gootenberg JS, Mori H, Oura S, Holmes B, Tanaka M, Seki M, Hirano H, Aburatani H, Ishitani R, Ikawa M, Yachie N, Zhang F, Nureki O. 2018. Engineered CRISPR-Cas9 nuclease with expanded targeting space. *Science* **361**:1259–1262.

63. Kleinstiver BP, Prew MS, Tsai SQ, Topkar VV, Nguyen NT, Zheng Z, Gonzales AP, Li Z, Peterson RT, Yeh JR, Aryee MJ, Joung JK. 2015. Engineered CRISPR-Cas9 nucleases with altered PAM specificities. *Nature* **523**:481–485.

64. Kleinstiver BP, Prew MS, Tsai SQ, Nguyen NT, Topkar VV, Zheng Z, Joung JK. 2015. Broadening the targeting range of *Staphylococcus aureus* CRISPR-Cas9 by modifying PAM recognition. *Nat Biotechnol* **33**:1293–1298.

65. Hu JH, Miller SM, Geurts MH, Tang W, Chen L, Sun N, Zeina CM, Gao X, Rees HA, Lin Z, Liu DR. 2018. Evolved Cas9 variants with broad PAM compatibility and high DNA specificity. *Nature* **556**:57–63.

66. Esvelt KM, Carlson JC, Liu DR. 2011. A system for the continuous directed evolution of biomolecules. *Nature* **472**:499–503.

67. Bolotin A, Quinquis B, Sorokin A, Ehrlich SD. 2005. Clustered regularly interspaced short palindrome repeats (CRISPRs) have spacers of extrachromosomal origin. *Microbiol (Reading)* **151**:2551–2561.

68. Wang T, Wei JJ, Sabatini DM, Lander ES. 2014. Genetic screens in human cells using the CRISPR-Cas9 system. *Science* **343**:80–84.

69. Gagnon JA, Valen E, Thyme SB, Huang P, Akhmetova L, Pauli A, Montague TG, Zimmerman S, Richter C, Schier AF. 2014. Efficient mutagenesis by Cas9 protein-mediated oligonucleotide insertion and large-scale assessment of single-guide RNAs. *PLoS One* **9**:e98186.

70. Xu H, Xiao T, Chen CH, Li W, Meyer CA, Wu Q, Wu D, Cong L, Zhang F, Liu JS, Brown M, Liu XS. 2015. Sequence determinants of improved CRISPR sgRNA design. *Genome Res* **25**:1147–1157.

71. Jinek M, Jiang F, Taylor DW, Sternberg SH, Kaya E, Ma E, Anders C, Hauer M, Zhou K, Lin S, Kaplan M, Iavarone AT, Charpentier E, Nogales E, Doudna JA. 2014. Structures of Cas9 endonucleases reveal RNA-mediated conformational activation. *Science* **343**:1247997.

72. Xu X, Duan D, Chen SJ. 2017. CRISPR-Cas9 cleavage efficiency correlates strongly with target-sgRNA folding stability: from physical mechanism to off-target assessment. *Sci Rep* **7**:143.

73. Haeussler M, Schönig K, Eckert H, Eschstruth A, Mianné J, Renaud JB, Schneider-Maunoury S, Shkumatava A, Teboul L, Kent J, Joly JS, Concordet JP. 2016. Evaluation of off-target and on-target scoring algorithms and integration into the guide RNA selection tool CRISPOR. *Genome Biol* **17**:148.

74. Doench JG, Fusi N, Sullender M, Hegde M, Vaimberg EW, Donovan KF, Smith I, Tothova Z, Wilen C, Orchard R, Virgin HW, Listgarten J, Root DE. 2016. Optimized sgRNA design to maximize activity and minimize off-target effects of CRISPR-Cas9. *Nat Biotechnol* **34**:184–191.

75. Jensen KT, Fløe L, Petersen TS, Huang J, Xu F, Bolund L, Luo Y, Lin L. 2017. Chromatin accessibility and guide sequence secondary structure affect CRISPR-Cas9 gene editing efficiency. *FEBS Lett* **591**:1892–1901.

76. Uusi-Mäkelä MIE, Barker HR, Bäuerlein CA, Häkkinen T, Nykter M, Rämet M. 2018. Chromatin accessibility is associated with CRISPR-Cas9 efficiency in the zebrafish (*Danio rerio*). *PLoS One* **13**:e0196238.

77. Wiedenheft B, van Duijn E, Bultema JB, Waghmare SP, Zhou K, Barendregt A, Westphal W, Heck AJ, Boekema EJ, Dickman MJ, Doudna JA. 2011. RNA-guided complex from a bacterial immune system enhances target recognition through seed sequence interactions. *Proc Natl Acad Sci USA* **108**:10092–10097.

78. Semenova E, Jore MM, Datsenko KA, Semenova A, Westra ER, Wanner B, van der Oost J, Brouns SJ, Severinov K. 2011. Interference by clustered regularly interspaced short palindromic repeat (CRISPR) RNA is governed by a seed sequence. *Proc Natl Acad Sci USA* **108**:10098–10103.

79. Jiang W, Bikard D, Cox D, Zhang F, Marraffini LA. 2013. RNA-guided editing of bacterial genomes using CRISPR-Cas systems. *Nat Biotechnol* **31**:233–239.

80. Wu X, Scott DA, Kriz AJ, Chiu AC, Hsu PD, Dadon DB, Cheng AW, Trevino AE, Konermann S, Chen S, Jaenisch R, Zhang F, Sharp PA. 2014. Genome-wide binding of the CRISPR endonuclease Cas9 in mammalian cells. *Nat Biotechnol* **32**:670–676.

81. Fu Y, Foden JA, Khayter C, Maeder ML, Reyon D, Joung JK, Sander JD. 2013. High-frequency off-target mutagenesis induced by CRISPR-Cas nucleases in human cells. *Nat Biotechnol* **31**:822–826.

82. Pattanayak V, Lin S, Guilinger JP, Ma E, Doudna JA, Liu DR. 2013. High-throughput profiling of off-target DNA cleavage reveals RNA-programmed Cas9 nuclease specificity. *Nat Biotechnol* **31**:839–843.

83. Tycko J, Barrera LA, Huston NC, Friedland AE, Wu X, Gootenberg JS, Abudayyeh OO, Myer VE, Wilson CJ, Hsu PD. 2018. Pairwise library screen systematically interrogates *Staphylococcus aureus* Cas9 specificity in human cells. *Nat Commun* **9**:2962.

84. Tsai SQ, Zheng Z, Nguyen NT, Liebers M, Topkar VV, Thapar V, Wyvekens N, Khayter C, Iafrate AJ, Le LP, Aryee MJ, Joung JK. 2015. GUIDE-seq enables genome-wide profiling of off-target cleavage by CRISPR-Cas nucleases. *Nat Biotechnol* **33**:187–197.

85. Kim D, Bae S, Park J, Kim E, Kim S, Yu HR, Hwang J, Kim JI, Kim JS. 2015. Digenome-seq: genome-wide profiling of CRISPR-Cas9 off-target effects in human cells. *Nat Methods* **12**:237–243.

86. Ran FA, Hsu PD, Lin CY, Gootenberg JS, Konermann S, Trevino AE, Scott DA, Inoue A, Matoba S, Zhang Y, Zhang F. 2013. Double nicking by RNA-guided CRISPR Cas9 for enhanced genome editing specificity. *Cell* **154**:1380–1389.

87. Guilinger JP, Thompson DB, Liu DR. 2014. Fusion of catalytically inactive Cas9 to FokI nuclease improves the specificity of genome modification. *Nat Biotechnol* **32**:577–582.

88. Slaymaker IM, Gao L, Zetsche B, Scott DA, Yan WX, Zhang F. 2016. Rationally engineered Cas9 nucleases with improved specificity. *Science* **351**:84–88.

89. Kleinstiver BP, Pattanayak V, Prew MS, Tsai SQ, Nguyen NT, Zheng Z, Joung JK. 2016. High-fidelity CRISPR-Cas9 nucleases with no detectable genome-wide off-target effects. *Nature* **529**:490–495.

90. Chen JS, Dagdas YS, Kleinstiver BP, Welch MM, Sousa AA, Harrington LB, Sternberg SH, Joung JK, Yildiz A, Doudna JA. 2017. Enhanced proofreading governs CRISPR-Cas9 targeting accuracy. *Nature* **550**:407–410.

91. Vakulskas CA, Dever DP, Rettig GR, Turk R, Jacobi AM, Collingwood MA, Bode NM, McNeill MS, Yan S, Camarena J, Lee CM, Park SH, Wiebking V, Bak RO, Gomez-Ospina N, Pavel-Dinu M, Sun W, Bao G, Porteus MH, Behlke MA. 2018. A high-fidelity Cas9 mutant delivered as a ribonucleoprotein complex enables efficient gene editing in human hematopoietic stem and progenitor cells. *Nat Med* **24**:1216–1224.

92. Casini A, Olivieri M, Petris G, Montagna C, Reginato G, Maule G, Lorenzin F, Prandi D, Romanel A, Demichelis F, Inga A, Cereseto A. 2018. A highly specific SpCas9 variant is identified by in vivo screening in yeast. *Nat Biotechnol* **36**:265–271.

93. Chang HHY, Pannunzio NR, Adachi N, Lieber MR. 2017. Non-homologous DNA end joining and alternative pathways to double-strand break repair. *Nat Rev Mol Cell Biol* **18**:495–506.

94. Bétermier M, Bertrand P, Lopez BS. 2014. Is non-homologous end-joining really an inherently error-prone process? *PLoS Genet* **10**:e1004086.

95. Waters CA, Strande NT, Wyatt DW, Pryor JM, Ramsden DA. 2014. Nonhomologous end joining: a good solution for bad ends. *DNA Repair (Amst)* **17**:39–51.

96. Mateos-Gomez PA, Kent T, Deng SK, McDevitt S, Kashkina E, Hoang TM, Pomerantz RT, Sfeir A. 2017. The helicase domain of Polθ counteracts RPA to promote alt-NHEJ. *Nat Struct Mol Biol* **24**:1116–1123.

97. Sharma S, Javadekar SM, Pandey M, Srivastava M, Kumari R, Raghavan SC. 2015. Homology and enzymatic requirements of microhomology-dependent alternative end joining. *Cell Death Dis* **6**:e1697.

98. Guo T, Feng YL, Xiao JJ, Liu Q, Sun XN, Xiang JF, Kong N, Liu SC, Chen GQ, Wang Y, Dong MM, Cai Z, Lin H, Cai XJ, Xie AY. 2018. Harnessing accurate non-homologous end joining for efficient precise deletion in CRISPR/Cas9-mediated genome editing. *Genome Biol* **19**:170.

99. Hsu PD, Lander ES, Zhang F. 2014. Development and applications of CRISPR-Cas9 for genome engineering. *Cell* **157**:1262–1278.

100. Shalem O, Sanjana NE, Hartenian E, Shi X, Scott DA, Mikkelson T, Heckl D, Ebert BL, Root DE, Doench JG, Zhang F. 2014. Genome-scale CRISPR-Cas9 knockout screening in human cells. *Science* **343**:84–87.

101. Zhou Y, Zhu S, Cai C, Yuan P, Li C, Huang Y, Wei W. 2014. High-throughput screening of a CRISPR/Cas9 library for functional genomics in human cells. *Nature* **509**:487–491.

102. Mou H, Smith JL, Peng L, Yin H, Moore J, Zhang XO, Song CQ, Sheel A, Wu Q, Ozata DM, Li Y, Anderson DG, Emerson CP, Sontheimer EJ, Moore MJ, Weng Z, Xue W. 2017. CRISPR/Cas9-mediated genome editing induces exon skipping by alternative splicing or exon deletion. *Genome Biol* **18**:108.

103. Sui T, Song Y, Liu Z, Chen M, Deng J, Xu Y, Lai L, Li Z. 2018. CRISPR-induced exon skipping is dependent on premature termination codon mutations. *Genome Biol* **19**:164.

104. Giuliano CJ, Lin A, Girish V, Sheltzer JM. 2019. Generating single cell-derived knockout clones in mammalian cells with CRISPR/Cas9. *Curr Protoc Mol Biol* **128**:e100.

105. Shi J, Wang E, Milazzo JP, Wang Z, Kinney JB, Vakoc CR. 2015. Discovery of cancer drug targets by CRISPR-Cas9 screening of protein domains. *Nat Biotechnol* **33**:661–667.

106. Bae S, Kweon J, Kim HS, Kim JS. 2014. Microhomology-based choice of Cas9 nuclease target sites. *Nat Methods* **11**:705–706.

107. Popp MW, Maquat LE. 2016. Leveraging rules of nonsense-mediated mRNA decay for genome engineering and personalized medicine. *Cell* **165**:1319–1322.

108. Symington LS, Gautier J. 2011. Double-strand break end resection and repair pathway choice. *Annu Rev Genet* **45**:247–271.

109. Zhang H, Tombline G, Weber BL. 1998. BRCA1, BRCA2, and DNA damage response: collision or collusion? *Cell* **92**:433–436.

110. Miura H, Quadros RM, Gurumurthy CB, Ohtsuka M. 2018. Easi-CRISPR for creating knock-in and conditional knockout mouse models using long ssDNA donors. *Nat Protoc* **13**:195–215.

111. Ran FA, Hsu PD, Wright J, Agarwala V, Scott DA, Zhang F. 2013. Genome engineering using the CRISPR-Cas9 system. *Nat Protoc* **8**:2281–2308.

112. Richardson CD, Ray GJ, DeWitt MA, Curie GL, Corn JE. 2016. Enhancing homology-directed genome editing by catalytically active and inactive CRISPR-Cas9 using asymmetric donor DNA. *Nat Biotechnol* **34**:339–344.

113. Paix A, Folkmann A, Goldman DH, Kulaga H, Grzelak MJ, Rasoloson D, Paidemarry S, Green R, Reed RR, Seydoux G. 2017. Precision genome editing using synthesis-dependent repair of Cas9-induced DNA breaks. *Proc Natl Acad Sci USA* **114**:E10745–E10754.

114. Richardson CD, Kazane KR, Feng SJ, Zelin E, Bray NL, Schäfer AJ, Floor SN, Corn JE. 2018. CRISPR-Cas9 genome editing in human cells occurs via the Fanconi anemia pathway. *Nat Genet* **50**:1132–1139.

115. Mikuni T, Nishiyama J, Sun Y, Kamasawa N, Yasuda R. 2016. High-throughput, high-resolution mapping of protein localization in mammalian brain by in vivo genome editing. *Cell* **165**:1803–1817.

116. Nishiyama J, Mikuni T, Yasuda R. 2017. Virus-mediated genome editing via homology-directed repair in mitotic and postmitotic cells in mammalian brain. *Neuron* **96**:755–768.e5.

117. Heyer WD, Ehmsen KT, Liu J. 2010. Regulation of homologous recombination in eukaryotes. *Annu Rev Genet* **44**:113–139.

118. Maruyama T, Dougan SK, Truttmann MC, Bilate AM, Ingram JR, Ploegh HL. 2015. Increasing the efficiency of precise genome editing with CRISPR-Cas9 by inhibition of nonhomologous end joining. *Nat Biotechnol* **33**:538–542.

119. Chu VT, Weber T, Wefers B, Wurst W, Sander S, Rajewsky K, Kühn R. 2015. Increasing the efficiency of homology-directed repair for CRISPR-Cas9-induced precise gene editing in mammalian cells. *Nat Biotechnol* **33**:543–548.

120. Charpentier M, Khedher AHY, Menoret S, Brion A, Lamribet K, Dardillac E, Boix C, Perrouault L, Tesson L, Geny S, De Cian A, Itier JM, Anegon I, Lopez B, Giovannangeli C, Concordet JP. 2018. CtIP fusion to Cas9 enhances transgene integration by homology-dependent repair. *Nat Commun* **9**:1133.

121. Rees HA, Yeh WH, Liu DR. 2019. Development of hRad51-Cas9 nickase fusions that mediate HDR without double-stranded breaks. *Nat Commun* **10**:2212.

122. Nambiar TS, Billon P, Diedenhofen G, Hayward SB, Taglialatela A, Cai K, Huang JW, Leuzzi G, Cuella-Martin R, Palacios A, Gupta A, Egli D, Ciccia A. 2019. Stimulation of CRISPR-mediated homology-directed repair by an engineered RAD18 variant. *Nat Commun* **10**:3395.

123. Lin S, Staahl BT, Alla RK, Doudna JA. 2014. Enhanced homology-directed human genome engineering by controlled timing of CRISPR/Cas9 delivery. *Elife* **3**:e04766.

124. Sakuma T, Nakade S, Sakane Y, Suzuki KT, Yamamoto T. 2016. MMEJ-assisted gene knock-in using TALENs and CRISPR-Cas9 with the PITCh systems. *Nat Protoc* **11**:118–133.

125. Suzuki K, Tsunekawa Y, Hernandez-Benitez R, Wu J, Zhu J, Kim EJ, Hatanaka F, Yamamoto M, Araoka T, Li Z, Kurita M, Hishida T, Li M, Aizawa E, Guo S, Chen S, Goebl A, Soligalla RD, Qu J, Jiang T, Fu X, Jafari M, Esteban CR, Berggren WT, Lajara J, Nuñez-Delicado E, Guillen P, Campistol JM, Matsuzaki F, Liu GH, Magistretti P, Zhang K, Callaway EM, Zhang K, Belmonte JC. 2016. In vivo genome editing via CRISPR/Cas9 mediated homology-independent targeted integration. *Nature* **540**:144–149.

126. Gao Y, Hisey E, Bradshaw TWA, Erata E, Brown WE, Courtland JL, Uezu A, Xiang Y, Diao Y, Soderling SH. 2019. Plug-and-play protein modification using homology-independent universal genome engineering. *Neuron* **103**:583–597.e8.

127. Ihry RJ, Worringer KA, Salick MR, Frias E, Ho D, Theriault K, Kommineni S, Chen J, Sondey M, Ye C, Randhawa R, Kulkarni T, Yang Z, McAllister G, Russ C, Reece-Hoyes J, Forrester W, Hoffman GR, Dolmetsch R, Kaykas A. 2018. p53 inhibits CRISPR-Cas9 engineering in human pluripotent stem cells. *Nat Med* **24**:939–946.

128. Caldecott KW. 2008. Single-strand break repair and genetic disease. *Nat Rev Genet* **9**:619–631.

129. Davis L, Maizels N. 2014. Homology-directed repair of DNA nicks via pathways distinct from canonical double-strand break repair. *Proc Natl Acad Sci USA* **111**:E924–E932.

130. Bothmer A, Phadke T, Barrera LA, Margulies CM, Lee CS, Buquicchio F, Moss S, Abdulkerim HS, Selleck W, Jayaram H, Myer VE, Cotta-Ramusino C. 2017. Characterization of the interplay between DNA repair and CRISPR/Cas9-induced DNA lesions at an endogenous locus. *Nat Commun* **8**:13905.

131. Rees HA, Liu DR. 2018. Base editing: precision chemistry on the genome and transcriptome of living cells. *Nat Rev Genet* **19**:770–788.

132. Nishida K, Arazoe T, Yachie N, Banno S, Kakimoto M, Tabata M, Mochizuki M, Miyabe A, Araki M, Hara KY, Shimatani Z, Kondo A. 2016. Targeted nucleotide editing using hybrid prokaryotic and vertebrate adaptive immune systems. *Science* **353**:aaf8729.

133. Komor AC, Kim YB, Packer MS, Zuris JA, Liu DR. 2016. Programmable editing of a target base in genomic DNA without double-stranded DNA cleavage. *Nature* **533**:420–424.

134. Tang W, Liu DR. 2018. Rewritable multi-event analog recording in bacterial and mammalian cells. *Science* **360**:eaap8992.

135. Gaudelli NM, Komor AC, Rees HA, Packer MS, Badran AH, Bryson DI, Liu DR. 2017. Programmable base editing of A•T to G•C in genomic DNA without DNA cleavage. *Nature* **551**:464–471.

136. Komor AC, Zhao KT, Packer MS, Gaudelli NM, Waterbury AL, Koblan LW, Kim YB, Badran AH, Liu DR. 2017. Improved base excision repair inhibition and bacteriophage Mu Gam protein yields C:G-to-T:A base editors with higher efficiency and product purity. *Sci Adv* **3**:eaao4774.

137. Yeh WH, Chiang H, Rees HA, Edge ASB, Liu DR. 2018. In vivo base editing of post-mitotic sensory cells. *Nat Commun* **9**:2184.

138. Kim K, Ryu SM, Kim ST, Baek G, Kim D, Lim K, Chung E, Kim S, Kim JS. 2017. Highly efficient RNA-guided base editing in mouse embryos. *Nat Biotechnol* **35**:435–437.

139. Liu Z, Lu Z, Yang G, Huang S, Li G, Feng S, Liu Y, Li J, Yu W, Zhang Y, Chen J, Sun Q, Huang X. 2018. Efficient generation of mouse models of human diseases via ABE- and BE-mediated base editing. *Nat Commun* **9**:2338.

140. Farzadfard F, Lu TK. 2018. Emerging applications for DNA writers and molecular recorders. *Science* **361**:870–875.

141. Hess GT, Frésard L, Han K, Lee CH, Li A, Cimprich KA, Montgomery SB, Bassik MC. 2016. Directed evolution using dCas9-targeted somatic hypermutation in mammalian cells. *Nat Methods* **13**:1036–1042.

142. Ma Y, Zhang J, Yin W, Zhang Z, Song Y, Chang X. 2016. Targeted AID-mediated mutagenesis (TAM) enables efficient genomic diversification in mammalian cells. *Nat Methods* **13**:1029–1035.

143. Kim HS, Jeong YK, Hur JK, Kim JS, Bae S. 2019. Adenine base editors catalyze cytosine conversions in human cells. *Nat Biotechnol* **37**:1145–1148.

144. Kim D, Lim K, Kim ST, Yoon SH, Kim K, Ryu SM, Kim JS. 2017. Genome-wide target specificities of CRISPR RNA-guided programmable deaminases. *Nat Biotechnol* **35**:475–480.

145. Kim D, Kim DE, Lee G, Cho SI, Kim JS. 2019. Genome-wide target specificity of CRISPR RNA-guided adenine base editors. *Nat Biotechnol* **37**:430–435.

146. Zuo E, Sun Y, Wei W, Yuan T, Ying W, Sun H, Yuan L, Steinmetz LM, Li Y, Yang H. 2019. Cytosine base editor generates substantial off-target single-nucleotide variants in mouse embryos. *Science* **364**:289–292.

147. Jin S, Zong Y, Gao Q, Zhu Z, Wang Y, Qin P, Liang C, Wang D, Qiu JL, Zhang F, Gao C. 2019. Cytosine, but not adenine, base editors induce genome-wide off-target mutations in rice. *Science* **364**:292–295.

148. Grünewald J, Zhou R, Garcia SP, Iyer S, Lareau CA, Aryee MJ, Joung JK. 2019. Transcriptome-wide off-target RNA editing induced by CRISPR-guided DNA base editors. *Nature* **569**:433–437.

149. Zhou C, Sun Y, Yan R, Liu Y, Zuo E, Gu C, Han L, Wei Y, Hu X, Zeng R, Li Y, Zhou H, Guo F, Yang H. 2019. Off-target RNA mutation induced by DNA base editing and its elimination by mutagenesis. *Nature* 571:275–278.

150. Gaudelli NM, Lam DK, Rees HA, Solá-Esteves NM, Barrera LA, Born DA, Edwards A, Gehrke JM, Lee SJ, Liquori AJ, Murray R, Packer MS, Rinaldi C, Slaymaker IM, Yen J, Young LE, Ciaramella G. 2020. Directed evolution of adenine base editors with increased activity and therapeutic application. *Nat Biotechnol* 38:892–900.

151. Kim YB, Komor AC, Levy JM, Packer MS, Zhao KT, Liu DR. 2017. Increasing the genome-targeting scope and precision of base editing with engineered Cas9-cytidine deaminase fusions. *Nat Biotechnol* 35:371–376.

152. Huang TP, Zhao KT, Miller SM, Gaudelli NM, Oakes BL, Fellmann C, Savage DF, Liu DR. 2019. Circularly permuted and PAM-modified Cas9 variants broaden the targeting scope of base editors. *Nat Biotechnol* 37:626–631.

153. Zong Y, Song Q, Li C, Jin S, Zhang D, Wang Y, Qiu JL, Gao C. 2018. Efficient C-to-T base editing in plants using a fusion of nCas9 and human APOBEC3A. *Nat Biotechnol* 36:950–953.

154. Thuronyi BW, Koblan LW, Levy JM, Yeh WH, Zheng C, Newby GA, Wilson C, Bhaumik M, Shubina-Oleinik O, Holt JR, Liu DR. 2019. Continuous evolution of base editors with expanded target compatibility and improved activity. *Nat Biotechnol* 37:1070–1079.

155. Kurt IC, Zhou R, Iyer S, Garcia SP, Miller BR, Langner LM, Grünewald J, Joung JK. 2021. CRISPR C-to-G base editors for inducing targeted DNA transversions in human cells. *Nat Biotechnol* 39:41–46.

156. Zhao D, Li J, Li S, Xin X, Hu M, Price MA, Rosser SJ, Bi C, Zhang X. 2021. Glycosylase base editors enable C-to-A and C-to-G base changes. *Nat Biotechnol* 39:35–40.

157. Sakata RC, Ishiguro S, Mori H, Tanaka M, Tatsuno K, Ueda H, Yamamoto S, Seki M, Masuyama N, Nishida K, Nishimasu H, Arakawa K, Kondo A, Nureki O, Tomita M, Aburatani H, Yachie N. 2020. Base editors for simultaneous introduction of C-to-T and A-to-G mutations. *Nat Biotechnol* 38:865–869.

158. Gehrke JM, Cervantes O, Clement MK, Wu Y, Zeng J, Bauer DE, Pinello L, Joung JK. 2018. An APOBEC3A-Cas9 base editor with minimized bystander and off-target activities. *Nat Biotechnol* 36:977–982.

159. Rees HA, Wilson C, Doman JL, Liu DR. 2019. Analysis and minimization of cellular RNA editing by DNA adenine base editors. *Sci Adv* 5:eaax5717.

160. Grünewald J, Zhou R, Iyer S, Lareau CA, Garcia SP, Aryee MJ, Joung JK. 2019. CRISPR DNA base editors with reduced RNA off-target and self-editing activities. *Nat Biotechnol* 37:1041–1048.

161. Anzalone AV, Randolph PB, Davis JR, Sousa AA, Koblan LW, Levy JM, Chen PJ, Wilson C, Newby GA, Raguram A, Liu DR. 2019. Search-and-replace genome editing without double-strand breaks or donor DNA. *Nature* 576:149–157.

162. Aida T, Wilde JJ, Yang L, Hou Y, Li M, Xu D, Lin J, Qi P, Lu Z, Feng G. 2020. Prime editing primarily induces undesired outcomes in mice. *bioRxiv*.

163. Bosch JA, Birchak G, Perrimon N. 2021. Precise genome engineering in *Drosophila* using prime editing. *Proc Natl Acad Sci USA* 118:e2021996118.

164. Gao P, Lyu Q, Ghanam AR, Lazzarotto CR, Newby GA, Zhang W, Choi M, Slivano OJ, Holden K, Walker JA, II, Kadina AP, Munroe RJ, Abratte CM, Schimenti JC, Liu DR, Tsai SQ, Long X, Miano JM. 2021. Prime editing in mice reveals the essentiality of a single base in driving tissue-specific gene expression. *Genome Biol* 22:83.

165. eGTEx Project. 2017. Enhancing GTEx by bridging the gaps between genotype, gene expression, and disease. *Nat Genet* 49:1664–1670.

166. Lohmueller KE, Pearce CL, Pike M, Lander ES, Hirschhorn JN. 2003. Meta-analysis of genetic association studies supports a contribution of common variants to susceptibility to common disease. *Nat Genet* 33:177–182.

167. Jansen IE, et al. 2019. Genome-wide meta-analysis identifies new loci and functional pathways influencing Alzheimer's disease risk. *Nat Genet* 51:404–413.

168. Kunkle BW, et al, Genetic and Environmental Risk in AD/Defining Genetic, Polygenic and Environmental Risk for Alzheimer's Disease Consortium (GERAD/PERADES). 2019. Genetic meta-analysis of diagnosed Alzheimer's disease identifies new risk loci and implicates Aβ, tau, immunity and lipid processing. *Nat Genet* 51:414–430.

169. Freimer NB, Mohr DC. 2019. Integrating behavioural health tracking in human genetics research. *Nat Rev Genet* 20:129–130.

170. Sanchez-Roige S, Fontanillas P, Elson SL, Pandit A, Schmidt EM, Foerster JR, Abecasis GR, Gray JC, de Wit H, Davis LK, MacKillop J, Palmer AA, 23andMe Research Team. 2018. Genome-wide association study of delay discounting in 23,217 adult research participants of European ancestry. *Nat Neurosci* 21:16–18.

171. Nagel M, Jansen PR, Stringer S, Watanabe K, de Leeuw CA, Bryois J, Savage JE, Hammerschlag AR, Skene NG, Muñoz-Manchado AB, White T, Tiemeier H, Linnarsson S, Hjerling-Leffler J, Polderman TJC, Sullivan PF, van der Sluis S, Posthuma D, 23andMe Research Team. 2018. Meta-analysis of genome-wide association studies for neuroticism in 449,484 individuals identifies novel genetic loci and pathways. *Nat Genet* 50:920–927.

172. Belsky DW, Domingue BW, Wedow R, Arseneault L, Boardman JD, Caspi A, Conley D, Fletcher JM, Freese J, Herd P, Moffitt TE, Poulton R, Sicinski K, Wertz J, Harris KM. 2018. Genetic analysis of social-class mobility in five longitudinal studies. *Proc Natl Acad Sci USA* 115:E7275–E7284.

173. Lee JJ, et al, 23andMe Research Team, COGENT (Cognitive Genomics Consortium), Social Science Genetic Association Consortium. 2018. Gene discovery and polygenic prediction from a genome-wide association study of educational attainment in 1.1 million individuals. *Nat Genet* 50:1112–1121.

174. Altshuler D, Daly MJ, Lander ES. 2008. Genetic mapping in human disease. *Science* 322:881–888.

175. Tam V, Patel N, Turcotte M, Bossé Y, Paré G, Meyre D. 2019. Benefits and limitations of genome-wide association studies. *Nat Rev Genet* 20:467–484.

176. Paquet D, Kwart D, Chen A, Sproul A, Jacob S, Teo S, Olsen KM, Gregg A, Noggle S, Tessier-Lavigne M. 2016. Efficient introduction of specific homozygous and heterozygous mutations using CRISPR/Cas9. *Nature* 533:125–129.

177. Lin YT, Seo J, Gao F, Feldman HM, Wen HL, Penney J, Cam HP, Gjoneska E, Raja WK, Cheng J, Rueda R, Kritskiy O, Abdurrob F, Peng Z, Milo B, Yu CJ, Elmsaouri S, Dey D, Ko T, Yankner BA, Tsai LH. 2018. APOE4 causes widespread molecular and cellular alterations associated with Alzheimer's disease phenotypes in human iPSC-derived brain cell types. *Neuron* 98:1141–1154.e7.

178. van der Kant R, Langness VF, Herrera CM, Williams DA, Fong LK, Leestemaker Y, Steenvoorden E, Rynearson KD, Brouwers JF, Helms JB, Ovaa H, Giera M, Wagner SL, Bang AG, Goldstein LSB. 2019. Cholesterol metabolism is a druggable axis that independently regulates tau and amyloid-β in iPSC-derived Alzheimer's disease neurons. *Cell Stem Cell* 24:363–375.e9.

179. Kwart D, Gregg A, Scheckel C, Murphy EA, Paquet D, Duffield M, Fak J, Olsen O, Darnell RB, Tessier-Lavigne M. 2019. A large panel of isogenic APP and PSEN1 mutant human iPSC neurons reveals shared endosomal abnormalities mediated by APP β-CTFs, not Aβ. *Neuron* 104:256–270.e5.

180. Soldner F, Stelzer Y, Shivalila CS, Abraham BJ, Latourelle JC, Barrasa MI, Goldmann J, Myers RH, Young RA, Jaenisch R. 2016. Parkinson-associated risk variant in distal enhancer of α-synuclein modulates target gene expression. *Nature* 533:95–99.

181. Zeng H, Guo M, Zhou T, Tan L, Chong CN, Zhang T, Dong X, Xiang JZ, Yu AS, Yue L, Qi Q, Evans T, Graumann J, Chen S. 2016. An isogenic human ESC platform for functional evaluation of genome-wide-association-study-identified diabetes genes and drug discovery. *Cell Stem Cell* 19:326–340.

182. Warren CR, O'Sullivan JF, Friesen M, Becker CE, Zhang X, Liu P, Wakabayashi Y, Morningstar JE, Shi X, Choi J, Xia F, Peters DT, Florido MHC, Tsankov AM, Duberow E, Comisar L, Shay J, Jiang X, Meissner A, Musunuru K, Kathiresan S, Daheron L, Zhu J, Gerszten RE, Deo RC, Vasan RS, O'Donnell CJ, Cowan CA. 2017. Induced pluripotent stem cell differentiation enables functional validation of GWAS variants in metabolic disease. *Cell Stem Cell* 20:547–557.e7.

183. Pashos EE, Park Y, Wang X, Raghavan A, Yang W, Abbey D, Peters DT, Arbelaez J, Hernandez M, Kuperwasser N, Li W, Lian Z, Liu Y, Lv W, Lytle-Gabbin SL, Marchadier DH, Rogov P, Shi J, Slovik KJ, Stylianou IM, Wang L, Yan R, Zhang X, Kathiresan S, Duncan SA, Mikkelsen TS, Morrisey EE, Rader DJ, Brown CD, Musunuru K. 2017. Large, diverse population cohorts of hiPSCs and derived hepatocyte-like cells reveal functional genetic variation at blood lipid-associated loci. *Cell Stem Cell* 20:558–570.e10.

184. Shi Y, Lin S, Staats KA, Li Y, Chang WH, Hung ST, Hendricks E, Linares GR, Wang Y, Son EY, Wen X, Kisler K, Wilkinson B, Menendez L, Sugawara T, Woolwine P, Huang M, Cowan MJ, Ge B, Koutsodendris N, Sandor KP, Komberg J, Vangoor VR, Senthilkumar K, Hennes V, Seah C, Nelson AR, Cheng TY, Lee SJ, August PR, Chen JA, Wisniewski N, Hanson-Smith V, Belgard TG, Zhang A, Coba M, Grunseich C, Ward ME, van den Berg LH, Pasterkamp RJ, Trotti D, Zlokovic BV, Ichida JK. 2018. Haploinsufficiency leads to neurodegeneration in C9ORF72 ALS/FTD human induced motor neurons. *Nat Med* 24:313–325.

185. Holtzman DM, Herz J, Bu G. 2012. Apolipoprotein E and apolipoprotein E receptors: normal biology and roles in Alzheimer disease. *Cold Spring Harb Perspect Med* 2:a006312.

186. Meyer K, Feldman HM, Lu T, Drake D, Lim ET, Ling KH, Bishop NA, Pan Y, Seo J, Lin YT, Su SC, Church GM, Tsai LH, Yankner BA. 2019. REST and neural gene network dysregulation in iPSC models of Alzheimer's disease. *Cell Rep* 26:1112–1127.e9.

187. Capecchi MR. 1989. The new mouse genetics: altering the genome by gene targeting. *Trends Genet* 5:70–76.

188. Skarnes WC, Rosen B, West AP, Koutsourakis M, Bushell W, Iyer V, Mujica AO, Thomas M, Harrow J, Cox T, Jackson D, Severin J, Biggs P, Fu J, Nefedov M, de Jong PJ, Stewart AF, Bradley A. 2011. A conditional knockout resource for the genome-wide study of mouse gene function. *Nature* 474:337–342.

189. Gerlai R. 2016. Gene targeting using homologous recombination in embryonic stem cells: the future for behavior genetics? *Front Genet* 7:43.

190. Yang H, Wang H, Shivalila CS, Cheng AW, Shi L, Jaenisch R. 2013. One-step generation of mice carrying reporter and conditional alleles by CRISPR/Cas-mediated genome engineering. *Cell* 154:1370–1379.

191. Niu Y, Shen B, Cui Y, Chen Y, Wang J, Wang L, Kang Y, Zhao X, Si W, Li W, Xiang AP, Zhou J, Guo X, Bi Y, Si C, Hu B, Dong G, Wang H, Zhou Z, Li T, Tan T, Pu X, Wang F, Ji S, Zhou Q, Huang X, Ji W, Sha J. 2014. Generation of gene-modified cynomolgus monkey via Cas9/RNA-mediated gene targeting in one-cell embryos. *Cell* 156:836–843.

192. Quadros RM, Miura H, Harms DW, Akatsuka H, Sato T, Aida T, Redder R, Richardson GP, Inagaki Y, Sakai D, Buckley SM, Seshacharyulu P, Batra SK, Behlke MA, Zeiner SA, Jacobi AM, Izu Y, Thoreson WB, Urness LD, Mansour SL, Ohtsuka M, Gurumurthy CB. 2017. Easi-CRISPR: a robust method for one-step generation of mice carrying conditional and insertion alleles using long ssDNA donors and CRISPR ribonucleoproteins. *Genome Biol* 18:92.

193. Gurumurthy CB, et al. 2019. Reproducibility of CRISPR-Cas9 methods for generation of conditional mouse alleles: a multi-center evaluation. *Genome Biol* 20:171.

194. Chen S, Lee B, Lee AY, Modzelewski AJ, He L. 2016. Highly efficient mouse genome editing by CRISPR ribonucleoprotein electroporation of zygotes. *J Biol Chem* 291:14457–14467.

195. Gurumurthy CB, Sato M, Nakamura A, Inui M, Kawano N, Islam MA, Ogiwara S, Takabayashi S, Matsuyama M, Nakagawa S, Miura H, Ohtsuka M. 2019. Creation of CRISPR-based germline-genome-engineered mice without ex vivo handling of zygotes by i-GONAD. *Nat Protoc* 14:2452–2482.

196. Waaijers S, Boxem M. 2014. Engineering the Caenorhabditis elegans genome with CRISPR/Cas9. *Methods* 68:381–388.

197. Ren X, Sun J, Housden BE, Hu Y, Roesel C, Lin S, Liu LP, Yang Z, Mao D, Sun L, Wu Q, Ji JY, Xi J, Mohr SE, Xu J, Perrimon N, Ni JQ. 2013. Optimized gene editing technology for *Drosophila melanogaster* using germ line-specific Cas9. *Proc Natl Acad Sci USA* 110:19012–19017.

198. Hwang WY, Fu Y, Reyon D, Maeder ML, Tsai SQ, Sander JD, Peterson RT, Yeh JR, Joung JK. 2013. Efficient genome editing in zebrafish using a CRISPR-Cas system. *Nat Biotechnol* 31:227–229.

199. Li W, Teng F, Li T, Zhou Q. 2013. Simultaneous generation and germline transmission of multiple gene mutations in rat using CRISPR-Cas systems. *Nat Biotechnol* 31:684–686.

200. Hai T, Teng F, Guo R, Li W, Zhou Q. 2014. One-step generation of knockout pigs by zygote injection of CRISPR/Cas system. *Cell Res* 24:372–375.

201. Yoshimatsu S, Okahara J, Sone T, Takeda Y, Nakamura M, Sasaki E, Kishi N, Shiozawa S, Okano H. 2019. Robust and efficient knock-in in embryonic stem cells and early-stage embryos of the common marmoset using the CRISPR-Cas9 system. *Sci Rep* 9:1528.

202. Ma H, Marti-Gutierrez N, Park SW, Wu J, Lee Y, Suzuki K, Koski A, Ji D, Hayama T, Ahmed R, Darby H, Van Dyken C, Li Y, Kang E, Park AR, Kim D, Kim ST, Gong J, Gu Y, Xu X, Battaglia D, Krieg SA, Lee DM, Wu DH, Wolf DP, Heitner SB, Belmonte JCI, Amato P, Kim JS, Kaul S, Mitalipov S. 2017. Correction of a pathogenic gene mutation in human embryos. *Nature* 548:413–419.

203. Gu B, Swigut T, Spencley A, Bauer MR, Chung M, Meyer T, Wysocka J. 2018. Transcription-coupled changes in nuclear mobility of mammalian cis-regulatory elements. *Science* 359:1050–1055.

204. Fonfara I, Richter H, Bratovič M, Le Rhun A, Charpentier E. 2016. The CRISPR-associated DNA-cleaving enzyme Cpf1 also processes precursor CRISPR RNA. *Nature* 532:517–521.

205. Liao C, Ttofali F, Slotkowski RA, Denny SR, Cecil TD, Leenay RT, Keung AJ, Beisel CL. 2019. Modular one-pot assembly of CRISPR arrays enables library generation and reveals factors influencing crRNA biogenesis. *Nat Commun* 10:2948.

206. Campa CC, Weisbach NR, Santinha AJ, Incarnato D, Platt RJ. 2019. Multiplexed genome engineering by Cas12a and CRISPR arrays encoded on single transcripts. *Nat Methods* 16:887–893.

207. Zetsche B, Heidenreich M, Mohanraju P, Fedorova I, Kneppers J, DeGennaro EM, Winblad N, Choudhury SR, Abudayyeh OO, Gootenberg JS, Wu WY, Scott DA, Severinov K, van der Oost J, Zhang F. 2017. Multiplex gene editing by CRISPR-Cpf1 using a single crRNA array. *Nat Biotechnol* 35:31–34.

208. Wang M, Mao Y, Lu Y, Tao X, Zhu JK. 2017. Multiplex gene editing in rice using the CRISPR-Cpf1 system. *Mol Plant* 10:1011–1013.

209. Sanson KR, DeWeirdt PC, Sangree AK, Hanna RE, Hegde M, Teng T, Borys SM, Strand C, Joung JK, Kleinstiver BP, Pan X, Huang A, Doench JG. 2019. Optimization of AsCas12a for combinatorial genetic screens in human cells. *bioRxiv*.

210. Sanjana NE, Shalem O, Zhang F. 2014. Improved vectors and genome-wide libraries for CRISPR screening. *Nat Methods* 11:783–784.

211. Tian R, Gachechiladze MA, Ludwig CH, Laurie MT, Hong JY, Nathaniel D, Prabhu AV, Fernandopulle MS, Patel R, Abshari M, Ward ME, Kampmann M. 2019. CRISPR interference-based platform for multimodal genetic screens in human iPSC-derived neurons. *Neuron* 104:239–255.e12.

212. Koike-Yusa H, Li Y, Tan EP, Velasco-Herrera MC, Yusa K. 2014. Genome-wide recessive genetic screening in mammalian cells with a lentiviral CRISPR-guide RNA library. *Nat Biotechnol* 32:267–273.

213. Aza-Blanc P, Cooper CL, Wagner K, Batalov S, Deveraux QL, Cooke MP. 2003. Identification of modulators of TRAIL-induced apoptosis via RNAi-based phenotypic screening. *Mol Cell* 12:627–637.

214. Kittler R, Putz G, Pelletier L, Poser I, Heninger AK, Drechsel D, Fischer S, Konstantinova I, Habermann B, Grabner H, Yaspo ML, Himmelbauer H, Korn B, Neugebauer K, Pisabarro MT, Buchholz F. 2004. An endoribonuclease-prepared siRNA screen in human cells identifies genes essential for cell division. *Nature* 432:1036–1040.

215. Luo B, Cheung HW, Subramanian A, Sharifnia T, Okamoto M, Yang X, Hinkle G, Boehm JS, Beroukhim R, Weir BA, Mermel C, Barbie DA, Awad T, Zhou X, Nguyen T, Piqani B, Li C, Golub TR, Meyerson M, Hacohen N, Hahn WC, Lander ES, Sabatini DM, Root DE. 2008. Highly parallel identification of essential genes in cancer cells. *Proc Natl Acad Sci USA* 105:20380–20385.

216. Luo J, Emanuele MJ, Li D, Creighton CJ, Schlabach MR, Westbrook TF, Wong KK, Elledge SJ. 2009. A genome-wide RNAi screen identifies multiple synthetic lethal interactions with the Ras oncogene. *Cell* 137:835–848.

217. Franceschini A, Meier R, Casanova A, Kreibich S, Daga N, Andritschke D, Dilling S, Rämö P, Emmenlauer M, Kaufmann A, Conde-Álvarez R, Low SH, Pelkmans L, Helenius A, Hardt WD, Dehio C, von Mering C. 2014. Specific inhibition of diverse pathogens in human cells by synthetic microRNA-like oligonucleotides inferred from RNAi screens. *Proc Natl Acad Sci USA* 111:4548–4553.

218. Behan FM, Iorio F, Picco G, Gonçalves E, Beaver CM, Migliardi G, Santos R, Rao Y, Sassi F, Pinnelli M, Ansari R, Harper S, Jackson DA, McRae R, Pooley R, Wilkinson P, van der Meer D, Dow D, Buser-Doepner C, Bertotti A, Trusolino L, Stronach EA, Saez-Rodriguez J, Yusa K, Garnett MJ. 2019. Prioritization of cancer therapeutic targets using CRISPR-Cas9 screens. *Nature* 568:511–516.

219. Lieb S, Blaha-Ostermann S, Kamper E, Rippka J, Schwarz C, Ehrenhöfer-Wölfer K, Schlattl A, Wernitznig A, Lipp JJ, Nagasaka K, van der Lelij P, Bader G, Koi M, Goel A, Neumüller RA, Peters JM, Kraut N, Pearson MA, Petronczki M, Wöhrle S. 2019. Werner syndrome helicase is a selective vulnerability of microsatellite instability-high tumor cells. *Elife* 8:e43333.

220. Chan EM, Shibue T, McFarland JM, Gaeta B, Ghandi M, Dumont N, Gonzalez A, McPartlan JS, Li T, Zhang Y, Bin Liu J, Lazaro JB, Gu P, Piett CG, Apffel A, Ali SO, Deasy R, Keskula P, Ng RWS, Roberts EA, Reznichenko E, Leung L, Alimova M, Schenone M, Islam M, Maruvka YE, Liu Y, Roper J, Raghavan S, Giannakis M, Tseng YY, Nagel ZD, D'Andrea A, Root DE, Boehm JS, Getz G, Chang S, Golub TR, Tsherniak A, Vazquez F, Bass AJ. 2019. WRN helicase is a synthetic lethal target in microsatellite unstable cancers. *Nature* 568:551–556.

221. Liu Y, Cao Z, Wang Y, Guo Y, Xu P, Yuan P, Liu Z, He Y, Wei W. 2018. Genome-wide screening for functional long noncoding RNAs in human cells by Cas9 targeting of splice sites. *Nat Biotechnol* 36:1203–1210.

222. Zhu S, Li W, Liu J, Chen CH, Liao Q, Xu P, Xu H, Xiao T, Cao Z, Peng J, Yuan P, Brown M, Liu XS, Wei W. 2016. Genome-scale deletion screening of human long non-coding RNAs using a paired-guide RNA CRISPR-Cas9 library. *Nat Biotechnol* 34:1279–1286.

223. Liu SJ, Horlbeck MA, Cho SW, Birk HS, Malatesta M, He D, Attenello FJ, Villalta JE, Cho MY, Chen Y, Mandegar MA, Olvera MP, Gilbert LA, Conklin BR, Chang HY, Weissman JS, Lim DA. 2017. CRISPRi-based genome-scale identification of functional long noncoding RNA loci in human cells. *Science* 355:eaah7111.

224. Joung J, Engreitz JM, Konermann S, Abudayyeh OO, Verdine VK, Aguet F, Gootenberg JS, Sanjana NE, Wright JB, Fulco CP, Tseng YY, Yoon CH, Boehm JS, Lander ES, Zhang F. 2017. Genome-scale activation screen identifies a lncRNA locus regulating a gene neighbourhood. *Nature* 548:343–346.

225. Bester AC, Lee JD, Chavez A, Lee YR, Nachmani D, Vora S, Victor J, Sauvageau M, Monteleone E, Rinn JL, Provero P, Church GM, Clohessy JG, Pandolfi PP. 2018. An integrated genome-wide CRISPRa approach to functionalize lncRNAs in drug resistance. *Cell* 173:649–664.e20.

226. Findlay GM, Boyle EA, Hause RJ, Klein JC, Shendure J. 2014. Saturation editing of genomic regions by multiplex homology-directed repair. *Nature* 513:120–123.

227. Findlay GM, Daza RM, Martin B, Zhang MD, Leith AP, Gasperini M, Janizek JD, Huang X, Starita LM, Shendure J. 2018. Accurate classification of BRCA1 variants with saturation genome editing. *Nature* 562:217–222.

228. Kircher M, Xiong C, Martin B, Schubach M, Inoue F, Bell RJA, Costello JF, Shendure J, Ahituv N. 2019. Saturation mutagenesis of twenty disease-associated regulatory elements at single base-pair resolution. *Nat Commun* 10:3583.

229. Hanna RE, Hegde M, Fagre CR, DeWeirdt PC, Sangree AK, Szegletes Z, Griffith A, Feeley MN, Sanson KR, Baidi Y, Koblan LW, Liu DR, Neal JT, Doench JG. 2021. Massively parallel assessment of human variants with base editor screens. *Cell* 184:1064–1080.e20.

230. Cuella-Martin R, Hayward SB, Fan X, Chen X, Huang JW, Taglialatela A, Leuzzi G, Zhao J, Rabadan R, Lu C, Shen Y, Ciccia A. 2021. Functional interrogation of DNA damage response variants with base editing screens. *Cell* 184:1081–1097.e19.

231. Gantz VM, Bier E. 2016. The dawn of active genetics. *Bioessays* 38:50–63.

232. Gantz VM, Jasinskiene N, Tatarenkova O, Fazekas A, Macias VM, Bier E, James AA. 2015. Highly efficient Cas9-mediated gene drive for population modification of the malaria vector mosquito *Anopheles stephensi*. *Proc Natl Acad Sci USA* 112:E6736–E6743.

233. Hammond A, Galizi R, Kyrou K, Simoni A, Siniscalchi C, Katsanos D, Gribble M, Baker D, Marois E, Russell S, Burt A, Windbichler N, Crisanti A, Nolan T. 2016. A CRISPR-Cas9 gene drive system targeting female reproduction in the malaria mosquito vector *Anopheles gambiae*. *Nat Biotechnol* 34:78–83.

234. Kyrou K, Hammond AM, Galizi R, Kranjc N, Burt A, Beaghton AK, Nolan T, Crisanti A. 2018. A CRISPR-Cas9 gene drive targeting doublesex causes complete population suppression in caged *Anopheles gambiae* mosquitoes. *Nat Biotechnol* 36:1062–1066.

235. Grunwald HA, Gantz VM, Poplawski G, Xu XS, Bier E, Cooper KL. 2019. Super-Mendelian inheritance mediated by CRISPR-Cas9 in the female mouse germline. *Nature* 566:105–109.

236. Esvelt KM, Smidler AL, Catteruccia F, Church GM. 2014. Concerning RNA-guided gene drives for the alteration of wild populations. *Elife* 3:e03401.

237. Gurwitz D. 2014. Gene drives raise dual-use concerns. *Science* 345:1010.

238. Akbari OS, Bellen HJ, Bier E, Bullock SL, Burt A, Church GM, Cook KR, Duchek P, Edwards OR, Esvelt KM, Gantz VM, Golic KG, Gratz SJ, Harrison MM, Hayes KR, James AA, Kaufman TC, Knoblich J, Malik HS, Matthews KA, O'Connor-Giles KM, Parks AL, Perrimon N, Port F, Russell S, Ueda R, Wildonger J. 2015. BIOSAFETY. *Safeguarding gene drive experiments in the laboratory*. *Science* 349:927–929.

239. Noble C, Adlam B, Church GM, Esvelt KM, Nowak MA. 2018. Current CRISPR gene drive systems are likely to be highly invasive in wild populations. *Elife* 7:e33423.

240. Champer J, Chung J, Lee YL, Liu C, Yang E, Wen Z, Clark AG, Messer PW. 2019. Molecular safeguarding of CRISPR gene drive experiments. *Elife* 8:e41439.

241. Wu B, Luo L, Gao XJ. 2016. Cas9-triggered chain ablation of cas9 as a gene drive brake. *Nat Biotechnol* 34:137–138.

242. Marshall JM. 2010. The Cartagena Protocol and genetically modified mosquitoes. *Nat Biotechnol* 28:896–897.

243. Oye KA, Esvelt K, Appleton E, Catteruccia F, Church G, Kuiken T, Lightfoot SB, McNamara J, Smidler A, Collins JP. 2014. Biotechnology. *Regulating gene drives*. *Science* 345:626–628.

244. Antoniani C, Meneghini V, Lattanzi A, Felix T, Romano O, Magrin E, Weber L, Pavani G, El Hoss S, Kurita R, Nakamura Y, Cradick TJ, Lundberg AS, Porteus M, Amendola M, El Nemer W, Cavazzana M, Mavilio F, Miccio A. 2018. Induction of fetal hemoglobin

synthesis by CRISPR/Cas9-mediated editing of the human β-globin locus. *Blood* **131**:1960–1973.

245. Psatha N, Reik A, Phelps S, Zhou Y, Dalas D, Yannaki E, Levasseur DN, Urnov FD, Holmes MC, Papayannopoulou T. 2018. Disruption of the BCL11A erythroid enhancer reactivates fetal hemoglobin in erythroid cells of patients with β-thalassemia major. *Mol Ther Methods Clin Dev* **10**:313–326.

246. Canver MC, Smith EC, Sher F, Pinello L, Sanjana NE, Shalem O, Chen DD, Schupp PG, Vinjamur DS, Garcia SP, Luc S, Kurita R, Nakamura Y, Fujiwara Y, Maeda T, Yuan GC, Zhang F, Orkin SH, Bauer DE. 2015. BCL11A enhancer dissection by Cas9-mediated in situ saturating mutagenesis. *Nature* **527**:192–197.

247. Samulski RJ, Zhu X, Xiao X, Brook JD, Housman DE, Epstein N, Hunter LA. 1991. Targeted integration of adeno-associated virus (AAV) into human chromosome 19. *EMBO J* **10**:3941–3950.

248. Eyquem J, Mansilla-Soto J, Giavridis T, van der Stegen SJ, Hamieh M, Cunanan KM, Odak A, Gönen M, Sadelain M. 2017. Targeting a CAR to the TRAC locus with CRISPR/Cas9 enhances tumour rejection. *Nature* **543**:113–117.

249. Hoban MD, Cost GJ, Mendel MC, Romero Z, Kaufman ML, Joglekar AV, Ho M, Lumaquin D, Gray D, Lill GR, Cooper AR, Urbinati F, Senadheera S, Zhu A, Liu PQ, Paschon DE, Zhang L, Rebar EJ, Wilber A, Wang X, Gregory PD, Holmes MC, Reik A, Hollis RP, Kohn DB. 2015. Correction of the sickle cell disease mutation in human hematopoietic stem/progenitor cells. *Blood* **125**:2597–2604.

250. Osborn MJ, Gabriel R, Webber BR, DeFeo AP, McElroy AN, Jarjour J, Starker CG, Wagner JE, Joung JK, Voytas DF, von Kalle C, Schmidt M, Blazar BR, Tolar J. 2015. Fanconi anemia gene editing by the CRISPR/Cas9 system. *Hum Gene Ther* **26**:114–126.

251. Román-Rodríguez FJ, Ugalde L, Álvarez L, Díez B, Ramírez MJ, Risueño C, Cortón M, Bogliolo M, Bernal S, March F, Ayuso C, Hanenberg H, Sevilla J, Rodríguez-Perales S, Torres-Ruiz R, Surrallés J, Bueren JA, Río P. 2019. NHEJ-mediated repair of CRISPR-Cas9-induced DNA breaks efficiently corrects mutations in HSPCs from patients with Fanconi anemia. *Cell Stem Cell* **25**:607–621.e7.

252. Schwank G, Koo BK, Sasselli V, Dekkers JF, Heo I, Demircan T, Sasaki N, Boymans S, Cuppen E, van der Ent CK, Nieuwenhuis EE, Beekman JM, Clevers H. 2013. Functional repair of CFTR by CRISPR/Cas9 in intestinal stem cell organoids of cystic fibrosis patients. *Cell Stem Cell* **13**:653–658.

253. Pauling L, Itano HA, Singer SJ, Wells IC. 1949. Sickle cell anemia, a molecular disease. *Science* **110**:543–548.

254. DeWitt MA, Magis W, Bray NL, Wang T, Berman JR, Urbinati F, Heo SJ, Mitros T, Muñoz DP, Boffelli D, Kohn DB, Walters MC, Carroll D, Martin DI, Corn JE. 2016. Selection-free genome editing of the sickle mutation in human adult hematopoietic stem/progenitor cells. *Sci Transl Med* **8**:360ra134.

255. Schumann K, Lin S, Boyer E, Simeonov DR, Subramaniam M, Gate RE, Haliburton GE, Ye CJ, Bluestone JA, Doudna JA, Marson A. 2015. Generation of knock-in primary human T cells using Cas9 ribonucleoproteins. *Proc Natl Acad Sci USA* **112**:10437–10442.

256. Roth TL, Puig-Saus C, Yu R, Shifrut E, Carnevale J, Li PJ, Hiatt J, Saco J, Krystofinski P, Li H, Tobin V, Nguyen DN, Lee MR, Putnam AL, Ferris AL, Chen JW, Schickel JN, Pellerin L, Carmody D, Alkorta-Aranburu G, Del Gaudio D, Matsumoto H, Morell M, Mao Y, Cho M, Quadros RM, Gurumurthy CB, Smith B, Haugwitz M, Hughes SH, Weissman JS, Schumann K, Esensten JH, May AP, Ashworth A, Kupfer GM, Greeley SAW, Bacchetta R, Meffre E, Roncarolo MG, Romberg N, Herold KC, Ribas A, Leonetti MD, Marson A. 2018. Reprogramming human T cell function and specificity with non-viral genome targeting. *Nature* **559**:405–409.

257. Shahbazi R, Sghia-Hughes G, Reid JL, Kubek S, Haworth KG, Humbert O, Kiem HP, Adair JE. 2019. Targeted homology-directed repair in blood stem and progenitor cells with CRISPR nanoformulations. *Nat Mater* **18**:1124–1132.

258. Kim S, Kim D, Cho SW, Kim J, Kim JS. 2014. Highly efficient RNA-guided genome editing in human cells via delivery of purified Cas9 ribonucleoproteins. *Genome Res* **24**:1012–1019.

259. Hendel A, Bak RO, Clark JT, Kennedy AB, Ryan DE, Roy S, Steinfeld I, Lunstad BD, Kaiser RJ, Wilkens AB, Bacchetta R, Tsalenko A, Dellinger D, Bruhn L, Porteus MH. 2015. Chemically modified guide RNAs enhance CRISPR-Cas genome editing in human primary cells. *Nat Biotechnol* **33**:985–989.

260. Seki A, Rutz S. 2018. Optimized RNP transfection for highly efficient CRISPR/Cas9-mediated gene knockout in primary T cells. *J Exp Med* **215**:985–997.

261. Hultquist JF, Hiatt J, Schumann K, McGregor MJ, Roth TL, Haas P, Doudna JA, Marson A, Krogan NJ. 2019. CRISPR-Cas9 genome engineering of primary CD4⁺ T cells for the interrogation of HIV-host factor interactions. *Nat Protoc* **14**:1–27.

262. Wu Y, Zeng J, Roscoe BP, Liu P, Yao Q, Lazzarotto CR, Clement K, Cole MA, Luk K, Baricordi C, Shen AH, Ren C, Esrick EB, Manis JP, Dorfman DM, Williams DA, Biffi A, Brugnara C, Biasco L, Brendel C, Pinello L, Tsai SQ, Wolfe SA, Bauer DE. 2019. Highly efficient therapeutic gene editing of human hematopoietic stem cells. *Nat Med* **25**:776–783.

263. Kalos M, June CH. 2013. Adoptive T cell transfer for cancer immunotherapy in the era of synthetic biology. *Immunity* **39**:49–60.

264. Qasim W, Zhan H, Samarasinghe S, Adams S, Amrolia P, Stafford S, Butler K, Rivat C, Wright G, Somana K, Ghorashian S, Pinner D, Ahsan G, Gilmour K, Lucchini G, Inglott S, Mifsud W, Chiesa R, Peggs KS, Chan L, Farzaneh F, Thrasher AJ, Vora A, Pule M, Veys P. 2017. Molecular remission of infant B-ALL after infusion of universal TALEN gene-edited CAR T cells. *Sci Transl Med* **9**:eaaj2013.

265. Genovese P, Schiroli G, Escobar G, Tomaso TD, Firrito C, Calabria A, Moi D, Mazzieri R, Bonini C, Holmes MC, Gregory PD, van der Burg M, Gentner B, Montini E, Lombardo A, Naldini L. 2014. Targeted genome editing in human repopulating haematopoietic stem cells. *Nature* **510**:235–240.

266. Schiroli G, Ferrari S, Conway A, Jacob A, Capo V, Albano L, Plati T, Castiello MC, Sanvito F, Gennery AR, Bovolenta C, Palchaudhuri R, Scadden DT, Holmes MC, Villa A, Sitia G, Lombardo A, Genovese P, Naldini L. 2017. Preclinical modeling highlights the therapeutic potential of hematopoietic stem cell gene editing for correction of SCID-X1. *Sci Transl Med* **9**:eaan0820.

267. De Ravin SS, Li L, Wu X, Choi U, Allen C, Koontz S, Lee J, Theobald-Whiting N, Chu J, Garofalo M, Sweeney C, Kardava L, Moir S, Viley A, Natarajan P, Su L, Kuhns D, Zarember KA, Peshwa MV, Malech HL. 2017. CRISPR-Cas9 gene repair of hematopoietic stem cells from patients with X-linked chronic granulomatous disease. *Sci Transl Med* **9**:eaah3480.

268. Humbert O, Radtke S, Samuelson C, Carrillo RR, Perez AM, Reddy SS, Lux C, Pattabhi S, Schefter LE, Negre O, Lee CM, Bao G, Adair JE, Peterson CW, Rawlings DJ, Scharenberg AM, Kiem HP. 2019. Therapeutically relevant engraftment of a CRISPR-Cas9-edited HSC-enriched population with HbF reactivation in nonhuman primates. *Sci Transl Med* **11**:eaaw3768.

269. Wolf DP, Mitalipov PA, Mitalipov SM. 2019. Principles of and strategies for germline gene therapy. *Nat Med* **25**:890–897.

270. Torikai H, Reik A, Soldner F, Warren EH, Yuen C, Zhou Y, Crossland DL, Huls H, Littman N, Zhang Z, Tykodi SS, Kebriaei P, Lee DA, Miller JC, Rebar EJ, Holmes MC, Jaenisch R, Champlin RE, Gregory PD, Cooper LJ. 2013. Toward eliminating HLA class I expression to generate universal cells from allogeneic donors. *Blood* **122**:1341–1349.

271. Stadtmauer EA, Fraietta JA, Davis MM, Cohen AD, Weber KL, Lancaster E, Mangan PA, Kulikovskaya I, Gupta M, Chen F, Tian L, Gonzalez VE, Xu J, Jung IY, Melenhorst JJ, Plesa G, Shea J, Matlawski T, Cervini A, Gaymon AL, Desjardins S, Lamontagne A, Salas-Mckee J, Fesnak A, Siegel DL, Levine BL, Jadlowsky JK, Young RM, Chew A, Hwang WT, Hexner EO, Carreno BM, Nobles CL, Bushman FD, Parker KR, Qi Y, Satpathy AT, Chang HY, Zhao Y, Lacey SF, June CH. 2020. CRISPR-engineered T cells in patients with refractory cancer. *Science* **367**:eaba7365.

272. Lu Y, Xue J, Deng T, Zhou X, Yu K, Deng L, Huang M, Yi X, Liang M, Wang Y, Shen H, Tong R, Wang W, Li L, Song J, Li J,

Su X, Ding Z, Gong Y, Zhu J, Wang Y, Zou B, Zhang Y, Li Y, Zhou L, Liu Y, Yu M, Wang Y, Zhang X, Yin L, Xia X, Zeng Y, Zhou Q, Ying B, Chen C, Wei Y, Li W, Mok T. 2020. Safety and feasibility of CRISPR-edited T cells in patients with refractory non-small-cell lung cancer. *Nat Med* **26**:732–740.

273. Haapaniemi E, Botla S, Persson J, Schmierer B, Taipale J. 2018. CRISPR-Cas9 genome editing induces a p53-mediated DNA damage response. *Nat Med* **24**:927–930.

274. Schiroli G, Conti A, Ferrari S, Della Volpe L, Jacob A, Albano L, Beretta S, Calabria A, Vavassori V, Gasparini P, Salataj E, Ndiaye-Lobry D, Brombin C, Chaumeil J, Montini E, Merelli I, Genovese P, Naldini L, Di Micco R. 2019. Precise gene editing preserves hematopoietic stem cell function following transient p53-mediated DNA damage response. *Cell Stem Cell* **24**:551–565.e8.

275. Milone MC, O'Doherty U. 2018. Clinical use of lentiviral vectors. *Leukemia* **32**:1529–1541.

276. Asokan A, Schaffer DV, Samulski RJ. 2012. The AAV vector toolkit: poised at the clinical crossroads. *Mol Ther* **20**:699–708.

277. Bainbridge JW, Smith AJ, Barker SS, Robbie S, Henderson R, Balaggan K, Viswanathan A, Holder GE, Stockman A, Tyler N, Petersen-Jones S, Bhattacharya SS, Thrasher AJ, Fitzke FW, Carter BJ, Rubin GS, Moore AT, Ali RR. 2008. Effect of gene therapy on visual function in Leber's congenital amaurosis. *N Engl J Med* **358**:2231–2239.

278. Kotterman MA, Schaffer DV. 2014. Engineering adeno-associated viruses for clinical gene therapy. *Nat Rev Genet* **15**:445–451.

279. Kessler PD, Podsakoff GM, Chen X, McQuiston SA, Colosi PC, Matelis LA, Kurtzman GJ, Byrne BJ. 1996. Gene delivery to skeletal muscle results in sustained expression and systemic delivery of a therapeutic protein. *Proc Natl Acad Sci USA* **93**:14082–14087.

280. Chirmule N, Propert K, Magosin S, Qian Y, Qian R, Wilson J. 1999. Immune responses to adenovirus and adeno-associated virus in humans. *Gene Ther* **6**:1574–1583.

281. Mingozzi F, High KA. 2011. Therapeutic in vivo gene transfer for genetic disease using AAV: progress and challenges. *Nat Rev Genet* **12**:341–355.

282. Nelson CE, Hakim CH, Ousterout DG, Thakore PI, Moreb EA, Castellanos Rivera RM, Madhavan S, Pan X, Ran FA, Yan WX, Asokan A, Zhang F, Duan D, Gersbach CA. 2016. In vivo genome editing improves muscle function in a mouse model of Duchenne muscular dystrophy. *Science* **351**:403–407.

283. Tabebordbar M, Zhu K, Cheng JKW, Chew WL, Widrick JJ, Yan WX, Maesner C, Wu EY, Xiao R, Ran FA, Cong L, Zhang F, Vandenberghe LH, Church GM, Wagers AJ. 2016. In vivo gene editing in dystrophic mouse muscle and muscle stem cells. *Science* **351**:407–411.

284. Villiger L, Grisch-Chan HM, Lindsay H, Ringnalda F, Pogliano CB, Allegri G, Fingerhut R, Häberle J, Matos J, Robinson MD, Thöny B, Schwank G. 2018. Treatment of a metabolic liver disease by in vivo genome base editing in adult mice. *Nat Med* **24**:1519–1525.

285. Swiech L, Heidenreich M, Banerjee A, Habib N, Li Y, Trombetta J, Sur M, Zhang F. 2015. In vivo interrogation of gene function in the mammalian brain using CRISPR-Cas9. *Nat Biotechnol* **33**:102–106.

286. Long C, Amoasii L, Mireault AA, McAnally JR, Li H, Sanchez-Ortiz E, Bhattacharyya S, Shelton JM, Bassel-Duby R, Olson EN. 2016. Postnatal genome editing partially restores dystrophin expression in a mouse model of muscular dystrophy. *Science* **351**:400–403.

287. Wu Z, Yang H, Colosi P. 2010. Effect of genome size on AAV vector packaging. *Mol Ther* **18**:80–86.

288. Chew WL, Tabebordbar M, Cheng JK, Mali P, Wu EY, Ng AH, Zhu K, Wagers AJ, Church GM. 2016. A multifunctional AAV-CRISPR-Cas9 and its host response. *Nat Methods* **13**:868–874.

289. Levy JM, Yeh WH, Pendse N, Davis JR, Hennessey E, Butcher R, Koblan LW, Comander J, Liu Q, Liu DR. 2020. Cytosine and adenine base editing of the brain, liver, retina, heart and skeletal muscle of mice via adeno-associated viruses. *Nat Biomed Eng* **4**:97–110.

290. Kim E, Koo T, Park SW, Kim D, Kim K, Cho HY, Song DW, Lee KJ, Jung MH, Kim S, Kim JH, Kim JH, Kim JS. 2017. In vivo genome editing with a small Cas9 orthologue derived from *Campylobacter jejuni*. *Nat Commun* **8**:14500.

291. Jo DH, Koo T, Cho CS, Kim JH, Kim JS, Kim JH. 2019. Long-term effects of in vivo genome editing in the mouse retina using *Campylobacter jejuni* Cas9 expressed via adeno-associated virus. *Mol Ther* **27**:130–136.

292. Friedland AE, Baral R, Singhal P, Loveluck K, Shen S, Sanchez M, Marco E, Gotta GM, Maeder ML, Kennedy EM, Kornepati AV, Sousa A, Collins MA, Jayaram H, Cullen BR, Bumcrot D. 2015. Characterization of *Staphylococcus aureus* Cas9: a smaller Cas9 for all-in-one adeno-associated virus delivery and paired nickase applications. *Genome Biol* **16**:257.

293. Yu W, Mookherjee S, Chaitankar V, Hiriyanna S, Kim JW, Brooks M, Ataeijannati Y, Sun X, Dong L, Li T, Swaroop A, Wu Z. 2017. Nrl knockdown by AAV-delivered CRISPR/Cas9 prevents retinal degeneration in mice. *Nat Commun* **8**:14716.

294. Maeder ML, Stefanidakis M, Wilson CJ, Baral R, Barrera LA, Bounoutas GS, Bumcrot D, Chao H, Ciulla DM, DaSilva JA, Dass A, Dhanapal V, Fennell TJ, Friedland AE, Giannoukos G, Gloskowski SW, Glucksmann A, Gotta GM, Jayaram H, Haskett SJ, Hopkins B, Horng JE, Joshi S, Marco E, Mepani R, Reyon D, Ta T, Tabbaa DG, Samuelsson SJ, Shen S, Skor MN, Stetkiewicz P, Wang T, Yudkoff C, Myer VE, Albright CF, Jiang H. 2019. Development of a gene-editing approach to restore vision loss in Leber congenital amaurosis type 10. *Nat Med* **25**:229–233.

295. Shen CC, Hsu MN, Chang CW, Lin MW, Hwu JR, Tu Y, Hu YC. 2019. Synthetic switch to minimize CRISPR off-target effects by self-restricting Cas9 transcription and translation. *Nucleic Acids Res* **47**:e13.

296. Charlesworth CT, Deshpande PS, Dever DP, Camarena J, Lemgart VT, Cromer MK, Vakulskas CA, Collingwood MA, Zhang L, Bode NM, Behlke MA, Dejene B, Cieniewicz B, Romano R, Lesch BJ, Gomez-Ospina N, Mantri S, Pavel-Dinu M, Weinberg KI, Porteus MH. 2019. Identification of preexisting adaptive immunity to Cas9 proteins in humans. *Nat Med* **25**:249–254.

297. Simhadri VL, McGill J, McMahon S, Wang J, Jiang H, Sauna ZE. 2018. Prevalence of pre-existing antibodies to CRISPR-associated nuclease Cas9 in the USA population. *Mol Ther Methods Clin Dev* **10**:105–112.

298. Wagner DL, Amini L, Wendering DJ, Burkhardt LM, Akyüz L, Reinke P, Volk HD, Schmueck-Henneresse M. 2019. High prevalence of *Streptococcus pyogenes* Cas9-reactive T cells within the adult human population. *Nat Med* **25**:242–248.

299. Staahl BT, Benekareddy M, Coulon-Bainier C, Banfal AA, Floor SN, Sabo JK, Urnes C, Munares GA, Ghosh A, Doudna JA. 2017. Efficient genome editing in the mouse brain by local delivery of engineered Cas9 ribonucleoprotein complexes. *Nat Biotechnol* **35**:431–434.

300. Zuris JA, Thompson DB, Shu Y, Guilinger JP, Bessen JL, Hu JH, Maeder ML, Joung JK, Chen ZY, Liu DR. 2015. Cationic lipid-mediated delivery of proteins enables efficient protein-based genome editing in vitro and in vivo. *Nat Biotechnol* **33**:73–80.

301. Lee K, Conboy M, Park HM, Jiang F, Kim HJ, Dewitt MA, Mackley VA, Chang K, Rao A, Skinner C, Shobha T, Mehdipour M, Liu H, Huang WC, Lan F, Bray NL, Li S, Corn JE, Kataoka K, Doudna JA, Conboy I, Murthy N. 2017. Nanoparticle delivery of Cas9 ribonucleoprotein and donor DNA *in vivo* induces homology-directed DNA repair. *Nat Biomed Eng* **1**:889–901.

302. Lee B, Lee K, Panda S, Gonzales-Rojas R, Chong A, Bugay V, Park HM, Brenner R, Murthy N, Lee HY. 2018. Nanoparticle delivery of CRISPR into the brain rescues a mouse model of fragile X syndrome from exaggerated repetitive behaviours. *Nat Biomed Eng* **2**:497–507.

303. Gao X, Tao Y, Lamas V, Huang M, Yeh WH, Pan B, Hu YJ, Hu JH, Thompson DB, Shu Y, Li Y, Wang H, Yang S, Xu Q, Polley DB, Liberman MC, Kong WJ, Holt JR, Chen ZY, Liu DR. 2018.

Treatment of autosomal dominant hearing loss by in vivo delivery of genome editing agents. *Nature* 553:217–221.

304. Yin H, Song CQ, Dorkin JR, Zhu LJ, Li Y, Wu Q, Park A, Yang J, Suresh S, Bizhanova A, Gupta A, Bolukbasi MF, Walsh S, Bogorad RL, Gao G, Weng Z, Dong Y, Koteliansky V, Wolfe SA, Langer R, Xue W, Anderson DG. 2016. Therapeutic genome editing by combined viral and non-viral delivery of CRISPR system components in vivo. *Nat Biotechnol* 34:328–333.

305. Yin H, Song CQ, Suresh S, Wu Q, Walsh S, Rhym LH, Mintzer E, Bolukbasi MF, Zhu LJ, Kauffman K, Mou H, Oberholzer A, Ding J, Kwan SY, Bogorad RL, Zatsepin T, Koteliansky V, Wolfe SA, Xue W, Langer R, Anderson DG. 2017. Structure-guided chemical modification of guide RNA enables potent non-viral in vivo genome editing. *Nat Biotechnol* 35:1179–1187.

306. Kocak DD, Josephs EA, Bhandarkar V, Adkar SS, Kwon JB, Gersbach CA. 2019. Increasing the specificity of CRISPR systems with engineered RNA secondary structures. *Nat Biotechnol* 37:657–666.

307. Wang HX, Song Z, Lao YH, Xu X, Gong J, Cheng D, Chakraborty S, Park JS, Li M, Huang D, Yin L, Cheng J, Leong KW. 2018. Nonviral gene editing via CRISPR/Cas9 delivery by membrane-disruptive and endosomolytic helical polypeptide. *Proc Natl Acad Sci USA* 115:4903–4908.

308. Rouet R, Thuma BA, Roy MD, Lintner NG, Rubitski DM, Finley JE, Wisniewska HM, Mendonsa R, Hirsh A, de Oñate L, Compte Barrón J, McLellan TJ, Bellenger J, Feng X, Varghese A, Chrunyk BA, Borzilleri K, Hesp KD, Zhou K, Ma N, Tu M, Dullea R, McClure KF, Wilson RC, Liras S, Mascitti V, Doudna JA. 2018. Receptor-mediated delivery of CRISPR-Cas9 endonuclease for cell-type-specific gene editing. *J Am Chem Soc* 140:6596–6603.

309. Wang XW, Hu LF, Hao J, Liao LQ, Chiu YT, Shi M, Wang Y. 2019. A microRNA-inducible CRISPR-Cas9 platform serves as a microRNA sensor and cell-type-specific genome regulation tool. *Nat Cell Biol* 21:522–530,

310. Hanewich-Hollatz MH, Chen Z, Hochrein LM, Huang J, Pierce NA. 2019. Conditional guide RNAs: programmable conditional regulation of CRISPR/Cas function in bacterial and mammalian cells via dynamic RNA nanotechnology. *ACS Cent Sci* 5:1241–1249.

311. Ferdosi SR, Ewaisha R, Moghadam F, Krishna S, Park JG, Ebrahimkhani MR, Kiani S, Anderson KS. 2019. Multifunctional CRISPR-Cas9 with engineered immunosilenced human T cell epitopes. *Nat Commun* 10:1842.

312. Moreno AM, Palmer N, Alemán F, Chen G, Pla A, Jiang N, Leong Chew W, Law M, Mali P. 2019. Immune-orthogonal orthologues of AAV capsids and of Cas9 circumvent the immune response to the administration of gene therapy. *Nat Biomed Eng* 3:806–816.

313. Shmakov SA, Makarova KS, Wolf YI, Severinov KV, Koonin EV. 2018. Systematic prediction of genes functionally linked to CRISPR-Cas systems by gene neighborhood analysis. *Proc Natl Acad Sci USA* 115:E5307–E5316.

314. Strecker J, Ladha A, Gardner Z, Schmid-Burgk JL, Makarova KS, Koonin EV, Zhang F. 2019. RNA-guided DNA insertion with CRISPR-associated transposases. *Science* 365:48–53.

315. Klompe SE, Vo PLH, Halpin-Healy TS, Sternberg SH. 2019. Transposon-encoded CRISPR-Cas systems direct RNA-guided DNA integration. *Nature* 571:219–225.

316. Saito M, Ladha A, Strecker J, Faure G, Neumann E, Altae-Tran H, Macrae RK, Zhang F. 2021. Dual modes of CRISPR-associated transposon homing. *Cell* 184:2441–2453.e18.

317. Abudayyeh OO, Gootenberg JS, Konermann S, Joung J, Slaymaker IM, Cox DB, Shmakov S, Makarova KS, Semenova E, Minakhin L, Severinov K, Regev A, Lander ES, Koonin EV, Zhang F. 2016. C2c2 is a single-component programmable RNA-guided RNA-targeting CRISPR effector. *Science* 353:aaf5573.

318. Smargon AA, Cox DBT, Pyzocha NK, Zheng K, Slaymaker IM, Gootenberg JS, Abudayyeh OA, Essletzbichler P, Shmakov S, Makarova KS, Koonin EV, Zhang F. 2017. Cas13b is a type VI-B CRISPR-associated RNA-guided RNase differentially regulated by accessory proteins Csx27 and Csx28. *Mol Cell* 65:618–630.e7.

319. Konermann S, Lotfy P, Brideau NJ, Oki J, Shokhirev MN, Hsu PD. 2018. Transcriptome engineering with RNA-targeting type VI-D CRISPR effectors. *Cell* 173:665–676.e14.

320. Yan WX, Chong S, Zhang H, Makarova KS, Koonin EV, Cheng DR, Scott DA. 2018. Cas13d is a compact RNA-targeting type VI CRISPR effector positively modulated by a WYL-domain-containing accessory protein. *Mol Cell* 70:327–339.e5.

321. Abudayyeh OO, Gootenberg JS, Essletzbichler P, Han S, Joung J, Belanto JJ, Verdine V, Cox DBT, Kellner MJ, Regev A, Lander ES, Voytas DF, Ting AY, Zhang F. 2017. RNA targeting with CRISPR-Cas13. *Nature* 550:280–284.

322. Cox DBT, Gootenberg JS, Abudayyeh OO, Franklin B, Kellner MJ, Joung J, Zhang F. 2017. RNA editing with CRISPR-Cas13. *Science* 358:1019–1027.

323. Abudayyeh OO, Gootenberg JS, Franklin B, Koob J, Kellner MJ, Ladha A, Joung J, Kirchgatterer P, Cox DBT, Zhang F. 2019. A cytosine deaminase for programmable single-base RNA editing. *Science* 365:382–386.

324. Esvelt KM, Mali P, Braff JL, Moosburner M, Yaung SJ, Church GM. 2013. Orthogonal Cas9 proteins for RNA-guided gene regulation and editing. *Nat Methods* 10:1116–1121.

325. Amrani N, Gao XD, Liu P, Edraki A, Mir A, Ibraheim R, Gupta A, Sasaki KE, Wu T, Donohoue PD, Settle AH, Lied AM, McGovern K, Fuller CK, Cameron P, Fazzio TG, Zhu LJ, Wolfe SA, Sontheimer EJ. 2018. NmeCas9 is an intrinsically high-fidelity genome-editing platform. *Genome Biol* 19:214.

326. Teng F, Cui T, Feng G, Guo L, Xu K, Gao Q, Li T, Li J, Zhou Q, Li W. 2018. Repurposing CRISPR-Cas12b for mammalian genome engineering. *Cell Discov* 4:63.

327. Liu JJ, Orlova N, Oakes BL, Ma E, Spinner HB, Baney KLM, Chuck J, Tan D, Knott GJ, Harrington LB, Al-Shayeb B, Wagner A, Brötzmann J, Staahl BT, Taylor KL, Desmarais J, Nogales E, Doudna JA. 2019. CasX enzymes comprise a distinct family of RNA-guided genome editors. *Nature* 566:218–223.

Genetic and Epigenetic Modulation of Gene Expression by CRISPR-dCas Systems

Jasprina N. Noordermeer[1], Crystal Chen[2], and Lei S. Qi[1,3,4]

[1]Department of Bioengineering, Stanford University, Stanford, CA 94305
[2]Department of Chemical Engineering, Stanford University, Stanford, CA 94305
[3]Department of Chemical and Systems Biology, Stanford University, Stanford, CA 94305
[4]ChEM-H Institute, Stanford University, Stanford, CA 94305

Introduction

Multicellular organism development requires intricate spatial and temporal coordination of complex and dynamic gene regulatory programs. In the last decades, researchers have identified many transcription factors that regulate these regulatory programs. Sequencing of the genomes of humans, as well as those of many other organisms, has greatly accelerated these studies, supporting work towards a better understanding of the molecular logic of how cells assemble and communicate to form complex tissues (reviewed in reference 1). The great progress made in DNA sequencing technologies has also significantly increased our ability to characterize and diagnose genetic diseases; to date, approximately 4,000 genetic disorders and their molecular abnormalities have been identified (2; reviewed in reference 3). With these advances come the scientific challenge and desire to precisely manipulate or perturb gene expression to better understand function or, in the case of genetic disorders, correct expression.

Until recently, few robust technologies were available for programmable control of gene expression, but several breakthroughs in the fields of genome editing and gene regulation have created unprecedented possibilities for genetic engineering. Interestingly, at the roots of several of these discoveries are painstaking fundamental studies into the mechanisms of bacterial and lower eukaryotic immune responses against the invasion of foreign genetic materials. In an approach called RNA interference (RNAi), researchers modulated an adaptive immune strategy of the worm *Caenorhabditis elegans* into a flexible and high-throughput method for sequence-specific gene silencing using small interfering RNAs or short hairpin RNAs (reviewed in reference 4). While RNAi has been a powerful approach for studying gene function in many different organisms, including eukaryotes, some serious drawbacks concerning low efficiency and high off-target effects have limited its efficacy and utility (5).

CRISPR: Biology and Applications, First Edition. Edited by Rodolphe Barrangou, Erik J. Sontheimer, and Luciano A. Marraffini.
© 2022 American Society for Microbiology. DOI: 10.1128/9781683673798.ch12

In the past decade, other approaches that use zinc finger nucleases (ZFNs) (6, 7) and transcription activator-like effector nucleases (TALENs) (8, 9) have been designed for genome editing in a variety of organisms. At the core of these technologies lies the use of engineered nucleases that are composed of sequence-specific DNA-binding domains fused to a nonspecific DNA cleavage module. This nuclease enables precise and efficient genetic modifications by cellular repair mechanisms such as error-prone nonhomologous end joining (NHEJ) and homology-directed repair (HDR). The programmability of the DNA-binding domains of ZFNs and TALENs to specifically target different DNA sequences makes these highly effective DNA editing tools. Their DNA binding modules have also been used in combination with a range of other effector domains, such as transcriptional activators/repressors, recombinases, methyltransferases, etc., generating a modular and efficient system for gene regulation (10–12). Despite their versatility, ZFN and TALEN methods for gene editing or gene regulation require separate construction and delivery of each DNA-binding protein, making high-throughput analysis and simultaneous multilocus targeting technically challenging, expensive, and time-consuming.

A major step forward in the ability to manipulate gene expression was inspired by the discovery of an elegant and efficient system of RNA-guided DNA endonucleases that bacteria and archaea evolved to defend themselves against invading viruses and other mobile elements. These RNA-mediated adaptive immune responses have been named clustered regularly interspaced short palindromic repeat (CRISPR)-CRISPR-associated (Cas) systems (13–25). CRISPR systems have been categorized into several distinct classes and types (reviewed in chapters 1 to 7).

Within a few short years, a series of groundbreaking findings opened up this system for use not only in gene editing but also for manipulating gene expression (reviewed in references 26 to 28). The first crucial findings were the demonstration that the CRISPR endonuclease Cas9 can be programmed to target almost any desired DNA sequence by the use of a single, short chimeric guide RNA to confer DNA binding specificity (14) and that the system can be adapted for genome editing in eukaryotic cells (29–31). Another major advance showed that the Cas9 genome editing tool can be repurposed into an engineered, nuclease-deficient Cas9 (dCas9) that targets specific DNA sequences without cleaving them (32). dCas9 thus became an RNA-guided DNA recognition platform, providing a precise, flexible, and scalable tool for many applications, such as transcriptional activation/repression, epigenetic modifications, chromatin repositioning, and genome imaging, depending on the effector module(s) linked to dCas9 (reviewed in references 33 and 34). In this chapter, we describe the bioengineering toolbox developed based on dCas9 or like molecules (e.g., dCas12), its applications for programmable gene expression in diverse organisms, and potential future clinical applications in regenerative medicine and gene therapy.

Nuclease-Deficient CRISPR-Associated 9 Systems for Programmable Gene Regulation

The type II RNA-guided endonuclease Cas9 cleaves target DNA as guided by a duplex of two RNAs: CRISPR RNA (crRNA), which recognizes the invading DNA via an approximately 20-bp region, and transactivating crRNA (tracrRNA), which is unique to the type II CRISPR class (17, 18). This CRISPR system requires only one Cas protein (Cas9), and after researchers (14) demonstrated that the crRNA-tracrRNA duplex can be replaced by a chimeric single guide RNA (sgRNA), the *Streptococcus pyogenes* Cas9 became the first and also most widely used Cas protein to be customized for genetic engineering. Importantly, the Cas9-sgRNA protein complex can be targeted via base pairing with the sgRNA to any genomic site that is adjacent to a protospacer-adjacent motif (PAM) sequence, inducing a DNA double-strand break (DSB) at a specific site within the base pairing region (35). This simple system was quickly developed into a highly efficient and reliable gene editing tool. First, a chimeric Cas9 associated with a single 20-bp sgRNA directing it to a specific genomic locus of interest would create a DSB at that site, triggering the second step, host DNA repair through NHEJ or HDR.

The use of Cas9 as a sequence-specific, nonmutagenic gene regulation tool only became possible after mutation of the two nuclease domains of the *S. pyogenes* Cas9, HNH and RuvC, that together mediate DNA DSBs (25, 32) (Fig. 12.1A and B). Each domain cleaves one DNA strand, creating a DSB proximal to the PAM sequence at the target site. A single point mutation in either of the two sites produces a nickase enzyme, and mutations in both domains (D10A and H840A for SpCas9) generate a complete loss of DNA cleavage activity (14). The resulting catalytically inactive dCas9 (32) can now be used to directly manipulate transcription of targeted genes or to recruit various transcriptional-level effector proteins, such as transcriptional repressors and activators.

Recently, Cas nucleases other than Cas9 have been identified and mutated to generate additional dCas molecules. For example, a point mutation in the RuvC domain of the

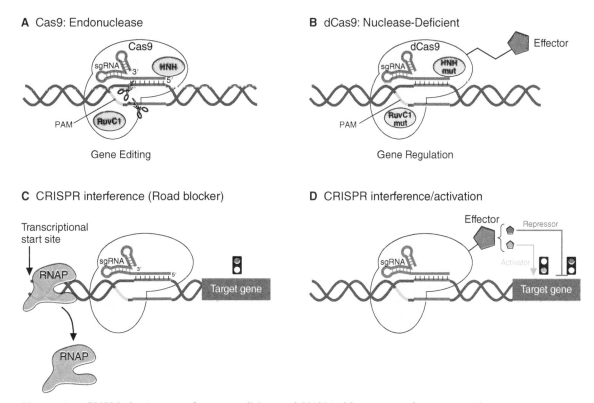

Figure 12.1 CRISPR-Cas9 system for gene editing and CRISPR-dCas9 system for gene regulation. **(A)** The Cas9 endonuclease targets specific DNA sequences by direct pairing of the sgRNA with the target site guided by the presence of a 5′ PAM. Cas9 DNA binding mediates cleavage of the target sequence by two nuclease domains, RuvC1 and HNH. **(B)** Nuclease-deficient dSpyCas9 protein contains mutations in its RuvC1 (RuvC1 mut; D10A) and HNH (HNH mut; H841A) domains, inactivating its nuclease activity. However, dCas9 maintains the ability to target specific DNA sequences through the sgRNA and PAM. dCas9 coupled to an effector of transcription can regulate specific gene expression. **(C)** dCas9 binds the DNA target through the sgRNA and sterically blocks the transcriptional elongation by the RNA polymerase (RNAP), thereby preventing target gene expression. This process is called CRISPRi. **(D)** In CRISPRi/a, a transcriptional activator or repressor can be coupled with dCas9 to achieve target gene activation or repression.

type V endonuclease Cas12a (previously named Cpf1) resulted in dCas12a (36, 37), which was subsequently used for logic-gated transcriptional control of gene expression in *Escherichia coli* (38). The identification of other Cas nucleases that are smaller or distinct in PAM requirements (for example, CasX [39]) will greatly expand the available RNA-guided genome editing and gene expression platforms. For example, through both RNA and protein engineering of a miniature type V-F CRISPR (Cas12f) that naturally fails to function in mammalian cells, we found that the reengineered nuclease and its nuclease-dead versions, termed CasMINI, can be used for effective gene editing, base editing, and gene activation and silencing, as detailed below (40).

Repression of Transcription via CRISPR Interference

Strong binding of catalytically inactive dCas9 to its DNA target, as guided by the sgRNA, can sterically hinder the RNA polymerase machinery, directly interfering with transcription elongation or blocking transcription initiation by blocking binding of endogenous transcription factors (Fig. 12.1C). This ability has been used to develop CRISPR interference (CRISPRi), a simple platform for targeted inhibition of gene expression in pro- and eukaryotes (32, 41–43). This method can suppress gene expression in bacteria with high efficiency. Furthermore, as multiple genes can be simultaneously repressed using different sgRNAs, CRISPRi has provided an alternative method for repressing bacterial genes expression on a genome-wide scale. It has been the method of choice for systematically assessing the roles of essential genes for bacterial growth and survival, cell shape, and morphology (44).

While silencing gene expression when just using dCas9 and sgRNA is highly efficient in prokaryotes, with some guides leading to 99% repression of target genes (32), steric hindrance of RNA polymerase activity by the dCas9-sgRNA complex is often less efficient for inhibiting transcription in eukaryotic cells. In mammalian cells, this strategy results in no to moderate repression (up to 80% repression) of endogenous genes, such as the transferrin receptor CD71, the tumor repressor protein TP53, or the cytokine receptor CXCR4 (45, 46).

Moderate repression has its own utility. Shariati and coworkers have used the steric hindrance property of dCas9-sgRNA to specifically block interactions between transcription factors and their binding sites, a method termed CRISPR disruption (CRISPRd) (47). CRISPRd offers a novel perturbation method in mammalian cells to map transcription factor binding sites and infer phenotypic function from the binding events, such as for the transcription factor Oct4 in murine embryonic stem cells. This offers a high-resolution approach to analyze functional genomics.

Many applications require more efficient repression. For this purpose, dCas9 has been fused to a number of transcriptional or epigenetic repression effector domains, such as the Kruppel-associated box (KRAB) domain of Kox1, the WPRW domain of Hes1, four concatenated copies of the mSin3 interaction domain (SID4X) or the chromoshadow domain of HP1α (Fig. 12.1D) (45, 48, 49).

The dCas9-KRAB fusion is an efficient and robust repressor that can downregulate expression of endogenous genes by more than 90% (45, 48). dCas9-KRAB can efficiently silence individual protein-coding genes as well as noncoding RNA genes by targeting promoter regions, 5′ untranslated regions, and proximal and distal enhancer elements. The KRAB-containing family of transcriptional repressors recruits additional corepressors, such as KRAB box-associated protein 1, and epigenetic readers, such as heterochromatin protein 1, that are associated with a loss of histone H3 acetylation and an increase in H3 lysine 9 trimethylation (H3K9me3) at the repressed promoters (50, 51). Similarly, the dCas9-KRAB fusion complex leads to reduced chromatin accessibility and increased levels of H3K9me3 at the target site (51). An improvement in the level of repression was found when dCas9 was fused to a bipartite repressor that contains both the KRAB repressor domain and that of methyl-CpG-binding protein 2 (52).

Large-scale genomic screens have demonstrated that the level of dCas9-KRAB repression is highly dependent on sgRNA binding site location (48). Stronger repression is often achieved by selecting the sgRNA target sequence from the region within −50 to +300 bp relative to the transcriptional start site (TSS) of the target gene, with maximum efficiency from approximately 50 to 100 bp downstream of the TSS. The optimal sgRNA length for repression is between 18 and 21 nucleotides for *S. pyogenes* dCas9, while the guanine/cytosine content and DNA strand targeted do not appear to be important factors for sgRNA repression efficiency (48).

Activation of Transcription via CRISPR Activation

For the development of CRISPR-dCas9-mediated activation of transcription (CRISPRa), a strategy similar to that for CRISPRi was used, exploiting the modularity of the dCas9 platform (Fig. 12.1D) (41, 45, 48, 53). Instead of applying repressor domains, activating effector domains were fused to dCas9, including the transactivation domain of NF-kB p65P65AD, the 16-amino-acid-long transactivation domain of the herpes simplex viral protein 16 (VP16), or multiple tandem copies of VP16, such as VP64 (4 copies of VP16) or VP160 (10 copies of VP16) (48, 54, 55). The approach of fusing dCas9 to various transcription-activating effectors yielded a modest (2- to 5-fold) increase in expression levels in eukaryotic cells; in bacteria, fusing dCas9 with the ω subunit of RNA polymerase increases reporter gene expression up to 3-fold (41). While the efficacy of activation can be significantly increased by tiling multiple sgRNAs at the promoter site (54, 56), many other second- and third-generation CRISPRa approaches were designed to improve the effectiveness, specificity, inducibility, and modularity of the synthetic gene activation system.

Improvements of dCas9-Coupled Activators of Gene Expression

To increase its effectiveness above the dCas9-VP64 fusion, two groups published a strategy in 2014 (43, 48) in which they fused dCas9 with a recently developed multipeptide array called SunTag. A SunTag array was fused to dCas9 at the carboxy terminus, consisting of a series of small polypeptides (GCN4s) that can recruit multiple copies of its cognate single-chain variable fragment (scFv, part of an anti-GCN4 antibody), each coupled to VP64. This approach resulted in a transcriptional effector that bound multiple copies of VP64 (10 or 24 copies) to a single copy of dCas9 (Fig. 12.2A). The SunTag system resulted in significant amplification of gene activation with a single sgRNA (43, 48). Besides the SunTag system, other systems, such as SpyTag, have been used to recruit multiple activators (57).

In contrast to the SunTag system, which increases gene activation by recruiting multiple copies of the same transactivator, the VP64-p65-Rta (VPR) method fuses dCas9 to a tripartite transactivation complex composed of VP64, the transactivation domain of NF-kB p65P65AD, and the Epstein-Barr virus R transactivator (Rta) proteins in tandem (Fig. 12.2B) (58). This dCas9-VPR fusion resulted in an improved level of activation of endogenous gene expression compared to the earlier dCas9-VP64 system and was able to also activate gene loci in *Drosophila melanogaster*, *Saccharomyces cerevisiae*, and *Mus musculus* (58). This strategy of synergistically recruiting multiple distinct activators to dCas9 aimed to resemble more closely the naturally occurring cellular gene activation systems that act via the intricate coordination of a collection of transcriptional effectors.

A SunTag

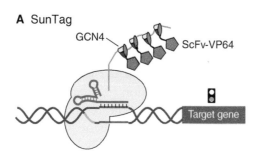

B VPR

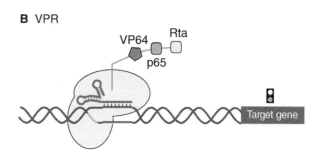

C SAM

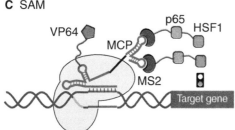

D scRNA

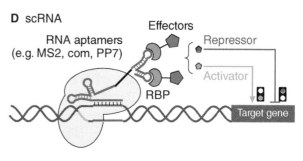

E Dimerization Systems

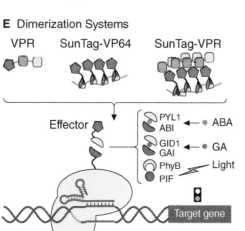

F dCas9-KAL or CRISPRoff

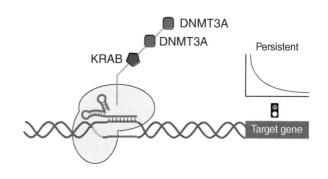

G Split dCas9 systems

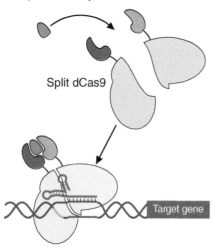

H Receptor-Coupled Systems

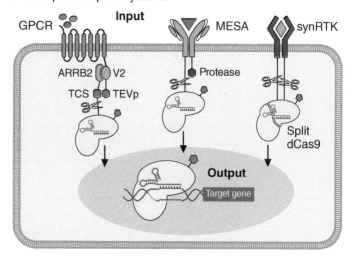

The third CRISPRa strategy, called the synergistic activation mediator (SAM) approach, uses the dCas9 and sgRNA as scaffolds to recruit transcription activation effectors to increase the gene activation level (53). It employs dCas9-VP64 with a modified sgRNA that contains two MS2 RNA aptamers (Fig. 12.2C). Each of these MS2 aptamers recruits a pair of cognate RNA-binding proteins, MS2 coat proteins, that are fused to the transactivator domain of p65 and HSF1 (53), resulting in increased gene activation.

The above-mentioned approaches were devised to optimize efficiency, but dCas9 has also been redesigned for realizing multiplexed, multimodular gene regulation, such as to enable simultaneous gene activation and repression in the same cell. A modification of the SAM approach was used for this application, where the sgRNA was fused to orthogonally acting protein-binding RNA aptamers (MS2, PP7, or com), creating an aptamer-modified sgRNA, or scaffold RNA (scRNA) (Fig. 12.2D) (59). This scRNA can recruit specific cognate RNA-binding proteins fused to, for example, a gene activator (VP64) or repressor (KRAB). When these activators and repressors are coupled to distinct sgRNAs associated with different genes, simultaneous multimodular gene regulation by the dCas9 complex can be achieved (60).

These CRISPRa methods are more efficient at activating target genes with a single sgRNA than the original dCas9-VP64 fusion in a range between 10- and 1,000-fold, depending on the specific gene and method examined (61). Direct comparisons with some of these methods show that VPR, SunTag, and SAM can activate a wide range of endogenous genes, and while it is likely that some systems are most optimally used in a cell-type- and context-dependent manner, all systems tested were active in a number of model organisms, such as *Drosophila*, murine, and human cells (61, 62). In addition, to reach even higher levels of activation, variants of these systems have been developed, such as by combining elements of SunTag, SAM, and VPR (61). However, the results of these studies have generally been disappointing, suggesting that recruiting more of the same transactivators does not necessarily lead to higher transcriptional activation efficiency above a certain level (reviewed in reference 34). Future CRISPRa strategies should include incorporating alternate transcriptional, epigenetic, and chromatin regulators, such as the Mediator complex, to increase transcriptional levels.

Figure 12.2 CRISPR-dCas9 based tools for manipulation of gene expression. (A) The SunTag approach leads to strong activation of gene expression. The tandem repeats of a small peptide GCN4 recruit multiple copies of a single-chain variable fragment (scFv) fused to the transcriptional activator VP64. (B) The VPR approach is a strategy for gene activation. dCas9 is fused to the combinatory tripartite transcriptional activator VP64-p65-Rta (VPR) to amplify the activation of transcription. (C) The SAM approach is a strategy for gene activation. dCas9 is fused to VP64 and the sgRNA has been modified to contain two MS2 RNA aptamers to recruit the MS2 bacteriophage coat protein (MCP), which is fused to the transcriptional activators p65 and heat shock factor 1 (HSF1). (D) The scRNA approach is a strategy for simultaneous gene activation and repression. A hybrid RNA scaffold coupling an sgRNA and an RNA aptamer (e.g., MS2, com, or PP7) can recruit RNA-binding proteins (e.g., MCP, COM, or PCP) tethered to either a transcriptional activator or repressor. (E) Dimerization systems consist of chemical- and light-controlled CRISPR-dCas9 systems for inducible gene regulation. Chemical- or light-induced dimerization systems (e.g., PYL1::ABI, GID::GAI, and PhyB::PIF) have been fused to dCas9 and transcriptional effectors, respectively. The addition of corresponding chemical (e.g., abscisic acid [ABA] or gibberellin [GA]) or light can induce gene regulation. (F) dCas9-KAL and CRISPRoff systems are strategies for long-term epigenetic silencing. The dCas9 is fused to the combinatory tripartite transcriptional repressor KRAB-DNMT3A-DNMT3L (KAL) to modify H3K9me3 and DNA methylation at CpG sites, which causes long-term silencing of nearby genes. (G) Only when both parts of split dCas9-effectors are fused via the presence of an appropriate induction signal can they form an active dCas9-effector fusion protein (H) Input/output molecular devices for gene regulations. GPCR, MESA, and synRTK (synthetic receptor tyrosine kinases) systems have been introduced for dCas-based gene regulation.

Inducible CRISPRi/a systems

Further development of CRISPRi/a approaches also focused on inducible and reversible gene expression in a spatial and temporal fashion to develop dynamic systems to better mimic native gene circuits and networks. In addition, adaptations were made to enable simultaneous regulation of multiple genes. The modularity and flexibility of the CRISPR approach have presented a range of strategies by which this level of regulation can be achieved. Inducible gene expression was first realized by coupling dCas9 to optogenetically or chemically inducible peptide dimerization domains. Conditional CRISPRa activity was also achieved by using split variants of dCas9, engineered allosteric proteins, or RNA switches (Fig. 12.2E to G). Lastly, native cellular signaling receptors were coupled to the dCas9 activation system, creating a chimeric input/output molecular device to control endogenous genes by diverse cellular inputs simultaneously in the same cell. In this section, several examples of these newly developed conditional CRISPRa systems are discussed (Fig. 12.2G).

Several studies have been published that used light-inducible, dCas9-based peptide heterodimerization to activate rapid and reversible endogenous gene expression when cells were exposed to blue light (Fig. 12.2E) (63–66). These authors fused either of the blue light-sensing heterodimerizing proteins CIB1 and CRY2 to dCas9 and either VP64 or p65AD, respectively. After illuminating the transfected cells with blue light, CIB1-CRY2 dimerization recruits the transcriptional activator to the dCas9 and sgRNA complex at its DNA target, leading to activation of gene expression. The highest activation levels in this system were comparable to those reported for dCas9-VP64-mediated transcriptional activation in HEK293 cells (55). In addition to the CIB1-CRY2 systems, a phytochrome-based red light-inducible PhyB-PIF (67) and the blue light heterodimer pMAG-nMAG (63–66) systems were developed.

A second, ligand-inducible approach uses small molecules instead of light as the input signal (Fig. 12.2E). Chemical inputs such as rapamycin employ a two-component system, consisting of dCas9 and VP64 fused to the rapamycin-binding FK506 binding protein 12 (FKBP) and the FKBP-rapamycin binding (FRB) domains, respectively (68, 69). In the presence of rapamycin, the dCas9-FRB-FKBP-VP64 complex is assembled and can activate robust gene expression. Gene activation is reversible and ceases after rapamycin is removed. Various other chemical ligands and their dimerization domains have been used, such as the abscisic acid (ABA)-inducible ABI-PYL1 (69–71) and the gibberellin (GA)-inducible GID1-GAI24 (71) domains. Endogenous human genes were shown to be efficiently activated by both rapamycin and GA-mediated dimerization (69).

Instead of dCas9, dCas12a has also been used in these systems, fused to the rapamycin analog A/C heterodimerization domains (72). Interestingly, combining orthogonal dCas9 with either ABA or GA inducer systems separately can independently control gene expression of multiple different genes in the same cell. In this way, Boolean logic-gated driven control of dCas9 by multiple cellular inputs can be realized to achieve multiplexed, synthetic modulation of gene expression.

An alternative approach for an additional level of control to inducible dCas9 regulation is using split dCas9 halves that can be reassembled into active dCas9 when the appropriate induction signals are received (Fig. 12.2G). Zetsche et al. used split Cas9 for DNA editing experiments, whereby Cas9 was split into N- and C-terminal halves and fused to the rapamycin-binding domains FRB and FKBP, respectively (68). Addition of rapamycin stimulates dimerization of both halves into an active Cas9 protein. To prevent autoassembly, the two split Cas9 fragments are kept in different cellular compartments with nuclear export and nuclear localization sequences. Rapamycin-induced dimerization transports the complex to the nucleus, where it can target DNA, guided by the Cas9-associated sgRNA. A similar strategy is used for split dCas9 that binds VP64 and activates transcription (73). In this study, split dCas9 was fused to the ligand-binding domain of the estrogen receptor (ERT) that interacts with the chaperone Hsp90, sequestering it to the cytosol. When the ligand 4-hydroxyamoxifen is added, it dissociates the Hsp90 and ERT interaction such that dCas9 reassembles and is translocated to the nucleus, where it can promote gene regulation (73). Gene activation can be fine-tuned through controlling dimerization in a ligand dose-dependent and gated-control manner for systems with multiple distinct inputs (74). Lastly, split dCas9 allows for smaller proteins to be packaged into the adeno-associated virus (AAV) vector delivery system, thereby potentially bypassing the size limitations of this delivery system.

In the section above, we discussed the scRNA strategy by which orthogonal regulation of dCas9 can induce both activation and repression of endogenous genes simultaneously in the same cell. By using an scRNA that can recruit multiple cognate RNA-binding proteins bound to an activator (VP64) or repressor (KRAB), multiplexed transcriptional regulation is achieved. More recently, a variation on this approach was developed in which the RNA-binding proteins are modified with small-molecule-mediated protein degradation domains (degrons) that control activation complex stability (75). The degrons induce rapid proteasome-mediated degradation; addition of small molecules can bind and stabilize these domains, enabling transcriptional activation by the complex. For example, unstable protein domains such as estrogen receptor or dihydrofolate reductase were linked to an aptamer-binding protein (e.g., MS2-binding protein MCP or PP7-binding protein PCP) and a transcriptional activator (e.g., VP64). The presence of these degrons results in rapid degradation of the chimeric fusion protein. However, when small molecules such as 4OHT or TMP bind these domains, they are stabilized, allowing protein complex binding to the MS2 or PP7 binding sites in the sgRNA for activation of transcription (75).

The transcriptional regulator activity of the dCas9 complex can also be controlled by converting the sgRNA into a biosensor by coupling it to RNA aptamers that respond to internal and external riboswitch-responsive signals (76). In the absence of the signal, the guide region of the sgRNA pairs with an antisense stem loop, preventing it from binding its target DNA site. When a signal is present and sensed by the RNA aptamers, a conformational change makes the sgRNA guide region accessible to recruit the dCas9 effector complex to the target DNA site, whereby gene expression can be activated (76).

Lastly, several types of chimeric receptors have been coupled with the dCas9 dimerization and split dCas9 systems to allow for multi-input AND/OR/NOT logic gate control of gene expression (Fig. 12.2H). These systems combine ligand dose-dependent regulation of gene expression with spatial and temporal control, thereby more closely resembling endogenous signal transduction pathways in mammalian cells. Generally, native ligands interacting with their cognate receptors are recruited in conjunction with synthetic chimeric receptors to reprogram cell fate in a predictable and productive manner. For example, cancer cells can be repurposed by altering oncogenic signal transduction pathways into antioncogenic pathways to fight tumorigenesis (reviewed in reference 77). The chimeric receptors consist of a ligand-sensing extracellular receptor domain (input) and an intracellular gene regulatory domain (output). The modularity of many of these domains ultimately makes this strategy suitable for achieving a high level of specificity for inducible control of gene expression. Here, we discuss chimeric G-protein-coupled receptors (GPCRs) (78, 79) used for this strategy. However, other studies have been published that combine dCas9 approaches with chimeric synthetic receptors such as chimeric antigen receptors (77), synthetic Notch receptors (80), and modular extracellular sensor architecture system receptors (81). These and other receptor systems have been adapted as general platforms to design custom response programs, such as primary T cell responses to a diversity of antigens, including cytokine secretion, T cell differentiation, and antibody delivery.

Chimeric GPCRs were used to develop a synthetic input/output signaling system called Tango (82, 83), where the C-terminal intracellular domain of the GPCR was replaced with a proteolytically cleavable synthetic transcription factor. In addition, an adaptor protein, beta arrestin (ARRB2), was fused to the tobacco etch virus (TEV) protease and coexpressed in the same cell. Upon ligand binding to the GPCR, ARRB2 binds the GPCR and cleaves and releases the transcription factor, which translocates to the nucleus, where it can activate transcription. Numerous ligands and transcriptional effectors have been customized to conditionally drive gene expression with this system, including the split dCas9 based transcription factor fused to the intracellular GPCR domain. In an approach called ChaCha (84), the original Tango design was reversed such that the dCas9-effector was coupled to the C terminus of the ARRB2 adaptor instead of the GPCR via a proteolytically cleavable linker, and the intracellular GPCR domain was replaced with the TEV protease (Fig. 12.2G). Upon ligand activation, the GPCR-TEV fusion protein proteolytically cleaves the ARRB2 adaptor, thereby releasing the dCas9 effector, which translocates to the nucleus and modulates transcription. Mathematical modeling suggests that the ChaCha design can release multiple dCas9 effector molecules per GPCR molecule (84). Moreover, the ChaCha approach resulted in transcriptional regulation that is dose dependent, reversible, and inducible by numerous ligands, including chemokines, fatty acids, hormones, mitogens, and synthetic compounds, potentially simultaneously activating multiple endogenous genes (84).

While major progress has been made to generate toolkits for purposeful manipulation of cellular signaling pathways, most of the described approaches exhibit considerable ligand-independent activity. In the coming years, many technical advances are expected that will combine elements of the above-described strategies to improve specificity, sensitivity, and efficiency and reduce ligand-independent activity.

Epigenetic Control of Transcription by dCas9 Effector Fusion Proteins

It is well established that locus-specific transcription is dependent on the state of the local chromatin structure, which, in turn, is maintained in a heritable manner by patterns of DNA methylation and histone methylation and acetylation (reviewed in reference 85). Transcriptionally active genes are generally present in decondensed euchromatin, while inactive genes are localized in condensed heterochromatin. Broadly speaking, euchromatin is characterized by hypomethylation of cytosine within CpG motifs, while cytosines in these motifs in heterochromatin are methylated (5-methylcytosine). Transcriptionally active loci in euchromatin are bound by nucleosomes bearing acetylated histone H4, and heterochromatic nucleosomal histone H4 is hypomethylated. Conversely, a specific lysine in histone H3, H3K27, is hypermethylated in heterochromatic nucleosomes but hypomethylated in euchromatic nucleosomes. Furthermore, histone H3 methylated at the position H3K9 is enriched in euchromatic repeat regions (86). Histone methylation at two other sites, H3K4 and H3K79, is also associated with euchromatin. Importantly, these processes can be manipulated *in vivo*, and many studies have proven a causal link between the presence of these epigenetic markers and the transcriptional activity of genes in a given locus.

Many groups have recently taken advantage of the sgRNA-mediated chromosomal targeting specificity of dCas9, along with the identification of enzymes that install, remove, or respond to specific epigenetic marks, to generate fusion proteins that can be used to manipulate locus-specific epigenetic codes. This has allowed control of gene transcription within specified loci. Here, we discuss these studies grouped by the specific dCas9 fusion approach.

Manipulation of locus-specific cytosine methylation, either by increasing or decreasing methylation via methylase or demethylase dCas9 fusion partners, respectively, has confirmed our understanding of cytosine methylation as an inhibitor of transcription. These approaches seem to have largely locus-independent results. Initial studies characterizing dCas9-methylase fusion protein activity *in vivo* (87–90) employed catalytic methylase domains from DNMT3A and a DNMT3A/DNMT3L chimera. These studies demonstrated the expected methylation pattern (which in these early studies appeared relatively confined to the target site [±~50 bp]), the transcriptional repression of the targeted genes, and, in some cases, the heritability of epigenetic 5-methylcytosine marks and the repressed state (87). Furthermore, these effects were demonstrated at a variety of loci, indicating a likely genome-wide utility for this approach.

Recently, a few studies showed that a tripartite combination of KRAB, DNMT3A, and DNMT3L can be fused onto a single dCas9 protein for long-term epigenetic silencing. Known as CRISPRoff (91) or dCas-KAL (92) technology, simultaneous rewriting of the H3K9me3 and CpG DNA methylation on DNA chromatin can lead to silencing at specific targeted gene loci in actively dividing cells on a timescale of months. Excitingly, these technologies could be expanded for whole-genome epigenetic screening, which may facilitate defining effective sgRNA binding sites as well as investigating epigenetic modulation for long-term silencing.

Conversely, *in vivo* expression of a dCas9-TET1 cytosine dioxygenase fusion, which indirectly catalyzes the demethylation of cytosines in the vicinity of targeted loci, resulted in the transcriptional derepression of targeted genes (88, 93). A similar dCas9-TET1 fusion, targeted to the methylated expanded CGG motifs in the upstream region of the silenced *FMR* gene, which causes fragile X syndrome, resulted in transcriptional derepression of *FMR* (94).

Subsequent studies have added the prokaryotic M.SssI CpG methylase, in its native form (95) and when reconstituted from split domains (96), as a useful methylase. Interestingly, altering the linker between one of the split domains and dCas9 increased methylation near the target site without reducing more distal methylation (97). Mutations of this methylase's DNA binding site to weaken its affinity for the DNA, however, resulted in decreased methylation of sites farther from the Cas9/sgRNA binding site. These studies indicate the great promise of protein engineering to optimize target specificity and activity of dCas9 fusions. More generally, when using more sensitive techniques for off-target identification than those previously employed, two groups identified widespread off-target activity of direct dCas9-DNMT fusions (98, 99). Off-target effects were shown to be suppressed by multimerizing the fusion protein at the target site via a SunTag motif (99, 100). Lastly, dCas9 SunTag multimerization of TET1 led to increased local demethylation relative to monomeric fusions (101).

The "histone code" (102), namely, the contributions of acetylation and methylation of specific histone amino acids, has also been manipulated by dCas9 fusion protein transcriptional regulation. Targeting a fusion of the catalytic domain of the human histone acetylase p300 and dCas9 to specific promoters and enhancers led to increased acetylation of histone H3 and gene activation (103). Subsequent studies (70, 104, 105) reported the development of temporally regulatable systems of targeted locus acetylation. At present, it is less clear that solely altering locus-specific histone methylation levels is a robust method for changing gene regulation; such regulation likely requires other coordinate changes to the epigenome (106–108).

In summary, the use of dCas9-based fusion protein epigenetic writers and erasers has shown early promise. We expect that further contributions to this fast-growing area of research will more fully flesh out the histone code by characterizing the effects of simultaneous alterations to the variety of modifications discussed and will provide tools for both future research and clinical applications.

CRISPR-Mediated Chromatin Repositioning and Modulation of the Three-Dimensional Genome: CRISPR-Genome Organizer and CRISPR-Engineered Chromatin Organization

In addition to promoters, enhancers, epigenetic effectors, and higher-order chromatin structure, there is considerable evidence that transcription is also regulated by the localization of a gene to nuclear bodies or compartments. Reciprocally, other studies have demonstrated that the transcriptional status of a gene likely influences its nuclear sublocalization. A recent study demonstrated that dCas9/sgRNA-mediated relocalization of genes to a specific subset of well-studied intranuclear bodies resulted in transcriptional repression of targeted genes (109). This approach, termed CRISPR-genome organizer (CRISPR-GO), provides a novel means to probe the relationship between transcriptional regulation and nuclear sublocalization by examining the effects of changing the nuclear sublocalization of chromosomal loci. Furthermore, it provides another CRISPR-Cas9 method for regulating endogenous transcription for therapeutic goals.

The authors of this study exploited two different chemically inducible and reversible dimerization systems by fusing one of the dimerization partners to dCas9 and the other to a protein with a known nuclear sublocalization. The nuclear regions targeted were the nuclear periphery (via an emerin fusion protein), Cajal bodies (via coilin), and promyelocytic leukemia (PML) bodies (via PML). Repetitive and nonrepetitive loci were bound with dCas9 programmed by sequence-specific sgRNAs. Chemical induction of CRISPR-GO dimerization was shown by three-dimensional fluorescence *in situ* hybridization (3D-FISH) to relocalize targeted chromosomal loci to the nuclear subdomain dictated by the targeting fusion dimer halves. Removal of the inducer resulted in the disappearance of the repositioned loci. Targeting and its reversal exhibited different timescales, suggesting different dynamics between the formation and dissociation of the specific nuclear bodies.

The authors next demonstrated that relocalizing an integrated reporter gene to the nuclear periphery resulted in transcriptional repression. Surprisingly, relocalizing three endogenous loci to the periphery did not result in decreased transcription, indicating—at least for nuclear periphery localization—that repression may be locus dependent or require additional factors. Relocalizing the integrated reporter gene locus or one of the endogenous loci (Chr3q29) to Cajal bodies resulted in decreased transcription. Determination of the expression levels of other genes near the targeted Chr3q29 locus indicated that the transcriptionally repressed domain extended on both sides of the target site by at least some 35 kb; another gene, about 600 kb downstream, was also repressed by chromosomal relocalization.

Finally, to demonstrate the potential utility of CRISPR-GO for addressing basic biological questions, the authors observed that CRISPR-GO-mediated anchoring of telomeres to the nuclear membrane resulted in decreased cell viability, likely due in part to causing a G_0/G_1 cell cycle arrest (109). Conversely, localizing telomeres to Cajal bodies resulted in increased cell viability. Together, these results indicate that appropriate telomere subnuclear localization is required for cellular function and that CRISPR-GO may prove a powerful tool for more generally addressing the roles of and requirements for subnuclear localization of higher-order chromosomal structures.

In another study, CRISPR was used to manipulate and investigate large regions of heterochromatin hallmarked by heterochromatin protein 1 (HP1) (110). The technology, termed CRISPR-engineered chromatin organization (CRISPR-EChO), combines live-cell imaging with inducible and reversible recruitment of HP1 proteins to large-scale targeted genomic sites. Using sgRNAs that target endogenous repetitive elements in the human genome, CRISPR-EChO demonstrated unique capabilities for studying heterochromatin functions. For example, HP1a tiled across kilobase-scale genomic DNA can form novel contacts with natural heterochromatin in a process that resembles molecular condensation. Simultaneously targeting two distant regions can integrate these regions in a few hours. Formation of CRISPR-EChO-mediated heterochromatin can inducibly and reversibly change chromatin from a diffuse to a compact state, wherein the compact state exhibits delayed disassembly kinetics and likely leads to epigenetic memory formation. Furthermore, CRISPR-EChO-mediated heterochromatin represses gene transcription across over 600 kb on the same chromosome. These findings further highlight the unique applications of CRISPR-mediated 3D genome regulation in studying supranucleosomal chromatin organization in living cells.

Applications of CRISPR-dCas9-Mediated Manipulation of Gene Expression for Basic Cell Biology, Regenerative Medicine, and Gene Therapy

The CRISPR-dCas9 system has created exciting new applications for both discovery-oriented basic research and applied research, such as diagnostics, gene therapy, and regenerative medicine (reviewed in reference 33). Its usage for loss-of-function and gain-of-function genome-wide screens, dissection and manipulation of dynamic genetic circuits, and modulation of cell fates has yielded important new information about the genetic networks that control cell identity, reprogramming, and differentiation. In one example, authentic induced pluripotent stem cells were generated with CRISPR-based activation of the endogenous Oct4 or Sox2 loci using dCas9-Suntag-VP64/sgRNAs (111). These and other advances show the robust and versatile functionality of these next-generation technologies that use synthetic transcription factors to reprogram cell fate (reviewed in references 27 and 34). The dCas9 platform has also been used to design novel methods for chromatin and RNA immunoprecipitation (i.e., dCas9-ChIP) (112), which led to the identification of new genome regulatory elements.

CRISPR/Cas9a/i approaches have also been very effective for functional genetic studies in animal cells of various species, including *C. elegans*, *Drosophila*, mice, humans, and others (61). However, so far CRISPRi/a's *in vivo* use has been complicated by the combined size of the dCas9/effector open reading frame, the sgRNA, and their respective promoters, which often exceeds the packaging capacity of commonly used AAV vectors. Liao and colleagues circumvented this limitation by adapting the SAM module (described above) to develop a system in which the transcriptional activators were separated from Cas9 (113). They infected Cas9 transgenic mice with an AAV that expresses multimerized cotranscriptional activators (i.e., the MPH complex) and modified sgRNAs. This approach was used to ameliorate the phenotypes of several rare disease mouse models, such as for acute kidney injury, type 1 diabetes, and the *mdx* mouse model for Duchenne muscular dystrophy. This study provides a powerful example of the use of CRISPRa to increase expression of clinically relevant target genes *in vivo* (113).

In another study (114), this method for *in vivo* CRISPRa was significantly improved by using a single AAV particle carrying a VP64-dCas9-VP64 fusion and an sgRNA. Using a considerably smaller dCas9 protein variant derived from *Staphylococcus aureus*, modest upregulation of the LAMA1 protein (a few fold) significantly ameliorated the congenital muscular dystrophy type 1A (MDC1A) disease phenotype. This general strategy may be applicable to many other genetic disorders, including haploinsufficiency diseases.

Several methods have been developed that combine elements of existing DNA FISH technologies with CRISPR dCas9-based imaging, creating new ways for imaging chromatin and RNA dynamics in live cells (115–119). These studies will certainly also lead to improved imaging for medical diagnostics.

Off-Targeting Issues in dCas9 Fusion-Mediated Transcriptional Regulation

The targeting specificity of dCas9 fusion approaches for gene regulation by any of the methods described above is critical for drawing valid conclusions from their results. The targeting specificity of these approaches will be supremely important for their clinical application. For example, adventitious upregulation of a nontargeted gene could lead to oncogenesis or other disease. Since genome engineering with nuclease-active Cas9 was the first tool developed in the CRISPR-Cas9 toolbox, considerable efforts (reviewed in references 120 to 124) have been made to evaluate and minimize Cas9 off-targeting by a number of means, including improvement in off-target detection methods, use of alternative Cas9 species or rational engineering of better-characterized ones, and manipulation of sgRNA sequences.

Although dCas9-based approaches have only recently been developed, there is significant literature pertaining to the specificity of several dCas9 fusion proteins. However, true comparison among fusions requires laborious whole-genome approaches to quantify both gene-specific covalent modifications and transcription levels, as well as some means to normalize between different epigenetic marker assays. There is some evidence from a study comparing dCas9-KRAB to Cas9 that the dCas9 fusion is more affected by mismatches between the sgRNA and its genomic target and therefore likely to have fewer off-targets than Cas9 (48); however, the dCas9 fusion partner likely plays a significant role in determining the targeting specificity. A chromatin immunoprecipitation-sequencing (ChIP-seq) analysis of dCas9 binding specificity, performed with hemagglutinin-tagged dCas9, indicated relatively low off-targeting by dCas9 (112). However, earlier studies from two other groups (125, 126), using a similar ChIP-seq approach and the same tagged version of dCas9, found relatively high occupancy of off-target sites by dCas9. The reasons for this discrepancy remain unclear.

To our knowledge, only a single study of the targeting specificity of a dCas9-epigenetic writer/eraser fusion protein has been reported to date. Using dCas9-BFP fusions with the methylases DNMT3A and DNMT3B (127), this study reported that expression of these fusions resulted in significant numbers (hundreds to thousands) of off-target methylation sites, as assayed by whole-genome bisulfite sequencing. Interestingly, there was only a weak correlation between these hypermethylated sites and predicted off-target dCas9 binding sites. Taken together with the improvements in dCas9-methylase and -demethylase specificity observed upon multimerizing epigenetic writer/erasers with a single dCas9 species at the target site (99, 101) and the increased specificity of the dCas9-split M.SssI CpG methylase when mutated to decrease DNA binding affinity (128) discussed above, this study suggests that the specificity of dCas9 binding to DNA is not the sole determinant of the on- to off-target ratio. Thus, it seems likely that each novel dCas9 fusion will need to be evaluated for its targeting specificity.

Future Perspectives

CRISPR-Cas9 technology has revolutionized our ability to design innovative experimental strategies for basic and applied biosciences. Novel approaches are available for gene editing and gene regulation. dCas9-based CRISPRi/a, the topic of this chapter, has gradually evolved into a versatile, robust, and sophisticated method for inducible, reversible, and multiplexable regulation of gene expression for both loss- and gain-of-function studies and epigenetic modifications of genes of interest. Besides its applications for the study of genetic regulatory networks, cell fate determination, differentiation, and chromatin dynamics, it has become a powerful toolbox for gene therapy, regenerative medicine and medical diagnostics.

However, before CRISPR-Cas will reach its full potential, especially for its use in therapeutics, important issues still need to be resolved. These include off-target activity and potential human immunogenicity to CRISPR-Cas9 proteins. A recent study shows that more than half of the human population likely already have preexisting adaptive immune responses against the *S. aureus* and *S. pyogenes* Cas9 proteins (129). Therefore, it will be important to explore other Cas9 variants to which humans are not yet exposed. More generally, the difficulty of therapeutic delivery to their appropriate targets will also hamper clinical use. The limited capacity of the currently used delivery vectors for dCas9 complexes can be addressed by either exploring other delivery vehicles or using smaller or split functionally equivalent Cas9 variants. For example, the Cas9 gene from *S. aureus* is 25% smaller than the commonly used *S. pyogenes* variant (130). Lastly, and potentially most importantly, the field is ready for a comprehensive policy for ethical and responsible use of CRISPR-based genome technologies for future gene editing and manipulation in the clinic.

References

1. Hood L, Rowen L. 2013. The Human Genome Project: big science transforms biology and medicine. *Genome Med* 5:79.

2. Maurano MT, Humbert R, Rynes E, Thurman RE, Haugen E, Wang H, Reynolds AP, Sandstrom R, Qu H, Brody J, Shafer A, Neri F, Lee K, Kutyavin T, Stehling-Sun S, Johnson AK, Canfield TK, Giste E, Diegel M, Bates D, Hansen RS, Neph S, Sabo PJ, Heimfeld S, Raubitschek A, Ziegler S, Cotsapas C, Sotoodehnia N, Glass I, Sunyaev SR, Kaul R, Stamatoyannopoulos JA. 2012. Systematic localization of common disease-associated variation in regulatory DNA. *Science* 337:1190–1195.

3. Austin CP. 2004. The impact of the completed human genome sequence on the development of novel therapeutics for human disease. *Annu Rev Med* 55:1–13.

4. Mohr SE, Smith JA, Shamu CE, Neumüller RA, Perrimon N. 2014. RNAi screening comes of age: improved techniques and complementary approaches. *Nat Rev Mol Cell Biol* 15:591–600.

5. Jackson AL, Bartz SR, Schelter J, Kobayashi SV, Burchard J, Mao M, Li B, Cavet G, Linsley PS. 2003. Expression profiling reveals off-target gene regulation by RNAi. *Nat Biotechnol* 21:635–637.

6. Kim YG, Cha J, Chandrasegaran S. 1996. Hybrid restriction enzymes: zinc finger fusions to Fok I cleavage domain. *Proc Natl Acad Sci U S A* 93:1156–1160.

7. Urnov FD, Rebar EJ, Holmes MC, Zhang HS, Gregory PD. 2010. Genome editing with engineered zinc finger nucleases. *Nat Rev Genet* 11:636–646.

8. Boch J, Scholze H, Schornack S, Landgraf A, Hahn S, Kay S, Lahaye T, Nickstadt A, Bonas U. 2009. Breaking the code of DNA binding specificity of TAL-type III effectors. *Science* 326:1509–1512.

9. Christian M, Cermak T, Doyle EL, Schmidt C, Zhang F, Hummel A, Bogdanove AJ, Voytas DF. 2010. Targeting DNA double-strand breaks with TAL effector nucleases. *Genetics* 186:757–761.

10. Gaj T, Gersbach CA, Barbas CF, III. 2013. ZFN, TALEN, and CRISPR/Cas-based methods for genome engineering. *Trends Biotechnol* 31:397–405.

11. Zhang F, Cong L, Lodato S, Kosuri S, Church GM, Arlotta P. 2011. Efficient construction of sequence-specific TAL effectors for modulating mammalian transcription. *Nat Biotechnol* 29:149–153.

12. Sander JD, Dahlborg EJ, Goodwin MJ, Cade L, Zhang F, Cifuentes D, Curtin SJ, Blackburn JS, Thibodeau-Beganny S, Qi Y, Pierick CJ, Hoffman E, Maeder ML, Khayter C, Reyon D, Dobbs D, Langenau DM, Stupar RM, Giraldez AJ, Voytas DF, Peterson RT, Yeh J-RJ, Joung JK. 2011. Selection-free zinc-finger-nuclease engineering by context-dependent assembly (CoDA). *Nat Methods* 8:67–69.

13. Gasiunas G, Barrangou R, Horvath P, Siksnys V. 2012. Cas9-crRNA ribonucleoprotein complex mediates specific DNA cleavage for adaptive immunity in bacteria. *Proc Natl Acad Sci U S A* 109:E2579–E2586.

14. Jinek M, Chylinski K, Fonfara I, Hauer M, Doudna JA, Charpentier E. 2012. A programmable dual-RNA-guided DNA endonuclease in adaptive bacterial immunity. *Science* 337:816–821.

15. Sapranauskas R, Gasiunas G, Fremaux C, Barrangou R, Horvath P, Siksnys V. 2011. The *Streptococcus thermophilus* CRISPR/Cas system provides immunity in *Escherichia coli*. *Nucleic Acids Res* 39:9275–9282.

16. Deveau H, Garneau JE, Moineau S. 2010. CRISPR/Cas system and its role in phage-bacteria interactions. *Annu Rev Microbiol* 64:475–493.

17. Garneau JE, Dupuis M-È, Villion M, Romero DA, Barrangou R, Boyaval P, Fremaux C, Horvath P, Magadán AH, Moineau S. 2010. The CRISPR/Cas bacterial immune system cleaves bacteriophage and plasmid DNA. *Nature* 468:67–71.

18. Barrangou R, Fremaux C, Deveau H, Richards M, Boyaval P, Moineau S, Romero DA, Horvath P. 2007. CRISPR provides acquired resistance against viruses in prokaryotes. *Science* **315**:1709–1712.

19. Brouns SJJ, Jore MM, Lundgren M, Westra ER, Slijkhuis RJH, Snijders APL, Dickman MJ, Makarova KS, Koonin EV, van der Oost J. 2008. Small CRISPR RNAs guide antiviral defense in prokaryotes. *Science* **321**:960–964.

20. Pourcel C, Salvignol G, Vergnaud G. 2005. CRISPR elements in *Yersinia pestis* acquire new repeats by preferential uptake of bacteriophage DNA, and provide additional tools for evolutionary studies. *Microbiology (Reading)* **151**:653–663.

21. Shmakov S, Abudayyeh OO, Makarova KS, Wolf YI, Gootenberg JS, Semenova E, Minakhin L, Joung J, Konermann S, Severinov K, Zhang F, Koonin EV. 2015. Discovery and functional characterization of diverse class 2 CRISPR-Cas systems. *Mol Cell* **60**:385–397.

22. Rath D, Amlinger L, Rath A, Lundgren M. 2015. The CRISPR-Cas immune system: biology, mechanisms and applications. *Biochimie* **117**:119–128.

23. Mojica FJM, Díez-Villaseñor C, García-Martínez J, Soria E. 2005. Intervening sequences of regularly spaced prokaryotic repeats derive from foreign genetic elements. *J Mol Evol* **60**:174–182.

24. Bolotin A, Quinquis B, Sorokin A, Ehrlich SD. 2005. Clustered regularly interspaced short palindrome repeats (CRISPRs) have spacers of extrachromosomal origin. *Microbiology (Reading)* **151**:2551–2561.

25. Marraffini LA, Sontheimer EJ. 2008. CRISPR interference limits horizontal gene transfer in staphylococci by targeting DNA. *Science* **322**:1843–1845.

26. Wright AV, Nuñez JK, Doudna JA. 2016. Biology and applications of CRISPR systems: harnessing nature's toolbox for genome engineering. *Cell* **164**:29–44.

27. Adli M. 2018. The CRISPR tool kit for genome editing and beyond. *Nat Commun* **9**:1911.

28. Dominguez AA, Lim WA, Qi LS. 2016. Beyond editing: repurposing CRISPR-Cas9 for precision genome regulation and interrogation. *Nat Rev Mol Cell Biol* **17**:5–15.

29. Jinek M, East A, Cheng A, Lin S, Ma E, Doudna J. 2013. RNA-programmed genome editing in human cells. *eLife* **2**:e00471 http://dx.doi.org/10.7554/eLife.00471.

30. Cong L, Ran FA, Cox D, Lin S, Barretto R, Habib N, Hsu PD, Wu X, Jiang W, Marraffini LA, Zhang F. 2013. Multiplex genome engineering using CRISPR/Cas systems. *Science* **339**:819–823.

31. Mali P, Yang L, Esvelt KM, Aach J, Guell M, DiCarlo JE, Norville JE, Church GM. 2013. RNA-guided human genome engineering via Cas9. *Science* **339**:823–826.

32. Qi LS, Larson MH, Gilbert LA, Doudna JA, Weissman JS, Arkin AP, Lim WA. 2013. Repurposing CRISPR as an RNA-guided platform for sequence-specific control of gene expression. *Cell* **152**:1173–1183.

33. Lau C-H. 2018. Applications of CRISPR-Cas in bioengineering, biotechnology, and translational research. *CRISPR J* **1**:379–404.

34. Xu X, Qi LS. 2019. A CRISPR-dCas toolbox for genetic engineering and synthetic biology. *J Mol Biol* **431**:34–47.

35. Mojica FJM, Díez-Villaseñor C, García-Martínez J, Almendros C. 2009. Short motif sequences determine the targets of the prokaryotic CRISPR defence system. *Microbiology (Reading)* **155**:733–740.

36. Liu Y, Han J, Chen Z, Wu H, Dong H, Nie G. 2017. Engineering cell signaling using tunable CRISPR-Cpf1-based transcription factors. *Nat Commun* **8**:2095.

37. Magnusson JP, Rios AR, Wu L, Qi LS. 2021. Enhanced Cas12a multi-gene regulation using a CRISPR array separator. *eLife* **10**:e66406 http://dx.doi.org/10.7554/eLife.66406.

38. Oesinghaus L, Simmel FC. 2019. Switching the activity of Cas12a using guide RNA strand displacement circuits. *Nat Commun* **10**:2092.

39. Liu J-J, Orlova N, Oakes BL, Ma E, Spinner HB, Baney KLM, Chuck J, Tan D, Knott GJ, Harrington LB, Al-Shayeb B, Wagner A, Brötzmann J, Staahl BT, Taylor KL, Desmarais J, Nogales E, Doudna JA. 2019. CasX enzymes comprise a distinct family of RNA-guided genome editors. *Nature* **566**:218–223.

40. Xu X, Chemparathy A, Zeng L, Kempton HR, Shang S, Nakamura M, Qi LS. 2021. Engineered miniature CRISPR-Cas system for mammalian genome regulation and editing. *Mol Cell* **81**:4333–4345.e4.

41. Bikard D, Jiang W, Samai P, Hochschild A, Zhang F, Marraffini LA. 2013. Programmable repression and activation of bacterial gene expression using an engineered CRISPR-Cas system. *Nucleic Acids Res* **41**:7429–7437.

42. Larson MH, Gilbert LA, Wang X, Lim WA, Weissman JS, Qi LS. 2013. CRISPR interference (CRISPRi) for sequence-specific control of gene expression. *Nat Protoc* **8**:2180–2196.

43. Tanenbaum ME, Gilbert LA, Qi LS, Weissman JS, Vale RD. 2014. A protein-tagging system for signal amplification in gene expression and fluorescence imaging. *Cell* **159**:635–646.

44. Peters JM, Colavin A, Shi H, Czarny TL, Larson MH, Wong S, Hawkins JS, Lu CHS, Koo B-M, Marta E, Shiver AL, Whitehead EH, Weissman JS, Brown ED, Qi LS, Huang KC, Gross CA. 2016. A comprehensive, CRISPR-based functional analysis of essential genes in bacteria. *Cell* **165**:1493–1506.

45. Gilbert LA, Larson MH, Morsut L, Liu Z, Brar GA, Torres SE, Stern-Ginossar N, Brandman O, Whitehead EH, Doudna JA, Lim WA, Weissman JS, Qi LS. 2013. CRISPR-mediated modular RNA-guided regulation of transcription in eukaryotes. *Cell* **154**:442–451.

46. Lawhorn IEB, Ferreira JP, Wang CL. 2014. Evaluation of sgRNA target sites for CRISPR-mediated repression of TP53. *PLoS One* **9**:e113232.

47. Shariati SA, Dominguez A, Xie S, Wernig M, Qi LS, Skotheim JM. 2019. Reversible disruption of specific transcription factor-DNA interactions using CRISPR/Cas9. *Mol Cell* **74**:622–633.e4.

48. Gilbert LA, Horlbeck MA, Adamson B, Villalta JE, Chen Y, Whitehead EH, Guimaraes C, Panning B, Ploegh HL, Bassik MC, Qi LS, Kampmann M, Weissman JS. 2014. Genome-scale CRISPR-mediated control of gene repression and activation. *Cell* **159**:647–661.

49. Konermann S, Brigham MD, Trevino A, Hsu PD, Heidenreich M, Cong L, Platt RJ, Scott DA, Church GM, Zhang F. 2013. Optical control of mammalian endogenous transcription and epigenetic states. *Nature* **500**:472–476.

50. Groner AC, Meylan S, Ciuffi A, Zangger N, Ambrosini G, Dénervaud N, Bucher P, Trono D. 2010. KRAB-zinc finger proteins and KAP1 can mediate long-range transcriptional repression through heterochromatin spreading. *PLoS Genet* **6**:e1000869

51. Thakore PI, D'Ippolito AM, Song L, Safi A, Shivakumar NK, Kabadi AM, Reddy TE, Crawford GE, Gersbach CA. 2015. Highly specific epigenome editing by CRISPR-Cas9 repressors for silencing of distal regulatory elements. *Nat Methods* **12**:1143–1149.

52. Yeo NC, Chavez A, Lance-Byrne A, Chan Y, Menn D, Milanova D, Kuo C-C, Guo X, Sharma S, Tung A, Cecchi RJ, Tuttle M, Pradhan S, Lim ET, Davidsohn N, Ebrahimkhani MR, Collins JJ, Lewis NE, Kiani S, Church GM. 2018. An enhanced CRISPR repressor for targeted mammalian gene regulation. *Nat Methods* **15**:611–616.

53. Konermann S, Brigham MD, Trevino AE, Joung J, Abudayyeh OO, Barcena C, Hsu PD, Habib N, Gootenberg JS, Nishimasu H, Nureki O, Zhang F. 2015. Genome-scale transcriptional activation by an engineered CRISPR-Cas9 complex. *Nature* **517**:583–588.

54. Cheng AW, Wang H, Yang H, Shi L, Katz Y, Theunissen TW, Rangarajan S, Shivalila CS, Dadon DB, Jaenisch R. 2013. Multiplexed activation of endogenous genes by CRISPR-on, an RNA-guided transcriptional activator system. *Cell Res* **23**:1163–1171.

55. Perez-Pinera P, Kocak DD, Vockley CM, Adler AF, Kabadi AM, Polstein LR, Thakore PI, Glass KA, Ousterout DG, Leong KW, Guilak F, Crawford GE, Reddy TE, Gersbach CA. 2013. RNA-guided gene activation by CRISPR-Cas9-based transcription factors. *Nat Methods* **10**:973–976.

56. Maeder ML, Linder SJ, Cascio VM, Fu Y, Ho QH, Joung JK. 2013. CRISPR RNA-guided activation of endogenous human genes. *Nat Methods* 10:977–979.

57. Ma D, Peng S, Huang W, Cai Z, Xie Z. 2018. Rational design of mini-Cas9 for transcriptional activation. *ACS Synth Biol* 7:978–985.

58. Chavez A, Scheiman J, Vora S, Pruitt BW, Tuttle M, P R Iyer E, Lin S, Kiani S, Guzman CD, Wiegand DJ, Ter-Ovanesyan D, Braff JL, Davidsohn N, Housden BE, Perrimon N, Weiss R, Aach J, Collins JJ, Church GM. 2015. Highly efficient Cas9-mediated transcriptional programming. *Nat Methods* 12:326–328.

59. Zalatan JG, Lee ME, Almeida R, Gilbert LA, Whitehead EH, La Russa M, Tsai JC, Weissman JS, Dueber JE, Qi LS, Lim WA. 2015. Engineering complex synthetic transcriptional programs with CRISPR RNA scaffolds. *Cell* 160:339–350.

60. Briner AE, Donohoue PD, Gomaa AA, Selle K, Slorach EM, Nye CH, Haurwitz RE, Beisel CL, May AP, Barrangou R. 2014. Guide RNA functional modules direct Cas9 activity and orthogonality. *Mol Cell* 56:333–339.

61. Chavez A, Tuttle M, Pruitt BW, Ewen-Campen B, Chari R, Ter-Ovanesyan D, Haque SJ, Cecchi RJ, Kowal EJK, Buchthal J, Housden BE, Perrimon N, Collins JJ, Church G. 2016. Comparison of Cas9 activators in multiple species. *Nat Methods* 13:563–567.

62. Zhou H, Liu J, Zhou C, Gao N, Rao Z, Li H, Hu X, Li C, Yao X, Shen X, Sun Y, Wei Y, Liu F, Ying W, Zhang J, Tang C, Zhang X, Xu H, Shi L, Cheng L, Huang P, Yang H. 2018. in vivo simultaneous transcriptional activation of multiple genes in the brain using CRISPR-dCas9-activator transgenic mice. *Nat Neurosci* 21:440–446.

63. Polstein LR, Gersbach CA. 2015. A light-inducible CRISPR-Cas9 system for control of endogenous gene activation. *Nat Chem Biol* 11:198–200.

64. Nihongaki Y, Kawano F, Nakajima T, Sato M. 2015. Photoactivatable CRISPR-Cas9 for optogenetic genome editing. *Nat Biotechnol* 33:755–760.

65. Nihongaki Y, Yamamoto S, Kawano F, Suzuki H, Sato M. 2015. CRISPR-Cas9-based photoactivatable transcription system. *Chem Biol* 22:169–174.

66. Nihongaki Y, Furuhata Y, Otabe T, Hasegawa S, Yoshimoto K, Sato M. 2017. CRISPR-Cas9-based photoactivatable transcription systems to induce neuronal differentiation. *Nat Methods* 14:963–966.

67. Levskaya A, Weiner OD, Lim WA, Voigt CA. 2009. Spatiotemporal control of cell signalling using a light-switchable protein interaction. *Nature* 461:997–1001.

68. Zetsche B, Volz SE, Zhang F. 2015. A split-Cas9 architecture for inducible genome editing and transcription modulation. *Nat Biotechnol* 33:139–142.

69. Bao Z, Jain S, Jaroenpuntaruk V, Zhao H. 2017. Orthogonal genetic regulation in human cells using chemically induced CRISPR/Cas9 activators. *ACS Synth Biol* 6:686–693.

70. Chen T, Gao D, Zhang R, Zeng G, Yan H, Lim E, Liang F-S. 2017. Chemically controlled epigenome editing through an inducible dCas9 system. *J Am Chem Soc* 139:11337–11340.

71. Gao Y, Xiong X, Wong S, Charles EJ, Lim WA, Qi LS. 2016. Complex transcriptional modulation with orthogonal and inducible dCas9 regulators. *Nat Methods* 13:1043–1049.

72. Tak YE, Kleinstiver BP, Nuñez JK, Hsu JY, Horng JE, Gong J, Weissman JS, Joung JK. 2017. Inducible and multiplex gene regulation using CRISPR-Cpf1-based transcription factors. *Nat Methods* 14:1163–1166.

73. Nguyen DP, Miyaoka Y, Gilbert LA, Mayerl SJ, Lee BH, Weissman JS, Conklin BR, Wells JA. 2016. Ligand-binding domains of nuclear receptors facilitate tight control of split CRISPR activity. *Nat Commun* 7:12009.

74. Hill ZB, Martinko AJ, Nguyen DP, Wells JA. 2018. Human antibody-based chemically induced dimerizers for cell therapeutic applications. *Nat Chem Biol* 14:112–117.

75. Maji B, Moore CL, Zetsche B, Volz SE, Zhang F, Shoulders MD, Choudhary A. 2017. Multidimensional chemical control of CRISPR-Cas9. *Nat Chem Biol* 13:9–11.

76. Liu Y, Zhan Y, Chen Z, He A, Li J, Wu H, Liu L, Zhuang C, Lin J, Guo X, Zhang Q, Huang W, Cai Z. 2016. Directing cellular information flow via CRISPR signal conductors. *Nat Methods* 13:938–944.

77. Lim WA, June CH. 2017. The principles of engineering immune cells to treat cancer. *Cell* 168:724–740.

78. Rosenbaum DM, Rasmussen SGF, Kobilka BK. 2009. The structure and function of G-protein-coupled receptors. *Nature* 459:356–363.

79. Wacker D, Stevens RC, Roth BL. 2017. How ligands illuminate GPCR molecular pharmacology. *Cell* 170:414–427.

80. Morsut L, Roybal KT, Xiong X, Gordley RM, Coyle SM, Thomson M, Lim WA. 2016. Engineering customized cell sensing and response behaviors using synthetic Notch receptors. *Cell* 164:780–791.

81. Daringer NM, Dudek RM, Schwarz KA, Leonard JN. 2014. Modular extracellular sensor architecture for engineering mammalian cell-based devices. *ACS Synth Biol* 3:892–902.

82. Barnea G, Strapps W, Herrada G, Berman Y, Ong J, Kloss B, Axel R, Lee KJ. 2008. The genetic design of signaling cascades to record receptor activation. *Proc Natl Acad Sci U S A* 105:64–69.

83. Kroeze WK, Sassano MF, Huang X-P, Lansu K, McCorvy JD, Giguère PM, Sciaky N, Roth BL. 2015. PRESTO-Tango as an open-source resource for interrogation of the druggable human GPCRome. *Nat Struct Mol Biol* 22:362–369.

84. Kipniss NH, Dingal PCDP, Abbott TR, Gao Y, Wang H, Dominguez AA, Labanieh L, Qi LS. 2017. Engineering cell sensing and responses using a GPCR-coupled CRISPR-Cas system. *Nat Commun* 8:2212.

85. Allis CD, Jenuwein T. 2016. The molecular hallmarks of epigenetic control. *Nat Rev Genet* 17:487–500.

86. Nakayama J, Rice JC, Strahl BD, Allis CD, Grewal SI. 2001. Role of histone H3 lysine 9 methylation in epigenetic control of heterochromatin assembly. *Science* 292:110–113.

87. Amabile A, Migliara A, Capasso P, Biffi M, Cittaro D, Naldini L, Lombardo A. 2016. Inheritable silencing of endogenous genes by hit-and-run targeted epigenetic editing. *Cell* 167:219–232.e14.

88. Liu XS, Wu H, Ji X, Stelzer Y, Wu X, Czauderna S, Shu J, Dadon D, Young RA, Jaenisch R. 2016. Editing DNA methylation in the mammalian genome. *Cell* 167:233–247.e17.

89. McDonald JI, Celik H, Rois LE, Fishberger G, Fowler T, Rees R, Kramer A, Martens A, Edwards JR, Challen GA. 2016. Reprogrammable CRISPR/Cas9-based system for inducing site-specific DNA methylation. *Biol Open* 5:866–874.

90. Vojta A, Dobrinić P, Tadić V, Bočkor L, Korać P, Julg B, Klasić M, Zoldoš V. 2016. Repurposing the CRISPR-Cas9 system for targeted DNA methylation. *Nucleic Acids Res* 44:5615–5628.

91. Nuñez JK, Chen J, Pommier GC, Cogan JZ, Replogle JM, Adriaens C, Ramadoss GN, Shi Q, Hung KL, Samelson AJ, Pogson AN, Kim JYS, Chung A, Leonetti MD, Chang HY, Kampmann M, Bernstein BE, Hovestadt V, Gilbert LA, Weissman JS. 2021. Genome-wide programmable transcriptional memory by CRISPR-based epigenome editing. *Cell* 184:2503–2519.e17.

92. Nakamura M, Ivec AE, Gao Y, Qi LS. 2021. *Durable CRISPR-based epigenetic silencing*. *BioDesign Res* 2021:9815820 http://dx.doi.org/10.34133/2021/9815820.

93. Choudhury SR, Cui Y, Lubecka K, Stefanska B, Irudayaraj J. 2016. CRISPR-dCas9 mediated TET1 targeting for selective DNA demethylation at BRCA1 promoter. *Oncotarget* 7:46545–46556.

94. Liu XS, Wu H, Krzisch M, Wu X, Graef J, Muffat J, Hnisz D, Li CH, Yuan B, Xu C, Li Y, Vershkov D, Cacace A, Young RA, Jaenisch R. 2018. Rescue of fragile X syndrome neurons by DNA methylation editing of the FMR1 gene. *Cell* 172:979–992.e6.

95. Lei Y, Zhang X, Su J, Jeong M, Gundry MC, Huang Y-H, Zhou Y, Li W, Goodell MA. 2017. Targeted DNA methylation in vivo using an engineered dCas9-MQ1 fusion protein. *Nat Commun* 8:16026.

96. Xiong T, Meister GE, Workman RE, Kato NC, Spellberg MJ, Turker F, Timp W, Ostermeier M, Novina CD. 2017. Targeted DNA methylation in human cells using engineered dCas9-methyltransferases. *Sci Rep* 7:6732.

97. Xiong T, Rohm D, Workman RE, Roundtree L, Novina CD, Timp W, Ostermeier M. 2018. Protein engineering strategies for improving the selective methylation of target CpG sites by a dCas9-directed cytosine methyltransferase in bacteria. *PLoS One* 13:e0209408.

98. Galonska C, Charlton J, Mattei AL, Donaghey J, Clement K, Gu H, Mohammad AW, Stamenova EK, Cacchiarelli D, Klages S, Timmermann B, Cantz T, Schöler HR, Gnirke A, Ziller MJ, Meissner A. 2018. Genome-wide tracking of dCas9-methyltransferase footprints. *Nat Commun* 9:597.

99. Pflueger C, Tan D, Swain T, Nguyen T, Pflueger J, Nefzger C, Polo JM, Ford E, Lister R. 2018. A modular dCas9-SunTag DNMT3A epigenome editing system overcomes pervasive off-target activity of direct fusion dCas9-DNMT3A constructs. *Genome Res* 28:1193–1206.

100. Huang Y-H, Su J, Lei Y, Brunetti L, Gundry MC, Zhang X, Jeong M, Li W, Goodell MA. 2017. DNA epigenome editing using CRISPR-Cas SunTag-directed DNMT3A. *Genome Biol* 18:176.

101. Morita S, Noguchi H, Horii T, Nakabayashi K, Kimura M, Okamura K, Sakai A, Nakashima H, Hata K, Nakashima K, Hatada I. 2016. Targeted DNA demethylation in vivo using dCas9-peptide repeat and scFv-TET1 catalytic domain fusions. *Nat Biotechnol* 34:1060–1065.

102. Jenuwein T, Allis CD. 2001. Translating the histone code. *Science* 293:1074–1080.

103. Hilton IB, D'Ippolito AM, Vockley CM, Thakore PI, Crawford GE, Reddy TE, Gersbach CA. 2015. Epigenome editing by a CRISPR-Cas9-based acetyltransferase activates genes from promoters and enhancers. *Nat Biotechnol* 33:510–517.

104. Gao D, Liang F-S. 2018. Chemical inducible dCas9-guided editing of H3K27 acetylation in mammalian cells. *Methods Mol Biol* 1767:429–445.

105. Shrimp JH, Grose C, Widmeyer SRT, Thorpe AL, Jadhav A, Meier JL. 2018. Chemical control of a CRISPR-Cas9 acetyltransferase. *ACS Chem Biol* 13:455–460.

106. Cano-Rodriguez D, Gjaltema RAF, Jilderda LJ, Jellema P, Dokter-Fokkens J, Ruiters MHJ, Rots MG. 2016. Writing of H3K4Me3 overcomes epigenetic silencing in a sustained but context-dependent manner. *Nat Commun* 7:12284.

107. O'Geen H, Bates SL, Carter SS, Nisson KA, Halmai J, Fink KD, Rhie SK, Farnham PJ, Segal DJ. 2019. Ezh2-dCas9 and KRAB-dCas9 enable engineering of epigenetic memory in a context-dependent manner. *Epigenetics Chromatin* 12:26.

108. O'Geen H, Ren C, Nicolet CM, Perez AA, Halmai J, Le VM, Mackay JP, Farnham PJ, Segal DJ. 2017. dCas9-based epigenome editing suggests histone methylation is not sufficient for target gene repression. *Nucleic Acids Res* 45:9901–9916.

109. Wang H, Xu X, Nguyen CM, Liu Y, Gao Y, Lin X, Daley T, Kipniss NH, La Russa M, Qi LS. 2018. CRISPR-mediated programmable 3D genome positioning and nuclear organization. *Cell* 175:1405–1417.e14.

110. Gao Y, Han M, Shang S, Wang H, Qi LS. 2021. Interrogation of the dynamic properties of higher-order heterochromatin using CRISPR-dCas9. *Mol Cell* 81:4287–4299.e5.

111. Liu P, Chen M, Liu Y, Qi LS, Ding S. 2018. CRISPR-based chromatin remodeling of the endogenous Oct4 or Sox2 locus enables reprogramming to pluripotency. *Cell Stem Cell* 22:252–261.e4.

112. Polstein LR, Perez-Pinera P, Kocak DD, Vockley CM, Bledsoe P, Song L, Safi A, Crawford GE, Reddy TE, Gersbach CA. 2015. Genome-wide specificity of DNA binding, gene regulation, and chromatin remodeling by TALE- and CRISPR/Cas9-based transcriptional activators. *Genome Res* 25:1158–1169.

113. Liao H-K, Hatanaka F, Araoka T, Reddy P, Wu M-Z, Sui Y, Yamauchi T, Sakurai M, O'Keefe DD, Núñez-Delicado E, Guillen P, Campistol JM, Wu C-J, Lu L-F, Esteban CR, Izpisua Belmonte JC.

114. 2017. In vivo target gene activation via CRISPR/Cas9-mediated trans-epigenetic modulation. *Cell* 171:1495–1507.e15.

114. Kemaladewi DU, Bassi PS, Erwood S, Al-Basha D, Gawlik KI, Lindsay K, Hyatt E, Kember R, Place KM, Marks RM, Durbeej M, Prescott SA, Ivakine EA, Cohn RD. 2019. A mutation-independent approach for muscular dystrophy via upregulation of a modifier gene. *Nature* 572:125–130.

115. Anton T, Bultmann S, Leonhardt H, Markaki Y. 2014. Visualization of specific DNA sequences in living mouse embryonic stem cells with a programmable fluorescent CRISPR/Cas system. *Nucleus* 5:163–172.

116. Chen B, Gilbert LA, Cimini BA, Schnitzbauer J, Zhang W, Li G-W, Park J, Blackburn EH, Weissman JS, Qi LS, Huang B. 2013. Dynamic imaging of genomic loci in living human cells by an optimized CRISPR/Cas system. *Cell* 155:1479–1491.

117. Deng W, Shi X, Tjian R, Lionnet T, Singer RH. 2015. CASFISH: CRISPR/Cas9-mediated in situ labeling of genomic loci in fixed cells. *Proc Natl Acad Sci U S A* 112:11870–11875.

118. Ma H, Naseri A, Reyes-Gutierrez P, Wolfe SA, Zhang S, Pederson T. 2015. Multicolor CRISPR labeling of chromosomal loci in human cells. *Proc Natl Acad Sci U S A* 112:3002–3007.

119. Nelles DA, Fang MY, O'Connell MR, Xu JL, Markmiller SJ, Doudna JA, Yeo GW. 2016. Programmable RNA tracking in live cells with CRISPR/Cas9. *Cell* 165:488–496.

120. Jamal M, Ullah A, Ahsan M, Tyagi R, Habib Z, Rehman K. 2018. Improving CRISPR-Cas9 on-target specificity. *Curr Issues Mol Biol* 26:65–80.

121. Tadić V, Josipović G, Zoldoš V, Vojta A. 2019. CRISPR/Cas9 based epigenome editing: an overview of dCas9-based tools with special emphasis on off-target activity. *Methods* 164-165:109–119.

122. Tsai SQ, Joung JK. 2016. Defining and improving the genome-wide specificities of CRISPR-Cas9 nucleases. *Nat Rev Genet* 17:300–312.

123. Tycko J, Myer VE, Hsu PD. 2016. Methods for optimizing CRISPR-Cas9 genome editing specificity. *Mol Cell* 63:355–370.

124. Zhang J-H, Adikaram P, Pandey M, Genis A, Simonds WF. 2016. Optimization of genome editing through CRISPR-Cas9 engineering. *Bioengineered* 7:166–174.

125. Kuscu C, Arslan S, Singh R, Thorpe J, Adli M. 2014. Genome-wide analysis reveals characteristics of off-target sites bound by the Cas9 endonuclease. *Nat Biotechnol* 32:677–683.

126. Wu X, Scott DA, Kriz AJ, Chiu AC, Hsu PD, Dadon DB, Cheng AW, Trevino AE, Konermann S, Chen S, Jaenisch R, Zhang F, Sharp PA. 2014. Genome-wide binding of the CRISPR endonuclease Cas9 in mammalian cells. *Nat Biotechnol* 32:670–676.

127. Lin L, Liu Y, Xu F, Huang J, Daugaard TF, Petersen TS, Hansen B, Ye L, Zhou Q, Fang F, Yang L, Li S, Fløe L, Jensen KT, Shrock E, Chen F, Yang H, Wang J, Liu X, Xu X, Bolund L, Nielsen AL, Luo Y. 2018. Genome-wide determination of on-target and off-target characteristics for RNA-guided DNA methylation by dCas9 methyltransferases. *Gigascience* 7:giy011.

128. Xiong K, Zhou Y, Hyttel P, Bolund L, Freude KK, Luo Y. 2016. Generation of induced pluripotent stem cells (iPSCs) stably expressing CRISPR-based synergistic activation mediator (SAM). *Stem Cell Res (Amst)* 17:665–669.

129. Charlesworth CT, Deshpande PS, Dever DP, Camarena J, Lemgart VT, Cromer MK, Vakulskas CA, Collingwood MA, Zhang L, Bode NM, Behlke MA, Dejene B, Cieniewicz B, Romano R, Lesch BJ, Gomez-Ospina N, Mantri S, Pavel-Dinu M, Weinberg KI, Porteus MH. 2019. Identification of preexisting adaptive immunity to Cas9 proteins in humans. *Nat Med* 25:249–254.

130. Ran FA, Cong L, Yan WX, Scott DA, Gootenberg JS, Kriz AJ, Zetsche B, Shalem O, Wu X, Makarova KS, Koonin EV, Sharp PA, Zhang F. 2015. in vivo genome editing using *Staphylococcus aureus* Cas9. *Nature* 520:186–191.

CRISPR Screens

Jonathan D. D'Gama[1,2], Joseph S. Park[1,2], and Matthew K. Waldor[1,2,3]

[1]Department of Microbiology, Harvard Medical School, Boston, MA
[2]Division of Infectious Diseases, Brigham & Women's Hospital, Boston, MA
[3]Howard Hughes Medical Institute, Boston, MA

Introduction

Genetic screening in eukaryotes is being transformed by CRISPR/Cas9 (CRISPR) genome editing technology. This far-reaching application of CRISPR technology harnesses its robust, programmable, RNA-guided, DNA genome editing activity for the creation of libraries of mutant eukaryotic cells. Since CRISPR-based mutagenesis relies on easily synthesized oligonucleotides that are complementary to target sequences, mutant cell libraries containing inactivating mutations in thousands of genetic elements, such as all protein coding genes, can be created in a straightforward fashion. These genome-scale loss-of-function libraries enable genetic screens for phenotypes of interest to be carried out with accuracy and relative ease (1, 2). Moreover, the repurposing of Cas9 as a programmable DNA-binding protein that allows it to serve as a chassis for the delivery of proteins with various functions to specific loci has greatly expanded the potential scope of CRISPR screens.

Development of genome-scale CRISPR screening technology represented a fusion of several technologies, including CRISPR/Cas9 genome editing (3, 4), design and synthesis of large-scale guide RNA libraries (1, 2, 4), and massively parallel sequencing-based measurement of guide RNA frequencies by next-generation sequencing. Critical enabling advances in CRISPR/Cas9 genome editing that allowed for genome-scale screens were the development of single guide RNAs (sgRNAs) fusing the CRISPR RNA (crRNA) and *trans*-activating crRNA (see chapters 4 and 11) (5), the identification of the sgRNA scaffold architecture most effective for *in vivo* editing (6), and the use of a Cas9 variant with both high efficacy *in vivo* and a very small and nonspecific protospacer adjacent motif (PAM) requirement (NGG in the case of *Streptococcus pyogenes*), enabling targeting of large fractions of genomes (3, 4). Genome-scale screens also required computational algorithms for design of large numbers of unique sgRNAs that collectively target the entire coding genome (1, 2, 4). Physical creation of extensive sgRNAs libraries consisting of tens to hundreds of thousands of sgRNAs relies on array-based oligonucleotide synthesis technology that can cheaply generate massive numbers of diverse oligonucleotides.

CRISPR: Biology and Applications, First Edition. Edited by Rodolphe Barrangou, Erik J. Sontheimer, and Luciano A. Marraffini.
© 2022 American Society for Microbiology. DOI: 10.1128/9781683673798.ch13

Prior to the CRISPR revolution, methods for eukaryotic genome editing and genetic screening had important limitations. CRISPR screens improved on previous genetic screening technologies in three key ways. First, CRISPR/Cas9 genome editing yields more complete and long-lasting inactivation of protein function than RNA interference (RNAi) (3, 4). Second, it relies on a form of targeted mutagenesis, unlike the insertional mutagenesis used in haploid cell screens, that has fewer off-target effects than RNAi (7). Third, CRISPR-based approaches are more portable than other approaches and function similarly and effectively across cell lines and model organisms. These advantages coupled with the technology's relative ease of use and cost-effectiveness have led to its broad adoption. Hence, CRISPR screens have ushered in a new era in eukaryotic functional genomics, significantly impacting diverse fields and opening up new areas of research. In this chapter, we first discuss CRISPR screening technology and then highlight recent studies that applied it to elucidate interactions between host cells and bacterial pathogens.

Genetic Screens

Screens identify individuals (or cells) with phenotypes of interest in heterogeneous populations. Genetic screens permit the association of genetic elements or genotypes with a phenotype through the use of a population whose phenotypic heterogeneity arises from underlying genetic diversity. Forward genetic screens, through phenotypic selection, uncover the causative genotype(s) of a phenotype, while reverse genetic screens uncover the phenotype(s) produced by specific genotypes. Phenotypic selection, the identification of a phenotype of interest, can be either positive or negative, enriching for or against the phenotype and thus increasing or decreasing the abundance of the causative genotype(s). Genetic diversity in a population is typically achieved by mutating a population that was originally genetically homogenous. Mutagenesis methods vary in efficacy, specificity, and the range of phenotypic and genotypic diversity produced. Ideally, mutagenesis generates a wide range of genotypic and phenotypic diversity in a population by creating only a single genetic mutation per individual. Screens require large populations to query large domains of genotypic and phenotypic diversity. Traditional mutagenesis methods, such as chemical mutagenesis, generate mutations in random genomic locations. In contrast, CRISPR-based genome editing techniques permit the generation of targeted mutations, reducing the population sizes required for genome-scale screens.

Overview of CRISPR/Cas9 Screens

CRISPR screens consist of several steps, beginning with sgRNA design and ending with identification of genetic hits (Fig. 13.1). After custom sgRNA library design and synthesis, sgRNAs are cloned into sgRNA lentiviral vectors. There are also commercially available, premade genome-scale sgRNA libraries (e.g., GeCKO [1] and Avana/Brunello [8, 9]) that can be used in place of a custom library. A set of sgRNA-encoding lentiviral vectors is transduced into cells to generate a pooled lentivirus sgRNA library. Cell lines (often stably expressing the Cas9 enzyme) are transfected with the lentivirus library such that each individual cell generally receives only one sgRNA, ensuring that each cell has only one gene inactivated by CRISPR genome editing. Selection for cells containing the sgRNA vector results in the generation of a pooled library of mutant cells in which each cell has a single gene inactivated. CRISPR/Cas9 genome editing generally yields null alleles due to the mutagenic DNA repair of the double-strand breaks (DSBs) introduced by Cas9 cleavage at the locus specified by the sgRNA. Gene inactivation leads to various phenotypes in the mutant cells. For example, cells receiving sgRNAs targeting essential genes are lost from the population due to cell nonviability. As the pooled library is passaged or enriched for a

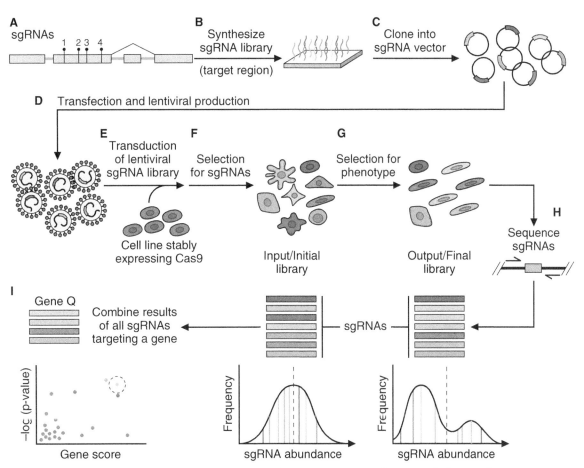

Figure 13.1 Overview of steps in CRISPR screens. (**A**) A CRISPR screen begins with design of sgRNAs targeting genes to be included in the library. (**B**) Oligonucleotides encoding the chosen targeting sequences are synthesized using array-based DNA synthesis technology. (**C**) These oligonucleotides are cloned into an appropriate vector to construct the sgRNA vector library in typical lab strains of *E. coli*. (**D**) The purified sgRNA vector library is transfected into 293T cells to produce a lentiviral sgRNA library. (**E**) A cell line in which the screen is to be performed is transduced with the lentiviral sgRNA library at a low titer to ensure that each cell receives only one sgRNA. The cell line chosen may constitutively express the Cas9 enzyme. (**F**) A mutant cell library is generated after selection for cells carrying integrated sgRNA constructs that mediate indel formation in their targets. (**G**) Selection enriches for mutant cells with the desired phenotype. (**H**) The distributions of sgRNA abundances in the initial and final (post-phenotypic selection) cell populations are determined using massively parallel next-generation sequencing of the sgRNA targeting region. The abundance of sgRNAs in the initial and final populations is compared. (**I**) Finally, utilizing one of several available computation-based statistical algorithms, the phenotypes produced by the sgRNAs targeting single genes are summed, generating a unified gene score that represents the overall enrichment or depletion of inactivating mutations in the gene after selection and a *P* value representing the likelihood of the result.

particular phenotype, the corresponding causative sgRNAs are amplified. Measuring the relative abundance of sgRNAs provides a gauge of the distribution of mutant cells in the population after selection. That is, the sgRNA targeting sequences also serve as molecular barcodes and sequencing sgRNA targeting sequences can be used to infer the distribution of inactivated genes. To increase the robustness and accuracy of screens, results from multiple sgRNAs targeting the same gene are amalgamated via computation-based statistical algorithms, yielding gene-level enrichment or depletion scores (10, 11). Subsequently, candidate genes can be picked for experimental validation and further study.

Screen Design, Execution, and Analysis

The first genome-scale CRISPR screens consisted of loss-of-function screens used to uncover genes required for cellular growth and susceptibility to drugs (1, 2). Subsequently, CRISPR screens have been applied to a wide range of fields (Fig. 13.2), such as cancer biology, in which the technology has been used to generate a detailed catalogue of the genetic dependencies of hundreds of cancer cell lines (12, 13) and genetic requirements for metastasis in vivo (14), and cell biology, in which it has been used to identify genes involved in basic cellular processes (15, 16). The flexibility and relative ease of CRISPR screening have led to many expansions on the original format, enabling complex screens for more nuanced phenotypes, enhancing the map of the functions of genetic elements in the genome.

CRISPR screens can be used to identify genetic elements undergoing positive and/or negative selection. In the former, selective pressure on the mutant cell library leads to enrichment of a subset of mutant cells along with their corresponding sgRNAs because most mutant cells are rapidly removed from the population. Since analysis of pooled screens consists of comparing sgRNA abundances at different time points, detecting strongly enriched sgRNAs is relatively straightforward. In contrast, in negative selection, or "dropout screens," selective pressure causes some mutants and sgRNAs to be depleted from the population, but the majority of mutant cells are still present. In addition, due to a reduction in the absolute magnitude of sgRNA abundance and an increased number of hits compared to positive selection screens, negative selection screens are more dependent on appropriate controls and deeper sequencing of sgRNAs to obtain accurate results. Despite these additional

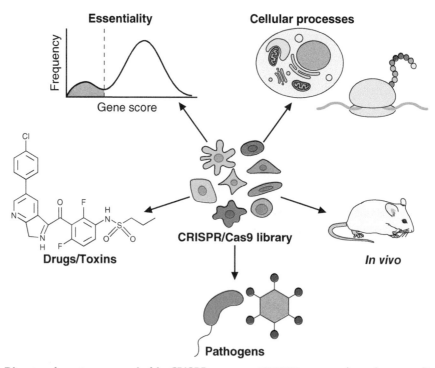

Figure 13.2 Diverse phenotypes queried in CRISPR screens. CRISPR screens have been applied to identify genes involved in a wide range of phenotypes, including genes essential for viability, genes whose inactivation confers resistance to a specific drug or toxin, genes required for pathogen infection of hosts (e.g., receptors) or pathogen life cycles, genes for *in vivo* growth of tumors, and genes for many fundamental cellular processes (e.g., autophagy and nonsense mediated decay).

considerations, this approach has been very successful in identifying essential genes in a variety of cell lines and cancers (12, 17–19), as well as genes with more modest effects on proliferation (20).

The magnitude of selective pressure determines the pace at which phenotypes are enriched or depleted. For example, screens with strong selective pressures, such as those performed with potent drugs, are ideal for positive selection screens. Strong selection leads to rapid changes in the abundance of sgRNAs and high levels of enrichment for true positive hits. However, strong selection may result in false negatives or positives, as mutant cells with weaker phenotypes may not become enriched. Weaker selective pressures can permit the detection of cells with wider ranges of phenotypes and mutant cells that are either enriched or depleted by selection. These screens, though, have additional requirements, including larger starting cell populations, the need for replicate screens, and deeper sequencing of sgRNAs.

Methods for Phenotypic Measurement and Enrichment

Several approaches for phenotypic measurement and enrichment have been employed in CRISPR screens. The methods can be broadly divided into two types, depending on whether or not the approach relies on viability-based selection (Fig. 13.3). Cellular proliferation rates are the default phenotype tracked in most CRISPR screens, since mutant cells in libraries divide over the course of the experiments. sgRNAs that target genes that modulate growth under the specific conditions in which the library is maintained alter the fitness of mutant cells. With several passages, fitness differentials are magnified, leading to enrichment and depletion of different sets of sgRNAs, which are detected via comparison of the sgRNA abundances in the pre- and postpassaged (selected) libraries. This approach has been used to identify essential genes (21) and cancer cell susceptibilities *in vitro* (12, 13) and *in vivo* in mouse models of tumor growth (14, 22), as well as genes conferring susceptibility to various toxins and drugs (1, 2, 23) or cytotoxic pathogens (8, 24).

An elaboration of the viability-based screen that extends the range of phenotypes that can be measured is to engineer a link between a phenotype of interest and cellular viability. For example, Gilbert et al. (25) conducted a CRISPR screen to study the retrograde trafficking of cholera toxin, which is not ordinarily cytotoxic, by utilizing a cholera toxin-diphtheria toxin fusion protein (26). The fusion protein contains a cytotoxic domain from diphtheria toxin, which relies on cholera toxin for internalization, retrotranslocation, and proper subcellular localization in the cytosol to be active. Hence, mutant cells in which trafficking of cholera toxin is aberrant will not be intoxicated and survive, leading to their enrichment in the population. Hits that alter cholera toxin trafficking can be distinguished from those that confer resistance to diphtheria toxin cytotoxicity by monitoring the subcellular localization of the fusion protein in cells, e.g., through fractionation. These approaches allow cellular viability to be used to facilitate the study of complex and subtle aspects of cellular physiology. However, prior to engaging in such screens, the experimental system used to couple a phenotype of interest to cell viability must be rigorously tested; furthermore, screen analysis can be confounded by the selective pressures imparted by the cytotoxic payload.

There are also enrichment methods that do not rely on cell viability. Some of these rely on tools to perform single-cell measurements for selection. For example, fluorescence-activated cell sorting (FACS) has been used to enrich for mutants with specific fluorescence intensities, associated with the phenotype of interest. For study of genes involved in expression of surface proteins, fluorescent-dye-conjugated antibodies specific to proteins of interest can be used to stain cells prior to FACS (27). Alternatively, to study infections and intracellular

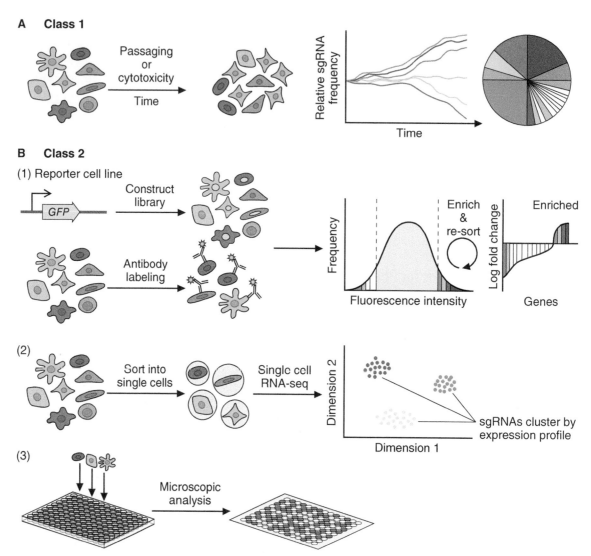

Figure 13.3 Approaches for phenotypic measurement and sgRNA enrichment. (**A**) Class 1 phenotypic measurements rely on viability and growth fitness to enrich for sgRNAs targeting specific genes. In these screens, the abundance of sgRNAs that create mutants with increased viability become enriched, whereas other sgRNAs become depleted. Typically, at the end of the screen, relatively few sgRNAs are high in abundance, dominating the distribution of sgRNAs (pie chart). Reporter cell lines in which a noncytotoxic phenotype is linked to a cytotoxin to create a viability-based screen also fall under this category. (**B**) Class 2 phenotypic measurements do not rely on cellular viability. Screens in this class often utilize single-cell measurements, including FACS-based screens in which mutant libraries are constructed in reporter cell lines in which expression of a particular phenotype is linked with expression of a fluorescent protein or binding of a fluorescent antibody (1). Cells with desired fluorescent intensities (typically very dim or very bright) are enriched via FACS. Enriched cells can be expanded and resorted to allow for multiple rounds of enrichment. After enrichment, sgRNAs of genes involved in the process queried will be enriched (2). Large-scale CRISPR mutagenesis can be coupled with single-cell RNA sequencing. The expression profile of each cell is determined and mutant cells with similar profiles cluster together, permitting identification of novel genes involved in a known process or novel associations between genes (3). Arrayed screening, unlike pooled screening, physically separates mutant cells prior to selection enabling screening of complex phenotypes, such as cell morphology and rapid association of phenotypes with genotypes.

processes, fluorescent invasive pathogens can be used or reporter systems can be created in which activation of a particular pathway leads to expression of a fluorescent protein, so increased activation corresponds to higher fluorescent intensities (28, 29). Technologies similar to FACS, like magnetic screening, have also been used in CRISPR screens (30).

Arrayed libraries, which consist of a set of defined CRISPR mutants, have also been used for screens. Arrayed screens enable immediate linkage of phenotype to genotype, since each well in the library contains cells with a single known sgRNA. Arrayed libraries can permit analyses of phenotypes not easily interrogated using pooled screens (31–33). For example, arrayed libraries have been used to perform microscopy-based screens to study genes required for coxsackievirus infection (33), cellular morphology, and the localization of subcellular complexes (32). This approach also mitigates *trans*-effects that can confound interpretation of pooled screens. In addition, this format may facilitate investigation of supracellular phenotypes such as the effect of gene inactivation on organoid development (34). However, there are barriers to setting up arrayed screens, since they often depend on access to automated robotic plate handlers and high-content, high-throughput imaging systems.

Cell Lines for CRISPR Screens

While initial CRISPR screens were primarily performed in widely used human cancer cell lines (e.g., A375, KBM7, and HL60) and embryonic stem cells (1, 2, 35), the approach was rapidly applied in many other cancer cell lines derived from different organs and cell types, facilitating study of cancer cell line dependencies, drug resistance, and other phenotypes. Recent studies have carried out CRISPR screens in more native contexts, such as human primary cells (36), induced pluripotent stem cells (37), or patient-derived cancer cell lines (38–40). While implementation of CRISPR screening in these contexts can be technically challenging, often requiring delivery of specialized, larger vectors containing both the Cas9 enzyme and sgRNA, the potential benefits can be substantial. Some of these limitations may be surmountable by utilizing smaller Cas9 variants (41, 42), to both decrease vector size and increase the efficiency of vector delivery, or smaller sgRNA libraries, to reduce the number of cells required for library generation.

Screen Analysis

Preexisting bioinformatic pipelines (43, 44) and new dedicated analytic tools (10, 11, 45–47) have been used to analyze the results of CRISPR screens. The goal of these algorithms is to provide quality control measures and combine the results obtained from multiple sgRNAs targeting the same gene into a gene-level score. Many sgRNA libraries include nontargeting, control sgRNAs, which serve as internal controls to assess screen performance and are utilized by some algorithms to determine the statistical significance of screen hits (2, 10). Cursory calculation of fold change values (final abundance/initial abundance) can be helpful, e.g., in positive selection screens, in which relatively few sgRNAs often become highly enriched. However, analytic methods provide statistical metrics enabling robust conclusions and aid in the identification of weakly enriched genes. The full list of screening "hits" can then be input into pathway enrichment algorithms (48–50) to aid in the discovery of cellular processes associated with the phenotype of interest. Hits and the pathways they implicate can also be utilized to construct custom sgRNA libraries targeting a reduced set of genes, potentially with more sgRNAs per gene, to perform secondary CRISPR screens (51, 52). These follow-up smaller-scale screens can improve overall screen accuracy.

Limitations and Improvements of CRISPR Screens

Despite their high efficacy and general reliability, it is important to recognize the limitations of CRISPR screens. High-quality sgRNA design is paramount for obtaining accurate screen results. The sgRNA targeting sequence is the primary determinant of both on- and off-target activities, i.e., the efficacy for generating mutations at the desired site and at other sites that are only partially similar in sequence. Even sgRNAs that target unique sequences can have widely varying efficacies and off-target activity that can confound the interpretation of screen results. To overcome this issue, initial libraries included up to 10 sgRNAs per gene in a library (2). Systematic analyses of sgRNA on- and off-target activity profiles, conducted by studying the effects of thousands of targeting and nontargeting sgRNAs, have led to substantial improvements in the design of sgRNAs with more consistent on-target activity and low off-target activity (2, 11, 53–55). Current genome-scale libraries contain more uniformly acting and specific sgRNAs per gene and can therefore rely on fewer sgRNAs per gene (9).

Even with proper sgRNA design, CRISPR/Cas9 genome editing can be ineffective in unusual cases. CRISPR/Cas9 nucleases create DSBs at the targeted locus that are repaired by nonhomologous end joining (NHEJ), a process that is typically highly mutagenic and produces loss-of-function alleles via the formation of insertions/deletions that result in frameshift mutations and premature stop codons. However, infrequently NHEJ repair may not alter gene activity (e.g., via a small in-frame mutation) or may even yield a gain-of-function allele (56). The formation of numerous DSBs by Cas9 nuclease activity can also activate a strong DNA damage response that results in cell cycle arrest and cell death (57), thus potentially leading to false-positive hits in some screens, particularly for genes with high copy numbers. Computational algorithms can now correct for this gene-independent response to CRISPR genome editing (12, 58–60).

The Cas9 enzyme used for genome editing constrains the potential target sites for editing. Due to the requirement for a PAM sequence to be immediately 3′ of the target sequence, there are limits to the number of targetable sequences in a genome. The *S. pyogenes* Cas9 PAM is NGG, which is found frequently in the human genome (61) but can pose an issue when working in organisms with low GC contents. Cas9 variants with altered PAMs have been engineered or evolved to expand targetable genome space (62–64), and Cas9 enzymes from species other than *S. pyogenes*, which use different PAM sequences, can be used for eukaryotic genome editing (3, 41, 62, 65, 66). Alternatively, additional CRISPR nucleases (e.g., Cas12a/Cpf1) (see chapters 6 and 11) will likely be substitutable for Cas9 in screens (67–71). The intrinsic specificity of different Cas9 enzymes impacts the on- and off-target activities of sgRNAs (64), and improved Cas9 variants have been evolved that have both improved on-target and reduced off-target activities (64, 72).

Expansion of the Scope of CRISPR Screens

Besides CRISPR knockout screens, repurposing Cas9 as a RNA-guided, DNA-binding protein lacking enzymatic activity has led to a major expansion of the scope of CRISPR screens (Fig. 13.4) (see chapter 12). By engineering inactivating mutations in the two nuclease domains of Cas9, an endonuclease dead version (dCas9) that retains its DNA binding, but not its cleaving or nicking activity, was created (73). Other proteins, protein domains, or enzymes can be fused to dCas9, enabling their targeting to specific loci by sgRNAs (74). Initial reports in eukaryotic cells utilized KRAB (Krüppel-associated box) or VP16/VP64 protein domains that inhibit and activate RNA polymerase, respectively, to inactivate (CRISPR inhibition [CRISPRi]) or activate (CRISPR activation [CRISPRa]) transcription of nearby genes, thereby converting CRISPR genome editing technology into a tool for programmable transcriptional regulation (74). Later CRISPRa technologies utilized alternative strategies for recruiting activation domains, including peptide array/SunTag (25) and MS2 RNA aptamer-based (75) approaches. Genome-scale screens with CRISPRi/a offer approaches complementary to CRISPR knockout screens for functional

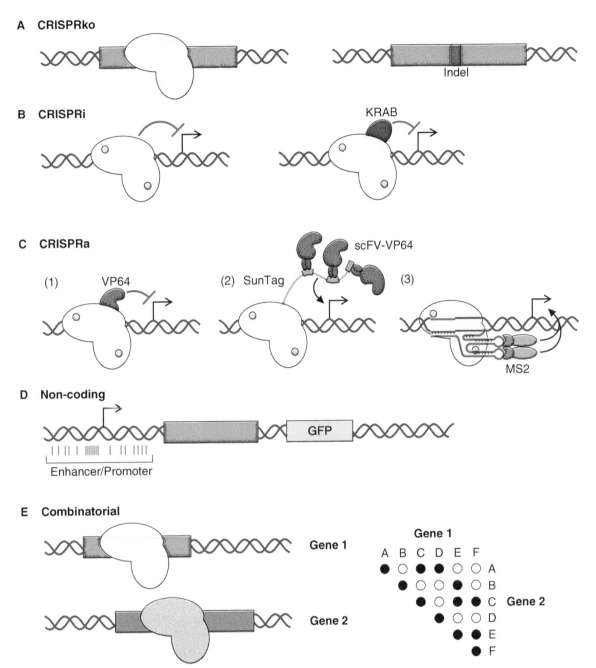

Figure 13.4 Extensions of CRISPR screens. (**A**) Originally, CRISPR genome-editing technology, yielding gene inactivations via generation of insertions/deletions (indels), was used. (**B to E**) Subsequently, development of catalytically inactive Cas9 enzyme (dCas9) enabled new CRISPR-based screens, including the following. (**B**) CRISPR interference (CRISPRi) is used for transcription inhibition through physical exclusion of RNA polymerase (left) or repression of RNA polymerase through fusion of dCas9 with a KRAB domain (right). (**C**) CRISPR activation (CRISPRa) permits overexpression of genes. Several strategies have been developed, including the following: fusing a VP64 domain (a potent, eukaryotic activator that is the minimal activation domain of the herpes simplex virus protein VP16 [131]) to dCas9 (1), fusing dCas9 to a peptide array that is bound by VP64 linked to a cognate single-chain variable fragment (scFv) antibody fragment specific to the epitope on the peptide (the SunTag approach) (25, 132) (2), and encoding an MS2 RNA aptamer on the sgRNA that is recognized by an MS2 protein that is fused to VP64 and other transcriptional activators (e.g., p65 and HSF1) (75) (3). (**D**) The function of the noncoding genome can be studied by tiling a regulatory region, such as an enhancer or promoter with sgRNAs, generating mutating indels via Cas9, and evaluating the effect on expression. (**E**) Combinatorial CRISPR screens provide opportunities to study genetic interactions on a large scale. One option is to utilize orthologous Cas9 enzymes from different organisms, which utilize unique sgRNAs. Genetic interactions can be detrimental (black) or neutral/alleviating (white).

genomics (25, 75). For example, CRISPRi permits investigation of the function of essential genes and gene dosage effects, while CRISPRa is a new, facile approach to study the effects of increased gene expression. Additional modifications of Cas9 include making its activity inducible, via chemical (76–78), genetic (79), or optical (80) means, permitting the study of spatiotemporal phenotypes in screens.

In addition to permitting high-throughput interrogation of protein-coding genes, CRISPR screens have become valuable tools for analyzing the function of the noncoding genome. For example, through construction of saturating sgRNA libraries that tile a particular locus, CRISPR screens have been used to delineate and study the importance of gene regulatory elements (81, 82). High-throughput analysis of another subset of the noncoding genome, long noncoding RNAs (lncRNAs), has also been enabled by CRISPR-based screens. CRISPRi has become a reliable method to study lncRNAs by repressing lncRNA transcription and enabled the first genome-scale screen of lncRNAs (83). The screen revealed that many lncRNAs are essential genetic elements or have growth-modulating properties, but only in specific cell types (83).

Analysis of genetic interactions is another emerging area enabled by CRISPR/Cas9 technology. Combinatorial CRISPR screens that modulate two distinct genetic elements in the same cell provide a means to perform large-scale studies of genetic interactions. Many types of screen designs are feasible due to the existence of various CRISPR screen types, orthologous Cas9 enzymes, and Cas9 multiplexing (84–87). While complete analysis of genetic interactions across the genome would require a prohibitively large mutant cell population size, analysis of gene subsets, e.g., genes involved in specific pathways, is feasible. Combining multiple CRISPR technologies, such as CRISPR/Cas9 with CRISPRa, is also possible (88).

The range of phenotypic measurements performed on mutant cell libraries is growing with technological improvements. As measurement of multidimensional single-cell (sc) phenotypes has become feasible, CRISPR screening followed by single-cell transcriptome sequencing (scRNA-seq) has been accomplished (89–91) using small-scale libraries, as well as coupling to mass spectrometry (92). Additional types of single-cell measurements such as other sequencing-based technologies (e.g., scATAC-seq [single-cell assay for transposase-accessible chromatin using sequencing] [93, 94], scChIP-seq [single-cell chromatin immunoprecipitation sequencing] [95, 96], or single-cell Hi-C [scHi-C] [97]) or metabolomics will likely also be linked with CRISPR screens (Fig. 13.3B) (2). Recent advances in imaging processing and *in situ* sequencing are enabling pooled imaged-based screens, circumventing the need for arrayed libraries to screen for morphological and other complex spatial cellular or subcellular phenotypes (98, 99).

CRISPR genome editing technology has been developed for many organisms, and screening platforms for mice (35), flies (100), *Saccharomyces cerevisiae* (101), and rice (102) have been reported. CRISPR-based screens are also enabling functional genomics in nonmodel organisms. For example, CRISPR screens in the human parasite *Toxoplasma gondii* have uncovered apicomplexan-specific essential genes (103) and hold promise for exploring many understudied organisms and cell lines, such as invertebrate insect cell lines (100). CRISPRi technology is being applied to conduct genome-scale screens in diverse bacteria (104–106) and may aid the study of the numerous exotic species in the human microbiota (107).

Illuminating Bacterial Host-Pathogen Interactions with CRISPR Screens

CRISPR screens have been instrumental in the investigation of host-pathogen interactions. Reviews of applications of CRISPR screens to study viral pathogens have been published recently (108, 109). Here we cover studies of bacterial pathogens. Several aspects of bacterial pathogenesis have been investigated using CRISPR screens, including studies to reveal host genes required for intoxication by exotoxins, the activity of bacterial secretion systems, bacterial attachment and invasion, and innate immune sensing of bacterial products (Fig. 13.5 and Table 13.1).

Exotoxins

Bacterial exotoxins are secreted proteins that can cause pathology in the host and are key virulence factors for numerous pathogens. Cytotoxic toxins, like potent drugs, provide strong selective pressures for screens, enabling the identification of genetic requirements for intoxication. Screens for identifying host targets of toxins published in early 2014 helped to provide proof of concept for CRISPR screens (23, 35). Koike-Yusa et al. (35) developed a genome-scale knockout library targeting all protein-coding genes in a murine cell line and identified genes necessary for intoxication by *Clostridium septicum* α-toxin, which

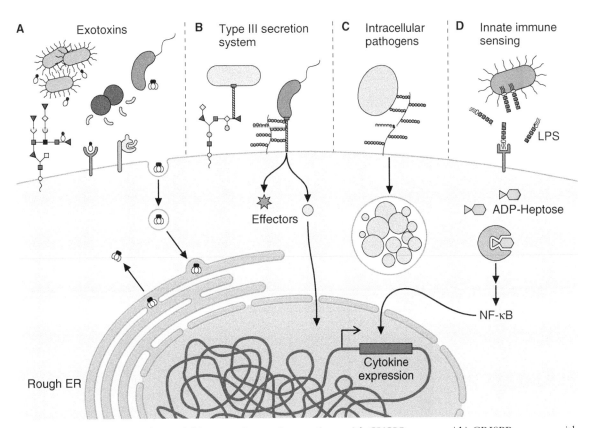

Figure 13.5 Illuminating bacterial host-pathogen interactions with CRISPR screens. (A) CRISPR screens with exotoxins have uncovered receptors for bacterial virulence factors, such as cell surface glycans and proteins, in addition to intracellular genes and pathways required for proper trafficking of toxins that act in the cytosol. **(B)** Bacterial type III secretion systems (T3SSs) translocate secreted effector proteins into host cells. CRISPR screens have uncovered differential requirements for binding and activity of various T3SSs. **(C)** CRISPR screens have revealed new host genetic requirements for pathogen intracellular life cycles, such as genes required for *Chlamydia trachomatis* invasion. **(D)** The innate immune system detects microbial associated molecular patterns (MAMPS), e.g., lipopolysaccharide (LPS) or the recently discovered metabolite ADP-heptose (130), and CRISPR screens have revealed host detection and response systems for these microbial products. CRISPR screens can elucidate the sensing pathways, which trigger activation of the inflammatory response (e.g., cytokine expression) and immune-related genes, often through activation of the transcription factor NF-κB.

Table 13.1 CRISPR screens investigating the interaction of host cells and bacterial pathogens

Immune pathway and pathogen	Gram stain	Virulence Factor	Phenotype	Cell line(s)	CRISPR technology	Top hits	Reference(s)
Exotoxin							
Clostridium septicum	Positive	α-Toxin	Viability	Murine embryonic cells	CRISPR ko	Cell surface receptor	35
Bacillus anthracis	Positive	Lethal factor	Viability	HeLa	CRISPR ko	Cell surface receptors and novel factors required for intoxication	23
Corynebacterium diphtheriae	Negative	Diphtheria toxin	Viability	HeLa	CRISPR ko	Cell surface receptors and novel factors required for intoxication	23
Clostridium difficile	Positive	Toxin A (TcdA)	Viability	HeLa	CRISPR ko	Colonic cell surface receptor	112
		Toxin B (TcdB)	Viability	HeLa	CRISPR ko	Cell surface receptor	113
Staphylococcus aureus	Positive	α-Hemolysin	Viability	U937	CRISPR ko	Cell surface receptor	115
		PVL	Viability	U937	CRISPR ko	Cell surface receptor	116
EHEC/*Shigella*	Negative	Shiga toxins (Stx1/2)	Viability	HT-29, 5637, HeLa	CRISPR ko	Cell surface receptor, glycolipid biosynthesis	117–119
EHEC	Negative	Subtilase toxin	Viability	HeLa	CRISPR ko	Cell surface receptor, glycan biosynthesis	121
Salmonella enterica serovars Paratyphi and Typhi	Negative	Typhoid toxin	Viability	HEK293T	CRISPR ko	Intracellular toxin trafficking	124
Vibrio cholerae	Negative	Cholera toxin	Viability	K562	CRISPRi and CRISPRa	Intracellular toxin trafficking	25
T3SS							
Vibrio parahaemolyticus	Negative	T3SS 1	Viability	HT-29	CRISPR ko	Cell surface receptor	8
		T3SS 2	Viability	HT-29	CRISPR ko	Cell surface receptor	8
EHEC	Negative	T3SS	Viability	HT-29	CRISPR ko	Cell surface receptor	117
Yersinia pestis	Negative	T3SS	Viability	U937	CRISPR ko	Cell surface receptor	126
Intracellular pathogen							
Chlamydia trachomatis	Negative		FACS	HT-29	CRISPR ko	Cell surface receptor, surface expression of GAGs	29
Immune sensing							
Pyroptosis (caspases)	Negative	LPS	Viability	Murine immortalized macrophages	CRISPR ko	Gasdermin D, the mediator of caspase-based pyroptosis	128
Pyroptosis (caspases)	Negative	LPS	Viability	RAW 264.7 murine macrophages	CRISPR ko	Regulators of pyroptosis	129
TNF (TLR)	Negative	LPS	FACS	Primary cells, mouse BMDC	CRISPR ko	TLR signaling pathway and regulators	51
NF-κB (ALPK1)	Negative/positive	ADP-heptose	FACS	HEK293T	CRISPR ko	Innate immune sensor for new bacterial metabolite	130

Abbreviations: FACS, fluorescence-activated cell sorting; CRISPR ko, CRISPR knockout screen; CRISPRi/a, CRISPR interference/activation screens; GAGs, glycosaminoglycans; T3SS, type III secretion system; TNF, tumor necrosis factor; TLR, Toll-like receptor; NF-κB, nuclear factor kappa-light-chain-enhancer of activated B cells; ALPK1, alpha-kinase 1.

contributes to the pathogenesis of myonecrosis. They identified expected genes, i.e., those involved in biosynthesis of glycosylphosphatidylinositol-anchored proteins, which serve as receptors for α-toxin (110), as well as genes involved in other previously unassociated pathways. Zhou et al. (23) carried out a screen for the receptors for the anthrax lethal factor (from *Bacillus anthracis*) and diphtheria (from *Corynebacterium diphtheriae*) toxins. Through use of a focused library targeting ~300 genes, they screened for mutant cells resistant to the two toxins and enriched for sgRNAs mediating the knockout of the known receptors (*ANTXR1/TEM8* and *HBEGF*) in both screens, in addition to novel factors required for cytotoxicity. Notably, due to cell line used, the screens did not enrich for the higher-affinity anthrax receptor gene (*ANTXR2*) (111), highlighting the importance of the specific cell line used for CRISPR screening.

Subsequently, genome-scale CRISPR screening technology was used to identify the receptor for one of the two exotoxins of *Clostridium difficile*, a major nosocomial pathogen. Through screening in a conventional epithelial cell line and extensive follow-up in a more physiologic cell line, organoids, and knockout mice, Tao et al. discovered that proteins of the Frizzled family serve as receptors for the *C. difficile* toxin B (TcdB) in colonic epithelial cells, resolving a long-standing question in the field (112). Recently, the same group has utilized a similar CRISPR knockout screening approach to identify receptors (sulfated glycosaminoglycans and low-density lipoprotein receptor) for *C. difficile* toxin A (TcdA) (113). Knowledge of the receptors for these potent toxins offers a valuable starting point for the development of novel therapeutics.

CRISPR screens have also been carried out to explore the mechanisms of pore-forming toxins. The *Staphylococcus aureus* α-hemolysin is known to use ADAM10 as a receptor (114), but a CRISPR screen by Virreira Winter et al. revealed several novel factors important for intoxication of host cells (115). Most of the previously unlinked genes contributed to intoxication by regulating ADAM10 expression, demonstrating how investigation of a host-pathogen interaction can reveal new aspects of host cell biological processes. A screen for another key *S. aureus* virulence factor toxin, Panton-Valentine leukocidin (PVL), a bicomponent pore-forming protein, identified a new host receptor for PVL. Tromp et al. (116) identified a novel human-specific host receptor for the toxin's pore-forming (F) component and demonstrated that the canonical receptor-binding (S) component and F component must both bind their respective surface receptors for host cell intoxication. These findings overturned the prevailing model of leukocidin function and explained the species specificity of PVL toxicity and the leukocidin's inefficacy in animal models.

The mechanism of action of Shiga toxin (Stx), an exotoxin primarily produced by enterohemorrhagic *Escherichia coli* (EHEC) and some *Shigella dysenteriae* strains, was explored in three recent studies. Each screen was performed in a different cell line, yet all three identified similar genes, demonstrating the robustness of CRISPR technology (117–119). These studies identified common novel genes implicated in cellular intoxication with Stx, whose receptor is known to be a specific glycolipid, glycosphingolipid globotriaosyl-ceramide (Gb3) (120). The genes (*LAPTM4A* and *TM9SF2*), whose functions were largely undefined, were demonstrated to be important for expression of Gb3, acting as a potential activator of a Gb3 biosynthetic enzyme (LAPTM4A) or being important for expression of all glycosphingolipids (TM9SF2). Another screen revealed genes required for intoxication by the EHEC subtilase toxin, which is only produced by a subset of strains, identifying new glycan receptors and a putative zinc transporter (SLC39A9/ZIP9) that serves as a general regulator of glycan biosynthesis (121).

Apart from identifying genes encoding pathways for synthesis of surface receptors, CRISPR screens have also been used to study the intracellular pathways required for toxin trafficking to intracellular targets. Typhoid toxin (TT; also called *Salmonella* cytolethal distending toxin) is an exotoxin produced by *Salmonella enterica* serovars Paratyphi and Typhi. TT is host adapted, binding to surface sialoglycans terminating in *N*-acetylneuraminic acid, which appear to be unique to humans (122). TT causes cytotoxicity by acting as a nuclease, inhibiting cellular proliferation (123). Chang et al. (124) conducted a CRISPR screen for resistance to TT and found that complexes involved in retrograde transport and the endoplasmic reticulum-associated protein degradation were required for proper TT trafficking. Another toxin that relies on similar pathways for intracellular transport is cholera toxin (CT), the cholera pathogen's principal virulence factor. High-resolution investigation of the intracellular pathways required for CT trafficking, which was known to rely on the retrograde pathway for cytosolic delivery, served as the basis for the initial demonstration of the utility of CRISPRi and CRISPRa (25). The complementary CRISPRi and CRISPRa screens recovered numerous genes known to be involved in the early steps of CT entry and trafficking as well as genes encoding specific complexes involved in Golgi transport, explaining previously unknown steps in CT trafficking within the Golgi (125).

Type III Secretion Systems

Many bacterial pathogens utilize type III secretion systems (T3SSs) to transfer proteins, called effectors, directly from the bacterial to the host cell cytoplasm. Inside host cells, effectors modulate diverse processes and can cause cell death. Three CRISPR screens have been performed to discover host cell genes required for T3SS-mediated cytotoxicity to date. Studying *Vibrio parahaemolyticus*, a cause of diarrheal disease arising from consumption of contaminated seafood, Blondel et al. dissected the differences in the host requirements for the cytotoxicity caused by the pathogen's two distinct T3SSs (8). The dual screens revealed that the two T3SSs rely on different host surface structures, heparan sulfated glycosaminoglycans and fucose, respectively, for delivery of effectors into host cells, highlighting the importance of glycans in T3SS activity and in determining pathogen species specificity. The genes identified in the screen primarily illuminated requirements for early steps in cytotoxicity, leaving unanswered the identity of the host partners of the effector proteins. A screen for cytotoxicity caused by EHEC, which encodes a single T3SS, also identified glycans as putative surface structures required for efficient T3SS effector delivery into host cells (117). This screen also simultaneously uncovered requirements for Stx intoxication, as the EHEC strain utilized encoded both the toxin and a hyperactive T3SS. Unexpectedly, both Stx- and T3SS-mediated cytotoxicity relied on TM9SF2 and LAPTM4A, which modulate expression of Gb3, suggesting that T3SS effector translocation depends on the glycolipid that acts as the Stx receptor. Thus, two distinct cytotoxic pathways in the same pathogen appear to rely on the same host factors. Recently, Osei-Owusu et al. performed a CRISPR screen to identify host factors required for intoxication and death of immune cells by the T3SS of *Yersinia pestis*, the causative agent of plague (126). *N*-Formylpeptide receptor (FPR1) was identified as the top hit, and follow-up experiments demonstrated its requirement for T3SS effector translocation and an interaction with the needle cap protein of the *Y. pestis* T3SS, which facilitated bacterial adhesion. Much remains to be explored regarding the interactions between T3SS structural components and effector proteins with host factors. Host genes required for additional pathogen secretion systems (e.g., type IV secretion) that translocate proteins to host cells could also be interrogated with CRISPR screens.

Intracellular Pathogens

Numerous bacterial pathogens have complex intracellular life cycles within host cells. However, only one CRISPR screen exploring host genes required for an obligate intracellular pathogen has been reported. Park et al. (29) studied the host requirements for invasion of *Chlamydia trachomatis*, a common cause of sexually transmitted disease and infectious blindness. Using a fluorescence-activated cell sorting (FACS)-based screen design, mutant host cells resistant to infection with a fluorescent *C. trachomatis* strain were identified. In addition to recovering many genes involved in the biosynthesis of heparan sulfate, a known host *C. trachomatis* receptor (127), the screen also identified components of the coat protein/coatomer complex (COPI). COPI proteins were shown to play roles both in surface expression of heparan sulfate and in promoting efficient *C. trachomatis* T3SS effector translocation. Most COPI proteins are essential (21), suggesting that hypomorphic alleles were created by CRISPR mutagenesis in this screen.

Immune Sensing of Bacteria

CRISPR screens have also yielded exciting insights into the mechanisms of host innate immune sensing of and defense against bacteria. An early application of CRISPR screening was utilized by Shi et al. (128) in a pioneering study to identify mediators of pyroptosis, a cell death pathway triggered by inflammatory caspases. Shi et al. electroporated lipopolysaccharide (LPS), which triggers caspase-11 (mouse)- or caspase-4/5 (human)-dependent pyroptosis, into a murine macrophage line and discovered that along with caspase-11, the previously uncharacterized protein gasdermin D is required for pyroptosis. A parallel screen also in murine macrophages for requirements of caspase-1-mediated pyroptosis, activated by addition of potent agonist (a structural component of the bacterial T3SS) to the library, also identified gasdermin D. These results also helped explain the genetic basis of specific human autoimmune disorders caused by gain-of-function mutations in a gasdermin family member. Working on a similar topic, Napier et al. (129) discovered through a CRISPR screen that the complement pathway is involved in caspase-11 expression and pyroptosis.

Parnas et al. (51) performed a first-of-its-kind screen in primary cells for genes involved in regulating activation of an immune response after LPS stimulation of dendritic cells. The cell mutant library was created in bone marrow-derived dendritic cells extracted from a "Cas9 mouse" that constitutively expressed the Cas9 enzyme. After LPS exposure, FACS was used to enrich for cells with high or low intracellular levels of tumor necrosis factor (TNF), a major inflammatory cytokine. Notably, the authors relied on a secondary, targeted screen based on the top hits from the initial genome-scale screen to reduce false positives and generate a more robust gene list. The screens uncovered many known and unknown genes involved in innate immune sensing of LPS and in the process established a workflow for CRISPR screens in primary cells. Recently, Zhou et al. (130) utilized a FACS and fluorescent reporter-based CRISPR screen to identify the innate immune receptor for a new microbe-associated molecular pattern that they discovered, the bacterial metabolite ADP-heptose, thereby revealing a mechanism for a new form of innate immune sensing. By harnessing the versatility of CRISPR screen designs, much remains to be discovered regarding immune responses to microorganisms.

Final Thoughts

Since their development a mere 5 years ago, CRISPR screens have transformed genetic analysis. The potency, accuracy, and versatility of CRISPR-based tools for genetic screens has ushered in a new era of genome-wide functional genomics. Given the large array of existing and emerging screen designs, CRISPR-based screens still hold untapped potential to further transform and disrupt myriad fields.

Acknowledgments
This study was supported by the National Institute of General Medical Sciences grant T32GM007753 (J.D.D.), HHMI Medical Research Fellowship (J.S.P.), and National Institute of Allergy and Infectious Diseases grant R01-AI-043247 and the HHMI (M.K.W.).

References

1. Shalem O, Sanjana NE, Hartenian E, Shi X, Scott DA, Mikkelson T, Heckl D, Ebert BL, Root DE, Doench JG, Zhang F. 2014. Genome-scale CRISPR-Cas9 knockout screening in human cells. *Science* 343:84–87.

2. Wang T, Wei JJ, Sabatini DM, Lander ES. 2014. Genetic screens in human cells using the CRISPR-Cas9 system. *Science* 343:80–84.

3. Cong L, Ran FA, Cox D, Lin S, Barretto R, Habib N, Hsu PD, Wu X, Jiang W, Marraffini LA, Zhang F. 2013. Multiplex genome engineering using CRISPR/Cas systems. *Science* 339:819–823.

4. Mali P, Yang L, Esvelt KM, Aach J, Guell M, DiCarlo JE, Norville JE, Church GM. 2013. RNA-guided human genome engineering via Cas9. *Science* 339:823–826.

5. Jinek M, Chylinski K, Fonfara I, Hauer M, Doudna JA, Charpentier E. 2012. A programmable dual-RNA-guided DNA endonuclease in adaptive bacterial immunity. *Science* 337:816–821.

6. Hsu PD, Scott DA, Weinstein JA, Ran FA, Konermann S, Agarwala V, Li Y, Fine EJ, Wu X, Shalem O, Cradick TJ, Marraffini LA, Bao G, Zhang F. 2013. DNA targeting specificity of RNA-guided Cas9 nucleases. *Nat Biotechnol* 31:827–832.

7. Smith I, Greenside PG, Natoli T, Lahr DL, Wadden D, Tirosh I, Narayan R, Root DE, Golub TR, Subramanian A, Doench JG. 2017. Evaluation of RNAi and CRISPR technologies by large-scale gene expression profiling in the Connectivity Map. *PLoS Biol* 15:e2003213.

8. Blondel CJ, Park JS, Hubbard TP, Pacheco AR, Kuehl CJ, Walsh MJ, Davis BM, Gewurz BE, Doench JG, Waldor MK. 2016. CRISPR/Cas9 screens reveal requirements for host cell sulfation and fucosylation in bacterial type III secretion system-mediated cytotoxicity. *Cell Host Microbe* 20:226–237.

9. Sanson KR, Hanna RE, Hegde M, Donovan KF, Strand C, Sullender ME, Vaimberg EW, Goodale A, Root DE, Piccioni F, Doench JG. 2018. Optimized libraries for CRISPR-Cas9 genetic screens with multiple modalities. *Nat Commun* 9:5416.

10. Li W, Xu H, Xiao T, Cong L, Love MI, Zhang F, Irizarry RA, Liu JS, Brown M, Liu XS. 2014. MAGeCK enables robust identification of essential genes from genome-scale CRISPR/Cas9 knockout screens. *Genome Biol* 15:554.

11. Doench JG, Fusi N, Sullender M, Hegde M, Vaimberg EW, Donovan KF, Smith I, Tothova Z, Wilen C, Orchard R, Virgin HW, Listgarten J, Root DE. 2016. Optimized sgRNA design to maximize activity and minimize off-target effects of CRISPR-Cas9. *Nat Biotechnol* 34:184–191.

12. Meyers RM, Bryan JG, McFarland JM, Weir BA, Sizemore AE, Xu H, Dharia NV, Montgomery PG, Cowley GS, Pantel S, Goodale A, Lee Y, Ali LD, Jiang G, Lubonja R, Harrington WF, Strickland M, Wu T, Hawes DC, Zhivich VA, Wyatt MR, Kalani Z, Chang JJ, Okamoto M, Stegmaier K, Golub TR, Boehm JS, Vazquez F, Root DE, Hahn WC, Tsherniak A. 2017. Computational correction of copy number effect improves specificity of CRISPR-Cas9 essentiality screens in cancer cells. *Nat Genet* 49:1779–1784.

13. Wang T, Yu H, Hughes NW, Liu B, Kendirli A, Klein K, Chen WW, Lander ES, Sabatini DM. 2017. Gene essentiality profiling reveals gene networks and synthetic lethal interactions with oncogenic Ras. *Cell* 168:890–903.e15.

14. Chen S, Sanjana NE, Zheng K, Shalem O, Lee K, Shi X, Scott DA, Song J, Pan JQ, Weissleder R, Lee H, Zhang F, Sharp PA. 2015. Genome-wide CRISPR screen in a mouse model of tumor growth and metastasis. *Cell* 160:1246–1260.

15. Stewart SE, Menzies SA, Popa SJ, Savinykh N, Petrunkina Harrison A, Lehner PJ, Moreau K. 2017. A genome-wide CRISPR screen reconciles the role of N-linked glycosylation in galectin-3 transport to the cell surface. *J Cell Sci* 130:3234–3247.

16. Yang J, Rajan SS, Friedrich MJ, Lan G, Zou X, Ponstingl H, Garyfallos DA, Liu P, Bradley A, Metzakopian E. 2019. Genome-scale CRISPRa screen identifies novel factors for cellular reprogramming. *Stem Cell Reports* 12:757–771.

17. Mair B, Tomic J, Masud SN, Tonge P, Weiss A, Usaj M, Tong AHY, Kwan JJ, Brown KR, Titus E, Atkins M, Chan KSK, Munsie L, Habsid A, Han H, Kennedy M, Cohen B, Keller G, Moffat J. 2019. Essential gene profiles for human pluripotent stem cells identify uncharacterized genes and substrate dependencies. *Cell Rep* 27:599–615.e12.

18. Wang X, Wang S, Troisi EC, Howard TP, Haswell JR, Wolf BK, Hawk WH, Ramos P, Oberlick EM, Tzvetkov EP, Ross A, Vazquez F, Hahn WC, Park PJ, Roberts CWM. 2019. BRD9 defines a SWI/SNF sub-complex and constitutes a specific vulnerability in malignant rhabdoid tumors. *Nat Commun* 10:1881.

19. Chen L, Alexe G, Dharia NV, Ross L, Iniguez AB, Conway AS, Wang EJ, Veschi V, Lam N, Qi J, Gustafson WC, Nasholm N, Vazquez F, Weir BA, Cowley GS, Ali LD, Pantel S, Jiang G, Harrington WF, Lee Y, Goodale A, Lubonja R, Krill-Burger JM, Meyers RM, Tsherniak A, Root DE, Bradner JE, Golub TR, Roberts CW, Hahn WC, Weiss WA, Thiele CJ, Stegmaier K. 2018. CRISPR-Cas9 screen reveals a MYCN-amplified neuroblastoma dependency on EZH2. *J Clin Invest* 128:446–462.

20. Hart T, Chandrashekhar M, Aregger M, Steinhart Z, Brown KR, MacLeod G, Mis M, Zimmermann M, Fradet-Turcotte A, Sun S, Mero P, Dirks P, Sidhu S, Roth FP, Rissland OS, Durocher D, Angers S, Moffat J. 2015. High-resolution CRISPR screens reveal fitness genes and genotype-specific cancer liabilities. *Cell* 163:1515–1526.

21. Wang T, Birsoy K, Hughes NW, Krupczak KM, Post Y, Wei JJ, Lander ES, Sabatini DM. 2015. Identification and characterization of essential genes in the human genome. *Science* 350:1096–1101.

22. Manguso RT, Pope HW, Zimmer MD, Brown FD, Yates KB, Miller BC, Collins NB, Bi K, LaFleur MW, Juneja VR, Weiss SA, Lo J, Fisher DE, Miao D, Van Allen E, Root DE, Sharpe AH, Doench JG, Haining WN. 2017. In vivo CRISPR screening identifies Ptpn2 as a cancer immunotherapy target. *Nature* 547:413–418.

23. Zhou Y, Zhu S, Cai C, Yuan P, Li C, Huang Y, Wei W. 2014. High-throughput screening of a CRISPR/Cas9 library for functional genomics in human cells. *Nature* 509:487–491.

24. Ma H, Dang Y, Wu Y, Jia G, Anaya E, Zhang J, Abraham S, Choi J-G, Shi G, Qi L, Manjunath N, Wu H. 2015. A CRISPR-based screen identifies genes essential for West-Nile-virus-induced cell death. *Cell Rep* 12:673–683.

25. Gilbert LA, Horlbeck MA, Adamson B, Villalta JE, Chen Y, Whitehead EH, Guimaraes C, Panning B, Ploegh HL, Bassik MC, Qi LS, Kampmann M, Weissman JS. 2014. Genome-scale CRISPR-mediated control of gene repression and activation. *Cell* 159:647–661.

26. Guimaraes CP, Carette JE, Varadarajan M, Antos J, Popp MW, Spooner E, Brummelkamp TR, Ploegh HL. 2011. Identification of host cell factors required for intoxication through use of modified cholera toxin. *J Cell Biol* 195:751–764.

27. Burr ML, Sparbier CE, Chan Y-C, Williamson JC, Woods K, Beavis PA, Lam EYN, Henderson MA, Bell CC, Stolzenburg S, Gilan O, Bloor S, Noori T, Morgens DW, Bassik MC, Neeson PJ, Behren A,

Darcy PK, Dawson S-J, Voskoboinik I, Trapani JA, Cebon J, Lehner PJ, Dawson MA. 2017. CMTM6 maintains the expression of PD-L1 and regulates anti-tumour immunity. *Nature* 549:101–105.

28. DeJesus R, Moretti F, McAllister G, Wang Z, Bergman P, Liu S, Frias E, Alford J, Reece-Hoyes JS, Lindeman A, Kelliher J, Russ C, Knehr J, Carbone W, Beibel M, Roma G, Ng A, Tallarico JA, Porter JA, Xavier RJ, Mickanin C, Murphy LO, Hoffman GR, Nyfeler B. 2016. Functional CRISPR screening identifies the ufmylation pathway as a regulator of SQSTM1/p62. *Elife* 5: e17290.

29. Park JS, Helble JD, Lazarus JE, Yang G, Blondel CJ, Doench JG, Starnbach MN, Waldor MK. 2019. A FACS-based genome-wide CRISPR screen reveals a requirement for COPI in Chlamydia trachomatis invasion. *iScience* 11:71–84.

30. Haney MS, Bohlen CJ, Morgens DW, Ousey JA, Barkal AA, Tsui CK, Ego BK, Levin R, Kamber RA, Collins H, Tucker A, Li A, Vorselen D, Labitigan L, Crane E, Boyle E, Jiang L, Chan J, Rincón E, Greenleaf WJ, Li B, Snyder MP, Weissman IL, Theriot JA, Collins SR, Barres BA, Bassik MC. 2018. Identification of phagocytosis regulators using magnetic genome-wide CRISPR screens. *Nat Genet* 50:1716–1727.

31. Metzakopian E, Strong A, Iyer V, Hodgkins A, Tzelepis K, Antunes L, Friedrich MJ, Kang Q, Davidson T, Lamberth J, Hoffmann C, Davis GD, Vassiliou GS, Skarnes WC, Bradley A. 2017. Enhancing the genome editing toolbox: genome wide CRISPR arrayed libraries. *Sci Rep* 7:2244.

32. de Groot R, Lüthi J, Lindsay H, Holtackers R, Pelkmans L. 2018. Large-scale image-based profiling of single-cell phenotypes in arrayed CRISPR-Cas9 gene perturbation screens. *Mol Syst Biol* 14:e8064.

33. Kim HS, Lee K, Kim S-J, Cho S, Shin HJ, Kim C, Kim J-S. 2018. Arrayed CRISPR screen with image-based assay reliably uncovers host genes required for coxsackievirus infection. *Genome Res* 28:859–868.

34. Gao X, Bali AS, Randell SH, Hogan BLM. 2015. GRHL2 coordinates regeneration of a polarized mucociliary epithelium from basal stem cells. *J Cell Biol* 211:669–682.

35. Koike-Yusa H, Li Y, Tan E-P, Velasco-Herrera MC, Yusa K. 2014. Genome-wide recessive genetic screening in mammalian cells with a lentiviral CRISPR-guide RNA library. *Nat Biotechnol* 32:267–273.

36. Tothova Z, Krill-Burger JM, Popova KD, Landers CC, Sievers QL, Yudovich D, Belizaire R, Aster JC, Morgan EA, Tsherniak A, Ebert BL. 2017. Multiplex CRISPR/Cas9-based genome editing in human hematopoietic stem cells models clonal hematopoiesis and myeloid neoplasia. *Cell Stem Cell* 21:547–555.e8.

37. Li X-L, Li G-H, Fu J, Fu Y-W, Zhang L, Chen W, Arakaki C, Zhang J-P, Wen W, Zhao M, Chen WV, Botimer GD, Baylink D, Aranda L, Choi H, Bechar R, Talbot P, Sun C-K, Cheng T, Zhang X-B. 2018. Highly efficient genome editing via CRISPR-Cas9 in human pluripotent stem cells is achieved by transient BCL-XL overexpression. *Nucleic Acids Res* 46:10195–10215.

38. Hong AL, Tseng Y-Y, Cowley GS, Jonas O, Cheah JH, Kynnap BD, Doshi MB, Oh C, Meyer SC, Church AJ, Gill S, Bielski CM, Keskula P, Imamovic A, Howell S, Kryukov GV, Clemons PA, Tsherniak A, Vazquez F, Crompton BD, Shamji AF, Rodriguez-Galindo C, Janeway KA, Roberts CWM, Stegmaier K, van Hummelen P, Cima MJ, Langer RS, Garraway LA, Schreiber SL, Root DE, Hahn WC, Boehm JS. 2016. Integrated genetic and pharmacologic interrogation of rare cancers. *Nat Commun* 7:11987.

39. Hong AL, Tseng Y-Y, Wala JA, Kim W-J, Kynnap BD, Doshi MB, Kugener G, Sandoval GJ, Howard TP, Li J, Yang X, Tillgren M, Ghandi M, Sayeed A, Deasy R, Ward A, McSteen B, Labella KM, Keskula P, Tracy A, Connor C, Clinton CM, Church AJ, Crompton BD, Janeway KA, Van Hare B, Sandak D, Gjoerup O, Bandopadhayay P, Clemons PA, Schreiber SL, Root DE, Gokhale PC, Chi SN, Mullen EA, Roberts CW, Kadoch C, Beroukhim R, Ligon KL, Boehm JS, Hahn WC. 2019. Renal medullary carcinomas depend upon SMARCB1 loss and are sensitive to proteasome inhibition. *Elife* 8: e44161.

40. MacLeod G, Bozek DA, Rajakulendran N, Monteiro V, Ahmadi M, Steinhart Z, Kushida MM, Yu H, Coutinho FJ, Cavalli FMG, Restall I, Hao X, Hart T, Luchman HA, Weiss S, Dirks PB, Angers S. 2019. Genome-wide CRISPR-Cas9 screens expose genetic vulnerabilities and mechanisms of temozolomide sensitivity in glioblastoma stem cells. *Cell Rep* 27:971–986.e9.

41. Burstein D, Harrington LB, Strutt SC, Probst AJ, Anantharaman K, Thomas BC, Doudna JA, Banfield JF. 2017. New CRISPR-Cas systems from uncultivated microbes. *Nature* 542:237–241.

42. Ran FA, Cong L, Yan WX, Scott DA, Gootenberg JS, Kriz AJ, Zetsche B, Shalem O, Wu X, Makarova KS, Koonin EV, Sharp PA, Zhang F. 2015. In vivo genome editing using Staphylococcus aureus Cas9. *Nature* 520:186–191.

43. Luo B, Cheung HW, Subramanian A, Sharifnia T, Okamoto M, Yang X, Hinkle G, Boehm JS, Beroukhim R, Weir BA, Mermel C, Barbie DA, Awad T, Zhou X, Nguyen T, Piqani B, Li C, Golub TR, Meyerson M, Hacohen N, Hahn WC, Lander ES, Sabatini DM, Root DE. 2008. Highly parallel identification of essential genes in cancer cells. *Proc Natl Acad Sci USA* 105:20380–20385.

44. König R, Chiang CY, Tu BP, Yan SF, DeJesus PD, Romero A, Bergauer T, Orth A, Krueger U, Zhou Y, Chanda SK. 2007. A probability-based approach for the analysis of large-scale RNAi screens. *Nat Methods* 4:847–849.

45. Dai Z, Sheridan JM, Gearing LJ, Moore DL, Su S, Wormald S, Wilcox S, O'Connor L, Dickins RA, Blewitt ME, Ritchie ME. 2014. edgeR: a versatile tool for the analysis of shRNA-seq and CRISPR-Cas9 genetic screens. *F1000 Res* 3:95.

46. Allen F, Behan F, Khodak A, Iorio F, Yusa K, Garnett M, Parts L. 2019. JACKS: joint analysis of CRISPR/Cas9 knockout screens. *Genome Res* 29:464–471.

47. Winter J, Schwering M, Pelz O, Rauscher B, Zhan T, Heigwer F, Boutros M. 2017. CRISPRAnalyzeR: interactive analysis, annotation and documentation of pooled CRISPR screens. *bioRxiv* 2017:109967.

48. Subramanian A, Tamayo P, Mootha VK, Mukherjee S, Ebert BL, Gillette MA, Paulovich A, Pomeroy SL, Golub TR, Lander ES, Mesirov JP. 2005. Gene set enrichment analysis: a knowledge-based approach for interpreting genome-wide expression profiles. *Proc Natl Acad Sci USA* 102:15545–15550.

49. Huang DW, Sherman BT, Zheng X, Yang J, Imamichi T, Stephens R, Lempicki RA. 2009. Extracting biological meaning from large gene lists with DAVID. *Curr Protoc Bioinformatics Chapter 13:Unit 13.11.

50. Mi H, Muruganujan A, Huang X, Ebert D, Mills C, Guo X, Thomas PD. 2019. Protocol update for large-scale genome and gene function analysis with the PANTHER classification system (v.14.0). *Nat Protoc* 14:703–721.

51. Parnas O, Jovanovic M, Eisenhaure TM, Herbst RH, Dixit A, Ye CJ, Przybylski D, Platt RJ, Tirosh I, Sanjana NE, Shalem O, Satija R, Raychowdhury R, Mertins P, Carr SA, Zhang F, Hacohen N, Regev A. 2015. A genome-wide CRISPR screen in primary immune cells to dissect regulatory networks. *Cell* 162:675–686.

52. Doench JG. 2018. Am I ready for CRISPR? A user's guide to genetic screens. *Nat Rev Genet* 19:67–80.

53. Doench JG, Hartenian E, Graham DB, Tothova Z, Hegde M, Smith I, Sullender M, Ebert BL, Xavier RJ, Root DE. 2014. Rational design of highly active sgRNAs for CRISPR-Cas9-mediated gene inactivation. *Nat Biotechnol* 32:1262–1267.

54. Morgens DW, Wainberg M, Boyle EA, Ursu O, Araya CL, Tsui CK, Haney MS, Hess GT, Han K, Jeng EE, Li A, Snyder MP, Greenleaf WJ, Kundaje A, Bassik MC. 2017. Genome-scale measurement of off-target activity using Cas9 toxicity in high-throughput screens. *Nat Commun* 8:15178.

55. Chuai G, Ma H, Yan J, Chen M, Hong N, Xue D, Zhou C, Zhu C, Chen K, Duan B, Gu F, Qu S, Huang D, Wei J, Liu Q. 2018. DeepCRISPR: optimized CRISPR guide RNA design by deep learning. *Genome Biol* 19:80.

56. Donovan KF, Hegde M, Sullender M, Vaimberg EW, Johannessen CM, Root DE, Doench JG. 2017. Creation of novel protein variants

with CRISPR/Cas9-mediated mutagenesis: turning a screening by-product into a discovery tool. *PLoS One* 12:e0170445.

57. Aguirre AJ, Meyers RM, Weir BA, Vazquez F, Zhang C-Z, Ben-David U, Cook A, Ha G, Harrington WF, Doshi MB, Kost-Alimova M, Gill S, Xu H, Ali LD, Jiang G, Pantel S, Lee Y, Goodale A, Cherniack AD, Oh C, Kryukov G, Cowley GS, Garraway LA, Stegmaier K, Roberts CW, Golub TR, Meyerson M, Root DE, Tsherniak A, Hahn WC. 2016. Genomic copy number dictates a gene-independent cell response to CRISPR/Cas9 targeting. *Cancer Discov* 6:914–929.

58. Iorio F, Behan FM, Gonçalves E, Bhosle SG, Chen E, Shepherd R, Beaver C, Ansari R, Pooley R, Wilkinson P, Harper S, Butler AP, Stronach EA, Saez-Rodriguez J, Yusa K, Garnett MJ. 2018. Unsupervised correction of gene-independent cell responses to CRISPR-Cas9 targeting. *BMC Genomics* 19:604.

59. de Weck A, Golji J, Jones MD, Korn JM, Billy E, McDonald ER III, Schmelzle T, Bitter H, Kauffmann A. 2018. Correction of copy number induced false positives in CRISPR screens. *PLOS Comput Biol* 14:e1006279.

60. Wu A, Xiao T, Fei T, Shirley Liu X, Li W. 2018. Reducing false positives in CRISPR/Cas9 screens from copy number variations. *bioRxiv* 2018:247031.

61. Hsu PD, Lander ES, Zhang F. 2014. Development and applications of CRISPR-Cas9 for genome engineering. *Cell* 157:1262–1278.

62. Kleinstiver BP, Prew MS, Tsai SQ, Topkar VV, Nguyen NT, Zheng Z, Gonzales APW, Li Z, Peterson RT, Yeh J-RJ, Aryee MJ, Joung JK. 2015. Engineered CRISPR-Cas9 nucleases with altered PAM specificities. *Nature* 523:481–485.

63. Kleinstiver BP, Prew MS, Tsai SQ, Nguyen NT, Topkar VV, Zheng Z, Joung JK. 2015. Broadening the targeting range of Staphylococcus aureus CRISPR-Cas9 by modifying PAM recognition. *Nat Biotechnol* 33:1293–1298.

64. Hu JH, Miller SM, Geurts MH, Tang W, Chen L, Sun N, Zeina CM, Gao X, Rees HA, Lin Z, Liu DR. 2018. Evolved Cas9 variants with broad PAM compatibility and high DNA specificity. *Nature* 556:57–63.

65. Esvelt KM, Mali P, Braff JL, Moosburner M, Yaung SJ, Church GM. 2013. Orthogonal Cas9 proteins for RNA-guided gene regulation and editing. *Nat Methods* 10:1116–1121.

66. Liu J-J, Orlova N, Oakes BL, Ma E, Spinner HB, Baney KLM, Chuck J, Tan D, Knott GJ, Harrington LB, Al-Shayeb B, Wagner A, Brötzmann J, Staahl BT, Taylor KL, Desmarais J, Nogales E, Doudna JA. 2019. CasX enzymes comprise a distinct family of RNA-guided genome editors. *Nature* 566:218–223.

67. Zetsche B, Gootenberg JS, Abudayyeh OO, Slaymaker IM, Makarova KS, Essletzbichler P, Volz SE, Joung J, van der Oost J, Regev A, Koonin EV, Zhang F. 2015. Cpf1 is a single RNA-guided endonuclease of a class 2 CRISPR-Cas system. *Cell* 163:759–771.

68. Kim HK, Song M, Lee J, Menon AV, Jung S, Kang Y-M, Choi JW, Woo E, Koh HC, Nam J-W, Kim H. 2017. In vivo high-throughput profiling of CRISPR-Cpf1 activity. *Nat Methods* 14:153–159.

69. Liu P, Luk K, Shin M, Idrizi F, Kwok S, Roscoe B, Mintzer E, Suresh S, Morrison K, Frazão JB, Bolukbasi MF, Ponnienselvan K, Luban J, Zhu LJ, Lawson ND, Wolfe SA. 2019. Enhanced Cas12a editing in mammalian cells and zebrafish. *Nucleic Acids Res* 47:4169–4180.

70. Fonfara I, Richter H, Bratovič M, Le Rhun A, Charpentier E. 2016. The CRISPR-associated DNA-cleaving enzyme Cpf1 also processes precursor CRISPR RNA. *Nature* 532:517–521.

71. Liu J, Srinivasan S, Li CY, Ho IL, Rose J, Shaheen MG, Wang G, Yao W, Deem A, Bristow C, Hart T, Draetta G. 2019. Pooled library screening with multiplexed Cpf1 library. *Nat Commun* 10:3144.

72. Kleinstiver BP, Pattanayak V, Prew MS, Tsai SQ, Nguyen NT, Zheng Z, Joung JK. 2016. High-fidelity CRISPR-Cas9 nucleases with no detectable genome-wide off-target effects. *Nature* 529:490–495.

73. Qi LS, Larson MH, Gilbert LA, Doudna JA, Weissman JS, Arkin AP, Lim WA. 2013. Repurposing CRISPR as an RNA-guided platform for sequence-specific control of gene expression. *Cell* 152:1173–1183.

74. Gilbert LA, Larson MH, Morsut L, Liu Z, Brar GA, Torres SE, Stern-Ginossar N, Brandman O, Whitehead EH, Doudna JA, Lim WA, Weissman JS, Qi LS. 2013. CRISPR-mediated modular RNA-guided regulation of transcription in eukaryotes. *Cell* 154:442–451.

75. Konermann S, Brigham MD, Trevino AE, Joung J, Abudayyeh OO, Barcena C, Hsu PD, Habib N, Gootenberg JS, Nishimasu H, Nureki O, Zhang F. 2015. Genome-scale transcriptional activation by an engineered CRISPR-Cas9 complex. *Nature* 517:583–588.

76. Davis KM, Pattanayak V, Thompson DB, Zuris JA, Liu DR. 2015. Small molecule-triggered Cas9 protein with improved genome-editing specificity. *Nat Chem Biol* 11:316–318.

77. Maji B, Gangopadhyay SA, Lee M, Shi M, Wu P, Heler R, Mok B, Lim D, Siriwardena SU, Paul B, Dančík V, Vetere A, Mesleh MF, Marraffini LA, Liu DR, Clemons PA, Wagner BK, Choudhary A. 2019. A high-throughput platform to identify small-molecule inhibitors of CRISPR-Cas9. *Cell* 177:1067–1079.e19.

78. Liu KI, Ramli MN, Woo CW, Wang Y, Zhao T, Zhang X, Yim GR, Chong BY, Gowher A, Chua MZ, Jung J, Lee JH, Tan MH. 2016. A chemical-inducible CRISPR-Cas9 system for rapid control of genome editing. *Nat Chem Biol* 12:980–987.

79. Pawluk A, Amrani N, Zhang Y, Garcia B, Hidalgo-Reyes Y, Lee J, Edraki A, Shah M, Sontheimer EJ, Maxwell KL, Davidson AR. 2016. Naturally occurring off-switches for CRISPR-Cas9. *Cell* 167:1829–1838.e9.

80. Polstein LR, Gersbach CA. 2015. A light-inducible CRISPR-Cas9 system for control of endogenous gene activation. *Nat Chem Biol* 11:198–200.

81. Canver MC, Smith EC, Sher F, Pinello L, Sanjana NE, Shalem O, Chen DD, Schupp PG, Vinjamur DS, Garcia SP, Luc S, Kurita R, Nakamura Y, Fujiwara Y, Maeda T, Yuan G-C, Zhang F, Orkin SH, Bauer DE. 2015. BCL11A enhancer dissection by Cas9-mediated in situ saturating mutagenesis. *Nature* 527:192–197.

82. Sanjana NE, Wright J, Zheng K, Shalem O, Fontanillas P, Joung J, Cheng C, Regev A, Zhang F. 2016. High-resolution interrogation of functional elements in the noncoding genome. *Science* 353:1545–1549.

83. Liu SJ, Horlbeck MA, Cho SW, Birk HS, Malatesta M, He D, Attenello FJ, Villalta JE, Cho MY, Chen Y, Mandegar MA, Olvera MP, Gilbert LA, Conklin BR, Chang HY, Weissman JS, Lim DA. 2017. CRISPRi-based genome-scale identification of functional long noncoding RNA loci in human cells. *Science* 355:321–331.

84. Wong ASL, Choi GCG, Cui CH, Pregernig G, Milani P, Adam M, Perli SD, Kazer SW, Gaillard A, Hermann M, Shalek AK, Fraenkel E, Lu TK. 2016. Multiplexed barcoded CRISPR-Cas9 screening enabled by CombiGEM. *Proc Natl Acad Sci USA* 113:2544–2549.

85. Najm FJ, Strand C, Donovan KF, Hegde M, Sanson KR, Vaimberg EW, Sullender ME, Hartenian E, Kalani Z, Fusi N, Listgarten J, Younger ST, Bernstein BE, Root DE, Doench JG. 2018. Orthologous CRISPR-Cas9 enzymes for combinatorial genetic screens. *Nat Biotechnol* 36:179–189.

86. Han K, Jeng EE, Hess GT, Morgens DW, Li A, Bassik MC. 2017. Synergistic drug combinations for cancer identified in a CRISPR screen for pairwise genetic interactions. *Nat Biotechnol* 35:463–474.

87. Du D, Roguev A, Gordon DE, Chen M, Chen S-H, Shales M, Shen JP, Ideker T, Mali P, Qi LS, Krogan NJ. 2017. Genetic interaction mapping in mammalian cells using CRISPR interference. *Nat Methods* 14:577–580.

88. Boettcher M, Tian R, Blau JA, Markegard E, Wagner RT, Wu D, Mo X, Biton A, Zaitlen N, Fu H, McCormick F, Kampmann M, McManus MT. 2018. Dual gene activation and knockout screen reveals directional dependencies in genetic networks. *Nat Biotechnol* 36:170–178.

89. Dixit A, Parnas O, Li B, Chen J, Fulco CP, Jerby-Arnon L, Marjanovic ND, Dionne D, Burks T, Raychowdhury R, Adamson B, Norman TM, Lander ES, Weissman JS, Friedman N, Regev A. 2016. Perturb-Seq: dissecting molecular circuits with scalable single-cell RNA profiling of pooled genetic screens. *Cell* 167:1853–1866.e17.

90. Datlinger P, Rendeiro AF, Schmidl C, Krausgruber T, Traxler P, Klughammer J, Schuster LC, Kuchler A, Alpar D, Bock C. 2017.

Pooled CRISPR screening with single-cell transcriptome readout. *Nat Methods* **14**:297–301.

91. Jaitin DA, Weiner A, Yofe I, Lara-Astiaso D, Keren-Shaul H, David E, Salame TM, Tanay A, van Oudenaarden A, Amit I. 2016. Dissecting immune circuits by linking CRISPR-pooled screens with single-cell RNA-Seq. *Cell* **167**:1883–1896.e15.

92. Wroblewska A, Dhainaut M, Ben-Zvi B, Rose SA, Park ES, Amir ED, Bektesevic A, Baccarini A, Merad M, Rahman AH, Brown BD. 2018. Protein barcodes enable high-dimensional single-cell CRISPR screens. *Cell* **175**:1141–1155.e16.

93. Buenrostro JD, Wu B, Litzenburger UM, Ruff D, Gonzales ML, Snyder MP, Chang HY, Greenleaf WJ. 2015. Single-cell chromatin accessibility reveals principles of regulatory variation. *Nature* **523**:486–490.

94. Cusanovich DA, Daza R, Adey A, Pliner HA, Christiansen L, Gunderson KL, Steemers FJ, Trapnell C, Shendure J. 2015. Multiplex single cell profiling of chromatin accessibility by combinatorial cellular indexing. *Science* **348**:910–914.

95. Grosselin K, Durand A, Marsolier J, Poitou A, Marangoni E, Nemati F, Dahmani A, Lameiras S, Reyal F, Frenoy O, Pousse Y, Reichen M, Woolfe A, Brenan C, Griffiths AD, Vallot C, Gérard A. 2019. High-throughput single-cell ChIP-seq identifies heterogeneity of chromatin states in breast cancer. *Nat Genet* **51**:1060–1066.

96. Rotem A, Ram O, Shoresh N, Sperling RA, Goren A, Weitz DA, Bernstein BE. 2015. Single-cell ChIP-seq reveals cell subpopulations defined by chromatin state. *Nat Biotechnol* **33**:1165–1172.

97. Ramani V, Deng X, Qiu R, Gunderson KL, Steemers FJ, Disteche CM, Noble WS, Duan Z, Shendure J. 2017. Massively multiplex single-cell Hi-C. *Nat Methods* **14**:263–266.

98. Wang C, Lu T, Emanuel G, Babcock HP, Zhuang X. 2019. Imaging-based pooled CRISPR screening reveals regulators of lncRNA localization. *Proc Natl Acad Sci USA* **116**:10842–10851.

99. Feldman D, Singh A, Schmid-Burgk JL, Carlson RJ, Mezger A, Garrity AJ, Zhang F, Blainey PC. 2019. Optical pooled screens in human cells. *Cell* **179**:787–799.

100. Viswanatha R, Li Z, Hu Y, Perrimon N. 2018. Pooled genome-wide CRISPR screening for basal and context-specific fitness gene essentiality in Drosophila cells. *Elife* **7**: e36333.

101. Bao Z, HamediRad M, Xue P, Xiao H, Tasan I, Chao R, Liang J, Zhao H. 2018. Genome-scale engineering of Saccharomyces cerevisiae with single-nucleotide precision. *Nat Biotechnol* **36**:505–508.

102. Meng X, Yu H, Zhang Y, Zhuang F, Song X, Gao S, Gao C, Li J. 2017. Construction of a genome-wide mutant library in rice using CRISPR/Cas9. *Mol Plant* **10**:1238–1241.

103. Sidik SM, Huet D, Ganesan SM, Huynh M-HH, Wang T, Nasamu AS, Thiru P, Saeij JPJ, Carruthers VB, Niles JC, Lourido S. 2016. A genome-wide CRISPR screen in Toxoplasma identifies essential apicomplexan genes. *Cell* **166**:1423–1435.e12.

104. Cui L, Vigouroux A, Rousset F, Varet H, Khanna V, Bikard D. 2018. A CRISPRi screen in E. coli reveals sequence-specific toxicity of dCas9. *Nat Commun* **9**:1912.

105. Wang T, Guan C, Guo J, Liu B, Wu Y, Xie Z, Zhang C, Xing X-H. 2018. Pooled CRISPR interference screening enables genome-scale functional genomics study in bacteria with superior performance. *Nat Commun* **9**:2475.

106. Lee HH, Ostrov N, Wong BG, Gold MA, Khalil AS, Church GM. 2019. Functional genomics of the rapidly replicating bacterium Vibrio natriegens by CRISPRi. *Nat Microbiol* **4**:1105–1113.

107. Peters JM, Koo B-M, Patino R, Heussler GE, Hearne CC, Qu J, Inclan YF, Hawkins JS, Lu CHS, Silvis MR, Harden MM, Osadnik H, Peters JE, Engel JN, Dutton RJ, Grossman AD, Gross CA, Rosenberg OS. 2019. Enabling genetic analysis of diverse bacteria with Mobile-CRISPRi. *Nat Microbiol* **4**:244–250.

108. Puschnik AS, Majzoub K, Ooi YS, Carette JE. 2017. A CRISPR toolbox to study virus-host interactions. *Nat Rev Microbiol* **15**:351–364.

109. Gebre M, Nomburg JL, Gewurz BE. 2018. CRISPR-Cas9 genetic analysis of virus-host interactions. *Viruses* **10**:55.

110. Gordon VM, Nelson KL, Buckley JT, Stevens VL, Tweten RK, Elwood PC, Leppla SH. 1999. Clostridium septicum alpha toxin uses glycosylphosphatidylinositol-anchored protein receptors. *J Biol Chem* **274**:27274–27280.

111. Liu S, Crown D, Miller-Randolph S, Moayeri M, Wang H, Hu H, Morley T, Leppla SH. 2009. Capillary morphogenesis protein-2 is the major receptor mediating lethality of anthrax toxin in vivo. *Proc Natl Acad Sci USA* **106**:12424–12429.

112. Tao L, Zhang J, Meraner P, Tovaglieri A, Wu X, Gerhard R, Zhang X, Stallcup WB, Miao J, He X, Hurdle JG, Breault DT, Brass AL, Dong M. 2016. Frizzled proteins are colonic epithelial receptors for C. difficile toxin B. *Nature* **538**:350–355.

113. Tao L, Tian S, Zhang J, Liu Z, Robinson-McCarthy L, Miyashita S-I, Breault DT, Gerhard R, Oottamasathien S, Whelan SPJ, Dong M. 2019. Sulfated glycosaminoglycans and low-density lipoprotein receptor contribute to Clostridium difficile toxin A entry into cells. *Nat Microbiol* **4**:1760–1769.

114. Wilke GA, Bubeck Wardenburg J. 2010. Role of a disintegrin and metalloprotease 10 in Staphylococcus aureus alpha-hemolysin-mediated cellular injury. *Proc Natl Acad Sci USA* **107**:13473–13478.

115. Virreira Winter S, Zychlinsky A, Bardoel BW. 2016. Genome-wide CRISPR screen reveals novel host factors required for Staphylococcus aureus α-hemolysin-mediated toxicity. *Sci Rep* **6**:24242.

116. Tromp AT, Van Gent M, Abrial P, Martin A, Jansen JP, De Haas CJC, Van Kessel KPM, Bardoel BW, Kruse E, Bourdonnay E, Boettcher M, McManus MT, Day CJ, Jennings MP, Lina G, Vandenesch F, Van Strijp JAG, Lebbink RJ, Haas PA, Henry T, Spaan AN. 2018. Human CD45 is an F-component-specific receptor for the staphylococcal toxin Panton-Valentine leukocidin. *Nat Microbiol* **3**:708–717.

117. Pacheco AR, Lazarus JE, Sit B, Schmieder S, Lencer WI, Blondel CJ, Doench JG, Davis BM, Waldor MK. 2018. CRISPR screen reveals that EHEC's T3SS and Shiga toxin rely on shared host factors for infection. *mbio* **9**:e01003-18.

118. Tian S, Muneeruddin K, Choi MY, Tao L, Bhuiyan RH, Ohmi Y, Furukawa K, Furukawa K, Boland S, Shaffer SA, Adam RM, Dong M. 2018. Genome-wide CRISPR screens for Shiga toxins and ricin reveal Golgi proteins critical for glycosylation. *PLoS Biol* **16**:e2006951.

119. Yamaji T, Sekizuka T, Tachida Y, Sakuma C, Morimoto K, Kuroda M, Hanada K. 2019. A CRISPR screen identifies LAPTM4A and TM9SF proteins as glycolipid-regulating factors. *iScience* **11**:409–424.

120. Jacewicz M, Clausen H, Nudelman E, Donohue-Rolfe A, Keusch GT. 1986. Pathogenesis of shigella diarrhea. XI. Isolation of a shigella toxin-binding glycolipid from rabbit jejunum and HeLa cells and its identification as globotriaosylceramide. *J Exp Med* **163**:1391–1404.

121. Yamaji T, Hanamatsu H, Sekizuka T, Kuroda M, Iwasaki N, Ohnishi M, Furukawa J, Yahiro K, Hanada K. 2019. A CRISPR screen using subtilase cytotoxin identifies SLC39A9 as a glycan-regulating factor. *iScience* **15**:407–420.

122. Deng L, Song J, Gao X, Wang J, Yu H, Chen X, Varki N, Naito-Matsui Y, Galán JE, Varki A. 2014. Host adaptation of a bacterial toxin from the human pathogen Salmonella Typhi. *Cell* **159**:1290–1299.

123. Miller RA, Betteken MI, Guo X, Altier C, Duhamel GE, Wiedmann M. 2018. The typhoid toxin produced by the nontyphoidal Salmonella enterica serotype Javiana is required for induction of a DNA damage response in vitro and systemic spread in vivo. *mbio* **9**:e00467-18.

124. Chang S-J, Jin SC, Jiao X, Galán JE. 2019. Unique features in the intracellular transport of typhoid toxin revealed by a genome-wide screen. *PLoS Pathog* **15**:e1007704.

125. Wernick NLB, Chinnapen DJ-F, Cho JA, Lencer WI. 2010. Cholera toxin: an intracellular journey into the cytosol by way of the endoplasmic reticulum. *Toxins (Basel)* **2**:310–325.

126. Osei-Owusu P, Charlton TM, Kim HK, Missiakas D, Schneewind O. 2019. FPR1 is the plague receptor on host immune cells. *Nature* **574:**57–62.

127. Rosmarin DM, Carette JE, Olive AJ, Starnbach MN, Brummelkamp TR, Ploegh HL. 2012. Attachment of Chlamydia trachomatis L2 to host cells requires sulfation. *Proc Natl Acad Sci USA* **109:**10059–10064.

128. Shi J, Zhao Y, Wang K, Shi X, Wang Y, Huang H, Zhuang Y, Cai T, Wang F, Shao F. 2015. Cleavage of GSDMD by inflammatory caspases determines pyroptotic cell death. *Nature* **526:**660–665.

129. Napier BA, Brubaker SW, Sweeney TE, Monette P, Rothmeier GH, Gertsvolf NA, Puschnik A, Carette JE, Khatri P, Monack DM. 2016. Complement pathway amplifies caspase-11-dependent cell death and endotoxin-induced sepsis severity. *J Exp Med* **213:**2365–2382.

130. Zhou P, She Y, Dong N, Li P, He H, Borio A, Wu Q, Lu S, Ding X, Cao Y, Xu Y, Gao W, Dong M, Ding J, Wang D-C, Zamyatina A, Shao F. 2018. Alpha-kinase 1 is a cytosolic innate immune receptor for bacterial ADP-heptose. *Nature* **561:**122–126.

131. Seipel K, Georgiev O, Schaffner W. 1992. Different activation domains stimulate transcription from remote ('enhancer') and proximal ('promoter') positions. *EMBO J* **11:**4961–4968.

132. Tanenbaum ME, Gilbert LA, Qi LS, Weissman JS, Vale RD. 2014. A protein-tagging system for signal amplification in gene expression and fluorescence imaging. *Cell* **159:**635–646.

CRISPR-Based Antimicrobials

Justen Russell and David Bikard

Synthetic Biology Group, Department of Microbiology, Institut Pasteur, Paris 75015, France

Introduction

While studying CRISPR-Cas function in prokaryotic immunity against foreign DNA, researchers realized that when programmed to target the bacteria's own chromosome, these systems would kill the cell (1–4). This observation has inspired the idea of using CRISPR-Cas systems as antimicrobials, a technology in stark contrast to current antimicrobial approaches in terms of mechanism and specificity.

The antimicrobial era truly began in antiquity. Skeletal remains suggest not only that the broad-spectrum antimicrobial tetracycline made its way into the 5th-century Roman and Nubian diet but also that its presence there protected against some bacterial infections (5–7). The Romans would not have understood that this protection came from tetracycline-producing *Streptomyces* growing in their grain stores (5) any more than the ancient inhabitants of Irbid and Ar-Ramtha in the northwest of Jordan would have understood that the curative properties of their celebrated Red Mediterranean Soil derived from its abundance of antimicrobial-producing bacteria (8). That understanding, and the industrial production of antimicrobial molecules in the 20th century, underpinned a paradigm shift in health, hygiene, and food production, but also the global emergence of today's multidrug-resistant pathogens (9).

CRISPR-Cas-based technologies may herald another fundamental shift, as they are employed both in support of existing antimicrobials and as a sequence-specific replacement in a world of personalized medicine.

The problem of antimicrobial resistance has been well documented, beginning only shortly after the implementation of the first chemical antimicrobials (9, 10). In a 1948 paper, Mary Barber and Mary Rozwadowska-Dowzenko lamented the increasing number of patients they observed in their practice from whom penicillin-resistant strains of *Staphylococcus* could be isolated (11). Since then, a gradual decline in the rate of development of new antimicrobial families coupled with increasing rates of multidrug resistance in humans and livestock has supported the conclusion that current antibiotic practices are unsustainable (12–15). A 2009 report commissioned by the European Union estimated that 25,000 deaths, 2.5 million hospital days, and an increased €1.5

CRISPR: Biology and Applications, First Edition. Edited by Rodolphe Barrangou, Erik J. Sontheimer, and Luciano A. Marraffini.
© 2022 American Society for Microbiology. DOI: 10.1128/9781683673798.ch14

billion ($1.7 billion in U.S. dollars) in annual costs were a direct result of antimicrobial-resistant bacteria in Europe (16). CRISPR-Cas antimicrobials targeted against antimicrobial resistance genes might provide novel strategies to fight drug-resistant infections and limit the spread of resistance.

Beyond sustainability, an increased understanding of the human microbiota suggests that potential health risks are associated with antimicrobial overuse in human populations (17–19). Microbial dysbiosis, an imbalance in the commensal human microbiota that can result from antimicrobial treatment, has been linked to issues in metabolism (20–22), immune inflammation (23, 24), and even mental health (25–28). Antimicrobial usage massively alters the microbiome, and these disruptions of the microbiota can take years to correct (reviewed in reference 18). Early-life antibiotic usage has been linked to autoimmune disorders later in life, such as inflammatory bowel disease (17, 29, 30). Even antimicrobials with a narrower spectrum of action result in collateral killing of commensal bacteria alongside their intended targets (reviewed in references 18 and 31). Since CRISPR-Cas systems target DNA in a sequence-specific manner, they may enable the removal of specific bacteria from complex communities without affecting the rest of the microbiome.

Multiple groups have proposed using CRISPR-Cas systems designed to target pathogenicity genes, toxins, or antimicrobial resistance genes (32–34). Bacteria deemed problematic could be targeted without causing the dysbiosis that leaves patients susceptible to other pathogens. Such discerning antimicrobials could also, in theory, be used to engineer the microbiome, removing undesirable bacterial strains in favor of others.

There are as many potential applications as there remain barriers to be overcome, but already CRISPR-Cas-based antimicrobials have sparked excitement. In this chapter, we explore CRISPR-Cas antimicrobials, looking at their mechanisms of action, methods of delivery, current limitations, and future prospects.

Mechanisms of CRISPR-Cas Antimicrobials

A CRISPR-Cas antimicrobial consists of two key components: the CRISPR-Cas system, which provides nuclease activity, and the vector to deliver the CRISPR-Cas payload to the target microbe. Delivery systems are discussed below (see "Delivery Systems").

The CRISPR-Cas system is the heart of the CRISPR-Cas antimicrobial. Once introduced into a microorganism, it is responsible for identifying a DNA sequence and degrading it (Fig. 14.1). For bacteria and archaea which already contain a CRISPR-Cas system, this part of the CRISPR-Cas antimicrobial may simply be a guide RNA that directs the host-encoded CRISPR-associated proteins. For microorganisms lacking a functional CRISPR-Cas system, this includes the CRISPR array or guide RNA and all necessary associated genes.

In most cases, once the CRISPR antimicrobial has been introduced, microbicidal activity is dependent on the cleavage of DNA and the inability to repair the damage. If the DNA is in the chromosome, this will result in death of the microorganism, while if the DNA is on a plasmid, it will result in destruction of the plasmid. This may still be lethal if the targeted plasmid carries a toxin-antitoxin system (35). As a CRISPR-Cas antimicrobial that targets an antimicrobial resistance gene will cut the bacterial DNA only if this gene is present, it will only kill antibiotic-resistant bacteria that harbor the gene in their chromosomes. It will not kill any bacteria that do not carry the gene, avoiding collateral damage to the host microbiome.

CRISPR-Cas antimicrobials can be developed using different types of CRISPR systems, each presenting its own unique challenges and advantages, examined below. The differences between CRISPR systems are examined in more detail in chapter 1.

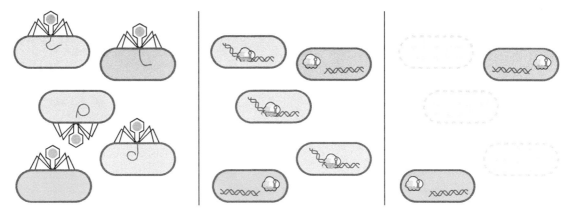

Figure 14.1 CRISPR-Cas antimicrobials can be used to selectively target bacteria that harbor a specific genetic element, i.e., an antibiotic resistance gene or virulence factor. Initially, DNA encoding Cas proteins and one or more guide RNAs targeting the genetic element would be delivered to many bacteria by a phage vector (left diagram) or conjugative plasmid. The newly expressed Cas nuclease would then scan the bacterial genome, cutting only if the targeted DNA is present (central diagram). If unrepaired, this cut results in the death of those bacteria without affecting the remainder (right diagram).

Type I CRISPR Systems

Type I CRISPR systems are the most abundant type of CRISPR systems in bacteria (36). They utilize the Cas3 exonuclease to degrade DNA (Fig. 14.2). Cas3 is recruited at the target site by an RNA-guided effector complex which consists of 3 to 5 proteins present in different stoichiometries (37). Type I systems have often been overshadowed by the single effector protein and clean cutting type II systems in biotechnological applications (38). However, for antimicrobial applications, the processive degradation of the target DNA by Cas3 could present an advantage.

In the typical type I CRISPR system, the multisubunit CRISPR-associated complex for antiviral defense (Cascade) scans and associates with the protospacer adjacent motif (PAM) and facilitates displacement of one strand of DNA in favor of the CRISPR RNA (crRNA) guide. If there is a full match between the crRNA and the DNA, this induces a conformation change in the DNA and cascade forming the R loop state (39, 40). Cas3 is recruited to nick the displaced DNA strand and processively degrades the target DNA, rendering it irreparable (39, 40). If the degraded region is in the chromosome, this typically results in cell death (39, 41). An inducible type I CRISPR system designed to target genomic DNA was used as a strategy of programmable cell death and was able to reduce the number of viable cells 10^8-fold (41).

Type II CRISPR Systems

The type II CRISPR system is simpler than the type I system, requiring only one gene, Cas9, and a single guide RNA (sgRNA) to locate and cleave target DNA. DNA is cut once to produce a clean single double-strand break (42–44). Thus, efficient killing of the bacteria, or curing of a plasmid, relies on this break not being repaired (Fig. 14.2). Provided all copies of the DNA are cut simultaneously, a double-strand break is sufficient to kill most bacteria when the target region is in the chromosome (45). Similarly, if all copies of a plasmid are cut, the plasmid will be destroyed, as homologous recombination and template-assisted repair will not be possible. Differences in the efficiency of guide RNAs have been observed, with weaker guides being unable to cut all chromosome copies simultaneously (45). This results in a series of cleavage and repair events that leads to an SOS response until either the CRISPR-Cas system is inactivated or the target is deleted.

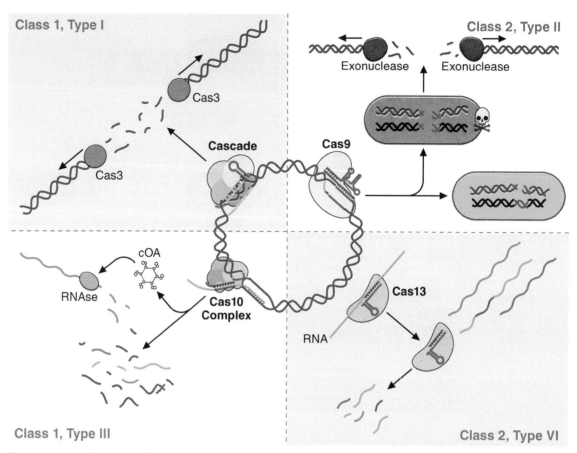

Figure 14.2 Mechanisms of potential CRISPR-Cas antimicrobials. There are a variety of CRISPR-Cas systems that can be utilized in a CRISPR-Cas antimicrobial. In the type I CRISPR system (top left), the multisubunit Cascade scans DNA for the PAM before attempting to displace one strand of DNA with its crRNA guide. If there is a match, the DNA is nicked and then processively degraded by Cas3 and other host proteins, resulting in large DNA gaps and, potentially, cell death if the damage cannot be repaired. In the type II system (top right), DNA is scanned and cut by Cas9. Natively Cas9 utilizes a crRNA guide paired with a tracrRNA (depicted here), though a single sgRNA can be engineered to fill the same purpose. If the RNA guide matches the DNA strand, Cas9 will cleave the DNA. If all copies of DNA are cut at the same time (provided an efficient NHEJ repair system is not present), then degradation of the DNA by host exonucleases will result in cell death. If only some copies of the DNA are cut, it can be repaired and the bacteria will survive the CRISPR-Cas antimicrobial. In the type VI system (bottom right), the effector protein Cas13 scans RNA instead of DNA. After target recognition, Cas13 is activated and degrades all cellular RNA in a nonspecific manner. This results in the arrest of cell growth and cellular dormancy. The type III system scans RNA instead of DNA (bottom left). Following recognition of target RNA by the guide sequence, the Cas10 complex degrades both DNA and RNA. Cyclic oligoadenylate messengers (cOAs) are also produced, leading to the activation of RNases and other antiviral effectors.

In the absence of a repair template, double-strand breaks can be repaired through non-homologous end-joining (NHEJ) systems, which, it is estimated, can be found in close to 25% of bacteria and many eukaryotes (46; reviewed in reference 47). If the blunt end break can be repaired, it will not result in bacterial killing and may introduce small insertions and deletions at the target sequence that will prevent further cutting. This can also be a beneficial outcome if the target gene is inactivated but is problematic when the gene is not. Interestingly, NHEJ repair of Cas9 breaks appears to be inefficient in bacteria that carry NHEJ systems, at least under laboratory conditions (46).

Strategies have been developed to block DNA repair after Cas9 cleavage in the chromosome of bacteria, ensuring that targeted bacteria will be killed. This was accomplished in one study using the Gam protein of phage Mu, which binds the loose DNA ends and prevents repair (45). Introduction of a dominant negative allele of *recA* (*recA56*) can also be used to block homologous recombination (48). Interestingly, the Csn2 protein of type IIA CRISPR-Cas systems, which is usually not included in genetic constructs for technological applications, was shown to block NHEJ repair, presumably through competitive inhibition with Ku (an early NHEJ protein responsible for binding the double-strand break and recruiting a ligase) (46). Other means may need to be found to block the repair pathways in eukaryotic pathogens if a type II system is used.

Other Types of CRISPR Systems

While the majority of CRISPR-Cas antimicrobial strategies reported so far have used type I or type II systems, other CRISPR-Cas systems display unique properties that might be of interest for the technology. For instance, type V systems carry the protein Cas12, which, like Cas9, introduces a double-strand break at the target position. As opposed to Cas9, which relies on a transactivating CRISPR RNA (tracrRNA) and host RNases to process the primary transcript of the CRISPR array into mature crRNAs, the Cas12a nuclease itself processes the primary transcript of the CRISPR array (49). As a result, multiplexing—targeting multiple protospacers at the same time—might be more convenient with nucleases like Cas12a (50).

In some types of CRISPR-Cas systems, the effector complex is guided to bind RNA instead of DNA. This is the case with the type III and type VI systems. In type III systems, recognition of the target RNA leads to the activation of both DNase and RNase activities, resulting in the degradation of both (51–53). Additionally, recognition of the target leads to the production of cyclic oligoadenylate messengers that activate RNases and other antiviral effectors (54–56). These systems have proven to be extremely tolerant to mutations in the target sequence, making them good candidates to limit bacterial escape, with the potential caveat that similar sequences may also be unintentionally targeted (57, 58). Type III systems are also more flexible in the sequences they can target, as they do not rely on a PAM sequence and can thus be guided more easily to any position of interest (59). The complexity of these systems, which rely on 6 or 7 Cas proteins, is a hurdle for their introduction into heterologous hosts, but reprogramming them for self-targeting in bacteria that naturally carry them remains a promising strategy.

Type VI systems carry the effector protein Cas13, which targets exclusively RNA and not DNA thanks to two HEPN (higher eukaryotes and prokaryotes nucleotide-binding) domains (60, 61). After target recognition, Cas13 is activated and degrades cellular RNA in a nonspecific manner, leading to the arrest of cell growth (62). This limits the potential for bacterial escape following Cas13 activation (62). If genomic DNA is cut, there is an opportunity to acquire protective mutations in the target through error-prone repair processes. This is not possible with mRNA degradation. Importantly, as DNA cutting does not provide the mechanism of cell death, bacterial killing does not rely on the presence of the target sequence in the chromosome. Experimental Cas13a-based antimicrobials have reportedly killed bacteria even when the targeted antimicrobial resistance genes were located on plasmids (63).

Nonmicrobicidal Activity

Not all applications of CRISPR-Cas antimicrobials rely on microbial killing. If a targeted DNA sequence is encoded on a plasmid rather than in the bacterial chromosome, the bacteria will survive, provided that no essential genes or antitoxin genes are also carried on the same plasmid. If this is the case, the plasmid can be cured from the bacterial population without any direct impact on the microbial community.

Anti-plasmid CRISPR activity could also be used to prevent the rise of antibiotic resistance. One major consideration with antibiotic resistance genes is horizontal transmission, in which genetic material is spread from one bacterium to another rather than through genetic transmission to daughter cells (64–69). The human microbiome is a repository of antimicrobial resistance genes (69, 70). While exact numbers are hard to discern, resistance genes against most known antibiotics can be found by sequencing the commensal flora. This on its own is not a problem, until these genes are spread into pathogenic or opportunistic species via horizontal gene transfer.

CRISPR-Cas antimicrobial systems could be delivered to microbes which do not currently harbor the target gene. These cells would not be killed but, provided that they retained the CRISPR system, would be "immunized" against acquiring the resistance genes in the future (32, 71). This could prevent the establishment of an antimicrobial resistance reservoir in potentially problematic bacterial communities. An immunized bacterium would encode a functional CRISPR-Cas system and guide RNA capable of degrading an antimicrobial resistance gene, for example, a beta-lactamase gene. It would respond to the future introduction of a beta-lactamase gene as it would to the DNA of any invading bacteriophage, i.e., by destroying it. The same response could also be targeted against virulence genes, thus preventing commensal bacteria from becoming pathogenic.

The genome of *Pseudomonas aeruginosa* strains bearing active CRISPR systems arc, on average, 300 kb smaller than those of strains lacking these loci, which might be explained by the ability of the CRISPR system to prevent the uptake of targeted mobile elements (72, 73). However, the abundance of organisms with active CRISPR systems that encode at least one guide RNA targeting the host chromosome shows that this barrier is not absolute (72, 74) (see "Considerations" below).

Typically, targeting of endogenous mobile genetic elements such as plasmids and prophages results in their removal, but due to strong selective pressure to retain resistance genes, it can also drive the inactivation of the CRISPR system of an immunized bacterium (72). This may be why in *P. aeruginosa*, despite preventing the acquisition of numerous mobile genetic elements, the presence of active CRISPR systems in genomes is not negatively correlated with the presence of antibiotic resistance genes that are typically spread via such mobile elements (73). In many cases, techniques to introduce additional selective pressure to maintain the CRISPR antimicrobial system may be needed to ensure that microorganisms remain "immunized." Moreover, this is not a problem limited to "immunized" microorganisms. Serious consideration will need to be given to the problem of promoting the growth of antibiotic-susceptible CRISPR-Cas treated microbes over antibiotic-resistant clones. After all, a single usage of an antibiotic could completely restore resistance by wiping out the susceptible population.

To circumvent this issue, some have proposed pairing the CRISPR-Cas antimicrobial with resistance against a second threat (34). For instance, a phage modified to carry the target sequence can be used to kill bacteria that either have not received or have lost the CRISPR system (34). In this study, a CRISPR array carrying spacers targeting beta-lactamase genes was delivered to *Escherichia coli* using phage lambda with the purpose of generating lysogens that would cure plasmids carrying the resistance genes. Subsequent challenge with a modified phage T7 carrying the target sequences eliminated those bacteria lacking the prophage, leaving only antibiotic-susceptible but phage-resistant clones (34).

Delivery Systems

The second part of any CRISPR-Cas antimicrobial is the delivery system. In eukaryotic cells, CRISPR-Cas systems have been delivered as plasmid DNA, as mRNA ready for translation, and even as guide RNA-Cas ribonucleoprotein complexes (75, 76). So far, only the delivery of plasmid DNA has been explored in bacteria. This plasmid could consist of a lone guide RNA capable of directing endogenous Cas proteins, or it could carry an array of Cas genes constituting the entire CRISPR-Cas system.

In principle, it would be ideal to efficiently deliver the CRISPR-Cas antimicrobial to all bacteria in the microbiome, resulting in the complete elimination of the targeted genetic element and "immunization" of the rest of the population. There is currently no known vector capable of introducing DNA into the majority of intestinal bacteria, let alone all human-associated microorganisms. In fact, some level of specificity of the delivery vector will likely be an important feature in order to avoid diluting the payload among too many unnecessary bacteria and to avoid off-target effects.

Therefore, CRISPR-Cas antimicrobials will use purpose-designed delivery vectors, delivering the DNA to a subset of bacteria in a specified niche. The requirements of a delivery method will depend on the local environment of the target microorganism and its traits. Targeting skin bacteria with topical formulations might be easier than targeting the complex, and difficult-to-reach, gut microbiome. While most intestinal bacteria live in the lumen, many are embedded in the mucus barrier or adherent to the epithelium, while some even hide inside cells. On the other hand, the skin microenvironment has its own complexities, such as hair follicles and abscesses where bacteria may hide. Different requirements will be demanded of a delivery vector that must survive the stomach to enter the upper intestine, penetrate bacterial biofilms in the colon, or reach the inner glands of a hair follicle. In the case of infection, delivery vectors may need to be capable of circulating through veins, arteries, lymph nodes, and various barriers to reach bacteria deep in infected tissue.

Conjugative Plasmids

One method proposed has been the use of conjugative plasmids (35, 77). These are replicating plasmids capable of self-transmission, carrying genes that enable their host bacterium to connect and transmit genetic material to neighboring bacteria. The guide RNA, with or without the accompanying CRISPR-Cas system, would be encoded along with conjugation elements on a single plasmid and transformed into a bacterial carrier. Bacteria carrying the plasmid would then be used to deliver the CRISPR-Cas antimicrobial to sister cells. Topical application of the bacteria could result in conjugation with skin flora, while their inclusion in a pill or suppository could allow the bacteria to reach the intestinal microbiome.

This method has benefits, as the conjugative carrier could persist on its own and may naturally migrate to the proper niche in the microbiome. For instance, intestinal bacteria carrying the plasmid may establish themselves in the small intestine alongside their conjugative target. As the plasmid spreads, bacteria harboring the target gene, for example, a beta-lactamase or enterotoxin gene, would be killed or cured of the undesired genetic element. Recipient bacteria that do not contain the target gene would, in turn, become carriers and, through conjugation, would spread the plasmid further. To encourage bacteria to retain the plasmid, they could contain a gene conferring a selective advantage; for example, it could provide protection against a bacteriophage (discussed above in "Nonmicrobicidal Activity").

While conjugation can occur between very distant bacterial species, making it an interesting strategy to develop CRISPR antimicrobials against a wide range of target species, this is not typically an efficient process. Efficiency can be further reduced by host factors; secretions from some intestinal epithelial cells have been observed to reduce conjugation efficiency 2-fold in nearby bacteria (78). When the goal is to eradicate or decolonize specific bacteria, it is critical to deliver the CRISPR-Cas system to the vast majority of target cells. Despite the limited efficiency of this strategy, if the donor strain can be maintained for a sufficient amount of time, it might be able to progressively drive out the target bacteria by applying a constant selective pressure to eliminate them. Nonetheless, inoculation of new bacteria into an established microbiome is not an easy prospect. Bacteria carrying the conjugative plasmid may struggle to establish themselves or to spread the conjugative plasmid to difficult-to-reach bacteria, such as those at the center of a biofilm.

Bacteriophages

The most common CRISPR-Cas antimicrobial design envisions the use of bacteriophage-derived delivery vectors. A bacteriophage consists of viral proteins, sometimes a phospholipid membrane, and viral nucleic acid. As viruses, they have no metabolic activity on their own. To reproduce they must find, bind, and deliver this genetic material into susceptible microbial hosts, which are then co-opted to produce more phages. In nature, bacteriophages have been known to carry their own CRISPR-Cas systems complete with arrays targeting other bacteriophages or bacteriophage defense systems (79–83). Whether directly co-opted from their hosts or acquired by horizontal gene transfer from another bacteriophage, these systems are active and can acquire new guide RNA (79–81). However, most synthetic CRISPR-Cas antimicrobials would replace replicating phage DNA with a synthetic plasmid containing a custom CRISPR-Cas system (Fig. 14.3). Such plasmids, also known as phagemids, encode a packaging signal enabling encapsulation in phage virions when cultured with a helper phage. The encapsulated phagemid is then transmitted into a susceptible host, where it replicates as a plasmid. This is a terminal state; without coinfection by the helper phage, the plasmid does not encode virion genes and is unable to transmit further.

Phage-inducible pathogenicity islands (PICIs), which also rely on a helper phage, have been suggested as an alternative to phagemids in the delivery of CRISPR-Cas antimicrobials (84). In nature, PICIs are genetic regions that encode a phage packaging signal and recombinases which integrate these mobile genetic elements into the host chromosome (85). During lytic phage infection, the entire genetic island is packaged into the virion of a helper phage and transmitted in a manner similar to that of a phagemid. After transmission, it then integrates into the genome of the target host. Though currently limited to bacterial species from which PICIs can be identified, in *Staphylococcus aureus*, replacing existing pathogenicity genes with CRISPR-Cas antimicrobial genes to generate a CRISPR-Cas PICI resulted in an antimicrobial that was more stable and transmitted at a higher efficiency than a comparable phagemid (84).

Phage Host Range

Host range is a current limitation of bacteriophage-derived delivery vectors. No single phage can target more than a narrow cohort of bacteria or archaea (reviewed in reference 86). While this limits the spread of resistance and off-target effects, this also means that a large number of bacteriophages need to be identified and developed as vectors, a difficult and often labor-intensive task.

To circumvent this problem, significant effort has been focused on modifying and expanding existing phage host ranges. Phages naturally acquire mutations, and it has been demonstrated that modest selective pressure can shift phage receptor binding. For example,

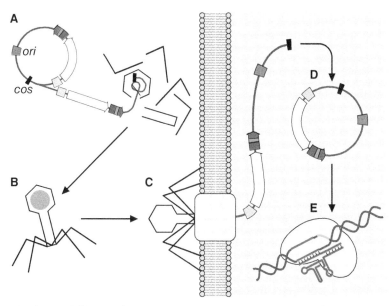

Figure 14.3 Bacteriophage delivery of CRISPR-Cas antimicrobials. Bacteriophages are a potential delivery vector of CRISPR-Cas antimicrobials. They may be used to deliver guide RNA, simple CRISPR arrays, or an entire CRISPR-Cas system to target bacteria. This DNA is encoded on a phagemid (**A**) containing a cohesive end site (*cos*) for phage packaging. The phagemid is replicated and packaged in forming viral particles (**B**) instead of the phage genome. This phagemid can then be delivered by the phage capsid (**C**), whereupon it recircularizes and replicates as a plasmid (**D**) that expresses the encoded CRISPR-Cas genes (**E**), leading to bacterial killing, plasmid curing, or immunization.

under low-glucose conditions, *E. coli* limits expression of LamB, the canonical receptor for phage lambda. These conditions lead to the rise of phage mutants that instead bind to OmpF, an alternative outer membrane protein (87–89). To improve on this concept, several approaches have been devised that pair either random or directed mutagenesis with high-throughput screening to select phage variants that have similarly "evolved" to bind to new proteins or hosts.

Many bacteriophages of the order *Caudovirales* initially bind their bacterial hosts using thin projections called tail fibers. One approach to expand phage host ranges has been to replace the tail fibers of one bacteriophage with those of another (90–92) (Fig. 14.4A).

By replacing the tail fibers of T7 with the tail fibers of related bacteriophages, one group was able to shift the host range from *E. coli* to *Klebsiella* or *Yersinia pseudotuberculosis* (92). Tail fiber/component swapping would greatly improve the effectiveness of many phage vectors (91).

Yosef and colleagues proposed a directed evolution setup in which rather than evolving the entire phage, a phagemid encoding a library of tail fiber variants would be paired with a helper phage lacking tail fiber genes (90). The resulting phages, identical except in their tail fibers, could then be tested on a range of bacteria. An antibiotic resistance gene carried on the phagemid allows for the selection of successful transformants. In this manner, mutations in the tail fiber genes that would extend the host range for phagemid transduction can be identified even if the target bacteria carry defense mechanisms, such as abortive infection systems, that would block successful phage replication (90).

Other means to alter a phage host range include attempts to identify the core sites in the tail fiber involved in host receptor recognition and modify only those (93–95). Taking inspiration from antibody recombination, in which a large number of unique ligands can be targeted following recombination of the relatively limited immunoglobulin fold, Yehl et al.

identified four distal loops in the tail fiber of bacteriophage T3 (94, 95). They found that by randomizing the amino acids of these nonstructured loops, they could isolate bacteriophage clones capable of targeting previously unsusceptible bacterial mutants lacking the canonical phage receptor (94). Such nonstructured amino acid loops near attachments sites, whether on tail fibers or other binding proteins, are ideal candidates for mutagenesis to maximize receptor diversity while preserving the overall protein structure (93–95) (Fig. 14.4B).

Despite these techniques, in some situations it might be impossible to identify receptor mutants that can bind a desired host. This would be the case for bacteria that express exopolysaccharide capsules that mask receptors. Some phages carry tail fibers with capsule-degrading domains, enabling phages to infect encapsulated bacteria. These domains could be fused to the tail fibers of phages lacking such activity.

Scientists have been able to modify the host range of phages through evolution or engineering, sometimes jumping from one species to another, but the novel species targeted are usually close relatives of the original bacterial host. Changing the host of a phage between distant bacterial clades, such as from a member of *Proteobacteria* to a member of *Firmicutes*, remains a challenge beyond our reach and might require modifications to the whole injection apparatus.

Clinical Usage of Bacteriophages

Bacteriophages have been explored as therapeutic agents since 1919; prior to the dawn of the antibiotic era, phages were produced commercially in Europe and North America (96, 97). Bacteriophage therapy continued in Eastern Europe and the USSR throughout the Cold War, but research into clinical usage of phages was largely absent outside of the Communist

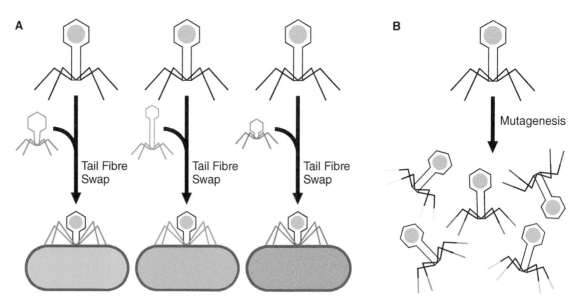

Figure 14.4 Altering bacteriophage host range. The host specificities of some bacteriophages are determined by their tail fibers. These important projections interact with and bind to lipopolysaccharides or proteins on the bacteria to facilitate infection. One method of altering phage host range is therefore to exchange the tail fibers with those of a related bacteriophage with a different host range (**A**). One well-characterized phage vector can be modified this way to deliver its DNA payload without having to completely engineer a novel viral vector for each new host. Another method for expanding host range relies on targeted mutagenesis to the tail fibers (or other bacterium-binding proteins) (**B**). Increasing the mutation rate of amino acid residues at the distal ends of the tail fiber can generate a library of bacteriophages to select from, some of which may have new host tropisms.

Bloc (96–98). Nonetheless, bacteriophage therapy has seen a reemergence of interest worldwide, with a large number of studies in recent years.

While phage therapy relies on phage infection and killing of microbial targets without the delivery of a CRISPR-Cas cassette, it provides important information for the clinical use of phagemid delivery vectors, such as the safety of application, and even intravenous injection, of bacteriophage particles, their expected half-life, and the resulting human immune response.

There is a large endemic population of bacteriophages in the human microbiome (99, 100). Though they do not infect human cells, bacteriophages are immunogenic. Bacteriophages are targeted by both cellular and humoral immune responses, including by bacteriophage-specific antibodies (101, 102). This antibody recognition does not appear to prevent large numbers of phage from binding to bacteria in infected tissue but does lead to opsonization, destruction, and immune signaling (103).

Macrophages also target bacteriophages, and when they phagocytose bacteriophages, they release a mix of immune signaling molecules called cytokines (99, 101). It was originally thought that the cytokines released were predominantly inflammatory, but this seems to have been a response to bacterial endotoxin contamination in the early experiments that originated from the bacterial host used to propagate the phages (99). Studies using purified bacteriophages lacking bacterial endotoxin have shown that bacteriophages alone provoke a weak antiviral immune response (104).

It is not known how this response can impact the ongoing microbial infection. If bacteriophages bias the immune response too strongly towards an antiviral response, this may impair the antimicrobial response (105). Indeed, it has been noted that some bacteriophages increase the rate of bacterial infiltration into tissue (99, 106).

Practical consideration must also be given to the effect the immune response has on bacteriophage availability. The majority of bacteriophages injected into mice are neutralized within minutes (101). This might be less of a hurdle for CRISPR-Cas antimicrobials delivered to the skin, gut, or other mucosal surfaces. As bacteriophages approach clinical usage, it will be imperative that researchers continue to examine the immunogenicity and half-life of bacteriophages in these settings as well.

In spite of these limitations, bacteriophages remain the best currently available vector for the delivery of CRISPR-Cas antimicrobials.

Considerations

Despite their promise, it is important to remember that CRISPR-Cas antimicrobials are not a panacea. There remain numerous hurdles to overcome before CRISPR-Cas antimicrobials can be used medically, and good design will be necessary in combination with good use practices to ensure that resistance does not develop against CRISPR-Cas antimicrobials themselves.

Resistance against CRISPR-Cas Antimicrobials

There is intense selective pressure to evolve resistance against any antimicrobial. CRISPR-Cas antimicrobials are no exception. Bacteria in nature can survive CRISPR targeting of their chromosomes. Indeed, self-targeting CRISPR guide RNAs are sometimes captured "by mistake" and can be found in CRISPR arrays of some bacteria (72, 107, 108). An early survey of genomes found self-targeting guide RNA in 18% of CRISPR-bearing organisms (109) Similar to a CRISPR-Cas antimicrobial, these self-targeting CRISPR systems should kill the bacteria (107, 108). What rescues these bacteria? In most cases, mutations are found in Cas genes, repeat sequences, guide RNA, or the target DNA that abrogate the CRISPR response (109). Similar mutations can inactivate CRISPR-Cas antimicrobials. Importantly, mutations in the CRISPR-Cas systems, while enabling the bacteria to survive, do not render the bacteria resistant to the CRISPR-Cas antimicrobial. Delivery of an intact CRISPR-Cas system would still be able to kill the target microbe.

In contrast, mutations in the target sequence make the bacteria resistant. One solution is to incorporate multiple guide RNAs to target several points in the same gene to ensure that a single mutation cannot prevent killing and to carefully design the delivery vector with the target bacteria in mind.

Apart from mutations in the target sequence, the recipient bacteria might carry anti-CRISPR proteins that can block the activity of Cas genes (110). Indeed, to evade bacterial CRISPR systems, many phages have genes that suppress or interfere with host CRISPR systems. There are many unique families of anti-CRISPRs (reviewed in references 77, 111, and 112), each with unique mechanisms of action. While we do not go into each here (see chapter 9), horizontal gene transfer between phages and bacteria is well documented, and in fact, bacteria which express anti-CRISPR genes have been identified (110). Bacteria targeted by CRISPR-Cas-based antimicrobials would face selective pressure to acquire and retain these genes. Again, designs to overcome this problem can be inspired by nature. Many bacteria have developed ways to counter anti-CRISPR proteins, including encoding two unique CRISPR systems (reviewed in reference 113).

Whether delivered by conjugative plasmids or phage capsids, the vectors carrying CRISPR-Cas systems will have to face host defenses against foreign DNA. Restriction-modification (RM) systems are maybe the most prevalent defense strategy against foreign DNA, with, on average, 2 RM systems per bacterial genome (114). These systems recognize specific sequences and cleave DNA unless the DNA is modified for protection, typically by methylation. In order to bypass these defenses, two strategies can be employed, either modifying DNA sequences so that they escape recognition by the restriction nucleases or eliminating targeted sequences altogether from the vectors.

Additionally, some bacteria carry xenogeneic silencing proteins which form nucleo-protein complexes on foreign DNA, for instance, by selectively targeting AT-rich DNA, and blocking its expression (115). Such systems might inhibit the expression of an anti-microbial CRISPR-Cas system but should be easily bypassed by ensuring that the vector sequence has characteristics (GC content, for example) similar to those of the host bacterial genome sequence.

What To Target

While CRISPR-Cas antimicrobials are highly discerning, sequence-specific killers, there is an abundance of unique antimicrobial resistance genes. For example, what would be the efficiency of targeting one beta-lactamase gene if there are more than 100 unique beta-lactamases remaining in the microbial population?

If the single beta-lactamase gene targeted is the only one used by a particularly virulent pathogen, targeting that gene alone will be sufficient. For other uses, it will be necessary to identify the minimum set of target sequences that covers the maximum number of variants of the resistance gene (33).

Another approach has been to look at conserved sequences found in common pathogenicity islands or on plasmids that bear pathogenicity factors or antimicrobial resistance genes (33). By targeting DNA sequences that are not part of any particular pathogenicity gene but are found only adjacent to pathogenicity genes, a large number of problematic genes can be purged from the microbiome at once (33).

Future Prospects

CRISPR-Cas based antimicrobials are in their infancy, and their impact on human and animal health remains to be determined. At this time, there are no currently available CRISPR-Cas antimicrobials for clinical usage, though at least four companies have been formed with the purpose of bringing the technology to market.

A number of studies have offered proof that, at least under controlled experimental conditions, CRISPR-Cas antimicrobials are capable of selectively killing pathogenic bacteria (32, 35) and targeting bacteria that harbor antibiotic resistance genes (33, 71).

This chapter has not, for the most part, addressed CRISPR-Cas antimicrobials designed to target fungal or protozoan pathogens. This is due not to inherent limitations of CRISPR-Cas systems but merely to a paucity of currently available research. CRISPR-Cas systems have been widely used in eukaryotic cells for genome editing purposes, and the systems can be adapted to work in a wide range of parasites (116–119). There will, however, be new challenges to address, including substantially different DNA repair mechanisms and completely new methods of DNA delivery.

We hope that in the coming decades researchers will identify novel solutions to the problems mentioned above, leading to the design of CRISPR-Cas antimicrobials that can efficiently be delivered to any microbe of interest in a complex ecosystem and ensuring that CRISPR-Cas antimicrobials will be safe and effective once they are administered.

Acknowledgment

David Bikard is a shareholder and Chief Scientific Officer of Eligo Bioscience.

References

1. Edgar R, Qimron U. 2010. The Escherichia coli CRISPR system protects from λ lysogenization, lysogens, and prophage induction. *J Bacteriol* **192**:6291–6294.

2. Bikard D, Hatoum-Aslan A, Mucida D, Marraffini LA. 2012. CRISPR interference can prevent natural transformation and virulence acquisition during in vivo bacterial infection. *Cell Host Microbe* **12**:177–186.

3. Gomaa AA, Klumpe HE, Luo ML, Selle K, Barrangou R, Beisel CL. 2014. Programmable removal of bacterial strains by use of genome-targeting CRISPR-Cas systems. *mBio* **5**:e00928-13.

4. Vercoe RB, Chang JT, Dy RL, Taylor C, Gristwood T, Clulow JS, Richter C, Przybilski R, Pitman AR, Fineran PC. 2013. Cytotoxic chromosomal targeting by CRISPR/Cas systems can reshape bacterial genomes and expel or remodel pathogenicity islands. *PLoS Genet* **9**:e1003454.

5. Cook M, Molto E, Anderson C. 1989. Fluorochrome labelling in Roman period skeletons from Dakhleh Oasis, Egypt. *Am J Phys Anthropol* **80**:137–143.

6. Bassett EJ, Keith MS, Armelagos GJ, Martin DL, Villanueva AR. 1980. Tetracycline-labeled human bone from ancient Sudanese Nubia (A.D. 350). *Science* **209**:1532–1534.

7. Nelson ML, Dinardo A, Hochberg J, Armelagos GJ. 2010. Brief communication: mass spectroscopic characterization of tetracycline in the skeletal remains of an ancient population from Sudanese Nubia 350–550 CE. *Am J Phys Anthropol* **143**:151–154.

8. Falkinham JO, III, Wall TE, Tanner JR, Tawaha K, Alali FQ, Li C, Oberlies NH. 2009. Proliferation of antibiotic-producing bacteria and concomitant antibiotic production as the basis for the antibiotic activity of Jordan's red soils. *Appl Environ Microbiol* **75**:2735–2741.

9. Levy SB. 1982. Microbial resistance to antibiotics. An evolving and persistent problem. *Lancet* **ii**:83–88.

10. Aminov RI. 2010. A brief history of the antibiotic era: lessons learned and challenges for the future. *Front Microbiol* **1**:134.

11. Barber M, Rozwadowska-Dowzenko M. 1948. Infection by penicillin-resistant staphylococci. *Lancet* **ii**:641–644.

12. Woolhouse M, Farrar J. 2014. Policy: an intergovernmental panel on antimicrobial resistance. *Nature* **509**:555–557.

13. Woolhouse M, Waugh C, Perry MR, Nair H. 2016. Global disease burden due to antibiotic resistance—state of the evidence. *J Glob Health* **6**:010306.

14. European Food Safety Authority, European Centre for Disease Prevention and Control. 2019. The European Union summary report on antimicrobial resistance in zoonotic and indicator bacteria from humans, animals and food in 2017. *EFSA J* **17**:e05598.

15. O'Neill J. 2014. *Antimicrobial Resistance: Tackling a Crisis for the Health and Wealth of nations. Review on Antimicrobial Resistance.* Wellcome Collection, London, United Kingdom.

16. Norrby R, Powell M, Aronsson B, Monnet DL, Lutsar I, Bocsan IS, Cars O, Giamarellou H, Gyssens IC. 2009. *The Bacterial Challenge: Time To React.* ECDC/EMEA, Stockholm, Sweden.

17. Round JL, Mazmanian SK. 2009. The gut microbiota shapes intestinal immune responses during health and disease. *Nat Rev Immunol* **9**:313–323.

18. Francino MP. 2016. Antibiotics and the human gut microbiome: dysbioses and accumulation of resistances. *Front Microbiol* **6**:1543.

19. Gilbert JA, Blaser MJ, Caporaso JG, Jansson JK, Lynch SV, Knight R. 2018. Current understanding of the human microbiome. *Nat Med* **24**:392–400.

20. Meijnikman AS, Gerdes VE, Nieuwdorp M, Herrema H. 2018. Evaluating causality of gut microbiota in obesity and diabetes in humans. *Endocr Rev* **39**:133–153.

21. Devaraj S, Hemarajata P, Versalovic J. 2013. The human gut microbiome and body metabolism: implications for obesity and diabetes. *Clin Chem* **59**:617–628.

22. Cani PD. 2019. Microbiota and metabolites in metabolic diseases. *Nat Rev Endocrinol* **15**:69–70.

23. Levy M, Kolodziejczyk AA, Thaiss CA, Elinav E. 2017. Dysbiosis and the immune system. *Nat Rev Immunol* 17:219–232.

24. Byrd AL, Belkaid Y, Segre JA. 2018. The human skin microbiome. *Nat Rev Microbiol* 16:143–155.

25. Yang Y, Tian J, Yang B. 2018. Targeting gut microbiome: a novel and potential therapy for autism. *Life Sci* 194:111–119.

26. Cuomo A, Maina G, Rosso G, Beccarini Crescenzi B, Bolognesi S, Di Muro A, Giordano N, Goracci A, Neal SM, Nitti M, Pieraccini F, Fagiolini A. 2018. The microbiome: a new target for research and treatment of schizophrenia and its resistant presentations? A systematic literature search and review. *Front Pharmacol* 9:1040.

27. Deans E. 2016. Microbiome and mental health in the modern environment. *J Physiol Anthropol* 36:1.

28. Valles-Colomer M, Falony G, Darzi Y, Tigchelaar EF, Wang J, Tito RY, Schiweck C, Kurilshikov A, Joossens M, Wijmenga C, Claes S, Van Oudenhove L, Zhernakova A, Vieira-Silva S, Raes J. 2019. The neuroactive potential of the human gut microbiota in quality of life and depression. *Nat Microbiol* 4:623–632.

29. Hildebrand H, Malmborg P, Askling J, Ekbom A, Montgomery SM. 2008. Early-life exposures associated with antibiotic use and risk of subsequent Crohn's disease. *Scand J Gastroenterol* 43:961–966.

30. Scott NA, Andrusaite A, Andersen P, Lawson M, Alcon-Giner C, Leclaire C, Caim S, Le Gall G, Shaw T, Connolly JPR, Roe AJ, Wessel H, Bravo-Blas A, Thomson CA, Kästele V, Wang P, Peterson DA, Bancroft A, Li X, Grencis R, Mowat AM, Hall LJ, Travis MA, Milling SWF, Mann ER. 2018. Antibiotics induce sustained dysregulation of intestinal T cell immunity by perturbing macrophage homeostasis. *Sci Transl Med* 10:eaao4755.

31. Langdon A, Crook N, Dantas G. 2016. The effects of antibiotics on the microbiome throughout development and alternative approaches for therapeutic modulation. *Genome Med* 8:39.

32. Bikard D, Euler CW, Jiang W, Nussenzweig PM, Goldberg GW, Duportet X, Fischetti VA, Marraffini LA. 2014. Exploiting CRISPR-Cas nucleases to produce sequence-specific antimicrobials. *Nat Biotechnol* 32:1146–1150.

33. Kim JS, Cho DH, Park M, Chung WJ, Shin D, Ko KS, Kweon DH. 2016. CRISPR/Cas9-mediated re-sensitization of antibiotic-resistant Escherichia coli harboring extended-spectrum β-lactamases. *J Microbiol Biotechnol* 26:394–401.

34. Yosef I, Manor M, Kiro R, Qimron U. 2015. Temperate and lytic bacteriophages programmed to sensitize and kill antibiotic-resistant bacteria. *Proc Natl Acad Sci USA* 112:7267–7272.

35. Citorik RJ, Mimee M, Lu TK. 2014. Sequence-specific antimicrobials using efficiently delivered RNA-guided nucleases. *Nat Biotechnol* 32:1141–1145.

36. Bernheim A, Bikard D, Touchon M, Rocha EPC. 2020. Atypical organizations and epistatic interactions of CRISPRs and cas clusters in genomes and their mobile genetic elements. *Nucleic Acids Res* 48:748–760.

37. Makarova KS, Wolf YI, Alkhnbashi OS, Costa F, Shah SA, Saunders SJ, Barrangou R, Brouns SJJ, Charpentier E, Haft DH, Horvath P, Moineau S, Mojica FJM, Terns RM, Terns MP, White MF, Yakunin AF, Garrett RA, van der Oost J, Backofen R, Koonin EV. 2015. An updated evolutionary classification of CRISPR-Cas systems. *Nat Rev Microbiol* 13:722–736.

38. Jiang W, Marraffini LA. 2015. CRISPR-Cas: new tools for genetic manipulations from bacterial immunity systems. *Annu Rev Microbiol* 69:209–228.

39. Westra ER, van Erp PB, Künne T, Wong SP, Staals RH, Seegers CL, Bollen S, Jore MM, Semenova E, Severinov K, de Vos WM, Dame RT, de Vries R, Brouns SJ, van der Oost J. 2012. CRISPR immunity relies on the consecutive binding and degradation of negatively supercoiled invader DNA by Cascade and Cas3. *Mol Cell* 46:595–605.

40. Xiao Y, Luo M, Dolan AE, Liao M, Ke A. 2018. Structure basis for RNA-guided DNA degradation by Cascade and Cas3. *Science* 361:eaat0839.

41. Caliando BJ, Voigt CA. 2015. Targeted DNA degradation using a CRISPR device stably carried in the host genome. *Nat Commun* 6:6989.

42. Garneau JE, Dupuis ME, Villion M, Romero DA, Barrangou R, Boyaval P, Fremaux C, Horvath P, Magadán AH, Moineau S. 2010. The CRISPR/Cas bacterial immune system cleaves bacteriophage and plasmid DNA. *Nature* 468:67–71.

43. Gasiunas G, Barrangou R, Horvath P, Siksnys V. 2012. Cas9-crRNA ribonucleoprotein complex mediates specific DNA cleavage for adaptive immunity in bacteria. *Proc Natl Acad Sci USA* 109:E2579–E2586.

44. Jinek M, Chylinski K, Fonfara I, Hauer M, Doudna JA, Charpentier E. 2012. A programmable dual-RNA-guided DNA endonuclease in adaptive bacterial immunity. *Science* 337:816–821.

45. Cui L, Bikard D. 2016. Consequences of Cas9 cleavage in the chromosome of *Escherichia coli*. *Nucleic Acids Res* 44:4243–4251.

46. Bernheim A, Calvo-Villamañán A, Basier C, Cui L, Rocha EPC, Touchon M, Bikard D. 2017. Inhibition of NHEJ repair by type II-A CRISPR-Cas systems in bacteria. *Nat Commun* 8:2094.

47. Shuman S, Glickman MS. 2007. Bacterial DNA repair by non-homologous end joining. *Nat Rev Microbiol* 5:852–861.

48. Moreb EA, Hoover B, Yaseen A, Valyasevi N, Roecker Z, Menacho-Melgar R, Lynch MD. 2017. Managing the SOS response for enhanced CRISPR-Cas-based recombineering in *E. coli* through transient inhibition of host RecA activity. *ACS Synth Biol* 6:2209–2218.

49. Fonfara I, Richter H, Bratovič M, Le Rhun A, Charpentier E. 2016. The CRISPR-associated DNA-cleaving enzyme Cpf1 also processes precursor CRISPR RNA. *Nature* 532:517–521.

50. Mapes AC, Trautner BW, Liao KS, Ramig RF. 2016. Development of expanded host range phage active on biofilms of multi-drug resistant *Pseudomonas aeruginosa*. *Bacteriophage* 6:e1096995.

51. Samai P, Pyenson N, Jiang W, Goldberg GW, Hatoum-Aslan A, Marraffini LA. 2015. Co-transcriptional DNA and RNA cleavage during type III CRISPR-Cas immunity. *Cell* 161:1164–1174.

52. Elmore JR, Sheppard NF, Ramia N, Deighan T, Li H, Terns RM, Terns MP. 2016. Bipartite recognition of target RNAs activates DNA cleavage by the type III-B CRISPR-Cas system. *Genes Dev* 30:447–459.

53. Estrella MA, Kuo FT, Bailey S. 2016. RNA-activated DNA cleavage by the type III-B CRISPR-Cas effector complex. *Genes Dev* 30:460–470.

54. Kazlauskiene M, Kostiuk G, Venclovas Č, Tamulaitis G, Siksnys V. 2017. A cyclic oligonucleotide signaling pathway in type III CRISPR-Cas systems. *Science* 357:605–609.

55. Niewoehner O, Garcia-Doval C, Rostøl JT, Berk C, Schwede F, Bigler L, Hall J, Marraffini LA, Jinek M. 2017. Type III CRISPR-Cas systems produce cyclic oligoadenylate second messengers. *Nature* 548:543–548.

56. Athukoralage JS, Rouillon C, Graham S, Grüschow S, White MF. 2018. Ring nucleases deactivate type III CRISPR ribonucleases by degrading cyclic oligoadenylate. *Nature* 562:277–280.

57. Manica A, Zebec Z, Steinkellner J, Schleper C. 2013. Unexpectedly broad target recognition of the CRISPR-mediated virus defence system in the archaeon *Sulfolobus solfataricus*. *Nucleic Acids Res* 41:10509–10517.

58. Maniv I, Jiang W, Bikard D, Marraffini LA. 2016. Impact of different target sequences on type III CRISPR-Cas immunity. *J Bacteriol* 198:941–950.

59. Pyenson NC, Gayvert K, Varble A, Elemento O, Marraffini LA. 2017. Broad targeting specificity during bacterial type III CRISPR-Cas immunity constrains viral escape. *Cell Host Microbe* 22:343–353.e3.

60. East-Seletsky A, O'Connell MR, Knight SC, Burstein D, Cate JHD, Tjian R, Doudna JA. 2016. Two distinct RNase activities of CRISPR-C2c2 enable guide-RNA processing and RNA detection. *Nature* 538:270–273.

61. Abudayyeh OO, Gootenberg JS, Konermann S, Joung J, Slaymaker IM, Cox DB, Shmakov S, Makarova KS, Semenova E, Minakhin L, Severinov K, Regev A, Lander ES, Koonin EV, Zhang F. 2016. C2c2 is a single-component programmable RNA-guided RNA-targeting CRISPR effector. *Science* 353:aaf5573.

62. Meeske AJ, Nakandakari-Higa S, Marraffini LA. 2019. Cas13-induced cellular dormancy prevents the rise of CRISPR-resistant bacteriophage. *Nature* 570:241–245.

63. Kiga K, Tan X-E, Ibarra-Chávez R, Watanabe S, Aiba Y, Sato'o Y, Li F-Y, Sasahara T, Cui B, Kawauchi M, Boonsiri T, Thitiananpakorn K, Taki Y, Azam AH, Suzuki M, Penadés JR, Cui L. 2019. Development of CRSIPR-Cas13a-based antimicrobials capable of sequence-specific killing of target bacteria. *bioRxiv* 10.1101/808741.

64. Watanabe T. 1963. Infective heredity of multiple drug resistance in bacteria. *Bacteriol Rev* 27:87–115.

65. Huddleston JR. 2014. Horizontal gene transfer in the human gastrointestinal tract: potential spread of antibiotic resistance genes. *Infect Drug Resist* 7:167–176.

66. Andam CP, Fournier GP, Gogarten JP. 2011. Multilevel populations and the evolution of antibiotic resistance through horizontal gene transfer. *FEMS Microbiol Rev* 35:756–767.

67. Wellington EM, Boxall AB, Cross P, Feil EJ, Gaze WH, Hawkey PM, Johnson-Rollings AS, Jones DL, Lee NM, Otten W, Thomas CM, Williams AP. 2013. The role of the natural environment in the emergence of antibiotic resistance in gram-negative bacteria. *Lancet Infect Dis* 13:155–165.

68. D'Costa VM, Griffiths E, Wright GD. 2007. Expanding the soil antibiotic resistome: exploring environmental diversity. *Curr Opin Microbiol* 10:481–489.

69. Schjørring S, Krogfelt KA. 2011. Assessment of bacterial antibiotic resistance transfer in the gut. *Int J Microbiol* 2011:312956.

70. Penders J, Stobberingh EE, Savelkoul PHM, Wolffs PFG. 2013. The human microbiome as a reservoir of antimicrobial resistance. *Front Microbiol* 4:87.

71. Goren M, Yosef I, Qimron U. 2017. Sensitizing pathogens to antibiotics using the CRISPR-Cas system. *Drug Resist Updat* 30:1–6.

72. Heussler GE, O'Toole GA. 2016. Friendly fire: biological functions and consequences of chromosomal targeting by CRISPR-Cas systems. *J Bacteriol* 198:1481–1486.

73. van Belkum A, Soriaga LB, LaFave MC, Akella S, Veyrieras JB, Barbu EM, Shortridge D, Blanc B, Hannum G, Zambardi G, Miller K, Enright MC, Mugnier N, Brami D, Schicklin S, Felderman M, Schwartz AS, Richardson TH, Peterson TC, Hubby B, Cady KC. 2015. Phylogenetic distribution of CRISPR-Cas systems in antibiotic-resistant *Pseudomonas aeruginosa*. *mBio* 6:e01796-15.

74. Buffie CG, Jarchum I, Equinda M, Lipuma L, Gobourne A, Viale A, Ubeda C, Xavier J, Pamer EG. 2012. Profound alterations of intestinal microbiota following a single dose of clindamycin results in sustained susceptibility to *Clostridium difficile*-induced colitis. *Infect Immun* 80:62–73.

75. Lino CA, Harper JC, Carney JP, Timlin JA. 2018. Delivering CRISPR: a review of the challenges and approaches. *Drug Deliv* 25:1234–1257.

76. Liu C, Zhang L, Liu H, Cheng K. 2017. Delivery strategies of the CRISPR-Cas9 gene-editing system for therapeutic applications. *J Control Release* 266:17–26.

77. Pursey E, Sünderhauf D, Gaze WH, Westra ER, van Houte S. 2018. CRISPR-Cas antimicrobials: challenges and future prospects. *PLoS Pathog* 14:e1006990.

78. Machado AMD, Sommer MOA. 2014. Human intestinal cells modulate conjugational transfer of multidrug resistance plasmids between clinical *Escherichia coli* isolates. *PLoS One* 9:e100739.

79. Box AM, McGuffie MJ, O'Hara BJ, Seed KD. 2015. Functional analysis of bacteriophage immunity through a type I-E CRISPR-Cas system in *Vibrio cholerae* and its application in bacteriophage genome engineering. *J Bacteriol* 198:578–590.

80. Seed KD, Lazinski DW, Calderwood SB, Camilli A. 2013. A bacteriophage encodes its own CRISPR/Cas adaptive response to evade host innate immunity. *Nature* 494:489–491.

81. Naser IB, Hoque MM, Nahid MA, Tareq TM, Rocky MK, Faruque SM. 2017. Analysis of the CRISPR-Cas system in bacteriophages active on epidemic strains of *Vibrio cholerae* in Bangladesh. *Sci Rep* 7:14880.

82. Al-Shayeb B, Sachdeva R, Chen L-X, Ward F, Munk P, Devoto A, Castelle CJ, Olm MR, Bouma-Gregson K, Amano Y, He C, Méheust R, Brooks B, Thomas A, Lavy A, Matheus-Carnevali P, Sun C, Goltsman DSA, Borton MA, Nelson TC, Kantor R, Jaffe AL, Keren R, Farag IF, Lei S, Finstad K, Amundson R, Anantharaman K, Zhou J, Probst AJ, Power ME, Tringe SG, Li W-J, Wrighton K, Harrison S, Morowitz M, Relman DA, Doudna JA, Lehours A-C, Warren L, Cate JHD, Santini JM, Banfield JF. 2019. Clades of huge phage from across Earth's ecosystems. *bioRxiv* 10.1101/572362:572362.

83. Hargreaves KR, Flores CO, Lawley TD, Clokie MR. 2014. Abundant and diverse clustered regularly interspaced short palindromic repeat spacers in *Clostridium difficile* strains and prophages target multiple phage types within this pathogen. *mBio* 5:e01045-13.

84. Ram G, Ross HF, Novick RP, Rodriguez-Pagan I, Jiang D. 2018. Conversion of staphylococcal pathogenicity islands to CRISPR-carrying antibacterial agents that cure infections in mice. *Nat Biotechnol* 36:971–976.

85. Fillol-Salom A, Martínez-Rubio R, Abdulrahman RF, Chen J, Davies R, Penadés JR. 2018. Phage-inducible chromosomal islands are ubiquitous within the bacterial universe. *ISME J* 12:2114–2128.

86. Hyman P, Abedon ST. 2010. Bacteriophage host range and bacterial resistance. *Adv Appl Microbiol* 70:217–248.

87. Burmeister AR, Lenski RE, Meyer JR. 2016. Host coevolution alters the adaptive landscape of a virus. *Proc Biol Sci* 283:20161528.

88. Maddamsetti R, Johnson DT, Spielman SJ, Petrie KL, Marks DS, Meyer JR. 2018. Gain-of-function experiments with bacteriophage lambda uncover residues under diversifying selection in nature. *Evolution* 72:2234–2243.

89. Meyer JR, Dobias DT, Weitz JS, Barrick JE, Quick RT, Lenski RE. 2012. Repeatability and contingency in the evolution of a key innovation in phage lambda. *Science* 335:428–432.

90. Yosef I, Goren MG, Globus R, Molshanski-Mor S, Qimron U. 2017. Extending the host range of bacteriophage particles for DNA transduction. *Mol Cell* 66:721–728.e3.

91. Goren MG, Yosef I, Qimron U. 2015. Programming bacteriophages by swapping their specificity determinants. *Trends Microbiol* 23:744–746.

92. Ando H, Lemire S, Pires DP, Lu TK. 2015. Engineering modular viral scaffolds for targeted bacterial population editing. *Cell Syst* 1:187–196.

93. Lu TK-T, Lemire S, Yang AC, Yehl KM, Van Amsterdam JR. October 2017. Synthetic bacteriophages and bacteriophage compositions. International patent PCT/US2017/058667.

94. Yehl K, Lemire S, Yang AC, Ando H, Mimee M, Torres MT, de la Fuente-Nunez C, Lu TK. 2019. Engineering phage host-range and suppressing bacterial resistance through phage tail fiber mutagenesis. *Cell* 179:459–469.e9.

95. Russell J, Bikard D. 2019. Learning from antibodies: phage host-range engineering. *Cell Host Microbe* 26:445–446.

96. Sulakvelidze A, Alavidze Z, Morris JG, Jr. 2001. Bacteriophage therapy. *Antimicrob Agents Chemother* 45:649–659.

97. Dublanchet A, Fruciano E. 2008. Brève histoire de la phagothérapie. *Med Mal Infect* 38:415–420.

98. Myelnikov D. 2018. An alternative cure: the adoption and survival of bacteriophage therapy in the USSR, 1922–1955. *J Hist Med Allied Sci* 73:385–411.

99. Van Belleghem JD, Dąbrowska K, Vaneechoutte M, Barr JJ, Bollyky PL. 2018. Interactions between bacteriophage, bacteria, and the mammalian immune system. *Viruses* 11:10.

100. **Shkoporov AN, Hill C.** 2019. Bacteriophages of the human gut: the "known unknown" of the microbiome. *Cell Host Microbe* 25:195–209.

101. **Merril CR, Biswas B, Carlton R, Jensen NC, Creed GJ, Zullo S, Adhya S.** 1996. Long-circulating bacteriophage as antibacterial agents. *Proc Natl Acad Sci USA* 93:3188–3192.

102. **Geier MR, Trigg ME, Merril CR.** 1973. Fate of bacteriophage lambda in non-immune germ-free mice. *Nature* 246:221–223.

103. **Żaczek M, Łusiak-Szelachowska M, Jończyk-Matysiak E, Weber-Dąbrowska B, Międzybrodzki R, Owczarek B, Kopciuch A, Fortuna W, Rogóż P, Górski A.** 2016. Antibody production in response to staphylococcal MS-1 phage cocktail in patients undergoing phage therapy. *Front Microbiol* 7:1681.

104. **Van Belleghem JD, Clement F, Merabishvili M, Lavigne R, Vaneechoutte M.** 2017. Pro- and anti-inflammatory responses of peripheral blood mononuclear cells induced by *Staphylococcus aureus* and *Pseudomonas aeruginosa* phages. *Sci Rep* 7:8004.

105. **Sweere JM, Van Belleghem JD, Ishak H, Bach MS, Popescu M, Sunkari V, Kaber G, Manasherob R, Suh GA, Cao X, de Vries CR, Lam DN, Marshall PL, Birukova M, Katznelson E, Lazzareschi DV, Balaji S, Keswani SG, Hawn TR, Secor PR, Bollyky PL.** 2019. Bacteriophage trigger antiviral immunity and prevent clearance of bacterial infection. *Science* 363:eaat9691.

106. **Bille E, Meyer J, Jamet A, Euphrasie D, Barnier J-P, Brissac T, Larsen A, Pelissier P, Nassif X.** 2017. A virulence-associated filamentous bacteriophage of *Neisseria meningitidis* increases host-cell colonisation. *PLoS Pathog* 13:e1006495.

107. **Paez-Espino D, Morovic W, Sun CL, Thomas BC, Ueda K, Stahl B, Barrangou R, Banfield JF.** 2013. Strong bias in the bacterial CRISPR elements that confer immunity to phage. *Nat Commun* 4:1430.

108. **Bikard D, Barrangou R.** 2017. Using CRISPR-Cas systems as antimicrobials. *Curr Opin Microbiol* 37:155–160.

109. **Stern A, Keren L, Wurtzel O, Amitai G, Sorek R.** 2010. Self-targeting by CRISPR: gene regulation or autoimmunity? *Trends Genet* 26:335–340.

110. **Rauch BJ, Silvis MR, Hultquist JF, Waters CS, McGregor MJ, Krogan NJ, Bondy-Denomy J.** 2017. Inhibition of CRISPR-Cas9 with Bacteriophage Proteins. *Cell* 168:150–158.e10.

111. **Pawluk A, Davidson AR, Maxwell KL.** 2018. Anti-CRISPR: discovery, mechanism and function. *Nat Rev Microbiol* 16:12–17.

112. **Borges AL, Davidson AR, Bondy-Denomy J.** 2017. The discovery, mechanisms, and evolutionary impact of anti-CRISPRs. *Annu Rev Virol* 4:37–59.

113. **Maxwell KL.** 2017. The anti-CRISPR story: a battle for survival. *Mol Cell* 68:8–14.

114. **Oliveira PH, Touchon M, Rocha EPC.** 2014. The interplay of restriction-modification systems with mobile genetic elements and their prokaryotic hosts. *Nucleic Acids Res* 42:10618–10631.

115. **Ali SS, Xia B, Liu J, Navarre WW.** 2012. Silencing of foreign DNA in bacteria. *Curr Opin Microbiol* 15:175–181.

116. **Ghorbal M, Gorman M, Macpherson CR, Martins RM, Scherf A, Lopez-Rubio JJ.** 2014. Genome editing in the human malaria parasite *Plasmodium falciparum* using the CRISPR-Cas9 system. *Nat Biotechnol* 32:819–821.

117. **Peng D, Kurup SP, Yao PY, Minning TA, Tarleton RL.** 2014. CRISPR-Cas9-mediated single-gene and gene family disruption in *Trypanosoma cruzi*. *mBio* 6:e02097-14.

118. **Shen B, Brown KM, Lee TD, Sibley LD.** 2014. Efficient gene disruption in diverse strains of *Toxoplasma gondii* using CRISPR/CAS9. *mBio* 5:e01114-14.

119. **Wagner JC, Platt RJ, Goldfless SJ, Zhang F, Niles JC.** 2014. Efficient CRISPR-Cas9-mediated genome editing in *Plasmodium falciparum*. *Nat Methods* 11:915–918.

Exploiting CRISPR-Cas Systems To Provide Phage Resistance in Industrial Bacteria

Rodolphe Barrangou and Avery Roberts

CRISPRlab, Department of Food, Bioprocessing and Nutrition Sciences, North Carolina State University, Raleigh, NC, 27695-7624

Introduction

A broad diversity of viruses ubiquitously distributed across the globe constitutes the most abundant biological entity on our planet. This predatory viral population primarily consists of bacteriophages, viruses of bacteria, which widely prey upon bacterial hosts that likewise occur broadly in virtually all environmental niches and habitats across our planet, notably oceans and soil (1). Recent advances in viral metagenomics have documented a remarkable phage genetic diversity, which we are only beginning to measure and for which we have a primitive understanding of the broad ecological impact (1). As a consequence, these bacterial hosts have developed an elaborate set of antiviral defense systems that match this phage abundance and diversity and constitute a pan-immune system widespread in bacterial communities (2, 3). Host-phage coevolutionary dynamics unfold over a complex molecular arms race in which diverse host immune systems drive elaborate countermeasures in viral predators. The diversified host defense arsenal is often redundant and subjected to horizontal gene transfer, matched by high mutation rates in adaptable and mosaic phage genomes (2–4).

The history of molecular biology has been defined and shaped by the deciphering of bacterial resistance systems, bookended by the repurposing of restriction enzymes from restriction-modification (R-M) systems and the reprogramming of Cas effector nucleases from CRISPR-Cas immune systems (5). The main antiviral defense systems in bacteria generally encompass abortive infection (Abi), R-M, toxin-antitoxin, adaptive immunity (CRISPR-Cas), and the recently discovered plethora of novel innate modalities such as BREX (bacteriophage exclusion), DISARM (defense island system associated with R-M), and yet-to-be-characterized mechanisms such as Gabija, Shedu, Zorya, and others (1, 2, 4). Remarkably, recent studies have revealed that numerous and diverse defense islands widespread in microbial genomes comprise potent and flexible immune systems that transcend phylogenetic boundaries to spread broadly in bacterial genomes, sometimes co-occurring, providing multiplexed resistance hinging on independent mechanisms of action (2, 3).

Over the past two decades, the chronicled discovery of CRISPR-Cas adaptive immune systems has revolutionized molecular biology, opening new avenues to provide and

CRISPR: Biology and Applications, First Edition. Edited by Rodolphe Barrangou, Erik J. Sontheimer, and Luciano A. Marraffini.
© 2022 American Society for Microbiology. DOI: 10.1128/9781683673798.ch15

enhance phage resistance in industrial bacteria (6). The humble beginnings of the CRISPR craze in industrial bacteria established the adaptive immunity role of CRISPR-Cas systems in dairy phage resistance (5), and the molecular mode of action was quickly defined as DNA-encoded (7), RNA-mediated (8) nucleic acid targeting (9). Early studies in the field identified CRISPR loci (10) and their associated *cas* genes (11), leading to the observation that CRISPR spacers share sequence homology to mobile genetic elements (12–14) and to their mechanistic characterization in 2007-2008 (6). As extensively detailed in the earlier chapters of this book, a diverse set of striking CRISPR-Cas systems broadly occur in bacteria and archaea, with shared mechanistic features encompassing adaptation, CRISPR RNA (crRNA) biogenesis, and interference. Adaptation enables the capture of invasive DNA through the acquisition of viral sequences as novel spacers at the leader end of the CRISPR array. This adaptive immunity mechanism constitutes a means to "vaccinate" the host and build up an immunization record over time that is captured iteratively within the CRISPR array. The crRNA biogenesis phase hinges on the transcription and processing of CRISPR-encoded repeats and spacers into the mature guide RNAs that provide sequence-specific targeting of complementary sequences. The interference process drives target nucleic acid elimination through Cas effector nucleases guided by crRNAs that drive sequence-specific nucleic acid destruction. The target nucleic acid cleavage varies across classes, types, and subtypes (15–17), notably with the CRISPR-associated complex for antiviral defense (Cascade) (8), the Cas9 endonuclease, and Cas12 dual nickase driving DNA targeting in types I, II, and V, respectively. In contrast, Csm/Cmr complexes and the Cas13 nuclease drive RNA targeting in types III and VI, respectively (18). A diversity of CRISPR-Cas systems occurs broadly in nature (15–17), with novel types and subtypes unearthed from metagenomic data sets on a regular basis (19). Together, these diverse CRISPR-Cas systems provide adaptive immunity against invasive genetic elements, such as bacteriophages, and can be exploited to specifically build up phage resistance in bacterial cultures widely used in industry.

CRISPR-Mediated Host-Phage Coevolutionary Dynamics in *Streptococcus thermophilus*

Dairy Cultures and Phage Resistance

Fermentation has been used for millennia to preserve fresh and raw food sources such as milk, grain, and vegetables and provide a safer and more stable food product to culturally and geographically diverse human communities. In particular, dairy fermentation has been broadly used to process milk into cheese and yoghurt, through the biotransformation of lactose into lactic acid, leading to the coagulation of milk by lactic acid bacteria such as *Streptococcus thermophilus* and *Lactococcus lactis*. Thousands of strains of *S. thermophilus* in particular have been domesticated by the dairy industry for the fast acidification of milk, and this workhorse has been deployed in all corners of the world for the speedy preservation of local milk supplied by various likewise-domesticated livestock species. Despite tremendous advances in the global food supply chain and the sophistication of food processing and fermentation, the organisms and processes driving the preservation of milk into fermented dairy products remain authentic. Occasionally, fermentation would appear delayed, or a batch would fail, due to unbeknownst phage predation.

Advances in microbiology over the 20th century, starting with the discovery and characterization of bacteriophages and culminating with the genomic characterization of bacteria and their viruses, have shed light on the molecular mechanisms providing phage resistance. The dairy industry was an early adopter of molecular biology to inform starter culture composition, particularly regarding genetic diversity and heterogeneity, as well as the use of culture rotation systems designed to circumvent product loss caused by widespread phages commonly found in milk. The establishment of the biological immune function of CRISPR-Cas systems in a *bona fide* industrialized strain of *S. thermophilus* fittingly laid a foundation for its commercial use in dairy cultures and the launch of CRISPR-enhanced starter cultures in 2011 by DuPont (20). These strains, and subsequently screened and formulated CRISPR-enhanced variants, have been used globally for the commercial manufacturing of dairy products for a decade, thus making virtually every pizza, cheeseburger, yoghurt, and cheese slice consumed a carrier of CRISPR technology. Besides the original proof of concept provided in industrial cultures, this approach has been independently replicated and used by others exploiting *S. thermophilus* dairy strains (21–25).

In *S. thermophilus* DGCC7710, the native type II-A CRISPR-Cas9 system can readily acquire CRISPR spacers when exposed to various phages, including φ858 and φ2972, both individually and as a phage cocktail (Fig. 15.1). Various phage genome regions can give rise to CRISPR immunity by copying and pasting of viral sequences as new spacers that are acquired by the CRISPR acquisition and adaptation machinery (26). When sequences unique to a phage family are acquired, resistance is specific to this subset of viruses. In contrast, in cases where sequences conserved between different phage families are acquired, CRISPR immunity can provide broad resistance against multiple phage families that share this sequence, enabling cross-immunity. In selected settings, the majority of bacterial colonies that grow following phage exposure, referred to as bacteriophage-insensitive mutants (BIMs), are CRISPR variants. Typically, CRISPR immunity affords these cultures 6 to 7 orders of magnitude of resistance, as defined by the efficiency of plaquing (one in ten million to one in a million phages retain the ability to infect the host). Early work involving a series of three experiments in *S. thermophilus* provided a proof of concept that CRISPR-Cas systems provide adaptive immunity in bacteria (7). First, it was shown that strains that survive phage exposure acquire a novel CRISPR spacer in a polarized manner at the leader end of the CRISPR array and that this spacer sequence is a perfect match to a phage genome segment. Second, it was demonstrated that adding a CRISPR spacer (inserting a phage sequence between CRISPR repeats) provides phage resistance *de novo*, whereas removing acquired spacers leads to phage sensitivity. Likewise, transplanting CRISPR content between two strains, each sensitive to one and resistant to another phage, actually swapped their phage sensitivity profiles, directly linking CRISPR genotype to phage resistance phenotype. Third, knocking out *cas* genes resulted in either loss of interference (*cas9* inactivation prevents targeting) or loss of adaptation (*cas4* inactivation prevents subsequent spacer acquisition). Deciphering the CRISPR-Cas mode of action in this model organism, containing the canonical type II-A system, enabled the discovery of the protospacer adjacent motif (PAM) in 2008 (27), the transplantation and reprogramming of CRISPR-Cas systems in other organisms (targeting of phage in *Escherichia coli*) (28), and the establishment that Cas9 is an endonuclease able to cleave phage and plasmid DNA (29) via two nickase motifs, namely, HNH and RuvC (30).

Phage Escape and Countermeasures

Given the arms race driving the coevolutionary dynamics between prey and predator, the buildup of phage resistance in the bacterial host predictably triggers a countermeasure in the lurking predator. In particular, given the selective pressure that CRISPR-encoded, sequence-specific immunity provides, phages have evolved various countermeasures that provide

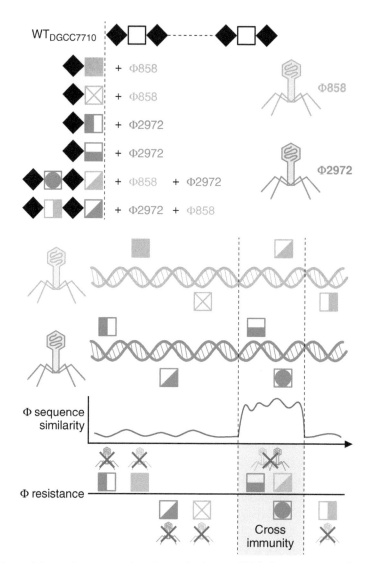

Figure 15.1 CRISPR provides resistance against bacteriophages. With *Streptococcus thermophilus* as a model species and the widely used commercial strain DGCC7710 as a model organism, upon exposure to phages of interest (e.g., φ858 and φ2972), novel spacers are acquired at the leader end of the CRISPR locus. Here, the DGCC7710 CRISPR locus is depicted with repeats as black diamonds and spacers as filled squares (white squares represent preexisting spacers). Potential new spacer acquisition events are depicted vertically below the CRISPR array, with filled squares of different patterns representing unique spacers specific to φ858 or φ2972 or both phages: spacers unique to one phage provide resistance against that specific virus, whereas spacers homologous to shared sequences can provide broad-spectrum resistance. The phage genomes are presented in the corresponding phage colors, and spacer squares are presented above or below the region of the corresponding phage genome for which their sequence matches. WT, wild type.

escape (Fig. 15.2). Given the complementarity between the CRISPR spacer sequence and the corresponding phage protospacer sequence, mutations, especially in the seed sequence and in the immediate vicinity of the cleavage site (e.g., –3 position from the 3′ edge of the protospacer in type II-A CRISPR-Cas systems), prevent targeting by CRISPR and cleavage by Cas nucleases such as Cas9 (31, 32). The seed sequence (32, 33) is a short stretch of nucleotides at the PAM-proximal edge of the spacer sequence that initiates and nucleates the formation of a stable RNA:DNA heteroduplex between the guide crRNA and the target DNA using Watson-Crick base-pairing complementarity that can extend the length of the protospacer to create an R-loop (34). A mutation in the seed sequence thus prevents extension of the

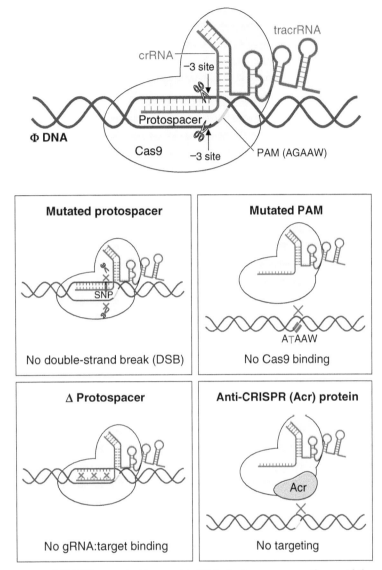

Figure 15.2 Phage escape of CRISPR-encoded immunity. With the *Streptococcus thermophilus* type II Sth1CRISPR-Cas9 interference complex as a model, the Cas9:crRNA:tracrRNA effector complex targets complementary double-stranded DNA flanked by a PAM (AGAAW). Various countermeasures enable bacteriophage to circumvent CRISPR targeting, including mutation of the protospacer sequence complementary to the crRNA, mutation of the PAM sequence targeted by Cas9, deletion of the targeted protospacer sequence, or blockage using anti-CRISPR proteins (Acrs).

crRNA:targetDNA heteroduplex, shortens phage DNA interrogation by the interference complex, and enables phage life cycle progression in the host.

Mechanistically, given the reliance on the PAM sequence for initiation of the crRNA:Cas effector interference complex, mutations in the conserved nucleotides in the PAM (31, 35, 36) prevent binding and consequently targeting by CRISPR-Cas immune systems. Preventing the initial, PAM-specific binding shields the phage DNA from molecular interrogation by the CRISPR interference complex.

Besides mutating the PAM and/or the protospacer, a deletion of this targeted sequence will preclude recognition by the interference complex. More recently, anti-CRISPR small proteins were discovered that specifically bind to Cas effectors and block recognition by the interference complex (37, 38). Together, these various escape mechanisms provide diversified avenues to counter adaptive immunity, albeit at an evolutionary cost, especially with regard to deletions and mutations of targeted chromosomal regions in viral genomes.

Expanding CRISPR Immunity in Industrial Bacteria

Characterization of CRISPR-Cas Systems

In order to exploit CRISPR-Cas immune systems to provide resistance against phages, it is necessary to first identify and characterize the various elements involved in the molecular mechanism of action to enable either the screening of natural variants or the engineering of potent spacers into the CRISPR array (Fig. 15.3) (39–42). First, the bacterial host of interest is microbiologically cultured and subjected to genome sequencing, with enough sequencing depth to ensure that repeated elements such as CRISPR arrays can be readily assembled. Then, a series of *in silico* analyses must be carried out to identify CRISPR loci and associated *cas* genes and then properly assign class, type, and subtype according to the occurrence of signature and accessory *cas* genes and known markers of *cas* operons (e.g., widespread acquisition genes *cas1* and *cas2*). Then, more in-depth computational analyses must be performed to evaluate the potential activity of a specific CRISPR-Cas system against invasive genetic elements by assessing the integrity of the CRISPR array and the matches between CRISPR spacers and corresponding protospacers in viral and plasmid sequences. Provided that good matches are identified, the sequences flanking the protospacers can then be queried to assess potentially conserved signatures

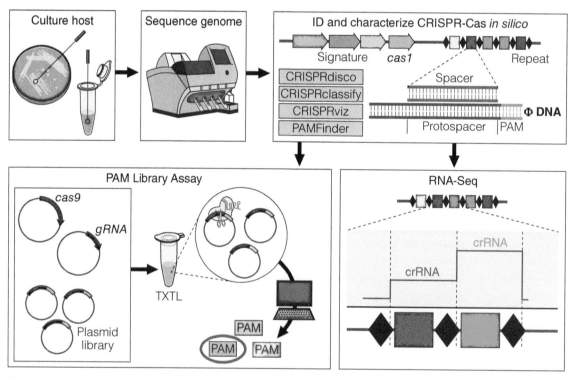

Figure 15.3 Characterization of CRISPR-Cas systems. To repurpose and exploit endogenous CRISPR-Cas systems in industrial bacteria, DNA is extracted from the cultured clone and subjected to genome sequencing. Various CRISPR pipelines can then be used to identify (ID) and predict the key features of CRISPR-Cas systems *in silico*, such as signature *cas* genes and spacer matches to protospacer sequences in various phage and plasmids. Key sequences and functional determinants, notably the PAM and guide RNA sequence, can then be validated using RNA-Seq and *in vivo* or, depicted here, *in vitro* PAM determination assays.

and predict the PAM (27, 43). The determination and deciphering of the *bona fide* PAM can hinge on validation either using targeting of plasmids carrying target sequences or via high-throughput screening of libraries that carry hypervariable sequences in the vicinity of the protospacer (44–46). Lastly, transcriptome sequencing (RNA-seq) analyses can be carried out to determine transcription levels for *cas* and CRISPR arrays and characterize the pre-crRNA processing patterns that yield the mature crRNAs that guide Cas effectors.

Altogether, the repurposing of CRISPR-Cas systems for strengthening phage resistance in industrial bacteria requires *a priori* knowledge to selectively screen for, select, or engineer bacteria with an enhanced CRISPR immunity repertoire.

Sequential CRISPR Vaccination Events Over Time

The dynamic nature of the interplay between host adaptive immunity and viral escape has practically impacted how industrial culture iterative vaccination occurs. Given these coevolutionary dynamics, there are rounds of vaccination that provide breadth and depth of CRISPR immunity, each of which gives rise to viral mutations in an attempt to circumvent immunity (Fig. 15.4). Iterative passages of host-phage cocultures for hundreds of passages, spanning thousands of generations, have been documented, and these typically enable the host to survive through a series of iterative vaccination events that target various phage chromosome areas and drive the demise of the viral population, presumably linked to the accumulation of mutations, some of which are detrimental to the virus life cycle. Besides single nucleotide polymorphisms (SNPs), insertions and deletions (indels) can span the PAM and protospacer and extend beyond the region targeted by CRISPR, ultimately damaging the integrity of the phage genome and consequently its ability to carry out its life cycle. In order to ascertain what would happen in a more biologically diverse environment, host-phage coevolutionary dynamics in the context of CRISPR have also encompassed multiple coexisting viral families. These conditions typically prolong the ability of phages to coexist with CRISPR-equipped hosts, but iterative spacer acquisition eventually provides resistance to phage cocktails and mixed viral communities.

Iterative Buildup of CRISPR Immunity in Industrial Starter Cultures

Microbiologically, the selection of natural phage-resistant variants following exposure to problematic phages has been used in the dairy industry for decades. The advent of molecular biology techniques and the deciphering of the molecular mechanisms underpinning various phage resistance mechanisms have enabled dairy microbiologists to selectively screen for variants that carry various phage resistance modalities, with Abi, cell surface resistance, and R-M historically in play and CRISPR-encoded resistance a more recent primary focus, i.e., in the past 15 years. There are well-documented protocols and techniques that enable the specific genesis of CRISPR BIMs and the generation and selection of CRISPR-immunized mutants that have acquired novel CRISPR spacers from problematic dairy phages (47). Actually, over 10 studies have been published on CRISPR BIM generation in *S. thermophilus* (47), and these protocols have been adapted to other genera and species of interest, such as *E. coli*, *Streptococcus mutans*, *Streptococcus agalactiae*, *Pseudomonas aeruginosa*, and *Sulfolobus solfataricus* (47). Typically, following host exposure to lytic phages, PCR analyses target the leader end of presumably active CRISPR loci to detect a size increase reflecting novel spacer integration. The PCR product is subsequently subjected to sequencing to confirm matches to viral sequences, and the ability of select BIMs to resist phage attack can be confirmed by plaque assays (47).

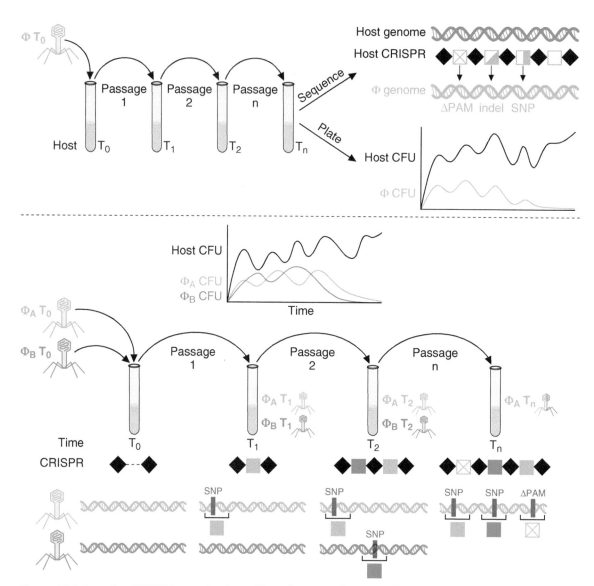

Figure 15.4 Iterative CRISPR immunizations. Host-phage cocultures can be passaged in broth culture in time series experiments and monitored using microbial techniques to quantify the host CFU and the phage PFU and, using genomic techniques to sequence the host genome, the phage genome and specifically determine the host CRISPR locus spacer content and the corresponding phage genomic mutations. Iterative passages over time in long-term experiments can encompass multiple phage species to determine the coevolutionary dynamics between host CRISPR-encoded immunity and various escape mechanisms in the phage populations. Here, spacers are depicted as squares filled with different patterns to represent unique spacer sequences. In the bottom panel, phages φA and φB are passaged alongside a host for multiple generations and accumulate mutations over time in response to spacer acquisition events in the host. Over time, DNA exchanges may cause phage populations to merge on a genomic level.

Based on the ability of CRISPR-Cas systems to sequentially acquire CRISPR spacers over time, it is possible by design to iteratively expand phage immunity by exposing a strain to a select series of phages for serial vaccination towards broad-spectrum immunity (Fig. 15.5). Using the dairy fermentative culture *Streptococcus thermophilus* as a major component of dairy cultures that is enhanced for phage resistance in the cheese and yoghurt industry, there are opportunities to naturally select for BIMs that acquire novel spacers in the two active CRISPR loci following exposure to phages commonly found in industrial

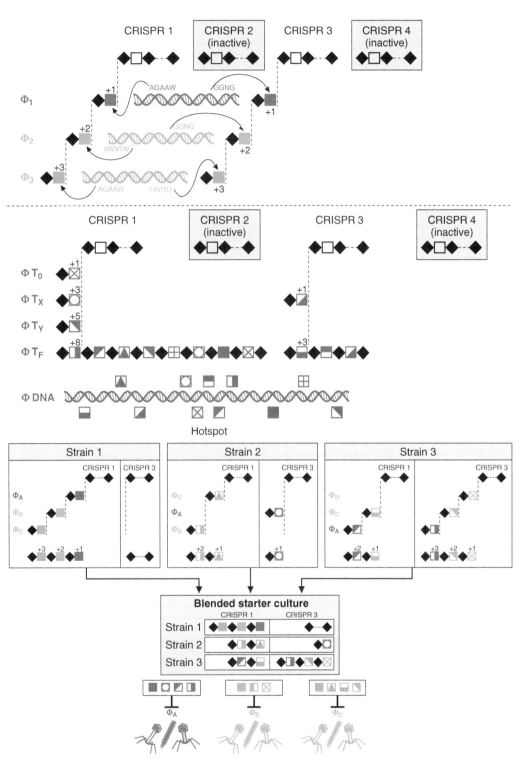

Figure 15.5 Iterative buildup of CRISPR immunity in industrial starter cultures. With *Streptococcus thermophilus* as a model system, in which up to 4 distinct CRISPR-Cas systems can occur concurrently, including active loci Sth1CRISPR-Cas9 and Sth3CRISPR-Cas9, a strain can be sequentially exposed to multiple phages to acquire spacers in both active CRISPR arrays (CRISPR1 and CRISPR3), iteratively. Eventually, multiple spacers can be acquired over time, with an example provided for 8 and 3 spacer acquisition events in CRISPR1 and CRISPR3, respectively, that target diverse and spread-out sequences in the phage genome, including high coverage of genomic hot spots highly susceptible to sampling by the CRISPR adaptation machinery. Eventually, iterative vaccination schemes can be developed in parallel in multiple strains to build broad and deep resistance against a comprehensive set of phages widely observed in industrial fermentations. These strains can then be combined in blended starter cultures to enhance phage resilience and extend the commercial life span of valuable cultures.

settings. Actually, there are multiple CRISPR-Cas systems typically present in the genomes of *Streptococcus thermophilus*, up to 4 concurrent systems spanning type I, type II, and type III (48). Previous studies have shown that multiple systems can be simultaneously active, notably with regard to spacer acquisition, and that these various systems are orthogonal and compatible (49–52). Accordingly, both type II CRISPR-Cas systems widely occurring in *S. thermophilus*, namely, Sth1CRISPR-Cas9 and Sth3CRISPR-Cas9, can independently acquire novel spacers from invasive phages (Fig. 15.5, top). Each is associated with its own PAM, thus enabling diversified targeting of and immunity to various sequences in phage genomes.

A noteworthy advantage of CRISPR-encoded immunity, besides sequence-specific targeting, is the ability to sequentially acquire novel spacers over time, following iterative exposure to and sampling from multiple phages (Fig. 15.5, middle). Impressively, both Sth1CRISPR-Cas9 and Sth3CRISPR-Cas9 have the ability to concurrently and iteratively acquire novel spacers from various industrial phages, providing broad and deep resistance. In combination with biases in sampling of select areas of the phage genome, this bacterium can thus build an immune potential repeatedly over time to highly cover genomic hot spots (Fig. 15.5, middle) (53, 54). Eventually, industrial microbiologists can use these vaccination approaches to create blends of BIMs that provide broad and deep resistance to the most commonly detected bacteriophages in the dairy industry (Fig. 15.5, bottom). As illustrated, different strains (e.g., strains 1, 2, and 3) can independently and in parallel be used to sequentially be exposed to multiple phages (e.g., red, blue, and green in Fig. 15.5), to each build resistance at up to two distinct type II CRISPR loci. Then, this combination of strains can be blended together. The resulting starter culture encompasses several distinct spacers in each immune system that comprehensively cover genome phages in general and can be selected based on spacers that specifically target highly conserved portions of the phage genomes. Presumably, this enables vaccination against previously unseen phage variants in which these conserved sequences occur. Also, based on conservation biases inherent to maintenance of functional attributes, heavy targeting of highly conserved sequences would also be more efficient, since mutations in these specific regions are presumed to be more deleterious. Altogether, this blended immunity approach has been heavily used in the dairy industry to naturally generate, screen, and combine phage-resistant strains that have been commercially available for a decade and utilized globally in the industry. Ironically, most of the first-generation CRISPR-related intellectual property, predating the 2012 single guide RNA technology (55) and 2013 genome editing era (56–58), focused on methods for using CRISPR-Cas immune systems against phages in dairy cultures.

When assessing the ecological context of CRISPR-mediated vaccination against phages, it is important to consider the genomic impact of iterative acquisition events. A study in 2013 determined the genomic impact of a series of 4 iterative vaccinations at two active *S. thermophilus* CRISPR loci, encompassing 8 total acquisition events ($n = 5$ in Sth1CRISPR-Cas9 and $n = 3$ in Sth3CRISPR-Cas9) from 4 distinct phages (59). The large majority of observed genomic differences (96%) did consist of CRISPR locus insertions, with 2 indels and 2 SNPs identified elsewhere, consistent with the baseline mutation rate in this species under this timeline. This is consistent with the maintenance of CRISPR loci integrity in long-term experiments and their broad conservation in this species. In contrast, the relative cost of circumventing CRISPR immunity in the phage population is deleterious at the genomic level, leading to extinction in long-term experiments and having a negative impact on viral fitness overall. The investigation of host-phage dynamics in the context of CRISPR does need more analyses, especially as it pertains to the relative cost(s) on fitness to the host compared to the phage (60–65). Thus, it appears that the relative cost of CRISPR immunity in the host is minimal compared to the cost of circumventing it in the phage (60, 65), an important argument in the industrial deployment of phage resistance in commercial cultures.

Using CRISPR Hypervariability for Genotyping

The serial acquisition of novel spacers over time in bacterial CRISPR loci forms a genetic record of immunization events. This hypervariable genetic watermark constitutes a valuable high-resolution chromosomal marker that can be exploited to capture the historical and evolutionary path of a given strain and chronicle the series of immunization events that occurred sequentially over time. Using these hypervariable genetic loci to detect, monitor, and track bacteria was the original application of CRISPR-Cas systems in bacteria, and also the very first filed and granted patent application in the field, filed back in 2004.

The iterative nature of the CRISPR adaptation process with sequential acquisition of new spacers in a polarized manner at the leader end of the locus is highly valuable for unraveling the historical exposures of a strain to invasive genetic elements, as well as comparing and contrasting clones derived from a common ancestor. The leader-distal end (occasionally referred to as the "trailer" end) remains conserved, with shared ancestral spacers preceding the terminal repeat (Fig. 15.6, top). Over time, various clones may acquire sequences from different phages, providing a point of divergence after which subsequent acquisition events add spacers that differ between clonal lines. Strains that share their full repertoire of spacers within a locus are often identical genetically. In contrast, strains that showcase unique leader end sets of spacers are considered unique. In cases where the difference consists of an additional acquisition event at the leader end of the CRISPR array, this is a recent acquisition event typical of a BIM that just underwent vaccination against a phage.

Since the 1990s, exploiting CRISPR spacer-based hypervariability for genotyping has been used in various industries to monitor bacteria of interest used in commercial settings, such as starter cultures in the dairy industry, probiotic strains in the dietary supplement industry, and food spoilage strains (66) and pathogenic strains (Fig. 15.6) (67). In some specific cases, CRISPR loci have been deciphered in hundreds and even thousands of strains to assess geographical origin, compare and contrast commercial products, and reconstitute disease history. For instance, DuPont has been sequencing the CRISPR loci of *S. thermophilus* strains widely used in dairy cultures across the globe for nearly 2 decades (27), mapping strain genotypes across products, manufacturers, and starter culture providers. To a lesser extent, this has been used in the dairy industry to monitor *Lactobacillus casei* and other widely used lactobacilli, as well as probiotic strains encompassing *Bifidobacterium animalis* subsp. *lactis* (68) and *Bifidobacterium longum* (see selected CRISPR spacer genotypes in Fig. 15.6).

Importantly, many studies have focused on using CRISPR hypervariability for typing human pathogens. Given the historical focus on bacterial species associated with infectious disease and their involvement in documented human infections and outbreaks, several studies have illustrated the potential of CRISPR-based genotyping for pathogen detection and monitoring and chronicled the phylogenetics of several organisms of medical interest, such as *Salmonella* (69, 70), *E. coli* (71), and *Clostridioides difficile* (72). Several pathogenic species have been subjected to extensive genome sequencing projects in the past 30 years and to CRISPR-focused genotyping analyses documenting the high resolution with which species that carry active CRISPR-Cas systems can be typed over space and time. Several studies have documented the congruence between CRISPR and serotyping in *Salmonella*

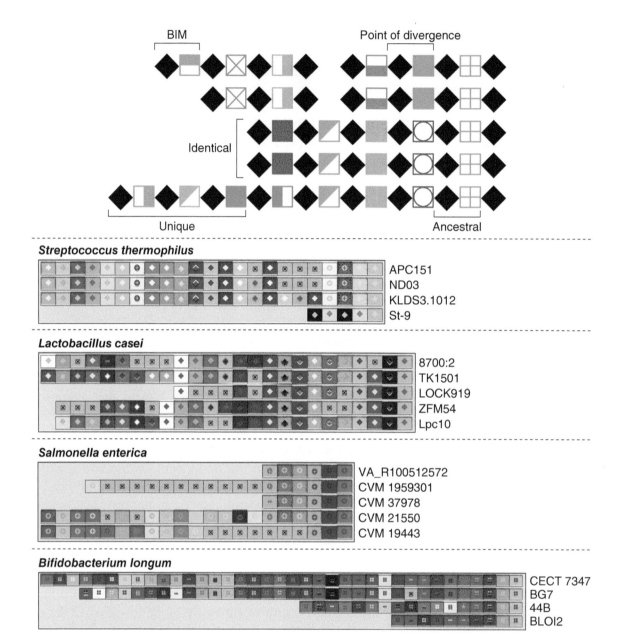

Figure 15.6 CRISPR-based bacterial genotyping. The CRISPR locus spacer content can be used to decipher phylogenetic relationships between related strains, with conserved spacers that align at the ancestral end and divergence over time as novel spacers get acquired, leading to clusters of identical strains, unique combinations of spacers in selected variants, and the visible addition of a recently acquired spacer in a bacteriophage-insensitive mutant (BIM). This scheme can be used to analyze commercial and product isolates in dairy cultures (e.g., *Streptococcus thermophilus* and *Lactobacillus casei*), in human pathogens (e.g., *Salmonella enterica*), and in probiotics (e.g., *Bifidobacterium longum*).

and *E. coli*, as well as the ability to implicate select CRISPR genotypes with specific food products, consumers, patients, and even hospitals. For instance, several studies have consistently documented that CRISPR-based typing in *Salmonella*, especially *Salmonella enterica*, is highly valuable and congruent with serovar clustering and other genotyping schemes such as multi-virulence-locus sequence typing and pulsed-field gel electrophoresis (70, 73–76).

Besides unique strains and isolates that have been microbiologically isolated and industrially manufactured, the CRISPR-based strain genotyping approach can also be implemented in mixed bacterial communities and applied to reconstitute strain-level genotypes in microbiomes (77). Conveniently, CRISPR identification tools (78) (Fig. 15.3) can be applied to metagenomic data to retrieve reads or contigs encompassing CRISPR repeats and reconstitute arrays with overlapping spacer content to reconstruct microbial populations and even assess genetic diversity at the strain level, including conserved ancestral spacers and recent vaccination events (Fig. 15.6). These approaches have been used in various metagenomic data sets to investigate bacterial population composition and variability in space and time. For instance, these loci have provided insights into the interplay between microbial host populations and corresponding viromes over time in environmental microbial communities (79–82). Furthermore, matches between CRISPR spacers and complementary protospacers in invasive elements also allow the study of the relationship between viral and bacterial populations and deciphering of the coevolutionary dynamics between viromes and microbiomes in various habitats (83–86).

Applications of CRISPR Immunity

While the advent of CRISPR-based genome editing technologies has revolutionized genetics and enabled the manipulation of genomes, transcriptomes, and epigenomes across the tree of life, the use of various CRISPR-based technologies in bacteria is on the rise. The phage resistance approaches discussed here in the context of dairy cultures can be implemented in a plethora of industrial bacteria and used to build resistance against various undesirable genetic elements and vaccinate diverse bacteria against a broad range of invasive nucleic acids (Fig. 15.7). For instance, antibiotic resistance genes and virulence cassettes can be targeted in pathogenic bacteria to reduce pathogenicity and virulence (e.g., in methicillin-resistant *Staphylococcus aureus*) and address both infectious disease in general and the rise of antibiotic resistance in particular. It was actually shown over a decade ago that spacer acquisition from plasmids carrying antibiotic resistance genes can provide vaccination against such undesirable mobile genetic elements (29), and this approach has been used to some extent to provide proof of concept for resensitization to antibiotics in bacteria. In industrial

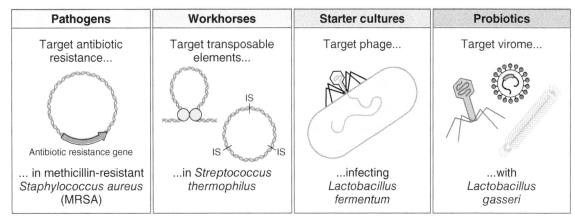

Figure 15.7 CRISPR immunity applications. CRISPR-Cas systems can be repurposed to provide adaptive or engineered immunity in various organisms for diverse purposes. In pathogens such as methicillin-resistant *Staphylococcus aureus*, antibiotic resistance genes can be targeted. In industrial workhorses used in biomanufacturing, transposable elements such as insertion sequence elements can be targeted to promote genomic stability in dairy cultures such as *Streptococcus thermophilus*. Predatory phages can be targeted in commercial starter cultures used in vegetable fermentations, such as *Lactobacillus fermentum*. In various probiotics, bacteriophages, such as the *Myoviridae*, are prevalent in the virome and can be targeted to enhance host gut colonization of *Lactobacillus gasseri*.

workhorses used for biomanufacturing, immunity against transposable elements could be built to remove insertion sequences and cure genomes out of insertion sequences and transposons to provide genomic stability. The aforementioned phage resistance scheme used to blend dairy cultures can be implemented in various bacteria widely used in the food supply chain to carry out fermentation processes, encompassing dairy, meats, vegetables, and grains (87). This approach can also be used to engineer next-generation beneficial bacteria such as probiotics used in the food supply chain and dietary supplement industry and, increasingly so, in bacteria formulated and engineered by microbiome companies to address various intestinal diseases and associated conditions (88). Aptly, our understanding of the virome now complements our appreciation for the importance of the microbiome in human health and disease, and as novel biotherapeutics are designed and engineered to promote human health, it is anticipated that enabling colonization and expanding life span *in vivo* should involve CRISPR-based vaccination against bacteriophages common to the niche in which these bacterial strains operate. Likewise, this concept can be implemented *ex vivo* in environmental habitats where bacteria are used for bioremediation, to prevent the uptake of undesirable genetic material or fend off widespread bacteriophages that are ubiquitously occurring across the planet.

One particular phylogenetic clade in which CRISPR-Cas systems have been extensively characterized in their native hosts, and also encompass genera of broad industrial interest (e.g., *Lactobacillus* and *Bifidobacterium*), is the lactic acid bacteria, which hold tremendous potential at the intersection of CRISPR biology and industrial microbiology (89, 90). Indeed, several comprehensive studies have determined the occurrence, diversity, and activity of CRISPR-Cas systems in bifidobacteria (91–93) and lactobacilli (94, 95). This is concurrent with substantial progress in genomic characterization of food bacteria (96). Thus, future CRISPR-based immunization strategies are anticipated to enhance colonization and extend life span *in vivo*, and CRISPR-enhanced probiotics, CRISPRbiotics, will continue to be developed for next-generation beneficial bacteria with enhanced functionalities.

Besides the harnessing of natural acquisition of phage spacers, our advancing understanding of primed adaptation (97, 98) can enable the nudging of the spacer integration process, and our ability to prompt the integration of select spacers from engineered plasmids (99) can also trigger the acquisition of designed spacers for industrial applications.

Conclusions and Outlook

The biological role of CRISPR-Cas systems as providers of adaptive immunity in bacteria can be exploited for the natural selection of vaccinated cells resistant to phages or the engineering of immunization against a broad spectrum of invasive genetic elements. The original proof of concept obtained in the dairy bacterium *Streptococcus thermophilus* has actually been exploited to build up enhanced broad-spectrum phage resistance in blended dairy starter cultures that have been made commercially available globally for a decade in the manufacturing of cheese and yoghurt. This approach can now be applied to various bacteria of industrial interest for biomanufacturing of enzymes and bioproducts, for cultures used in food fermentations, and in individual strains formulated to manipulate microbiomes associated with health and disease. There are opportunities to repurpose both endogenous and exogenous systems that are programmed to specifically target phages of interest and enhance phage resistance in cultures of interest, especially with proof of concept that several endogenous CRISPR-Cas systems work in industrial lactobacilli (94, 100).

Recent advances in microbiome research, in combination with the developments of CRISPR-based biotechnologies and novel synthetic biology techniques, altogether usher in a new era for the engineering of enhanced bacteria that shape the composition and function of microbiomes of interest. It is thus anticipated that these powerful technologies, encompassing CRISPR-encoded resistance against ubiquitous bacteriophages, will be widely used for the development of next-generation industrial bacteria with enhanced robustness, fitness, competitiveness, and longevity, to be used in various food, agriculture, biotechnological, and medical applications. Besides CRISPR, recently identified phage defense systems that lurk in the uncharacterized dark matter of microbial pan-genomes could also be used to enhance phage resistance in industrial cultures in the future. Likewise, beyond phage resistance, other applications of CRISPR-Cas systems can be used to enhance industrial cultures, including genome editing, screening, antibacterials (101), and molecular recordings, as discussed elsewhere in this volume.

Acknowledgments
We acknowledge invaluable discussions with and insights from the many collaborators and colleagues who provided input over the last two decades of CRISPR research, especially coinventors and coauthors.

Rodolphe Barrangou is a shareholder of DuPont, Caribou Biosciences, Intellia Therapeutics, Locus Biosciences, Inari Agriculture, TreeCo, Ancilia Therapeutics, and CRISPR Biotechnologies.

References

1. Dion MB, Oechslin F, Moineau S. 2020. Phage diversity, genomics and phylogeny. *Nat Rev Microbiol* **18**:125–138.

2. Bernheim A, Sorek R. 2020. The pan-immune system of bacteria: antiviral defence as a community resource. *Nat Rev Microbiol* **18**:113–119.

3. Gao L, Altae-Tran H, Böhning F, Makarova KS, Segel M, Schmid-Burgk JL, Koob J, Wolf YI, Koonin EV, Zhang F. 2020. Diverse enzymatic activities mediate antiviral immunity in prokaryotes. *Science* **369**:1077–1084.

4. Doron S, Melamed S, Ofir G, Leavitt A, Lopatina A, Keren M, Amitai G, Sorek R. 2018. Systematic discovery of antiphage defense systems in the microbial pangenome. *Science* **359**:eaar4120.

5. Sontheimer EJ, Barrangou R. 2015. The bacterial origins of the CRISPR genome-editing revolution. *Hum Gene Ther* **26**:413–424.

6. Barrangou R, Horvath P. 2017. A decade of discovery: CRISPR functions and applications. *Nat Microbiol* **2**:17092.

7. Barrangou R, Fremaux C, Deveau H, Richards M, Boyaval P, Moineau S, Romero DA, Horvath P. 2007. CRISPR provides acquired resistance against viruses in prokaryotes. *Science* **315**:1709–1712.

8. Brouns SJ, Jore MM, Lundgren M, Westra ER, Slijkhuis RJ, Snijders AP, Dickman MJ, Makarova KS, Koonin EV, van der Oost J. 2008. Small CRISPR RNAs guide antiviral defense in prokaryotes. *Science* **321**:960–964.

9. Marraffini LA, Sontheimer EJ. 2008. CRISPR interference limits horizontal gene transfer in staphylococci by targeting DNA. *Science* **322**:1843–1845.

10. Ishino Y, Shinagawa H, Makino K, Amemura M, Nakata A. 1987. Nucleotide sequence of the iap gene, responsible for alkaline phosphatase isozyme conversion in Escherichia coli, and identification of the gene product. *J Bacteriol* **169**:5429–5433.

11. Jansen R, Embden JD, Gaastra W, Schouls LM. 2002. Identification of genes that are associated with DNA repeats in prokaryotes. *Mol Microbiol* **43**:1565–1575.

12. Bolotin A, Quinquis B, Sorokin A, Ehrlich SD. 2005. Clustered regularly interspaced short palindrome repeats (CRISPRs) have spacers of extrachromosomal origin. *Microbiology (Reading)* **151**:2551–2561.

13. Pourcel C, Salvignol G, Vergnaud G. 2005. CRISPR elements in Yersinia pestis acquire new repeats by preferential uptake of bacteriophage DNA, and provide additional tools for evolutionary studies. *Microbiology (Reading)* **151**:653–663.

14. Mojica FJ, Díez-Villaseñor C, García-Martínez J, Soria E. 2005. Intervening sequences of regularly spaced prokaryotic repeats derive from foreign genetic elements. *J Mol Evol* **60**:174–182.

15. Makarova KS, Haft DH, Barrangou R, Brouns SJ, Charpentier E, Horvath P, Moineau S, Mojica FJ, Wolf YI, Yakunin AF, van der Oost J, Koonin EV. 2011. Evolution and classification of the CRISPR-Cas systems. *Nat Rev Microbiol* **9**:467–477.

16. Makarova KS, Wolf YI, Alkhnbashi OS, Costa F, Shah SA, Saunders SJ, Barrangou R, Brouns SJ, Charpentier E, Haft DH, Horvath P, Moineau S, Mojica FJ, Terns RM, Terns MP, White MF, Yakunin AF, Garrett RA, van der Oost J, Backofen R, Koonin EV. 2015. An updated evolutionary classification of CRISPR-Cas systems. *Nat Rev Microbiol* **13**:722–736.

17. Makarova KS, Wolf YI, Iranzo J, Shmakov SA, Alkhnbashi OS, Brouns SJJ, Charpentier E, Cheng D, Haft DH, Horvath P, Moineau S, Mojica FJM, Scott D, Shah SA, Siksnys V, Terns MP, Venclovas Č, White MF, Yakunin AF, Yan W, Zhang F, Garrett RA, Backofen R, van der Oost J, Barrangou R, Koonin EV. 2020. Evolutionary classification of CRISPR-Cas systems: a burst of class 2 and derived variants. *Nat Rev Microbiol* **18**:67–83.

18. Hale CR, Zhao P, Olson S, Duff MO, Graveley BR, Wells L, Terns RM, Terns MP. 2009. RNA-guided RNA cleavage by a CRISPR RNA-Cas protein complex. *Cell* **139**:945–956.

19. Burstein D, Harrington LB, Strutt SC, Probst AJ, Anantharaman K, Thomas BC, Doudna JA, Banfield JF. 2017. New CRISPR-Cas systems from uncultivated microbes. *Nature* **542**:237–241.

20. Barrangou R, Horvath P. 2012. CRISPR: new horizons in phage resistance and strain identification. *Annu Rev Food Sci Technol* **3**:143–162.

21. Hao M, Cui Y, Qu X. 2018. Analysis of CRISPR-Cas system in Streptococcus thermophilus and its application. *Front Microbiol* **9**:257.

22. Mills S, Griffin C, Coffey A, Meijer WC, Hafkamp B, Ross RP. 2010. CRISPR analysis of bacteriophage-insensitive mutants (BIMs)

of industrial Streptococcus thermophilus—implications for starter design. *J Appl Microbiol* **108**:945–955.

23. Delorme C, Legravet N, Jamet E, Hoarau C, Alexandre B, El-Sharoud WM, Darwish MS, Renault P. 2017. Study of Streptococcus thermophilus population on a world-wide and historical collection by a new MLST scheme. *Int J Food Microbiol* **242**:70–81.

24. Hynes AP, Lemay ML, Trudel L, Deveau H, Frenette M, Tremblay DM, Moineau S. 2017. Detecting natural adaptation of the Streptococcus thermophilus CRISPR-Cas systems in research and classroom settings. *Nat Protoc* **12**:547–565.

25. McDonnell B, Mahony J, Hanemaaijer L, Kouwen TRHM, van Sinderen D. 2018. Generation of bacteriophage-insensitive mutants of Streptococcus thermophilus via an antisense RNA CRISPR-Cas silencing approach. *Appl Environ Microbiol* **84**:e01733-17.

26. McGinn J, Marraffini LA. 2019. Molecular mechanisms of CRISPR-Cas spacer acquisition. *Nat Rev Microbiol* **17**:7–12.

27. Horvath P, Romero DA, Coûté-Monvoisin AC, Richards M, Deveau H, Moineau S, Boyaval P, Fremaux C, Barrangou R. 2008. Diversity, activity, and evolution of CRISPR loci in Streptococcus thermophilus. *J Bacteriol* **190**:1401–1412.

28. Sapranauskas R, Gasiunas G, Fremaux C, Barrangou R, Horvath P, Siksnys V. 2011. The Streptococcus thermophilus CRISPR/Cas system provides immunity in Escherichia coli. *Nucleic Acids Res* **39**:9275–9282.

29. Garneau JE, Dupuis ME, Villion M, Romero DA, Barrangou R, Boyaval P, Fremaux C, Horvath P, Magadán AH, Moineau S. 2010. The CRISPR/Cas bacterial immune system cleaves bacteriophage and plasmid DNA. *Nature* **468**:67–71.

30. Gasiunas G, Barrangou R, Horvath P, Siksnys V. 2012. Cas9-crRNA ribonucleoprotein complex mediates specific DNA cleavage for adaptive immunity in bacteria. *Proc Natl Acad Sci USA* **109**:E2579–E2586.

31. Deveau H, Barrangou R, Garneau JE, Labonté J, Fremaux C, Boyaval P, Romero DA, Horvath P, Moineau S. 2008. Phage response to CRISPR-encoded resistance in Streptococcus thermophilus. *J Bacteriol* **190**:1390–1400.

32. Semenova E, Jore MM, Datsenko KA, Semenova A, Westra ER, Wanner B, van der Oost J, Brouns SJ, Severinov K. 2011. Interference by clustered regularly interspaced short palindromic repeat (CRISPR) RNA is governed by a seed sequence. *Proc Natl Acad Sci USA* **108**:10098–10103.

33. Wiedenheft B, van Duijn E, Bultema JB, Waghmare SP, Zhou K, Barendregt A, Westphal W, Heck AJ, Boekema EJ, Dickman MJ, Doudna JA. 2011. RNA-guided complex from a bacterial immune system enhances target recognition through seed sequence interactions. *Proc Natl Acad Sci USA* **108**:10092–10097.

34. Szczelkun MD, Tikhomirova MS, Sinkunas T, Gasiunas G, Karvelis T, Pschera P, Siksnys V, Seidel R. 2014. Direct observation of R-loop formation by single RNA-guided Cas9 and Cascade effector complexes. *Proc Natl Acad Sci USA* **111**:9798–9803.

35. Paez-Espino D, Sharon I, Morovic W, Stahl B, Thomas BC, Barrangou R, Banfield JF. 2015. CRISPR immunity drives rapid phage genome evolution in Streptococcus thermophilus. *mBio* **6**:e00262-15.

36. Sun CL, Barrangou R, Thomas BC, Horvath P, Fremaux C, Banfield JF. 2013. Phage mutations in response to CRISPR diversification in a bacterial population. *Environ Microbiol* **15**:463–470.

37. Bondy-Denomy J, Pawluk A, Maxwell KL, Davidson AR. 2013. Bacteriophage genes that inactivate the CRISPR/Cas bacterial immune system. *Nature* **493**:429–432.

38. Bondy-Denomy J, Garcia B, Strum S, Du M, Rollins MF, Hidalgo-Reyes Y, Wiedenheft B, Maxwell KL, Davidson AR. 2015. Multiple mechanisms for CRISPR-Cas inhibition by anti-CRISPR proteins. *Nature* **526**:136–139.

39. Nethery MA, Barrangou R. 2019. Predicting and visualizing features of CRISPR-Cas systems. *Methods Enzymol* **616**:1–25.

40. Nethery MA, Barrangou R. 2019. CRISPR Visualizer: rapid identification and visualization of CRISPR loci via an automated high-throughput processing pipeline. *RNA Biol* **16**:577–584.

41. Briner AE, Barrangou R. 2016. Guide RNAs: a glimpse at the sequences that drive CRISPR-Cas systems. *Cold Spring Harb Protoc* **2016**:pdb.top090902.

42. Briner AE, Henriksen ED, Barrangou R. 2016. Prediction and validation of native and engineered Cas9 guide sequences. *Cold Spring Harb Protoc* **2016**:pdb.prot086785.

43. Mojica FJM, Díez-Villaseñor C, García-Martínez J, Almendros C. 2009. Short motif sequences determine the targets of the prokaryotic CRISPR defence system. *Microbiol Read* **155**:733–740.

44. Karvelis T, Gasiunas G, Young J, Bigelyte G, Silanskas A, Cigan M, Siksnys V. 2015. Rapid characterization of CRISPR-Cas9 protospacer adjacent motif sequence elements. *Genome Biol* **16**:253.

45. Leenay RT, Maksimchuk KR, Slotkowski RA, Agrawal RN, Gomaa AA, Briner AE, Barrangou R, Beisel CL. 2016. Identifying and visualizing functional PAM diversity across CRISPR-Cas systems. *Mol Cell* **62**:137–147.

46. Maxwell CS, Jacobsen T, Marshall R, Noireaux V, Beisel CL. 2018. A detailed cell-free transcription-translation-based assay to decipher CRISPR protospacer-adjacent motifs. *Methods* **143**:48–57.

47. Dupuis ME, Barrangou R, Moineau S. 2015. Procedures for generating CRISPR mutants with novel spacers acquired from viruses or plasmids. *Methods Mol Biol* **1311**:195–222.

48. Horvath P, Barrangou R. 2010. CRISPR/Cas, the immune system of bacteria and archaea. *Science* **327**:167–170.

49. Briner AE, Donohoue PD, Gomaa AA, Selle K, Slorach EM, Nye CH, Haurwitz RE, Beisel CL, May AP, Barrangou R. 2014. Guide RNA functional modules direct Cas9 activity and orthogonality. *Mol Cell* **56**:333–339.

50. Carte J, Christopher RT, Smith JT, Olson S, Barrangou R, Moineau S, Glover CV, III, Graveley BR, Terns RM, Terns MP. 2014. The three major types of CRISPR-Cas systems function independently in CRISPR RNA biogenesis in Streptococcus thermophilus. *Mol Microbiol* **93**:98–112.

51. Karvelis T, Gasiunas G, Miksys A, Barrangou R, Horvath P, Siksnys V. 2013. crRNA and tracrRNA guide Cas9-mediated DNA interference in Streptococcus thermophilus. *RNA Biol* **10**:841–851.

52. Young JC, Dill BD, Pan C, Hettich RL, Banfield JF, Shah M, Fremaux C, Horvath P, Barrangou R, Verberkmoes NC. 2012. Phage-induced expression of CRISPR-associated proteins is revealed by shotgun proteomics in Streptococcus thermophilus. *PLoS One* **7**:e38077.

53. Levy A, Goren MG, Yosef I, Auster O, Manor M, Amitai G, Edgar R, Qimron U, Sorek R. 2015. CRISPR adaptation biases explain preference for acquisition of foreign DNA. *Nature* **520**:505–510.

54. Paez-Espino D, Morovic W, Sun CL, Thomas BC, Ueda K, Stahl B, Barrangou R, Banfield JF. 2013. Strong bias in the bacterial CRISPR elements that confer immunity to phage. *Nat Commun* **4**:1430.

55. Jinek M, Chylinski K, Fonfara I, Hauer M, Doudna JA, Charpentier E. 2012. A programmable dual-RNA-guided DNA endonuclease in adaptive bacterial immunity. *Science* **337**:816–821.

56. Cong L, Ran FA, Cox D, Lin S, Barretto R, Habib N, Hsu PD, Wu X, Jiang W, Marraffini LA, Zhang F. 2013. Multiplex genome engineering using CRISPR/Cas systems. *Science* **339**:819–823.

57. Mali P, Yang L, Esvelt KM, Aach J, Guell M, DiCarlo JE, Norville JE, Church GM. 2013. RNA-guided human genome engineering via Cas9. *Science* **339**:823–826.

58. Jiang W, Bikard D, Cox D, Zhang F, Marraffini LA. 2013. RNA-guided editing of bacterial genomes using CRISPR-Cas systems. *Nat Biotechnol* **31**:233–239.

59. Barrangou R, Coûté-Monvoisin AC, Stahl B, Chavichvily I, Damange F, Romero DA, Boyaval P, Fremaux C, Horvath P. 2013. Genomic impact of CRISPR immunization against bacteriophages. *Biochem Soc Trans* **41**:1383–1391.

60. Chabas H, Nicot A, Meaden S, Westra ER, Tremblay DM, Pradier L, Lion S, Moineau S, Gandon S. 2019. Variability in the durability of CRISPR-Cas immunity. *Philos Trans R Soc Lond B Biol Sci* **374**:20180097.

61. Chevallereau A, Meaden S, van Houte S, Westra ER, Rollie C. 2019. The effect of bacterial mutation rate on the evolution of CRISPR-Cas adaptive immunity. *Philos Trans R Soc Lond B Biol Sci* 374:20180094.

62. Common J, Morley D, Westra ER, van Houte S. 2019. CRISPR-Cas immunity leads to a coevolutionary arms race between Streptococcus thermophilus and lytic phage. *Philos Trans R Soc Lond B Biol Sci* 374:20180098.

63. Common J, Walker-Sünderhauf D, van Houte S, Westra ER. 2020. Diversity in CRISPR-based immunity protects susceptible genotypes by restricting phage spread and evolution. *J Evol Biol* 33:1097–1108.

64. Meaden S, Capria L, Alseth E, Gandon S, Biswas A, Lenzi L, van Houte S, Westra ER. 2020. Phage gene expression and host responses lead to infection-dependent costs of CRISPR immunity. *ISME J*.

65. van Houte S, Ekroth AK, Broniewski JM, Chabas H, Ashby B, Bondy-Denomy J, Gandon S, Boots M, Paterson S, Buckling A, Westra ER. 2016. The diversity-generating benefits of a prokaryotic adaptive immune system. *Nature* 532:385–388.

66. Briner AE, Barrangou R. 2014. Lactobacillus buchneri genotyping on the basis of clustered regularly interspaced short palindromic repeat (CRISPR) locus diversity. *Appl Environ Microbiol* 80:994–1001.

67. Barrangou R, Dudley EG. 2016. CRISPR-based typing and next-generation tracking technologies. *Annu Rev Food Sci Technol* 7:395–411.

68. Briczinski EP, Loquasto JR, Barrangou R, Dudley EG, Roberts AM, Roberts RF. 2009. Strain-specific genotyping of Bifidobacterium animalis subsp. lactis by using single-nucleotide polymorphisms, insertions, and deletions. *Appl Environ Microbiol* 75:7501–7508.

69. DiMarzio M, Shariat N, Kariyawasam S, Barrangou R, Dudley EG. 2013. Antibiotic resistance in Salmonella enterica serovar Typhimurium associates with CRISPR sequence type. *Antimicrob Agents Chemother* 57:4282–4289.

70. Shariat N, DiMarzio MJ, Yin S, Dettinger L, Sandt CH, Lute JR, Barrangou R, Dudley EG. 2013. The combination of CRISPR-MVLST and PFGE provides increased discriminatory power for differentiating human clinical isolates of Salmonella enterica subsp. enterica serovar Enteritidis. *Food Microbiol* 34:164–173.

71. Yin S, Jensen MA, Bai J, Debroy C, Barrangou R, Dudley EG. 2013. The evolutionary divergence of Shiga toxin-producing Escherichia coli is reflected in clustered regularly interspaced short palindromic repeat (CRISPR) spacer composition. *Appl Environ Microbiol* 79:5710–5720.

72. Andersen JM, Shoup M, Robinson C, Britton R, Olsen KE, Barrangou R. 2016. CRISPR diversity and microevolution in Clostridium difficile. *Genome Biol Evol* 8:2841–2855.

73. Liu F, Barrangou R, Gerner-Smidt P, Ribot EM, Knabel SJ, Dudley EG. 2011. Novel virulence gene and clustered regularly interspaced short palindromic repeat (CRISPR) multilocus sequence typing scheme for subtyping of the major serovars of Salmonella enterica subsp. enterica. *Appl Environ Microbiol* 77:1946–1956.

74. Shariat N, Kirchner MK, Sandt CH, Trees E, Barrangou R, Dudley EG. 2013. Subtyping of Salmonella enterica serovar Newport outbreak isolates by CRISPR-MVLST and determination of the relationship between CRISPR-MVLST and PFGE results. *J Clin Microbiol* 51:2328–2336.

75. Shariat N, Sandt CH, DiMarzio MJ, Barrangou R, Dudley EG. 2013. CRISPR-MVLST subtyping of Salmonella enterica subsp. enterica serovars Typhimurium and Heidelberg and application in identifying outbreak isolates. *BMC Microbiol* 13:254.

76. Shariat N, Timme RE, Pettengill JB, Barrangou R, Dudley EG. 2015. Characterization and evolution of Salmonella CRISPR-Cas systems. *Microbiology* 161:374–386.

77. Briner AE, Barrangou R. 2016. Deciphering and shaping bacterial diversity through CRISPR. *Curr Opin Microbiol* 31:101–108.

78. Crawley AB, Henriksen JR, Barrangou R. 2018. CRISPRdisco: an automated pipeline for the discovery and analysis of CRISPR-Cas systems. *CRISPR J* 1:171–181.

79. Andersson AF, Banfield JF. 2008. Virus population dynamics and acquired virus resistance in natural microbial communities. *Science* 320:1047–1050.

80. Hidalgo-Cantabrana C, Sanozky-Dawes R, Barrangou R. 2018. Insights into the human virome using CRISPR spacers from microbiomes. *Viruses* 10:479.

81. Levin BR, Moineau S, Bushman M, Barrangou R. 2013. The population and evolutionary dynamics of phage and bacteria with CRISPR-mediated immunity. *PLoS Genet* 9:e1003312.

82. Tyson GW, Banfield JF. 2008. Rapidly evolving CRISPRs implicated in acquired resistance of microorganisms to viruses. *Environ Microbiol* 10:200–207.

83. Heidelberg JF, Nelson WC, Schoenfeld T, Bhaya D. 2009. Germ warfare in a microbial mat community: CRISPRs provide insights into the co-evolution of host and viral genomes. *PLoS One* 4:e4169.

84. Pride DT, Sun CL, Salzman J, Rao N, Loomer P, Armitage GC, Banfield JF, Relman DA. 2011. Analysis of streptococcal CRISPRs from human saliva reveals substantial sequence diversity within and between subjects over time. *Genome Res* 21:126–136.

85. Naidu M, Robles-Sikisaka R, Abeles SR, Boehm TK, Pride DT. 2014. Characterization of bacteriophage communities and CRISPR profiles from dental plaque. *BMC Microbiol* 14:175.

86. Held NL, Whitaker RJ. 2009. Viral biogeography revealed by signatures in Sulfolobus islandicus genomes. *Environ Microbiol* 11:457–466.

87. Barrangou R, Notebaart RA. 2019. CRISPR-directed microbiome manipulation across the food supply chain. *Trends Microbiol* 27:489–496.

88. Brandt K, Barrangou R. 2019. Applications of CRISPR technologies across the food supply chain. *Annu Rev Food Sci Technol* 10:133–150.

89. Horvath P, Coûté-Monvoisin AC, Romero DA, Boyaval P, Fremaux C, Barrangou R. 2009. Comparative analysis of CRISPR loci in lactic acid bacteria genomes. *Int J Food Microbiol* 131:62–70.

90. Roberts A, Barrangou R. 2020. Applications of CRISPR-Cas systems in lactic acid bacteria. *FEMS Microbiol Rev* 44:523–537.

91. Briner AE, Lugli GA, Milani C, Duranti S, Turroni F, Gueimonde M, Margolles A, van Sinderen D, Ventura M, Barrangou R. 2015. Occurrence and diversity of CRISPR-Cas systems in the genus Bifidobacterium. *PLoS One* 10:e0133661.

92. Hidalgo-Cantabrana C, Crawley AB, Sanchez B, Barrangou R. 2017. Characterization and exploitation of CRISPR loci in Bifidobacterium longum. *Front Microbiol* 8:1851.

93. Pan M, Nethery MA, Hidalgo-Cantabrana C, Barrangou R. 2020. Comprehensive mining and characterization of CRISPR-Cas systems in Bifidobacterium. *Microorganisms* 8:720.

94. Crawley AB, Henriksen ED, Stout E, Brandt K, Barrangou R. 2018. Characterizing the activity of abundant, diverse and active CRISPR-Cas systems in lactobacilli. *Sci Rep* 8:11544.

95. Sun Z, Harris HM, McCann A, Guo C, Argimón S, Zhang W, Yang X, Jeffery IB, Cooney JC, Kagawa TF, Liu W, Song Y, Salvetti E, Wrobel A, Rasinkangas P, Parkhill J, Rea MC, O'Sullivan O, Ritari J, Douillard FP, Paul Ross R, Yang R, Briner AE, Felis GE, de Vos WM, Barrangou R, Klaenhammer TR, Caufield PW, Cui Y, Zhang H, O'Toole PW. 2015. Expanding the biotechnology potential of lactobacilli through comparative genomics of 213 strains and associated genera. *Nat Commun* 6:8322.

96. Pan M, Barrangou R. 2020. Combining omics technologies with CRISPR-based genome editing to study food microbes. *Curr Opin Biotechnol* 61:198–208.

97. Datsenko KA, Pougach K, Tikhonov A, Wanner BL, Severinov K, Semenova E. 2012. Molecular memory of prior infections activates the CRISPR/Cas adaptive bacterial immunity system. *Nat Commun* 3:945.

98. Fineran PC, Gerritzen MJ, Suárez-Diez M, Künne T, Boekhorst J, van Hijum SA, Staals RH, Brouns SJ. 2014. Degenerate target sites mediate rapid primed CRISPR adaptation. *Proc Natl Acad Sci USA* 111:E1629–E1638.

99. **Hynes AP, Labrie SJ, Moineau S.** 2016. Programming native CRISPR arrays for the generation of targeted immunity. *mBio* 7:e00202-16.

100. **Hidalgo-Cantabrana C, Goh YJ, Pan M, Sanozky-Dawes R, Barrangou R.** 2019. Genome editing using the endogenous type I CRISPR-Cas system in Lactobacillus crispatus. *Proc Natl Acad Sci USA* **116:**15774–15783.

101. **Gomaa AA, Klumpe HE, Luo ML, Selle K, Barrangou R, Beisel CL.** 2014. Programmable removal of bacterial strains by use of genome-targeting CRISPR-Cas systems. *mBio* **5:**e00928-13.

CHAPTER

Recording Biological Information with CRISPR-Cas Systems

Mariia Y. Cherepkova[1], Tanmay Tanna[1], and Randall J. Platt[1,2]

[1]Department of Biosystems Science and Engineering, ETH Zurich, 4058 Basel, Switzerland
[2]Department of Chemistry, University of Basel, 4003, Basel, Switzerland

Introduction

Biological systems—in all their diverse forms, ranging from macromolecules to multicellular organisms—are highly complex and dynamic. For instance, biological states for single cells are determined by an interplay of genetic and epigenetic regulation, transcriptional, translational, and posttranslational events, sensing mechanisms for external stimuli, signaling cascades, and microenvironments within the cell and cell niches. With such complexity, a stable yet evolvable medium for encoding, transmitting, and utilizing biological information is essential, and DNA has emerged as the evolutionarily preferred solution. Recent advances in genome editing, coupled with substantial reductions in DNA sequencing and synthesis costs, have opened new opportunities for harnessing this potential of DNA as an information storage system for biological applications (1).

One of the main challenges of biology is to understand how molecular events give rise to complex cellular behaviors and states and guide the development of multicellular systems. Most current technologies for studying these processes, however, are disruptive, therefore limiting observations to a few snapshots in time and resulting in a readout lacking in information that associates one state with the following. One potential solution to this challenge is to convert cells into biological recorders that continuously capture information about dynamic cellular processes and stably encode it in DNA (2, 3). This memorized information can then be read out from populations of cells to provide longitudinal insights into cellular events, such as the abundance of biomolecules over time or cellular lineage of multicellular systems such as organisms, organoids, or tumors.

A major step taken in this direction was the development of DNA writing, a process by which biological or artificial information is encoded through modifications in DNA using natural or engineered enzymes termed DNA writers. DNA modifications, in the form of mutations such as insertions, deletions, inversions, or base substitutions, serve as memory states corresponding to input signals (2, 4). The first DNA writers relied on synthetic gene circuits comprised of toggle switches (5) or site-specific DNA recombinases (6). While these advances established the principles of DNA writing in living cells, these systems lacked scalability and orthogonality, precluding their widespread use for interrogating biological systems.

CRISPR: Biology and Applications, First Edition. Edited by Rodolphe Barrangou, Erik J. Sontheimer, and Luciano A. Marraffini.
© 2022 American Society for Microbiology. DOI: 10.1128/9781683673798.ch16

Due to recent advancements in genome editing, fueled by the discovery and exploration of the microbial adaptive immune system CRISPR-Cas, a new generation of DNA writing tools was created and is enabling a diversity of novel applications for investigating and engineering biology in basic research, biotechnology, and medicine (3). In this chapter, we review the development and existing applications of CRISPR-Cas systems for DNA writing and discuss features and current limitations of these technologies for recording biological information in living cells.

DNA Writing with CRISPR-Cas Systems

The natural diversity of CRISPR-Cas systems provides us with a plethora of opportunities for developing DNA writing devices with diverse mutational signatures and response dynamics. The CRISPR-Cas immune response consists of three stages: adaptation, expression/ processing, and interference. During the adaptation stage, a CRISPR acquisition complex comprised of Cas1 and Cas2 proteins recognizes foreign genetic elements and integrates short sequence fragments from these nucleic acids as new "spacers" into the host CRISPR array (7). In the subsequent stages, the acquired spacers are transcribed and processed into small CRISPR RNAs, which can guide effector Cas proteins to the complementary sequence in hostile genetic elements during future invasions, triggering cleavage, degradation, or mutation of the target sequence. Thus, during the natural CRISPR-Cas immune response, both the adaptation and interference stages involve modification of DNA, which can be leveraged for DNA writing.

The molecular machinery of the adaptation module from several microorganisms has been harnessed for DNA writing via spacer acquisition from DNA or RNA (Fig. 16.1) (8–10). The information encoded in the form of spacers can subsequently be retrieved by sequencing CRISPR arrays, either genomic or plasmid borne. Since CRISPR arrays can store multiple spacers, such systems enable multiple rounds of writing at a unique locus and can therefore provide greater information storage capacity. Additionally, new spacers are almost always inserted between a leader sequence and the first direct repeat sequence of the CRISPR array, which allows for resolution of the order of molecular events during recording (7, 11).

DNA writing by one of the most widely used Cas effector proteins, Cas9 from *Streptococcus pyogenes*, has provided a flexible platform for engineering DNA writing tools. The RNA-guided DNA endonuclease Cas9 can be utilized to generate random mutational outcomes at a specific target locus (Fig. 16.1). It recognizes target loci via short RNAs (single guide RNAs [sgRNAs]) and generates double-strand breaks (DSBs). In eukaryotic cells, these DSBs are then repaired through either nonhomologous end joining (NHEJ), which results in the formation of insertions and deletions (together called "indels"), or, less often, through homology-directed repair. Cas9 target sites must contain the recognition protospacer sequence (complementary to sgRNA) and a protospacer adjacent motif (PAM) (12, 13). Thus, a DNA writing module based on Cas9 should include a target locus with the PAM and enable the expression of sgRNA and Cas9.

DNA writers based on a wild-type Cas9 protein are efficient but can generate only a limited number of mutational outcomes and are thus characterized by a short time of writing owing to saturation of target loci, requiring multiple target loci to achieve a high recording capacity. To overcome these limitations, alternative Cas9-based DNA writing modules utilize self-targeting guide RNAs (stgRNAs), also known as homing guide RNAs (hgRNAs) (14, 15) (Fig. 16.1). In these systems, the stgRNA scaffold sequence, which is necessary for Cas binding, carries a GUU-to-GGG mutation immediately downstream of

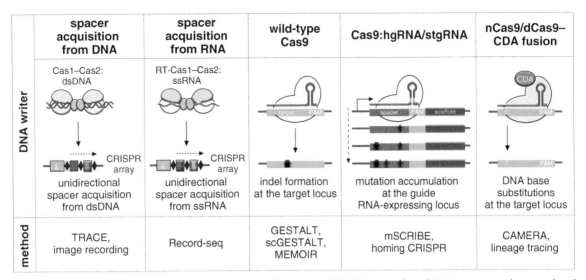

	spacer acquisition from DNA	spacer acquisition from RNA	wild-type Cas9	Cas9:hgRNA/stgRNA	nCas9/dCas9–CDA fusion
DNA writer	Cas1–Cas2: dsDNA CRISPR array unidirectional spacer acquisition from dsDNA	RT-Cas1–Cas2: ssRNA CRISPR array unidirectional spacer acquisition from ssRNA	indel formation at the target locus	mutation accumulation at the guide RNA-expressing locus	DNA base substitutions at the target locus
method	TRACE, image recording	Record-seq	GESTALT, scGESTALT, MEMOIR	mSCRIBE, homing CRISPR	CAMERA, lineage tracing

Figure 16.1 DNA writing by CRISPR-Cas systems. Diversity of CRISPR-Cas-based DNA writers, their mode of action, and applications in molecular recording and lineage tracing.

the spacer sequence, so that it matches the requisite NGG PAM sequence of Cas9. These synthetic stgRNAs direct Cas9 to target the DNA locus of the stgRNA itself, enabling retargeting and evolvability of the target locus and thus providing a higher diversity of mutations and extending the period of DNA writing.

Cas9-based DNA writers that generate mutational outcomes using DSBs are functional only in eukaryotic cells, since NHEJ repair pathways are rare in prokaryotes, and DSBs frequently cause cell death or a loss of extrachromosomal DNA (16).

By engineering Cas9 through the addition of different functional domains, the molecular arsenal of potential genetic modifications has been expanded. For instance, a "base editing" approach uses catalytically disabled (dCas9) or nicking Cas (nCas9) nucleases fused to nucleobase deaminases to insert mutations into DNA at single-base resolution without inducing DSBs, thus being functional in both eukaryotes and prokaryotes. As an example, a fusion of nCas9 to a cytidine deaminase domain (CDA) (Fig. 16.1) converts C:G base pairs to T:A base pairs in a 4- to 8-nucleotide window from the PAM site on the distal side of protospacer sequence (17).

In contrast to other DNA writers (such as site-specific DNA recombinases) that generally allow for only a limited number of predefined mutational outcomes, CRISPR-Cas systems provide a resource for the development of a new class of efficient, scalable, and easily programmable DNA writers. Furthermore, DNA writers that utilize CRISPR-Cas base editing or stgRNAs or leverage CRISPR spacer acquisition allow for multiple cycles of writing (4) onto a single DNA substrate, thus increasing recording capacity. The expanding toolbox of novel and engineered Cas proteins has found wide-ranging DNA writing applications, and here, we focus on two main applications: lineage tracing and molecular recording.

Lineage Tracing Using CRISPR-Cas9

In multicellular organisms, a single totipotent zygote gives rise to all cell types. This process is orchestrated by a plethora of cellular signals and involves multiple cell division and differentiation events that lead the zygote through cellular lineages resulting in different cell types. Lineage tracing methods aim to establish relationships between cell types by following the progeny of a single cell or groups of cells (Fig. 16.2A). Before the development of

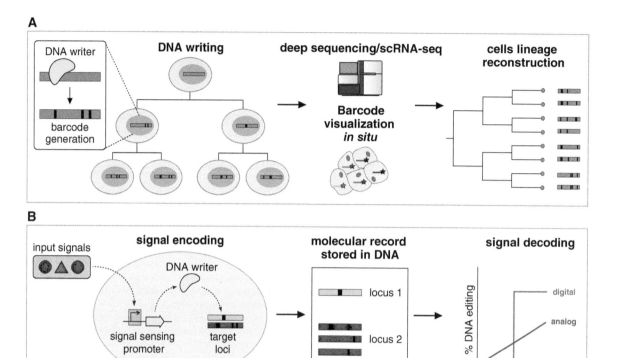

Figure 16.2 Applications of DNA writing. (**A**) Lineage tracing aims to establish relationships between cell types by following the progeny of a single cell or a group of cells. By introducing mutations at target loci, DNA writers generate barcodes over the course of cell division and differentiation. These barcodes can be subsequently retrieved by sequencing or using *in situ* hybridization with multiple fluorescently labeled probes. Subsequently, a lineage tree can be reconstructed using phylogenetic computational approaches. (**B**) Molecular recording aims to capture transient signals and stably encode them into DNA within cells. One example is depicted, wherein input signals activate transcription of a DNA writer from a signal-sensing promoter, leading to the accumulation of mutations at target loci. These molecular records can be sequenced and used to decode information about input signals. In a case of a digital recording regime (blue), molecular records represent the information about presence or absence of input signals, whereas in analog recording (red), mutations gradually accumulate reflecting differing magnitudes or durations of input signals.

CRISPR-Cas genome editing tools, lineage tracing techniques included dye-based markers, transplantation, nucleotide pulse-chase analysis, Cre-Lox, FLP-FRT recombination-based mutagenesis, and sequencing of somatic mutations (18). However, these methods do not provide a detailed lineage tree over time but rather allow cell labeling at single time points. Over the past few years, Cas9-mediated genome editing has proven to be a powerful tool for lineage tracing via the dynamic introduction of stochastic mutations in the genome, thus barcoding individual cells. These barcodes, continuously generated during cellular division and differentiation, facilitate reconstruction of the original lineage relationships. The lineage tree is built based on the diversity of the barcode sequences: the more distant these mutational signatures, the less related the cells.

Leveraging this idea, McKenna and colleagues developed genome editing of synthetic target arrays for lineage tracing (GESTALT), a lineage tracing method based on genome editing of a transgenic reporter sequence in the genome (19). By transfecting cells with a plasmid expressing Cas9 and sgRNAs, or injecting Cas9 ribonucleoproteins (RNPs) containing sgRNAs that target a reporter sequence into embryos, they were able to generate barcodes to record lineage information and demonstrate the reconstruction of lineage relationships in early embryonic development of the zebrafish (*Danio rerio*). The transgenic reporter in GESTALT contains 10 distinct sites in tandem, each of which is targeted by a different sgRNA, resulting in the generation of irreversible edits, subsequently retrieved by DNA sequencing of the GESTALT reporter barcode in adult tissues. Lineage trees are generated using patterns of shared mutations between barcodes, decoding relationships between different cell populations. Although GESTALT can provide the developmental history of a whole organism, this method has important limitations. First, the transient presence of Cas9 and sgRNAs restricts editing to a short time window, allowing for tracking of only a part of embryogenesis. Second, due to destruction of the cells during sample preparation, lineage trees generated using GESTALT lack information about specific cell types associated with identified barcodes.

To overcome these limitations, single-cell GESTALT (scGESTALT) (20) was developed, which combines the lineage recording capabilities of GESTALT with cell type identification by single-cell RNA sequencing (scRNA-seq). In scGESTALT, the edited barcodes are expressed from a reporter transgene consisting of nine distinct CRISPR loci in tandem, enabling detection at the RNA level. To extend the time frame of barcoding, different loci of the reporter are edited at different developmental stages. For barcoding cells during early embryogenesis, Cas9 RNPs containing sgRNAs targeting the first four loci of the reporter are injected at the one-cell stage. At later stages of development, expression of Cas9 is induced by heat shock, while sgRNAs targeting the other five loci are constitutively expressed. scRNA-seq of ~60,000 cells from the developing zebrafish brain identified >100 cell types. Using these data, lineage trees were generated with hundreds of branches that helped uncover relationships at the level of cell types, brain regions, and gene expression cascades during differentiation.

Recently, Chan and colleagues used a similar approach to map cell fates during early development in mouse embryos, from fertilization through gastrulation (21). In their system, a target locus with three Cas9 target sites and a unique barcode, as well as complementary sgRNAs, is delivered to oocytes using a single piggyBAC transposon vector, with multiple target sites being integrated into the genome of each cell. These three target loci are embedded into the 3′ untranslated region of a constitutively transcribed gene encoding a fluorescent protein to enable detection through RNA-seq. Further, the oocytes are fertilized with constitutive Cas9-green fluorescent protein-encoding sperm to initiate DNA writing. After gastrulation, the resulting mutational outcomes at target sites and cell's transcriptional phenotype are simultaneously detected using scRNA-seq. Applying this system, the authors reconstructed lineage identities and also identified phenotypic convergence in cells originating from different embryonic layers. Despite providing cell type information, the diversity of barcodes and time of editing provided by this approach, as well as by scGESTALT, are not sufficient to cover the number of cells during mammalian embryogenesis, limiting observations to a short period.

To increase the number of possible barcodes for lineage tracing, Kalhor and colleagues used an evolvable hgRNA, which, similarly to stgRNAs (Fig. 16.1), contains an NGG PAM sequence in the scaffold, enabling continuous mutagenesis of the hgRNA expression locus (14). hgRNAs generate barcode diversity more than eight times higher than that obtained with a conventional sgRNA design and do not require the integration of paired sgRNA and target elements into model organisms. This system can go through multiple rounds of mutagenesis until indels generated during DSB repair result in either the deletion of the PAM sequence or the formation of truncated nonfunctional hgRNAs. The recording capability of this system can be expanded by introducing multiple orthogonal hgRNA loci. Based on this approach, a mouse line carrying 60 independent hgRNA loci (22) has been created for *in vivo* barcoding and used to generate developmentally barcoded animals. Using this mouse line, lineage trees of early embryonic development and germ layers, neuroectoderm, and neural tube were reconstructed. Another substantial improvement to hgRNA-based lineage tracing could be achieved by enabling barcode recovery from the transcriptomes of single cells, thus preserving information about cell types.

One key limitation of DNA writers based on random mutagenesis using Cas9 is that DSBs are preferentially repaired with a deletion, rather than an insertion, which results in quick saturation or loss of target sites and barcodes. A recently described lineage tracing method named cell history recording by ordered insertion (CHYRON) addresses this by combining a Cas9-hgRNA system with a terminal deoxynucleotidyl transferase, which incorporates random nucleotides into Cas9-induced DSBs (23). This system allows the generation of continuous and ordered insertional mutations at the target locus and thus increases information encoding capacity. As a lineage tracer, this system proved powerful for reconstructing lineage relationships in human cell culture, setting the stage for future applications *in vivo*. Future improvements of CHYRON writing efficiency can potentially lead to a single-cell resolution of lineage tracing and molecular recording.

Base editing by cytidine deaminase-nCas9 fusion proteins can also be leveraged to address the problem of rapid saturation of barcodes due to deletions (17). The mutational activity of nCas9-CDA results in a high barcode diversity with single nucleotide resolution, without loss of barcodes due to deletions. This base editing approach was used to specifically target endogenous interspersed repeat regions in mammalian cell culture (24). The resulting genetic editing patterns in endogenous sites were used as cellular barcodes to reconstruct lineage trees. This method generates endogenous barcodes without making DSBs and also does not rely on the creation and insertion of complex target arrays. The base editing approach enables the editing of targets slowly but continuously, and it theoretically provides a diversity of cell barcodes sufficient to cover the approximately 37 trillion cells of the human body.

In contrast to deep sequencing, optical *in situ* detection of barcodes can provide spatial information about cell lineages. Another lineage tracing application of Cas9-based DNA writers called memory by engineered mutagenesis with optical *in situ* readout (MEMOIR) utilizes multiplexed single-molecule RNA fluorescence *in situ* hybridization to detect barcodes and cell-type-specific mRNA (25). Combining this optical readout with highly efficient DNA writers would be especially beneficial for defining developmental trajectories in embryos, tumors, and other systems.

Introduction to Molecular Recording

Current technologies to investigate cellular processes (such as transcriptomics, proteomics, and metabolomics) can only provide snapshots of molecular information, and therefore, the readout lacks information about a continuous sequence of biological events. Further, it is difficult to detect and monitor transient events using these snapshots. For addressing these limitations, DNA writers have been applied to capture biological, environmental, or artificial information within cells and encode it stably into DNA; this emergent application of DNA writing is termed molecular recording (Fig. 16.2B).

In general, the mechanism of molecular recorders involves three steps (2): (i) sensing of input stimuli such as specific cellular signals (e.g., presence or absence of small molecules) or stochastic information within the cell (e.g., abundance of mRNA transcripts), (ii) transformation of input signals or information into a standardized format for being encoded (e.g., through triggering the expression of DNA writing modules), and (iii) writing into a DNA medium (e.g., indel generation at predefined genomic loci) (4). Resulting "molecular records" encoded into the DNA of cell populations can be retrieved by sequencing or hybridization methods long after transient input signals have disappeared.

Molecular recording can be classified into digital and analog based on the encoding regime. Digital recording involves the encoding of information about presence or absence of input signals, whereas in analog recording, mutations or modifications gradually accumulate in response to differing magnitudes or durations of input signals. Thus, both the presence and concentration of biological molecules can serve as input signals for a recording system. Potential inputs can also include environmental stimuli such as temperature, light, pH, electricity, and other physiological characteristics capable of being sensed by cells and transformed into cellular signals (Fig. 16.2B). Further, in addition to detecting predefined signals, analog recording can also be used to encode information about cellular processes and cell states. A range of biological signals can be used as proxies for cell states; for example, transcription levels of mRNA and noncoding RNAs from different genes represent the transcriptional state of cells. Similarly, proteins and metabolites represent other markers of cell states.

One approach to sense input stimuli is to directly couple signals of interest to the expression or posttranscriptional/posttranslational activation of a DNA writer. Transcriptional activation can be achieved by using naturally occurring or engineered signal-responsive promoters (e.g., cAMP response element in mammalian cells), whereas signal-dependent conformation changes or modifications of DNA writers can be used for posttranslational activation (e.g., split-Cas9 nuclease) (4). Alternatively, engineered biosensor modules can be used to specifically detect the signal of interest and transform it into altered activation levels of the DNA writer (2). Another approach is to stochastically sample a broad range of input information, such as the pool of mRNA transcripts present within a cell, and continuously encode this sampled information using DNA writers.

The type of signal transformation, DNA writing module efficiency, and response time are other important features of molecular recorders, which determine whether an analog or digital recording regime can be achieved. Using synthetic biology and genome editing techniques, the final characteristics of the recorder can be optimized to match the desired application. For instance, transcription-based activation demonstrates slow signal transduction ($>10^2$ s), whereas enzyme-based posttranslational sensors respond to a signal much quicker ($<10^2$ s), making them suitable for tracking fast biological processes (26).

Molecular Recording Using CRISPR-Cas9

In a recent approach, named mammalian synthetic cellular recorders integrating biological events (mSCRIBE), Perli et al. leveraged stgRNAs and Cas9 for molecular recording in mammalian cells (15) (Fig. 16.3). By linking the expression of stgRNA or Cas9 to biological signals, the authors found an accumulation of stgRNA mutations upon increasing the magnitude and duration of input signals, thereby demonstrating analog memory being stored in a population of human cells. To demonstrate the power of this approach, they developed an inflammation recording cell line, where Cas9 expression is linked to the activation of NF-κB signaling that plays an essential role in coordinating a response to inflammation. These sentinel cells enabled the recording of different concentrations of tumor necrosis factor alpha (a potent activator of the NF-κB pathway) and upon implantation into mice were able to detect systemic inflammation caused by lipopolysaccharide injections, demonstrating *in vivo* recording of physiologically relevant biological signals. In the future, coupling stgRNA modules with the response elements of other signaling pathways could allow simultaneous tracking of multiple cellular processes. Although mSCRIBE enables

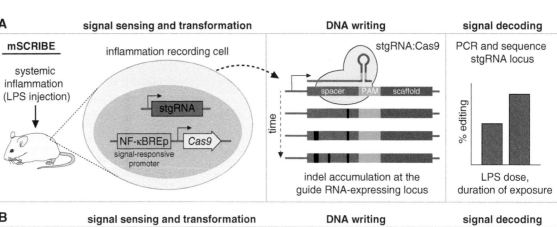

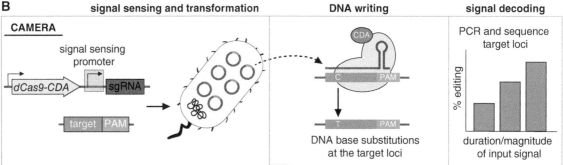

Figure 16.3 Examples of molecular recording by CRISPR-Cas9. (**A**) Leveraging Cas9 and self-targeting guide RNAs (stgRNAs), mSCRIBE enables detection of information about systemic inflammation. In this system, inflammation recording cells contain an NFκB-responsive promoter (NFκBEP) controlling the expression of Cas9. Systemic inflammation caused by lipopolysaccharide (LPS) injection into mice results in the accumulation of indels at the stgRNA loci. By quantifying mutations at the stgRNA loci, information about LPS dose and duration of exposure can be revealed. (**B**) Fusion of catalytically inactive dCas9 with a cytidine deaminase domain (dCas9-CDA) is utilized in a molecular recording approach named CAMERA. In this system, a signal sensing promoter triggers sgRNA expression and leads to base editing at target loci. The target loci are sequenced and editing efficiency is calculated to decode and quantify the input signals.

analog recording of biological signals, the recording is interrupted when the PAM sequence is deleted or the stgRNA is shortened below 16 bp, resulting in a limited recording time. Additionally, continuous self-mutation generates a diverse set of stgRNAs variants, leading to potential off-target effects and introducing noise into the recording process.

To avoid random DNA mutagenesis, base editing approaches utilizing dCas9 or nCas9 fusions with CDA have been utilized for molecular recording (27, 28). In a system called CRISPR-mediated analog multievent recording apparatus 2 (CAMERA 2), Tang and Liu used base editors to record various stimuli in bacteria and human cell cultures in the form of single-base mutations at defined genomic loci (27) (Fig. 16.3). In bacteria, CAMERA 2 utilizes a two-plasmid system with a "writing" plasmid expressing sgRNAs and dCas9-CDA fusion protein and multiple copies of a "recording" plasmid carrying different target sequences. The information is then recovered by sequencing the target loci within the recording plasmid. The percentage of edits per specific nucleotide position reflects the dosage, amplitude, and duration of input signals. Since a high-copy-number recording plasmid could have thousands of copies per cell, CAMERA 2 supports analog recording even in small populations (10 to 100 cells), in comparison to other recording technologies that enable recording only on a scale of 1,000 to 10,000 bacterial cells. Combining CAMERA 2 with different inducible promoters, the authors were able to record a wide range of signals, including exposure to antibiotics, nutrients, viruses, and light in bacteria. In a similar manner, CAMERA 2m exploits human safe harbor gene *CCR5* as a recording locus in mammalian cell culture. Controlling the expression of sgRNAs targeting *CCR5* by a synthetic inducible promoter allowed simultaneous recording of multiple stimuli in a population of cells. Furthermore, leveraging a Wnt-inducible T-cell factor/lymphoid enhancer factor promoter (LEF-TCF) to control the expression of the base editor, the authors were able to detect the activation of the Wnt pathway in human cells. Base editors provide predictable mutational outcomes with single-nucleotide resolution, which enables the direct association of input stimuli with mutations and allows for a high recording capacity. Recent works also leverage adenine base editors to introduce A:T-to-G:C mutations in a programmable manner (29). Combining cytidine and adenine base editors could be used to develop new powerful rewritable tools for molecular recording in the future.

Molecular Recording Using CRISPR Spacer Acquisition

CRISPR spacer acquisition evolved in bacteria and archaea as a natural mechanism to capture molecular records of genetic encounters and horizontal gene transfer events in the form of spacers stored within CRISPR arrays. Inspired by this, recent molecular recording approaches apply the molecular machinery of the CRISPR adaptation module as a tool for recording biological and arbitrary information into a DNA medium (Fig. 16.4). In addition to recording a diversity of input information, this approach can provide temporal resolution of recordings via the sequential and ordered spacer acquisition. This approach is also capable of directly sensing and encoding input information such as presence and abundance of transcripts or DNA sequences, precluding the need for signal transformation.

The pioneering example of applying CRISPR spacer acquisition for molecular recording utilizes the Cas1-Cas2 acquisition complex from *Escherichia coli* to integrate spacers directly into genomic CRISPR arrays. By overexpressing *E. coli* Cas1 and Cas2 proteins, Shipman and colleagues were able to generate records of specific DNA sequences in a population of bacterial genomes (8). Spacer acquisition was demonstrated from genomic and plasmid sources, as well as from synthetic double-stranded DNA sequences electroporated into the cell, and shown to be dependent on spacer sequence and the presence of a 5′ PAM. Since the spacers are integrated between a leader sequence and the first direct repeat sequence of the CRISPR array, this system was also shown to reconstruct the temporal sequence of events based on the order of acquisition. Further, the PAM recognition of this acquisition system can be modified through directed evolution, leading to the possibility of using multiple systems recognizing different PAMs orthogonally to enhance the range of input information. In a follow-up study, they encoded pixel information in the form of short fragments of synthetic DNA, which were then electroporated into recorder *E. coli* cells (30). These fragments served as artificial protospacers, being recognized by the CRISPR acquisition complex and integrated into CRISPR arrays in the bacterial genome. The images were thus permanently recorded into a population of cells and could subsequently be retrieved by sequencing the respective CRISPR arrays. Leveraging the unidirectional nature of CRISPR spacer acquisition, the authors also encoded multiple images over time within a population of bacteria, generating a short movie (Fig. 16.4A).

Besides recording synthetic inputs such as oligonucleotides, CRISPR spacer acquisition can be used to detect horizontal gene transfer events. Munck and colleagues leveraged the *E. coli* Cas1-Cas2 spacer acquisition complex to identify horizontally transferred elements in the human gut microbiome via exposing a recording *E. coli* strain to human clinical fecal samples (31). In the recording strain, upon overexpression of the Cas1-Cas2 acquisition complex, spacers were acquired from horizontally transferred elements. These exogenous spacers, permanently stored in genomic CRISPR arrays, were

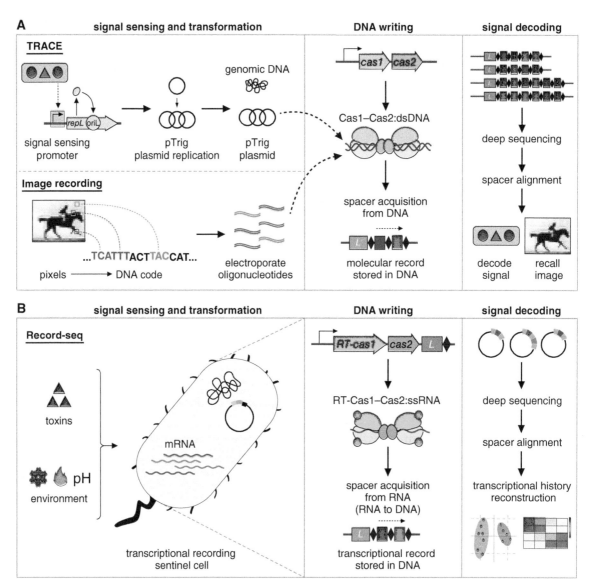

Figure 16.4 Molecular recording by CRISPR spacer acquisition. (**A**) Spacer acquisition from double-stranded DNA (dsDNA) by the Cas1-Cas2 protein complex from *E. coli* can be used for recording biological information or encoding digital information in a population of bacteria. Input digital information, such as an image or sequence of images, is encoded in DNA oligonucleotides (image recording), whereas in TRACE, an input biological signal is converted to a pTrig plasmid DNA abundance. The Cas1-Cas2 DNA writing module acquires spacers from input DNA, thus recording this information permanently in the form of spacers stored within CRISPR arrays. Deep sequencing of CRISPR arrays and aligning of the newly acquired spacers to reference sequences (i.e., *E. coli* genome, plasmid, or oligonucleotides) allows the original input signals to be decoded. (**B**) Record-seq leverages spacer acquisition from single-stranded RNA (ssRNA) for recording transcriptional information. Upon heterologous expression in *E. coli*, RT-Cas1–Cas2 acquires new spacers from RNA, based on transcript abundance. Thus, changes in the transcriptional state of a cell caused by a variety of input signals are reflected in the number of spacers aligning to the genes differentially expressed in response to the stimuli. Deep sequencing of CRISPR arrays and alignment of the newly acquired spacers to reference sequences allow the transcriptional history of a population of bacterial cells to be decoded.

then sequenced and used to assess their origins and transfer rates. This approach allows for detection and quantification of horizontal gene transfer events and demonstrates a potential application of molecular recording by CRISPR spacer acquisition to study the dynamics of complex microbial communities.

Analogously to Cas9-based molecular recording devices, the activity of CRISPR spacer acquisition can be coupled to a sensing module to detect input signals within cells. A technology termed temporal recording in arrays by CRISPR expansion (TRACE) employs *E. coli* spacer acquisition machinery to enable the recording of extracellular stimuli (Fig. 16.4A) (9). The input signal is recognized by an inducible synthetic promoter, which regulates the expression of RepL protein, essential for the replication of a copy number-inducible trigger plasmid. Consequently, a transcriptional signal can be converted to an altered abundance of the trigger plasmid in the intracellular DNA pool, providing a source of DNA protospacers for Cas1-Cas2 proteins expressed in the same cell. This signal can then be detected through an increase in spacers mapping to the corresponding trigger plasmid. TRACE has been successfully applied for recording multiple stimuli in a population of bacterial cells, such as the presence of isopropyl-β-D-thiogalactopyranoside (IPTG) or metabolites, and provided information regarding chronological order of input signals and magnitude of exposure. Despite providing a sequential resolution of recording, the input signals detected by TRACE are limited to a set of stimuli that need to be defined prior to the experiment, and further applications depend on the availability of inducible promoters that trigger spacer acquisition.

Although the prototypical Cas1-Cas2 integration complex from *E. coli* acquires spacers only from DNA, adaptation modules from other organisms are capable of acquiring spacers from RNA through naturally occurring fusions between a reverse transcriptase (RT) and Cas1 (RT-Cas1) (10, 32, 33). Such RT-Cas1–Cas2 protein complexes can be leveraged to directly record information about transcript abundance in the form of RNA-derived spacers, thus eliminating the need for coupling additional sensors to the adaptation module for recording biological stimuli.

Schmidt and colleagues recently demonstrated that an RT-Cas1–Cas2 complex derived from the human commensal bacterium *Fusicatenibacter saccharivorans* (*Fs*RT-Cas1–Cas2), upon heterologous expression in *E. coli*, acquires new spacers from RNA transcripts based on their abundance. This feature was harnessed to develop Record-seq (Fig. 16.4B), a method for recording RNA on a transcriptome scale (10). In a Record-seq experiment, the *Fs*RT-Cas1–Cas2 adaptation module records short spacer sequences from RNA templates into plasmid-borne CRISPR arrays, with spacer abundance serving as a proxy for transcript abundance. The plasmid DNA can then be extracted and deep sequenced to identify acquired spacers. By aligning acquired spacer sequences to the corresponding genes in the *E. coli* genome, cumulative gene expression can be quantified and transcriptional histories of the cells can be reconstructed. Thus, Record-seq allows for the direct recording of transcriptional responses without the need to create reporter strains dedicated to specific stimuli. Record-seq has been applied for detecting complex cellular responses to environmental stimuli, such as oxidative and acid stress, or exposure to different dosages of herbicides. Remarkably, in comparison to RNA sequencing, Record-seq is capable of capturing transcriptional response to transient stimuli and preserving this information even after the transcriptional state has changed. However, the extremely low efficiency of CRISPR spacer acquisition from RNA allows the recording of transcriptional events only on a population scale and hinders further applications of Record-seq, which would require higher sensitivity and better time resolution.

Outlook

CRISPR-Cas systems provide an advanced toolbox for the development of sophisticated DNA writing technologies, providing an entry point to recording multiple facets of dynamic biological processes through time. These CRISPR-based tools broaden the repertoire of potential DNA modifications, providing a highly flexible and efficient means to store information in DNA. This, in turn, provides a technology suite enabling the sensing of diverse stimuli, recording of multiple signals, and enhanced resolution of cellular lineage relationships. In this chapter, we described the diversity of CRISPR-Cas-based DNA writers and their emerging applications for cell lineage reconstruction and molecular recording.

Genome editing technology development is a dynamic area of research, and the next generation of DNA writing technologies will be realized through both rational engineering approaches and fundamentally new discoveries. For instance, one of the main limitations of Cas9 base editors is their off-target effects and the narrow range in which they can edit DNA. However, recent work on the continuous evolution of base editors resulted in the development of new protein variants with expanded target compatibility and improved activity (34), and the implementation of this new generation of base editors for DNA writing is imminent. Other CRISPR effector proteins with novel properties could also provide unique opportunities to encode information in DNA, such as Cas12a (35), which generates staggered cuts (in contrast to the blunt ends generated by Cas9), or Cas12k (36), which acts as an RNA-guided transposase, offering fundamentally new opportunities for DNA writing.

In addition to DNA writing modules based on Cas effector proteins, CRISPR spacer acquisition complexes present a resource for DNA writing with enormous information capture capacity and built-in temporal resolution. Moreover, molecular recording using CRISPR spacer acquisition from RNA eliminates the need for specific input signals or coupled sensing modules and enables us to directly encode a wide range of signals of interest. However, DNA writing modules based on CRISPR spacer acquisition are less efficient than those based on Cas9 and have only been demonstrated for prokaryotes, necessitating further development in this area before their full potential can be realized.

The application of CRISPR-Cas DNA writers to lineage tracing has enabled dynamic mutagenesis at target loci, increasing the time frame and resolution of cell lineage reconstruction. However, further application to mammalian *in vivo* studies is impeded by the low diversity of cellular barcodes created by existing DNA writers. In the future, DNA writing modules with higher information storage capacity, such as CRISPR spacer acquisition protein complexes or the new generation of improved base editors, can be harnessed for cellular lineage tracing. Another limitation of existing lineage tracing technology is a quick saturation of target loci available for mutagenesis, since current DNA writing modules continually introduce mutations upon being expressed. This could potentially be addressed by coupling DNA writing activity to cell division events, for instance, using cell cycle regulatory machinery of mammalian cells.

Analogously, molecular recording techniques have advanced with the expanded CRISPR-Cas toolbox, enabling the encoding of specific cellular signals or biological information into cell populations. A crucial limitation of current molecular recording technologies, especially those based on CRISPR spacer acquisition, is the low activity of DNA writing modules, which limits the amount of information retrieved per cell and necessitates population-level analyses. Enhancing the efficiency of these systems could potentially enable recordings at a single-cell level, in addition to improving temporal resolution, and facilitate the development of sophisticated approaches to understand dynamic cellular processes.

Besides experimental innovations, advances in analytical methods are needed to fully and efficiently leverage biological information encoded by DNA writers for recapitulating complex and dynamic biological processes. For instance, computational approaches for inferring the order of transcriptional states based on temporal information stored by molecular recorders could help reconstruct the sequence of molecular processes within the cell. Ultimately, we envision that combinatorial strategies integrating temporal and lineage information from DNA writing methods, with functional and spatial information from orthogonal technologies such as scRNA-seq, proteomics, metabolomics, topological mappings, and other multiplexed perturbation strategies, will enable single-cell-level analyses of the molecular mechanisms of development and disease. The continued development of novel DNA writing technologies will further expand our capacity to store information in DNA, and the application of these DNA writers will provide a powerful toolbox for unraveling the molecular events that orchestrate biology and disease.

References

1. Church GM, Gao Y, Kosuri S. 2012. Next-generation digital information storage in DNA. *Science* 337:1628.

2. Sheth RU, Wang HH. 2018. DNA-based memory devices for recording cellular events. *Nat Rev Genet* 19:718–732.

3. Schmidt F, Platt RJ. 2017. Applications of CRISPR-Cas for synthetic biology and genetic recording. *Curr Opin Syst Biol* 5:9–15.

4. Farzadfard F, Lu TK. 2018. Emerging applications for DNA writers and molecular recorders. *Science* 361:870–875.

5. Gardner TS, Cantor CR, Collins JJ. 2000. Construction of a genetic toggle switch in Escherichia coli. *Nature* 403:339–342.

6. Friedland AE, Lu TK, Wang X, Shi D, Church G, Collins JJ. 2009. Synthetic gene networks that count. *Science* 324:1199–1202.

7. Amitai G, Sorek R. 2016. CRISPR-Cas adaptation: insights into the mechanism of action. *Nat Rev Microbiol* 14:67–76.

8. Shipman SL, Nivala J, Macklis JD, Church GM. 2016. Molecular recordings by directed CRISPR spacer acquisition. *Science* 353:aaf1175.

9. Sheth RU, Yim SS, Wu FL, Wang HH. 2017. Multiplex recording of cellular events over time on CRISPR biological tape. *Science* 358:1457–1461.

10. Schmidt F, Cherepkova MY, Platt RJ. 2018. Transcriptional recording by CRISPR spacer acquisition from RNA. *Nature* 562:380–385.

11. Yosef I, Goren MG, Qimron U. 2012. Proteins and DNA elements essential for the CRISPR adaptation process in Escherichia coli. *Nucleic Acids Res* 40:5569–5576.

12. Ran FA, Cong L, Yan WX, Scott DA, Gootenberg JS, Kriz AJ, Zetsche B, Shalem O, Wu X, Makarova KS, Koonin EV, Sharp PA, Zhang F. 2015. In vivo genome editing using Staphylococcus aureus Cas9. *Nature* 520:186–191.

13. Jinek M, Chylinski K, Fonfara I, Hauer M, Doudna JA, Charpentier E. 2012. A programmable dual-RNA-guided DNA endonuclease in adaptive bacterial immunity. *Science* 337:816–821.

14. Kalhor R, Mali P, Church GM. 2017. Rapidly evolving homing CRISPR barcodes. *Nat Methods* 14:195–200.

15. Perli SD, Cui CH, Lu TK. 2016. Continuous genetic recording with self-targeting CRISPR-Cas in human cells. *Science* 353:aag0511.

16. Pitcher RS, Wilson TE, Doherty AJ. 2005. New insights into NHEJ repair processes in prokaryotes. *Cell Cycle* 4:675–678.

17. Komor AC, Kim YB, Packer MS, Zuris JA, Liu DR. 2016. Programmable editing of a target base in genomic DNA without double-stranded DNA cleavage. *Nature* 533:420–424.

18. Kretzschmar K, Watt FM. 2012. Lineage tracing. *Cell* 148:33–45.

19. McKenna A, Findlay GM, Gagnon JA, Horwitz MS, Schier AF, Shendure J. 2016. Whole-organism lineage tracing by combinatorial and cumulative genome editing. *Science* 353:aaf7907.

20. Raj B, Wagner DE, McKenna A, Pandey S, Klein AM, Shendure J, Gagnon JA, Schier AF. 2018. Simultaneous single-cell profiling of lineages and cell types in the vertebrate brain. *Nat Biotechnol* 36:442–450.

21. Chan MM, Smith ZD, Grosswendt S, Kretzmer H, Norman TM, Adamson B, Jost M, Quinn JJ, Yang D, Jones MG, Khodaverdian A,

Yosef N, Meissner A, Weissman JS. 2019. Molecular recording of mammalian embryogenesis. *Nature* 570:77–82.

22. Kalhor R, Kalhor K, Mejia L, Leeper K, Graveline A, Mali P, Church GM. 2018. Developmental barcoding of whole mouse via homing CRISPR. *Science* 361:eaat9804.

23. Loveless TB, Grotts JH, Schechter MW, Forouzmand E, Carlson CK, Agahi BS, Liang G, Ficht M, Liu B, Xie X, Liu CC. 2021. Lineage tracing and analog recording in mammalian cells by single-site DNA writing. *Nat Chem Biol* 17:739–747.

24. Hwang B, Lee W, Yum S-Y, Jeon Y, Cho N, Jang G, Bang D. 2019. Lineage tracing using a Cas9-deaminase barcoding system targeting endogenous L1 elements. *Nat Commun* 10:1234–1239.

25. Frieda KL, Linton JM, Hormoz S, Choi J, Chow KK, Singer ZS, Budde MW, Elowitz MB, Cai L. 2017. Synthetic recording and in situ readout of lineage information in single cells. *Nature* 541:107–111.

26. Olson EJ, Tabor JJ. 2012. Post-translational tools expand the scope of synthetic biology. *Curr Opin Chem Biol* 16:300–306.

27. Tang W, Liu DR. 2018. Rewritable multi-event analog recording in bacterial and mammalian cells. *Science* 360:eaap8992.

28. Farzadfard F, Gharaei N, Higashikuni Y, Jung G, Cao J, Lu TK. 2019. Single-nucleotide-resolution computing and memory in living cells. *Mol Cell* 75:769–780.e4.

29. Gaudelli NM, Komor AC, Rees HA, Packer MS, Badran AH, Bryson DI, Liu DR. 2018. Publisher Correction: Programmable base editing of A•T to G•C in genomic DNA without DNA cleavage. *Nature* 559:E8.

30. Shipman SL, Nivala J, Macklis JD, Church GM. 2017. CRISPR-Cas encoding of a digital movie into the genomes of a population of living bacteria. *Nature* 547:345–349.

31. Munck C, Sheth RU, Freedberg DE, Wang HH. 2020. Recording mobile DNA in the gut microbiota using an Escherichia coli CRISPR-Cas spacer acquisition platform. *Nat Commun* 11:95.

32. Silas S, Mohr G, Sidote DJ, Markham LM, Sanchez-Amat A, Bhaya D, Lambowitz AM, Fire AZ. 2016. Direct CRISPR spacer acquisition from RNA by a natural reverse transcriptase-Cas1 fusion protein. *Science* 351:aad4234.

33. González-Delgado A, Mestre MR, Martínez-Abarca F, Toro N. 2019. Spacer acquisition from RNA mediated by a natural reverse transcriptase-Cas1 fusion protein associated with a type III-D CRISPR-Cas system in Vibrio vulnificus. *Nucleic Acids Res* 47:10202–10211.

34. Thuronyi BW, Koblan LW, Levy JM, Yeh W-H, Zheng C, Newby GA, Wilson C, Bhaumik M, Shubina-Oleinik O, Holt JR, Liu DR. 2019. Continuous evolution of base editors with expanded target compatibility and improved activity. *Nat Biotechnol* 37:1070–1079.

35. Zetsche B, Gootenberg JS, Abudayyeh OO, Slaymaker IM, Makarova KS, Essletzbichler P, Volz SE, Joung J, van der Oost J, Regev A, Koonin EV, Zhang F. 2015. Cpf1 is a single RNA-guided endonuclease of a class 2 CRISPR-Cas system. *Cell* 163:759–771.

36. Strecker J, Ladha A, Gardner Z, Schmid-Burgk JL, Makarova KS, Koonin EV, Zhang F. 2019. RNA-guided DNA insertion with CRISPR-associated transposases. *Science* 365:48–53.

Afterword

The advent of CRISPR-based technologies has revolutionized genome editing, with a democratized technology now deployed in the clinic for human gene therapy, in the field for crop production, in the farm for livestock breeding, and even on our plates in the form of fermented dairy products. Yet the mysterious origin of these molecular machines in the dark matter of microbial genomes, as part of an intriguing adaptive immune system, illustrates the challenge of the scientific enterprise, aimed at understanding how life functions. Deciphering the molecular mechanisms underpinning CRISPR-encoded, RNA-mediated, Cas-protein-dependent targeting of nucleic acids has spawned an extensive biotechnological toolbox enabling the manipulation of virtually any genome, transcriptome, and epigenome on the planet, from microscopic viruses to giant trees, across all branches of life on earth. Unbelievably, it took less than 20 years to go from discovery to the clinic.

Reminiscent of the pace at which PCR was adopted and evoking the derivation of restriction-modification into restriction enzymes, the stories of CRISPR encompassed in these chapters illustrate how and why fundamental studies of bacteria lead to critical knowledge that can sometimes turn into valuable technologies that change the world. The basic focus on host-virus interplay has revealed much about the forces that drive predator-prey coevolutionary dynamics. Fittingly, the CRISPR phenomenon has prompted microbiologists to look for "the next CRISPR" and history has repeated itself, in a predictable *déjà vu*. Indeed, a plethora of novel defense systems has been identified and characterized recently, encompassing Thoeris, Zorya, Wadjet, Gabija, BREX, and distant relatives constituting hypervariable defense islands involved in the trafficking of nucleic acids into and between microbes and their viruses. At a time when we increasingly realize the critical importance of the microbiome and virome for human, animal, plant, and environmental health, this knowledge is crucial for understanding and manipulating bacteria and viruses towards healthier compositions and functions. Yet despite these advances, our knowledge of microbial genomes remains relatively limited, with a substantial proportion of the pangenome still comprising genes of unknown function. We still understand only a small portion of the overall genome-transcriptome-proteome dynamics and complex interplay in bacteria, let alone viruses. Considering the fact that the large majority of bacteria remains uncultured and the large majority of viruses remains unknown and invisible, we should contemplate how far we have to go as opposed to how far we have gone.

In light of this exciting future, we urge the readership to ponder what is next and how we may best deploy and exploit these technologies in nonmodel microorganisms and bacteriophages. Of course, there are noteworthy lessons to be learned from the CRISPR story, and we dare point out that we should balance strategic technology deployment and accidental scientific discovery. Of course, defining how bacteria interact with each other and with viruses and plasmids is unquestionably a worthwhile path. Another valuable lesson illustrated in the CRISPR stories covered here is the importance of collaborations and increased reliance on interdisciplinary efforts to tackle challenges as creative and synergistic teams. We advocate that this is both more productive and enjoyable, even amid intense competition. For us, it has been more fun and enjoyable to work with like-minded colleagues who become friends, and we thank the many authors of this book for their contributions to the field. This may be just the end of the beginning of the CRISPR era, and we hope it will usher in a new period in microbiology, in which we untangle the drivers of evolution and assemble the genomic puzzles responsible for phenotypes and functions of interest.

Index

CRISPR: Biology and Applications, First Edition. Edited by Rodolphe Barrangou, Erik J. Sontheimer,
and Luciano A. Marraffini.
© 2022 American Society for Microbiology. DOI: 10.1128/9781683673798.bindex